Informatics in Medical Imaging

IMAGING IN MEDICAL DIAGNOSIS AND THERAPY

William R. Hendee, Series Editor

Quality and Safety in Radiotherapy
Todd Pawlicki, Peter B. Dunscombe, Arno J. Mundt, and Pierre Scalliet, Editors
ISBN: 978-1-4398-0436-0

Adaptive Radiation Therapy
X. Allen Li, Editor
ISBN: 978-1-4398-1634-9

Quantitative MRI in Cancer
Thomas E. Yankeelov, David R. Pickens, and Ronald R. Price, Editors
ISBN: 978-1-4398-2057-5

Informatics in Medical Imaging
George C. Kagadis and Steve G. Langer, Editors
ISBN: 978-1-4398-3124-3

Forthcoming titles in the series

Image-Guided Radiation Therapy
Daniel J. Bourland, Editor
ISBN: 978-1-4398-0273-1

Informatics in Radiation Oncology
Bruce H. Curran and George Starkschall, Editors
ISBN: 978-1-4398-2582-2

Adaptive Motion Compensation in Radiotherapy
Martin Murphy, Editor
ISBN: 978-1-4398-2193-0

Image Processing in Radiation Therapy
Kristy Kay Brock, Editor
ISBN: 978-1-4398-3017-8

Proton and Carbon Ion Therapy
Charlie C.-M. Ma and Tony Lomax, Editors
ISBN: 978-1-4398-1607-3

Monte Carlo Techniques in Radiation Therapy
Jeffrey V. Siebers, Iwan Kawrakow, and David W. O. Rogers, Editors
ISBN: 978-1-4398-1875-6

Informatics in Medical Imaging
George C. Kagadis and Steve G. Langer, Editors
ISBN: 978-1-4398-3124-3

Stereotactic Radiosurgery and Radiotherapy
Stanley H. Benedict, Brian D. Kavanagh, and David J. Schlesinger, Editors
ISBN: 978-1-4398-4197-6

Cone Beam Computed Tomography
Chris C. Shaw, Editor
ISBN: 978-1-4398-4626-1

Handbook of Brachytherapy
Jack Venselaar, Dimos Baltas, Peter J. Hoskin, and Ali Soleimani-Meigooni, Editors
ISBN: 978-1-4398-4498-4

Targeted Molecular Imaging
Michael J. Welch and William C. Eckelman, Editors
ISBN: 978-1-4398-4195-0

IMAGING IN MEDICAL DIAGNOSIS AND THERAPY

William R. Hendee, Series Editor

Informatics in Medical Imaging

Edited by

George C. Kagadis
Steve G. Langer

CRC Press
Taylor & Francis Group
Boca Raton London New York

CRC Press is an imprint of the
Taylor & Francis Group, an **informa** business
A TAYLOR & FRANCIS BOOK

CRC Press
Taylor & Francis Group
6000 Broken Sound Parkway NW, Suite 300
Boca Raton, FL 33487-2742

Version Date: 2011909

International Standard Book Number: 978-1-4398-3124-3 (Hardback)

Visit the Taylor & Francis Web site at
http://www.taylorandfrancis.com

and the CRC Press Web site at
http://www.crcpress.com

To my son Orestis who has blessed me with love, continuously challenging me to become a better person, and my wife Voula who stands by me every day.

To George Nikiforidis and Bill Hendee for their continuous support and dear friendship.

George C. Kagadis

Of course I want to thank my mother (Betty Langer) and wife Sheryl for their support, but in addition I would like to dedicate this effort to my mentors . . .

My father Calvin Lloyd Langer, whose endless patience for a questioning youngster set a good example.

My graduate advisor Dr. Aaron Galonsky, who trusted a green graduate student in his lab and kindly steered him to a growing branch of physics.

My residency advisor, Dr. Joel Gray, who taught science ethics before that phrase became an oxymoron.

And to my precious Gabi, if her father can set half the example of his mentors, she will do well.

Steve G. Langer

Contents

Series Preface

Advances in the science and technology of medical imaging and radiation therapy are more profound and rapid than ever before, since their inception over a century ago. Further, the disciplines are increasingly cross-linked as imaging methods become more widely used to plan, guide, monitor, and assess the treatments in radiation therapy. Today, the technologies of medical imaging and radiation therapy are so complex and so computer-driven that it is difficult for the persons (physicians and technologists) responsible for their clinical use to know exactly what is happening at the point of care, when a patient is being examined or treated. The persons best equipped to understand the technologies and their applications are medical physicists, and these individuals are assuming greater responsibilities in the clinical arena to ensure that what is intended for the patient is actually delivered in a safe and effective manner.

The growing responsibilities of medical physicists in the clinical arenas of medical imaging and radiation therapy are not without their challenges, however. Most medical physicists are knowledgeable in either radiation therapy or medical imaging, and are experts in one or a small number of areas within their discipline. They sustain their expertise in these areas by reading scientific articles and attending scientific talks at meetings. In contrast, their responsibilities increasingly extend beyond their specific areas of expertise. To meet these responsibilities, medical physicists periodically must refresh their knowledge of advances in medical imaging or radiation therapy, and they must be prepared to function at the intersection of these two fields. How to accomplish these objectives is a challenge.

At the 2007 annual meeting of the American Association of Physicists in Medicine in Minneapolis, this challenge was the topic of conversation during a lunch hosted by Taylor & Francis Publishers and involving a group of senior medical physicists (Arthur L. Boyer, Joseph O. Deasy, C.-M. Charlie Ma, Todd A. Pawlicki, Ervin B. Podgorsak, Elke Reitzel, Anthony B. Wolbarst, and Ellen D. Yorke). The conclusion of this discussion was that a book series should be launched under the Taylor & Francis banner, with each volume in the series addressing a rapidly advancing area of medical imaging or radiation therapy of importance to medical physicists. The aim would be for each volume to provide medical physicists with the information needed to understand the technologies driving a rapid advance and their applications to safe and effective delivery of patient care.

Each volume in the series is edited by one or more individuals with recognized expertise in the technological area encompassed by the book. The editors are responsible for selecting the authors of individual chapters and ensuring that the chapters are comprehensive and intelligible to someone without such expertise. The enthusiasm of volume editors and chapter authors has been gratifying and reinforces the conclusion of the Minneapolis luncheon that this series of books addresses a major need of medical physicists.

Imaging in Medical Diagnosis and Therapy would not have been possible without the encouragement and support of the series manager, Luna Han of Taylor & Francis Publishers. The editors and authors, and most of all I, are indebted to her steady guidance of the entire project.

William R. Hendee
Series Editor
Rochester, Minnesota

Preface

The process of collecting and analyzing the data is critical in healthcare as it constitutes the basis for categorization of patient health problems. Data collected in medical practice ranges from free form text to structured text, numerical measurements, recorded signals, and imaging data. When admitted to the hospital, the patient often experiences additional tests varying from simple examinations such as blood tests, x-rays and electrocardiograms (ECGs), to more complex ones such as genetic tests, electromyograms (EMGs), computed tomography (CT), magnetic resonance imaging (MRI), and position emission tomography (PET). Historically, the demographics collected from all these tests were characterized by uncertainty because often there was not a single authoritative source for patient demographic information, and multiple points of human-entered data were not all in perfect agreement. The results from these tests are then archived in databases and subsequently retrieved (or not—if the "correct" demographic has been forgotten) upon requests by clinicians for patient management and analysis.

For these reasons, digital medical databases and, consequently, the Electronic Health Record (EHR) have emerged in healthcare. Today, these databases have the advantage of high computing power and almost infinite archiving capacity as well as Web availability. Access through the Internet has provided the potential for concurrent data sharing and relevant backup. This procedure of appropriate data acquisition, archiving, sharing, retrieval, and data mining is the focus of medical informatics. All this information is deemed vital for efficient provision of healthcare (Kagadis et al., 2008).

Medical imaging informatics is an important subcomponent of medical informatics and deals with aspects of image generation, manipulation, management, integration, storage, transmission, distribution, visualization, and security (Huang, 2005; Shortliffe and Cimino, 2006). Medical imaging informatics has advanced rapidly, and it is no surprise that it has evolved principally in radiology, the home of most imaging modalities. However, many other specialties (i.e., pathology, cardiology, dermatology, and surgery) have adopted the use of digital images; thus, imaging informatics is used extensively in these specialties as well.

Owing to continuous progress in image acquisition, archiving, and processing systems, the field of medical imaging informatics continues to rapidly change and there are many books written every year to reflect this evolution. While much reference material is available from the American Association of Physicists in Medicine (AAPM), the Society for Imaging Informatics in Medicine (SIIM) Task Group reports, European guidance documents, and the published literature, this book tries to fill a gap and provide an integrated publication dealing with the most essential and timely issues within the scope of informatics in medical imaging.

The target audience for this book is students, researchers, and professionals in medical physics and biomedical imaging with an interest in informatics. It may also be used as a reference guide for medical physicists and radiologists needing information on informatics in medical imaging. It provides a knowledge foundation of the state of the art in medical imaging informatics and points to major challenges of the future.

The book content is grouped into six sections. Section I deals with introductory material to informatics as it pertains to healthcare. Section II deals with the standard imaging informatics protocols, while Section III covers healthcare informatics based enabling technologies. In Section IV, key systems of radiology informatics are discussed and in Section V special focus is given to operational issues in medical imaging. Finally, Section VI looks at medical informatics issues outside the radiology department.

References

Huang, H.K. 2005. Medical imaging informatics research and development trends. *Comput. Med. Imag. Graph.*, 29, 91–3.

Kagadis, G.C., Nagy, P., Langer, S., Flynn, M., Starkschall, G. 2008. Anniversary paper: Roles of medical physicists and healthcare applications of informatics. *Med. Phys.*, 35, 119–27.

Shortliffe, E.H., Cimino, J.J. 2006. *Biomedical Informatics: Computer Applications in Healthcare and Biomedicine (Health Informatics)*. New York, NY: Springer.

George C. Kagadis
Steve G. Langer
Editors

Editors

George C. Kagadis, PhD is currently an assistant professor of medical physics and medical informatics at University of Patras, Greece. He received his Diploma in Physics from the University of Athens, Greece in 1996 and both his MSc and PhD in medical physics from the University of Patras, Greece in 1998 and 2002, respectively. He is a Greek State Scholarship Foundation grantee, a Fulbright Research Scholar, and a full AAPM member. He has authored approximately 70 journal papers and had presented over 20 talks at international meetings. Dr. Kagadis has been involved in European and national projects, including e-health. His current research interests focus on IHE, CAD applications, medical image processing and analysis as well as studies in molecular imaging. Currently, he is a member of the AAPM Molecular Imaging in Radiation Oncology Work Group, European Affairs Subcommittee, Work Group on Information Technology, and an associate editor to *Medical Physics*.

Steve G. Langer, PhD is currently a codirector of the radiology imaging informatics lab at the Mayo Clinic in Rochester, Minnesota and formerly served on the faculty of the University of Washington, Seattle. His formal training in nuclear physics at the University of Wisconsin, Madison and Michigan State has given way to a new mission: to design, enable, and guide into production high-performance computing solutions to implement next-generation imaging informatics analytics into the clinical practice. This includes algorithm design, validation, performance profiling, and deployment on vended or custom platforms as required. He also has extensive interests in validating the behavior and performance of human- and machine-based (CAD) diagnostic agents.

Editors

Contributors

Christos Alexakos
Department of Computer Engineering and Informatics
University of Patras
Rion, Greece

Luca Antiga
Biomedical Engineering Department
Mario Negri Institute
Bergamo, Italy

Peter Balter
Department of Radiation Physics
The University of Texas
Houston, Texas

Daniel J. Blezek
Department of Bioengineering
Mayo Clinic
Rochester, Minnesota

Anna Caroli
Biomedical Engineering Department
Mario Negri Institute
Bergamo, Italy
and
Laboratory of Epidemiology, Neuroimaging, and Telemedicine
IRCCS San Giovanni di Dio-FBF
Brescia, Italy

Lionel T. Cheng
Singapore Armed Forces Medical Corps
and
Singapore General Hospital
Singapore

John Drakos
Clinic of Haematology
University of Patras
Rion, Greece

Brad J. Erickson
Department of Radiology
Mayo Clinic
Rochester, Minnesota

William R. Hendee
Departments of Radiology, Radiation Oncology, Biophysics, and Population Health
Medical College of Wisconsin
Milwaukee, Wisconsin

Steven C. Horii
Department of Radiology
University of Pennsylvania Medical Center
Philadelphia, Pennsylvania

George C. Kagadis
Department of Medical Physics
University of Patras
Rion, Greece

Dimitris Karnabatidis
Department of Radiology
Patras University Hospital
Patras, Greece

Konstantinos Katsanos
Department of Radiology
Patras University Hospital
Patras, Greece

Shawn Kinzel
Information Systems
Mayo Clinic
Rochester, Minnesota

Helmut König
Siemens AG Healthcare Sector
Erlangen, Germany

Elizabeth A. Krupinski
Department of Radiology
University of Arizona
Tucson, Arizona

Steve G. Langer
Department of Diagnostic Radiology
Mayo Clinic
Rochester, Minnesota

Randy Luhta
CT Engineering
Philips Medical Systems
Cleveland, Ohio

Dirk Marwede
Department of Nuclear Medicine
University of Leipzig
Leipzig, Germany

Frank Meng
VA Greater Los Angles Healthcare System
Los Angeles, California

Craig Morioka
UCLA Medical Imaging Informatics
and
VA Greater Los Angles Healthcare System
Los Angeles, California

George C. Nikiforidis
Department of Medical Physics
University of Patras
Patras, Greece

Frédéric Noo
Department of Radiology
University of Utah
Salt Lake City, Utah

Herman Oosterwijk
OTech Inc.
Cross Roads, Texas

Katia Passera
Biomedical Engineering Department
Mario Negri Institute
Bergamo, Italy

George C. Sakellaropoulos
Department of Medical Physics
University of Patras
Patras, Greece

Ioannis Sechopoulos
Department of Radiology, Hematology, and Medical Oncology
Emory University
Atlanta, Georgia

George Starkschall
Department of Radiation Physics
The University of Texas
Houston, Texas

Scott Stekel
Department of Radiology
Mayo Clinic
Rochester, Minnesota

Brent K. Stewart
Department of Radiology
University of Washington School of Medicine
Seattle, Washington

Alisa Walz-Flannigan
Department of Radiology
Mayo Clinic
Rochester, Minnesota

John Yorkston
Carestream Health
Rochester, New York

Boris Zavalkovskiy
Enterprise Imaging
Lahey Clinic, Inc.
Burlington, Massachusetts

I

Introduction to Informatics in Healthcare

1

Ontologies in the Radiology Department

Dirk Marwede
University of Leipzig

Ontologies have become increasingly popular to structure knowledge and exchange information. In medicine, the main areas for the application of ontologies are the encoding of information with standardized terminologies and the use of formalized medical knowledge in expert systems for decision support. In medical imaging, the ever-growing number of imaging studies and digital data requires tools for comprehensive and effective information management. Ontologies provide human- and machine-readable information and bring the prospective of semantic data integration. As such, ontologies might enhance interoperability between systems and facilitate different tasks in the radiology department like patient management, structured reporting, decision support, and image retrieval.

1.1 Ontologies and Knowledge Representation

There have been many attempts to define what an ontology is. Originally, in the philosophical branch of metaphysics, an *ontology* deals with questions concerning the existence of entities in reality and how such entities relate to each other. In information and computer science, an ontology has been defined as a body of formally represented knowledge based on a conceptualization. Such a conceptualization is an explicit specification of objects, concepts, and other entities that are assumed to exist in some area of interest and the relations that hold among them (Genesereth and Nilsson, 1987; Gruber, 1993). Similarly, the term *knowledge representation* has been used in artificial intelligence, a branch of computer science, to describe a formal system representing knowledge by a set of rules that are used to infer (formalized reasoning) new knowledge within a specific domain.

Besides different definitions, the term ontology nowadays is often used to describe different levels of usage. These levels include (1) the definition of a common vocabulary, (2) the standardization of terms, concepts, or tasks, (3) conceptual schemas for transfer, reuse, and sharing of information, (4) organization and representation of knowledge, and (5) answering questions or queries. From those usages, some general benefits of ontologies in information management can be defined

- To enhance the interoperability between information systems
- To transmit, reuse, and share the structured data
- To facilitate the data aggregation and analysis
- To integrate the knowledge (e.g., a model) and data (e.g., patient data)

1.2 Ontology Components

1.2.1 Concepts and Instances

The main component of ontologies are *concepts* also called classes, entities, or elements. Concepts can be regarded as "unit of thoughts," that is, some conceptualization with a specific meaning whereas the meaning of concepts can be implicitly or explicitly defined. Concepts with implicit definitions are often called *primitive* concepts. In contrast, concepts with explicit definitions (i.e., *defined* concepts) are defined by relations to other concepts and sometimes restrictions (e.g., a value range). Concepts or classes can have *instances*, that is, individuals, for which all defined relations hold true. Concepts are components of a knowledge model whereas instances populate this model with individual data. For example, the concepts *Patient Name* and *Age* can have instances such as *John Doe* and *37*.

1.2.2 Relations

Relations are used to link concepts to each other or to attach attributes to concepts. Binary relations are used to relate concepts to each other. The hierarchical organization of concepts in an ontology is usually based on the *is_a* (i.e., is a subtype of) relation, which relates a parent concept to a child concept (e.g., "inflammation" *is_a* "disease"). The relation is also called *subsumption* as the relation subsumes *sub*-concepts under a *super*-concept. In the medical domain, many relations express structural (e.g., anatomy), spatial (e.g., location and position), functional (e.g., pathophysiological processes), or causative information (e.g., disease cause). For example, structural information can be described by partonomy relations like *part_of* or *has_part* (e.g., "liver vein" *part_of* "liver"), spatial information by the relation *located_in* (e.g., "cyst" *located_in* "liver"), or *contained_in* (e.g., "thrombus" *contained_in* "lumen of pulmonary artery"), and functional information by the relation *regulates* (e.g., "apoptosis" *regulates* "cell death"). Attributes can be attached to concepts by relations like *has_shape* or *has_density* (e.g., "pulmonary nodule" *has_shape* "round").

A relation can be defined by properties like transitivity, symmetry/antisymmetry, and reflexivity (Smith and Rosse, 2004; Smith et al., 2005). For example, a relation R over a class X is *transitive* if an element a is related to an element b, and b is in turn related to an element c, then a is also related to c (e.g., "pneumonia" *is_a* "inflammation" *is_a* "disease" denotes that "pneumonia" *is_a* "disease"). Relational properties are mathematical definitions from set theory, which can be explicitly defined in some ontology or representation languages (Baader et al., 2003; Levy, 2002).

1.2.3 Restrictions and Inheritance

Beside formal characteristics of relations, further logical statements can be attached to concepts. Such logical expressions are called restrictions or axioms, which explicitly define concepts. Basic restrictions include *domain* and *range* restrictions that define which concepts can be linked through a relation. Restrictions can be applied to the filler of a relation, for example, to a value, concept, or concept type and depend on the representation formalism used. In general, restrictions are commonly deployed in large ontologies to support reasoning tasks for checking consistency of the ontology (Baader et al., 2003; Rector et al., 1997). *Inheritance* is a mechanism deployed in most ontologies in which a child concept inherits all definitions of the parent concept. Some ontology languages support mechanism of multiple inheritance in which a child concept inherits definitions of different parent concepts.

1.3 Ontology Construction

The construction of an ontology usually starts with a *specification* to define the purpose and scope of an ontology. In a second step, concepts and relations in a domain are identified (*conceptualization*) often involving natural language processing (NLP) algorithms and domain experts. Afterwards, the description of concepts is transformed in a formal model by the use of restrictions (*formalization*) followed by the *implementation* of the ontology in a representation language. Finally, *maintenance* of the implemented ontology is achieved by testing, updating, and correcting the ontology. Many ontologies today, in particular controlled terminologies or basic symbolic knowledge models, do not support formalized reasoning. In fact, even if not all ontologies require reasoning support to execute specific tasks, reasoning techniques are useful during ontology construction to check consistency of the evolving ontology.

In most ontologies, concepts are precoordinated which means that primitive or defined concepts cannot be modified. However, in particular within large domains like medicine, some ontologies support postcoordination of concepts which allows to construct new concepts by the combination of primitive or defined concepts by the user (Rector and Nowlan, 1994). Postcoordination requires strict rules for concept definition to assure semantic and logical consistency within an ontology.

1.4 Representation Techniques

The expressivity of ontology languages to represent knowledge ranges from informal approaches with little or no specification of the meaning of terms to formal languages with strict logical definitions (Staab and Studer, 2009). In general, there is a tradeoff between logical expressivity of languages and computational efficiency, thus the appropriate ontology language or representation formalism needs to be chosen with regard to the domain of interest and the intent of the ontology.

First knowledge representation languages include *semantic networks* and *frame-based* approaches. Semantic networks represent semantic relations among concepts in a graph structure (Sowa, 1987). Within such networks, it is possible to represent logical description, for example existential graphs or conceptual graphs. Frame-based systems use a *frame* to represent an entity within a domain (Minsky, 1975). Frames are associated with a

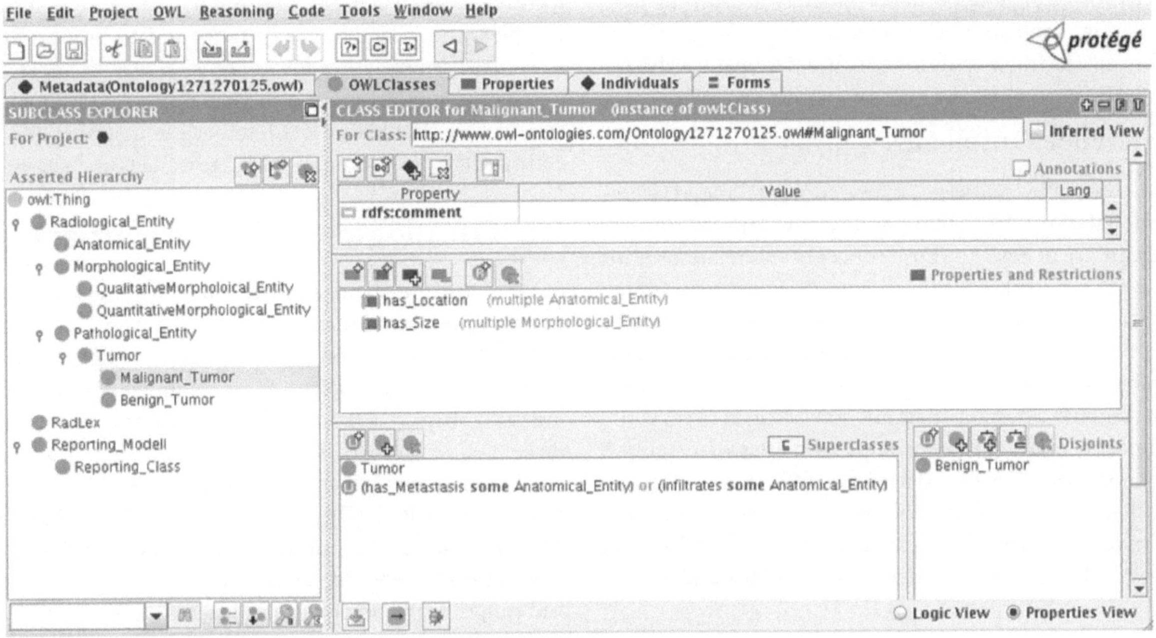

FIGURE 1.1 Definition of concepts in a frame-based ontology editor with OWL support (Protégé OWL).

number of *slots* that can be filled with *slot values* that are also frames. Protégé is a popular open-source ontology editor using frames, which is compatible to the open knowledge base connectivity protocol (OKBC) (Noy et al., 2003).

Description logics (DLs) are a family of representation languages using formal descriptions for concept definitions. In contrast to semantic networks and frame-based models, DLs use formal, *logic*-based semantics for knowledge representation. In addition to the description formalism, DLs are usually composed of two components—a terminological formalism describing names for complex descriptions (T-Box) and a assertional formalism used to state properties for individuals (A-Box) (Baader et al., 2009).

The resource description framework (RDF) is a framework for representing information about resources in a graph form. The Web Ontology Language (OWL), an extension of RDF, is a language for semantic representation of Web content. OWL adds more vocabulary for describing properties and classes, that is, relations between classes (e.g., disjointness), cardinality (e.g., "exactly one"), equality, richer typing of properties, characteristics of properties (e.g., symmetry), and enumerated classes.* OWL provides three sublanguages with increasing expressivity and reasoning power: *OWL Lite* supports users primarily concerned with classification hierarchies and simple constraints, *OWL DL* provides maximum expressiveness while retaining computational completeness, and *OWL Full* has maximum expressiveness and the syntactic freedom of RDF with no computational guarantees. Today, some frame-based ontology editors provide plug-ins for OWL support combining frame-based knowledge models with logical expressivity and reasoning capacities (Figure 1.1).

1.5 Types of Ontologies

Medicine is a very knowledge intensive area with a long tradition in structuring its information. First attempts focussed on the codification of medical terminology resulting in hierarchical organized controlled vocabularies and terminologies, for example, the International Classification of Disease (ICD). The introduction of basic relations between entries in different hierarchies resulted in more complex medical terminologies like the Systematized Nomenclature of Medicine (Snomed) (Spackman et al., 1997). However, in recent years, complex knowledge models with or without formal reasoning support have been constructed like the Foundational Model of Anatomy (FMA) (Rosse and Mejino, 2003) or the Generalized Architecture for Languages, Encyclopaedias, and Nomenclatures in Medicine (GALEN) (Rector and Nowlan, 1994).

1.5.1 Upper-Level Ontologies

A top- or upper-level ontology is a domain-independent representation of very basic concepts and relations (objects, space, time). In information and computer science, the main aim of such an ontology is to facilitate the integration and interoperability of domain-specific ontologies. Building a comprehensive upper-level ontology is a complex task and different upper-level ontologies have been developed with considerable differences in scope, syntax, semantics, and representational formalisms (Grenon and Smith, 2004; Herre et al., 2006; Masolo et al., 2003).

* http://www.w3.org/TR/owl-features/

Today, the use of a single upper-level ontology subsuming concepts and relations of all domain-specific ontologies is questioned and probably not desirable in terms of computational feasibility.

1.5.2 Reference Ontologies

In large domains like medicine, many concepts and relations are foundational in the sense that ontologies within the same or related domain use or refer to those concepts and relations. This observation has led to the notion of *Foundational* or *Reference Ontologies* that serve as a basis or reference for other ontologies (Burgun, 2006). The most-known reference ontology in medicine is the Foundational Model of Anatomy (FMA), a comprehensive ontology of structural human anatomy, consisting of over 70,000 different concepts and 170 relationships with approximately 1.5 million instantiations (Rosse and Mejino, 2003) (Figure 1.2). An important characteristic of reference ontologies is that they are developed independently of any particular purpose and should reflect the underlying reality (Bodenreider and Burgun, 2005).

1.5.3 Application Ontologies

Application ontologies are constructed with a specific context and target group in mind. In contrast to abstract concepts in upper-level ontologies or to the general and comprehensive knowledge in reference ontologies, concepts and relations represent a well-defined portion of knowledge to carry out a specific task. In medicine, many application ontologies are used for decision support, for example, for the representation of mammographic features of breast cancer. Those ontologies are designed to perform complex knowledge intensive tasks and to process and provide structured information for analysis. However, most application ontologies thus far do not adhere to upper-level ontologies or link to reference ontologies that hamper the mapping and interoperability between different knowledge models and systems.

1.6 Ontologies in Medical Imaging

Medical imaging and clinical radiology are knowledge intensive disciplines and there have been many efforts to capture this knowledge. Radiology departments are highly computerized environments using software for (1) image acquisition, processing, and display, (2) image evaluation and reporting, and (3) image and report archiving. Digital data are nowadays administered in different information systems, for example, patient and study data in Radiology Information Systems (RIS) and image data in Picture Archiving and Communication Systems (PACS).

Within radiology departments, knowledge is rather diverse and ranges from conceptual models for integrating information from different sources to expert knowledge models about diagnostic conclusions. A certain limitation of information processing within radiology departments today is that even if images and reports contain semantic information about anatomical and pathological structures, morphological features, and disease trends, there is no semantic link between images and reports. In addition, image and report data are administered in different systems (PACS, RIS) and communicated using different standards (DICOM, HL7), which impair the integration of semantic

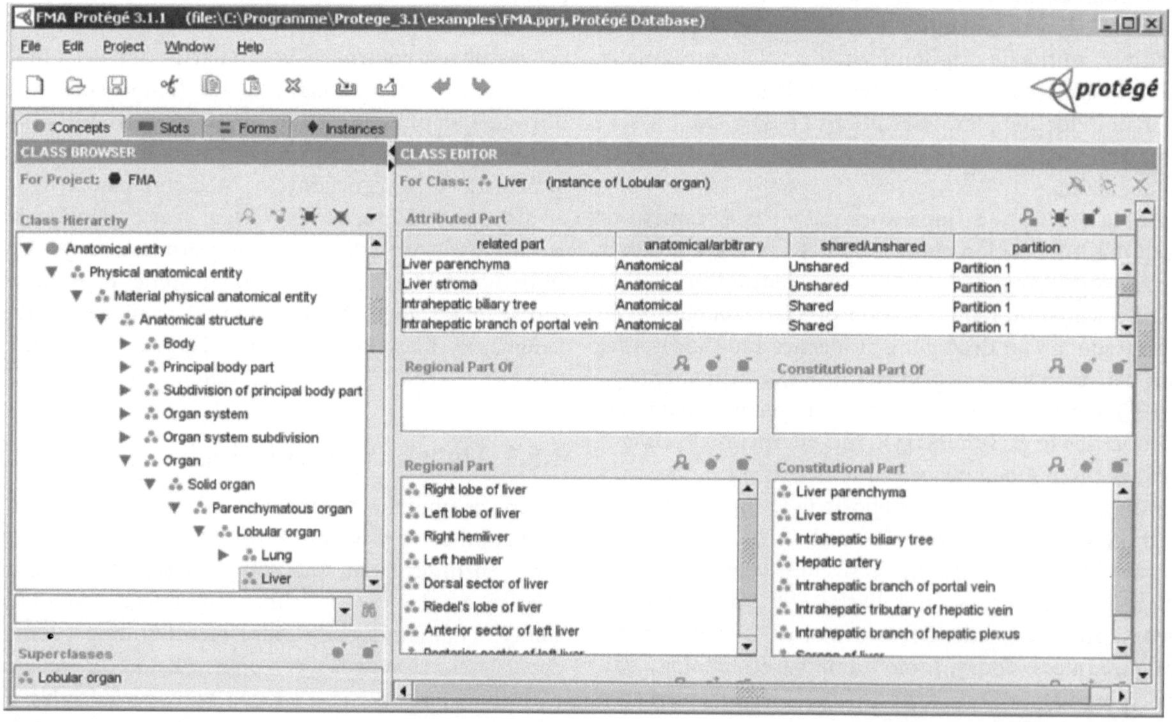

FIGURE 1.2 Hierarchical organization of anatomical concepts and symbolic relations in the Foundational Model of Anatomy (FMA).

radiological knowledge models and the interoperability between applications.

1.7 Foundational Elements and Principles

1.7.1 Terminologies in Radiology

In the past, several radiological lexicons have been developed such as the Fleischner Glossary of terms used in thoracic imaging (Tuddenham, 1984; Austin et al., 1996), the Breast Imaging Reporting and Data System (BIRADS) (Liberman and Menell, 2002), and the American College of Radiology Index (ACR) for diagnoses. As those lexicons represented only a small part of terms used in radiology and were not linked to other medical terminologies, the Radiological Society of North America (RSNA) started, in 2003, the development of a concise radiological lexicon called RadLex© (Langlotz, 2006).

RadLex was developed to unify terms in radiology and to facilitate indexing and retrieval of images and reports. The terminology can be accessed through an online term browser or downloaded for use. RadLex is a hierarchical, organized terminology consisting of approximately 12,000 terms grouped in 14 main term categories (Figure 1.3). Main categories are *anatomical entity* (e.g., "lung"), *imaging observation* (e.g., "pulmonary nodule"), *imaging observation characteristic* (e.g., "focality") and *modifiers* (e.g., "composition modifier"), *procedure steps* (e.g., "CT localizer radiograph") and *imaging procedure attributes* (e.g., modalities), *relationship* (e.g., is_a, part_of), and *teaching attributes* (e.g., "perceptual difficulty"). Thus far, the hierarchical organization of terms represents *is_a* and *part_of* relations between terms.

RadLex can be regarded as a hierarchical, organized, standardized terminology. RadLex thus far does not contain formal definitions or logical restrictions. However, evolving ontologies in radiology might use RadLex terms as a basis for concept definitions and different formal constructs for specific application tasks. In this manner, RadLex has been linked already to anatomical concepts of the Foundational Model of Anatomy to enrich the anatomical terms defined in RadLex with a comprehensive knowledge model of human anatomy (Mejino et al., 2008).

1.7.2 Interoperability

Ontologies affect different tasks in radiology departments like reporting, image retrieval, or patient management. To exchange and process the information between ontologies or systems, different levels of interoperability need to be distinguished (Tolk and Muguira, 2003; Turnitsa, 2005). The *technical* level is the most basic level assuring that a common protocol exists for data exchange. The *syntactic* level specifies a common data structure and format, and the *semantic* level defines the content and meaning of the exchanged information in terms of a reference model. *Pragmatic* interoperability specifies the context of the exchanged information making the processes explicit, which use the information in different systems. A *dynamic* level ensures that state changes of exchanged information are understood by the systems and on the highest level of interoperability, the *conceptual* level, a fully specified abstract concept model including constraints and assumptions is explicitly defined.

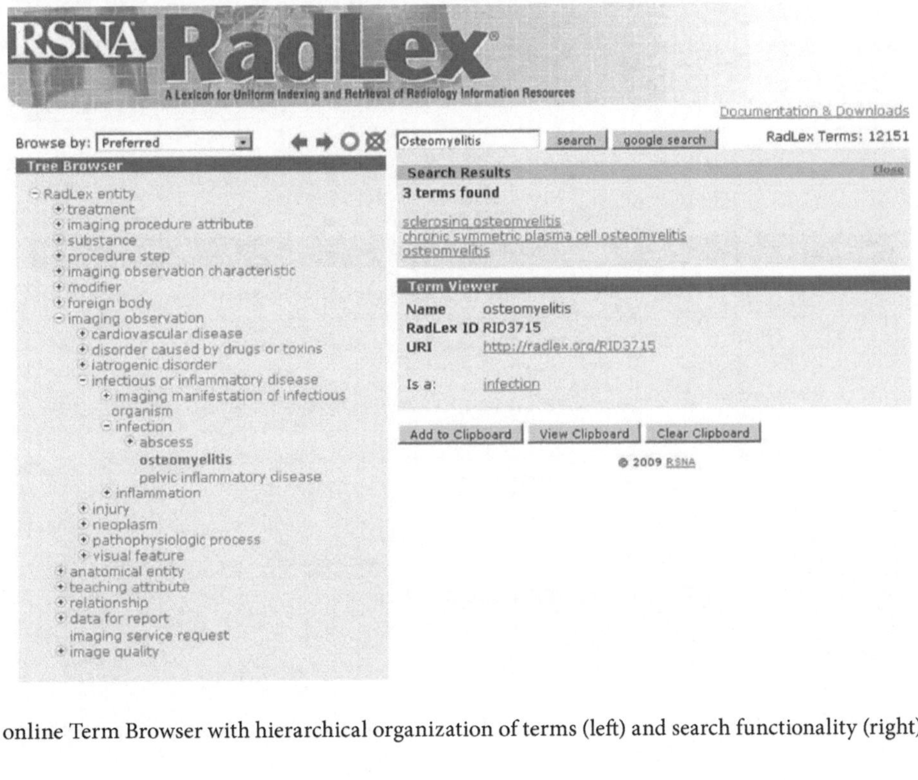

FIGURE 1.3 RadLex online Term Browser with hierarchical organization of terms (left) and search functionality (right).

1.8 Application Areas of Ontologies in Radiology

1.8.1 Imaging Procedure Appropriateness

Medical imaging procedures are performed to deliver accurate diagnostic and therapeutic information at the right moment. For each imaging study, an appropriate imaging technique and protocol are chosen depending on the medical context. In clinical practice, this context is defined by the patient condition, clinical question (indication), patient benefit, radiation exposure, and availability of imaging techniques determining the appropriateness of an imaging examination. During the 1990s, the American College of Radiology (ACR) has developed standardized criteria for the appropriate use of imaging technologies, the ACR Appropriateness Criteria (ACRAC).

The ACRAC represent specific clinical problems and associated imaging procedures with an appropriateness score ranging from 1 (not indicated) to 9 (most appropriate) (Figure 1.4). The ACRAC are organized in a relational database model and electronically available (Sistrom, 2008). A knowledge model of the ACRAC and online tools to represent, edit, and manage knowledge contained in the ACRAC were developed. This model was defined by the Appropriateness Criteria Model Encoding Language (ACME), which uses the Standard Generalized Mark-Up Language (SGML) to represent and interrelate the definitions of *conditions*, *procedures*, and *terms* in a semantic network (Kahn, 1998). To promote the application of appropriate criteria in clinical practice, an online system was developed to search, retrieve, and display ACRAC (Tjahjono and Kahn, 1999). However, to enhance the use of ACRAC criteria and its integration into different information systems (e.g., order entry), several additional requirements have been defined: a more formal representation syntax of clinical conditions, a standardized terminology or coding scheme for clinical concepts, and the representation of temporal information and uncertainty (Tjahjono and Kahn, 1999).

1.8.2 Clinical Practice Guidelines

"Clinical practice guidelines are systematically developed statements to assist the practitioners and patient decisions about appropriate healthcare for specific circumstances" (Field and Lohr, 1992). In the 1990s, early systems emerged representing originally paper-based clinical guidelines in a computable format. The most popular approaches were the GEODE-CM system for guidelines and data entry (Stoufflet et al., 1996), the Medial Logical Modules for alerts and reminders (Barrows et al., 1996; Hripcsak et al., 1996), the MBTA system for guidelines and reminders (Barnes and Barnett, 1995), and the EON architecture (Musen et al., 1996) and PRODIGY system (Purves, 1998) for guideline-based decision support. As those systems differed by representation technique, format, and functionality, the need for a common guideline representation format emerged.

In 1998, the Guideline Interchange Format (GLIF), a representation format for sharable computer-interpretable clinical practice guidelines, was developed. GLIF incorporates functionalities from former guideline systems and consists of three abstraction levels, a conceptual (human-readable) level for medical terms as free text represented in flow charts, a computable level with an

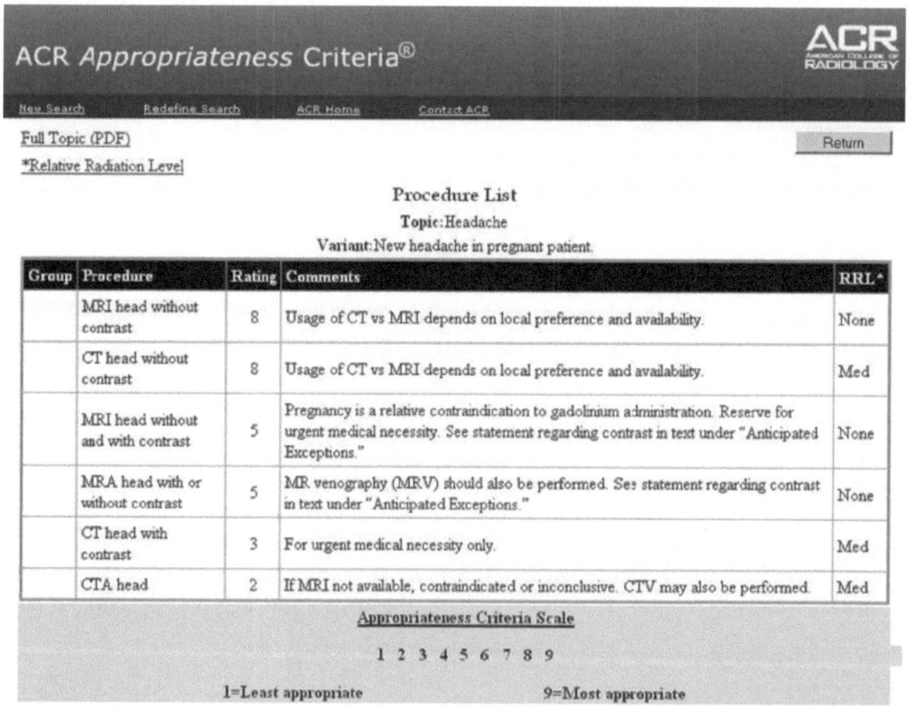
ACR *Appropriateness* Criteria®

New Search Redefine Search ACR Home Contact ACR

Full Topic (PDF)
*Relative Radiation Level

Return

Procedure List
Topic: Headache
Variant: New headache in pregnant patient

Group	Procedure	Rating	Comments	RRL*
	MRI head without contrast	8	Usage of CT vs MRI depends on local preference and availability.	None
	CT head without contrast	8	Usage of CT vs MRI depends on local preference and availability.	Med
	MRI head without and with contrast	5	Pregnancy is a relative contraindication to gadolinium administration. Reserve for urgent medical necessity. See statement regarding contrast in text under "Anticipated Exceptions."	None
	MRA head with or without contrast	5	MR venography (MRV) should also be performed. See statement regarding contrast in text under "Anticipated Exceptions."	None
	CT head with contrast	3	For urgent medical necessity only.	Med
	CTA head	2	If MRI not available, contraindicated or inconclusive. CTV may also be performed.	Med

Appropriateness Criteria Scale
1 2 3 4 5 6 7 8 9
1=Least appropriate 9=Most appropriate

FIGURE 1.4 Online access to the ACRAC: Detailed representation of clinical conditions, procedures, and appropriateness score.

expressive syntax to execute a guideline, and an implementation level to integrate guidelines in institutional clinical applications (Boxwala et al., 2004). The GLIF model represents guidelines as sets of classes for guideline entities, attributes, and data types. A flowchart is built by *Guideline_Steps*, which has the following subclasses: *Decision_Step* class for representing decision points, *Action_Step* class for modeling recommended actions or tasks, *Branch_Step* and *Synchronization_Step* classes for modeling concurrent guideline paths, and *Patient_State_Step* class for representing the patient state. In addition, the GLIF specification includes an expression and query language to access patient data and to map those data to variables defined as decision criteria.

In summary, computer-interpretable practice guidelines are able to use diverse medical data for diagnoses and therapy guidance. Integration of appropriate imaging criteria and imaging results in clinical guidelines is possible, but requires interoperability between information systems used in radiology and guideline systems. However, the successful implementation of computer-interpretable guidelines highly depends on the complexity of the guideline, the involvement of medical experts, the degree of interoperability with different information systems, and the integration in the clinical workflow.

1.8.3 Order Entry

In general, computer-based physician order entry (CPOE) refers to a variety of computer-based systems for medical orders (Sittig and Stead, 1994). For over 20 years, CPOE systems have been used mainly for ordering the medication and laboratory examinations; however, since, some years, radiology order entry systems (ROE) are emerging, enabling physicians are to order the image examinations electronically. CPOE and ROE systems assure standardized, legible, and complete orders and provide data for quality assurance and cost analysis.

There is no standard ROE system and many systems have been designed empirically according to the organizational and institutional demands. Physicians interact with the systems through a user interface, which typically is composed of order forms in which information can be typed in or selected from predefined lists. The ordering physician specifies the imaging modality or service and provides information about the patient like signs/symptoms and known diseases. Clinical information is usually encoded into a standardized terminology or classification schema like the International Classification of Diseases (ICD). Some systems incorporate the decision support in the order entry process, providing guidance for physicians which imaging study is the most appropriate (Rosenthal et al., 2006). There is evidence that those systems might change the ordering behavior of physicians and increase the quality of imaging orders (Sistrom et al., 2009).

Knowledge modeling of order entry and decision support elements is not trivial as relations between clinical information like signs and symptoms, suspected diseases, and appropriate imaging examinations are extensive and frequently complex. However, as standardized terminologies are implemented in most order entry systems and criteria for appropriate imaging have been defined, an ontology or knowledge model for the appropriate ordering of imaging examinations can be implemented and possibly shared across different institutions.

1.9 Image Interpretation

1.9.1 Structured Reporting

Structured reporting of imaging studies brings the prospect of unambiguous communication of exam results and automated report analysis for research, teaching, and quality improvement. In addition, structured reports address the major operational needs of radiology practices, including patient throughput, report turnaround time, documentation of service, and billing. As such, structured reports might serve as a basis for many other applications like decision support systems, reminder and notification programs, or electronic health records.

General requirements for structured reports are a controlled vocabulary or terminology and a standardized format and structure. Early structured reporting systems used data entry forms in which predefined terms or free-text was reported (Bell and Greenes, 1994; Kuhn et al., 1992). For the meaningful reporting of imaging observations, some knowledge models were developed to represent statements and diagnostic conclusions frequently found in radiology reports (Bell et al., 1994; Friedman et al., 1993; Marwede et al., 2007). However, integrating a controlled vocabulary with a knowledge model for reporting imaging findings in a user-friendly reporting system remains a challenging task.

In fact, the primary candidate for a controlled vocabulary is RadLex, the first comprehensive radiological terminology. There is some evidence that RadLex contains most terms present in radiology reports today, even if some terms need to be composed by terms from different hierarchies (Marwede et al., 2008). In 2008, the RSNA defined general requirements for structured radiology reports to provide a framework for the development of best practice reporting templates.* Those templates use standardized terms from RadLex and a simple knowledge representation scheme defined in extensible mark-up language (XML). Furthermore, a comprehensive model for image annotations like measurements or semantic image information has been developed using RadLex for structured annotations (Channin et al., 2009). In this model, annotations represent links between image regions and report items connecting semantic information in images with reports. Storage and export of annotations can be performed in different formats (Rubin et al., 2008).

Structured reporting applications today mainly use data entry forms in which the user types or selects terms from lists. Those forms provide static or dynamic menu-driven interfaces, which enable the radiologist to quickly select and report items. However, a promising approach to avoid distraction during review is to integrate speech recognition software into

* http://www.rsna.org/informatics/radreports.cfm

structured reporting applications (Liu et al., 2006). Such applications might provide new dimensions of interaction like the "talking template," which requests information or guides the radiologist through the structured report without interrupting the image review process (Sistrom, 2005).

1.9.2 Diagnostic Decision Support Systems

In radiology departments, diagnostic decision support systems (DSS) assist the radiologist during the image interpretation process in three ways: (1) to perceive image findings, (2) to interpret those findings to render a diagnosis, and (3) to make decisions and recommendations for patient management (Rubin, 2009). DSS systems are typically designed to integrate a medical knowledge base, patient data, and an inference engine to generate the specific advice.

In general, there are five main techniques used by DSS: *Rule-based reasoning* uses logical statements or rules to infer knowledge. Those systems acquire specific information about a case and then invoke appropriate rules by an inference engine. Similarly, *symbolic modeling* is an approach which defines knowledge by structured organization of concepts and relations. Concept definitions are explicitly stated and sometimes constrained by logical statements used to infer knowledge. An *artificial neural network* (ANN) is composed of a collection of interconnected elements whereas connections between elements are weighted and constitute the knowledge of the network. ANN does not require defined expert rules and can learn directly from observations. Training of the network is performed by presenting input variables and the observed dependent output variable. The network then determines internodal connections between elements and uses this knowledge for classification of new cases. *Bayesion Networks*—also called probalistic networks—reason about uncertain knowledge. They use diverse medical information (e.g., physical findings, laboratory exam results, image study findings) to determine the probability of a disease. Each variable in the network has two or more states with associated probability values summing up to 1 for each variable. Connections between variables are expressed as conditional probabilities such as sensitivity or specificity. In this manner, probabilistic networks can be constructed on the basis of published statistical study results. *Case-based Reasoning* (CBR) systems use knowledge from prior experiences to solve new problems. The systems contain cases indexed by associated features. Indexing of new cases is performed by retrieving similar cases from memory and adapting solutions from prior experiences to the new case (Kahn, 1994).

Applications concerned with the detection of imaging findings by quantitative analysis are called computer-aided diagnosis (CAD) systems. Those systems frequently use ANN for image analysis and were successfully deployed for the detection of breast lesions (Giger et al., 1994; Huo et al., 1998; Jiang et al., 1999; Xu et al., 1997), lung nodules (Giger et al., 1994; Xu et al., 1997) (REF DOI), and colon polyps (Yoshida and Dachman, 2004). DSS systems concerned with the diagnosis of a disease were developed at first for the diagnosis of lung diseases (Asada et al., 1990; Gross et al., 1990), bone tumors in skeletal radiography (Piraino et al., 1991), liver lesions (Maclin and Dempsey, 1992; Tombropoulus et al., 1993), and breast masses (Kahn et al., 1997; Wu et al., 1995). In recent years, applications have been developed using symbolic models for reasoning tasks (Alberdi et al., 2000; Rubin et al., 2006) and a composite approach of symbolic modeling and Bayesian networks for diagnostic decision support in mammography (Burnside et al., 2000).

Even if all techniques infer knowledge in some manner, symbolic modeling and rule-based reasoning approaches conform more precisely to what is understood by *ontologies* today. As inferred knowledge often is not trivial to understand by the user, those approaches tend to be more comprehensible to humans due to their representation formalism. In fact, this is besides workflow integration and speed of the reasoning process, one of the most important factors affecting the successful implementation of DSS systems (Bates et al., 2003).

1.9.3 Results Communication

1.9.3.1 DICOM-Structured Reporting

The use of structured reporting forms reduces the ambiguity of natural language reports and enhances the precision, clarity, and value of clinical documents (Hussein et al., 2004). DICOM-Structured Reporting (SR) is a supplement of the DICOM Standard developed to facilitate the encoding and exchange of report information. The Supplement defines a document architecture for storage and transmission of structured reports by using the DICOM hierarchical structure and services.

An SR document consists of *Content Items,* which are composed of name/value pairs. The name (concept name) is represented by a coded entry that uses an attribute triplet: (1) the code value (a computer readable identifier), (2) the code scheme designator (the coding organization), and (3) the code meaning (human-readable text). The value of content items is used to represent the diverse information like containers (e.g., headings, titles), text, names, time, date, or codes. For specific reporting applications and tasks, SR templates were developed, which describe and constrain content items, value types, relationship types, and value sets for SR documents.

By the use of content items, text strings or standardized terms can be used to encode and interrelate the image information. For example, a mass can be described by properties like margin or size, which is achieved by relating content items through the relationship *has_properties.* In this manner, a structured report represents some kind of knowledge model in which image findings are related to each other (Figure 1.5).

To unify the representation of radiological findings, a model integrating UMLS terms, radiological findings, and DICOM SR has been proposed (Bertaud et al., 2008). This is a promising approach to standardize and integrate the knowledge about imaging observations and their representation in structured format. However, as DICOM SR defines only few relations and allows basic constrains on document items, its semantic and logical expressivity is limited. In future applications, the use

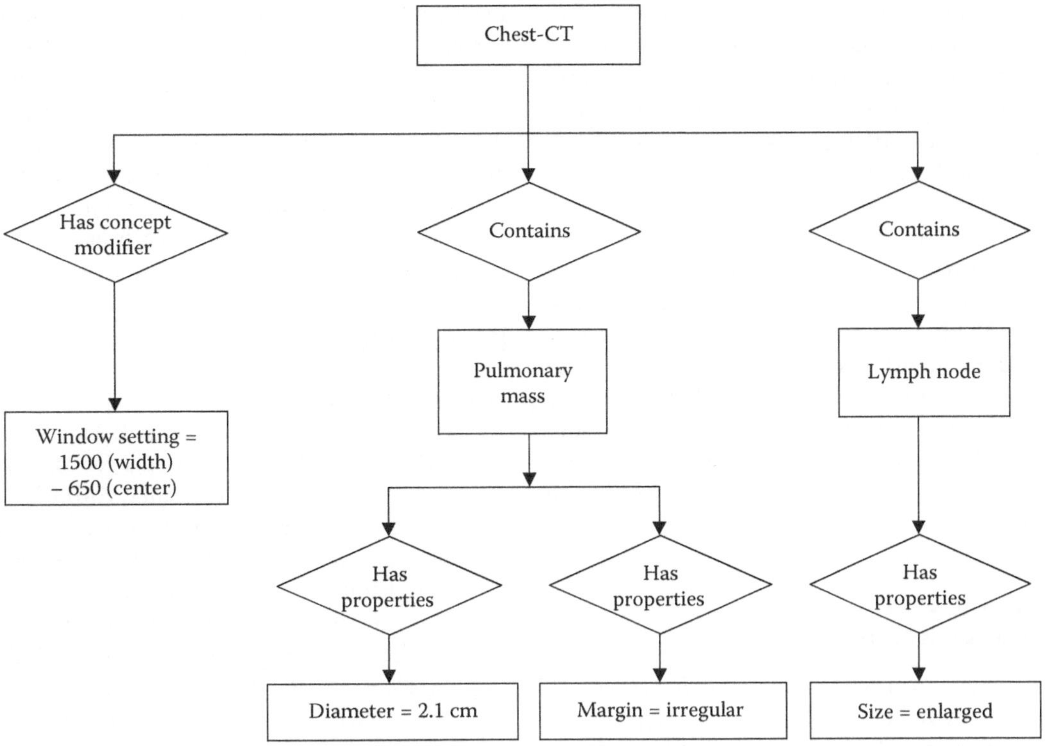

FIGURE 1.5 DICOM Structured Reporting Tree.

of a standardized radiological lexicons like RadLex and a more expressive representation formalism might increase the usefulness of structured reports and allows interoperability and analysis of imaging observations among different institutions.

1.9.3.2 Notification and Reminder

Notification and reminder systems track clinical data to issue alerts or inform physicians (Rubin, 2009). In radiology, such systems can be used to categorize the importance of findings and inform physicians about recommended actions. These systems facilitate the communication of critical results by assuring quick and appropriate communication (e.g., phone or email). In addition, systems can track the receipt of a message and send reminders if no appropriate action is taken. Communication and tracking of imaging results often are implemented in Web-based systems that have shown to improve the communication among radiologists, clinicians, and technologists (Halsted and Froehle, 2008; Johnson et al., 2005). As the primary basis for notification and reminder systems are imaging results, standardized terminologies and structured reports seem to be very useful as input for such systems in the future. However, definition of criteria for notification and reminder systems might benefit from ontologies capturing knowledge about imaging findings, clinical data, and recommended actions.

1.9.4 Semantic Image Retrieval

The number of digitally produced medical images is rising strongly and requires efficient strategies for management and access to those images. In radiology departments, access to image archives is usually based on patient identification or study characteristics (e.g., modality, study description) representing the underlying structure of data management.

Beginning in 1980, first systems were developed for querying images by content (Chang and Fu, 1980). With the introduction of digital imaging technologies, content-based image retrieval systems were developed using colors, textures, and shapes for image classification. Within radiology departments, applications executing classification and content-based search algorithms were introduced for mammography CT images of the lung, MRI and CT images of the brain, photon emission tomography (PET) images, and x-ray images of the spine (Muller et al., 2004).

Besides retrieving the images based on image content determined by segmentation algorithms or demographic and procedure information, the user often is interested in the context, that is, the meaning or interpretation of the image content (Kahn and Rubin, 2009; Lowe et al., 1998). One way to incorporate context in image retrieval applications is to index radiology reports or figure captions (Kahn and Rubin, 2009). Such approaches are encouraging if textual information is mapped to concept-based representations to reduce equivocal image retrieval results by lexical variants or ambiguous abbreviations.

Current context-based approaches for image retrieval use concepts like imaging technique (e.g., "chest x-ray"), anatomic region or field of view (e.g., "anterioposterior view"), major anatomic segments (e.g., "thorax"), image features (e.g., "density"), and findings (e.g., "pneumonia") for image retrieval. However, for a comprehensive semantic image retrieval application, a knowledge model of anatomical and pathological structures displayed on images and its image features would be desirable. For many diseases, however, image features are not unique and its presence or combination in a specific clinical context produces lists of possible diagnoses with different degrees of certainty. In this regard, criteria for diagnoses inferred from images are often imprecise and ill-defined and considerable intra- and interobserver variation is common (Tagare et al., 1997).

There have been some efforts to retrieve images based on semantic medical information. For example, indexing images by structured annotations using a standardized radiological lexicon (RadLex) allow the user to store such annotation together with images. Such annotations than can be queried and similar patients or images can be retrieved on the basis of the annotated information (Channin et al., 2009). Other approaches use automatic segmentation algorithms and concept-based annotations to label image content and use those concepts for image retrieval (Seifert et al., 2010).

1.9.5 Teaching Cases, Knowledge Bases, and E-Learning

There is a long tradition of collecting and archiving images for educational purposes in radiology. With the development of digital imaging techniques and PACS, images from interesting cases can be easily labeled or exported in collections. In recent years, many systems have been developed to archive, label, and retrieve images. Such systems often provide the possibility to attach additional clinical information to images or cases and share teaching files through the Web like the Medical Image Resource System (MIRC) (Siegel and Reiner, 2001). Today, many departments possess teaching archives that are continuously populated with cases encountered in the daily work routine. In fact, various comprehensive teaching archives exist on the Web providing extensive teaching cases (Scarsbrook et al., 2005).

One major challenge in the management of teaching files is the organization of cases for educational purposes. Most teaching archives label cases by examination type (e.g., "MRI"), body region (e.g., "abdominal imaging"), and diagnoses (e.g., "myxoid fibrosarcoma") using text strings. Even if many archives represent similar cases, such systems deploy their own information and organizational model and contain non uniform labels. One important aspect in usability and interoperability of teaching archives is the use of a standardized terminology and knowledge model for organization and retrieval of cases together with a strict guideline for labeling cases. An ontology- or concept-based organization of semantic image content would empower users to query cases by explicit criteria like combination of morphological features and classify cases according to additional attributes like analytical or perceptual difficulty.

The use of electronic educational material is called e-learning and many Web-based applications have been developed to present medical images together with additional educational material electronically. Most implementations deploy a learning management system to organize, publish, and maintain the material. Such systems usually encompass registration, delivery and tracking of multimedia courses and content, communication and interactions between students/residents and educators, and testing (Sparacia et al., 2007). Some e-learning applications for radiology are in use and such systems would certainly benefit from concept-based organization of semantic image content. In this way, cases and knowledge in existing teaching archives could be re-used within e-learning applications and interpretation and inference patterns frequently encountered in radiology could be used for the education of students and residents.

References

Alberdi, E., Taylor, P., Lee, R., Fox, J., Sordo, M., and Todd-Pokropek, A., 2000. Cadmium II: Acquisition and representation of radiological knowledge for computerized decision support in mammography. *Proc. AMIA Symp.*, pp. 7–11.

Asada, N., Doi, K., Macmahon, H., Montner, S. M., Giger, M. L., Abe, C., and Wu, Y., 1990. Potential usefulness of an artificial neural network for differential diagnosis of interstitial lung diseases: Pilot study. *Radiology*, 177, 857–60.

Austin, J. H., Muller, N. L., Friedman, P. J., Hansell, D. M., Naidich, D. P., Remy-jardin, M., Webb, W. R., and Zerhouni, E. A., 1996. Glossary of terms for CT of the lungs: Recommendations of the Nomenclature Committee of the Fleischner Society. *Radiology*, 200, 327–31.

Baader, F., Calvanese, D., Mcguiness, D., Nardi, D., and Patel-Schneider, P. (Eds.), 2003. *The Description Logics Handbook.* Cambridge University Press.

Baader, F., Horrocks, I., and Sattler, U. 2009. Description logics. In Staab, S. and Studer, R. (Eds.) *Handbook on Ontologies.* Berlin: Springer.

Barnes, M. and Barnett, G. O. 1995. An architecture for a distributed guideline server. *Proc. Ann. Symp. Comput. Appl. Med. Care*, pp. 233–7.

Barrows, R. C., Jr., Allen, B. A., Smith, K. C., Arni, V. V., and Sherman, E. 1996. A decision-supported outpatient practice system. *Proc. AMIA Ann. Fall Symp.*, pp. 792–6.

Bates, D. W., Kuperman, G. J., Wang, S., Gandhi, T., Kittler, A., Volk, L., Spurr, C., Khorasani, R., Tanasijevic, M., and Middleton, B. 2003. Ten commandments for effective clinical decision support: Making the practice of evidence-based medicine a reality. *J. Am. Med. Inform. Assoc.*, 10, 523–30.

Bell, D. S. and Greenes, R. A. 1994. Evaluation of UltraSTAR: Performance of a collaborative structured data entry system. *Proc. Ann. Symp. Comput. Appl. Med. Care*, pp. 216–22.

Bell, D. S., Pattison-Gordon, E., and Greenes, R. A. 1994. Experiments in concept modeling for radiographic image reports. *J. Am. Med. Inform. Assoc.*, 1, 249–62.

Bertaud, V., Lasbleiz, J., Mougin, F., Burgun, A., and Duvauferrier, R. 2008. A unified representation of findings in clinical radiology using the UMLS and DICOM. *Int. J. Med. Inform.*, 77, 621–9.

Bodenreider, O. and Burgun, A. 2005. Biomedical Ontologies. *Medical Informatics: Knowlegde Managment and Datamining in Biomedicine.* Berlin: Springer.

Boxwala, A. A., Peleg, M., Tu, S., Ogunyemi, O., Zeng, Q. T., Wang, D., Patel, V. L., Greenes, R. A., and Shortliffe, E. H. 2004. GLIF3: A representation format for sharable computer-interpretable clinical practice guidelines. *J. Biomed. Inform.*, 37, 147–61.

Burgun, A. 2006. Desiderata for domain reference ontologies in biomedicine. *J. Biomed. Inform.*, 39, 307–13.

Burnside, E., Rubin, D., and Shachter, R. 2000. A Bayesian network for mammography. *Proc. AMIA Symp.*, pp. 106–10.

Chang, N. S. and Fu, K. S. 1980. Query-by-pictorial-example. *IEEE Trans. Software Eng.*, 6, 519–24.

Channin, D. S., Mongkolwat, P., Kleper, V., and Rubin, D. L. 2009. The annotation and image mark-up project. *Radiology*, 253, 590–2.

Field, M. and Lohr, K. (Eds.) 1992. *Guidelines for Clinical Practice: From Development to Use.* Washington, DC: National Academy Press.

Friedman, C., Cimino, J. J., and Johnson, S. B. 1993. A conceptual model for clinical radiology reports. *Proc. Ann. Symp. Comput. Appl. Med. Care*, pp. 829–33.

Genesereth, M. and Nilsson, N. 1987. *Logical Foundations of Artificial Intelligence.* Los Altos, CA: Morgan Kaufmann.

Giger, M. L., Bae, K. T., and Macmahon, H. 1994. Computerized detection of pulmonary nodules in computed tomography images. *Invest. Radiol.*, 29, 459–65.

Grenon, P. and Smith, B. 2004. SNAP and SPAN: Towards dynamic spatial ontology. *Spatial Cogn. Computat.*, 4, 69–103.

Gross, G. W., Boone, J. M., Greco-Hunt, V., and Greenberg, B. 1990. Neural networks in radiologic diagnosis. II. Interpretation of neonatal chest radiographs. *Invest. Radiol.*, 25, 1017–23.

Gruber, T. 1993. A translation approach to portable ontology specifications. *Knowledge Acquisition*, 5, 199–220.

Halsted, M. J. and Froehle, C. M. 2008. Design, implementation, and assessment of a radiology workflow management system. *AJR Am. J. Roentgenol.*, 191, 321–7.

Herre, H., Heller, B., Burek, P., Hoehndorf, R., Loebe, F., and Michalek, H. 2006. General Formal Ontology (GFO): A Foundational Ontology Integrating Objects and Processes. Part I: Basic Principles (Version 1.0). *Onto-Med Report, Nr. 8. Research Group Ontologies in Medicine (Onto-Med), University of Leipzig.* http://www.onto-med.de/publications/#reports.

Hripcsak, G., Clayton, P. D., Jenders, R. A., Cimino, J. J., and Johnson, S. B. 1996. Design of a clinical event monitor. *Comput. Biomed. Res.*, 29, 194–221.

Huo, Z., Giger, M. L., Vyborny, C. J., Wolverton, D. E., Schmidt, R. A., and Doi, K. 1998. Automated computerized classification of malignant and benign masses on digitized mammograms. *Acad. Radiol.*, 5, 155–68.

Hussein, R., Engelmann, U., Schroeter, A., and Meinzer, H. P. 2004. Dicom structured reporting: Part 1. Overview and characteristics. *Radiographics*, 24, 891–6.

Jiang, Y., Nishikawa, R. M., Schmidt, R. A., Metz, C. E., Giger, M. L., and Doi, K. 1999. Improving breast cancer diagnosis with computer-aided diagnosis. *Acad. Radiol.*, 6, 22–33.

Johnson, A. J., Hawkins, H., and Applegate, K. E. 2005. Web-based results distribution: New channels of communication from radiologists to patients. *J. Am. Coll. Radiol.*, 2, 168–73.

Kahn, C. E., Jr. 1994. Artificial intelligence in radiology: Decision support systems. *Radiographics*, 14, 849–61.

Kahn, C. E., Jr. 1998. An Internet-based ontology editor for medical appropriateness criteria. *Comput. Methods Programs Biomed.*, 56, 31–6.

Kahn, C. E., Jr., Roberts, L. M., Shaffer, K. A., and Haddawy, P. 1997. Construction of a Bayesian network for mammographic diagnosis of breast cancer. *Comput. Biol. Med.*, 27, 19–29.

Kahn, C. E., Jr. and Rubin, D. L. 2009. Automated semantic indexing of figure captions to improve radiology image retrieval. *J. Am. Med. Inform. Assoc.*, 16, 380–6.

Kuhn, K., Gaus, W., Wechsler, J. G., Janowitz, P., Tudyka, J., Kratzer, W., Swobodnik, W., and Ditschuneit, H. 1992. Structured reporting of medical findings: Evaluation of a system in gastroenterology. *Methods Inf. Med.*, 31, 268–74.

Langlotz, C. P. 2006. RadLex: A new method for indexing online educational materials. *Radiographics*, 26, 1595–7.

Levy, A. 2002. *Basic Set Theory.* Dover Publications.

Liberman, L. and Menell, J. H. 2002. Breast imaging reporting and data system (BI-RADS). *Radiol. Clin. North Am.*, 40, 409–30.

Liu, D., Zucherman, M., and Tulloss, W. B., Jr. 2006. Six characteristics of effective structured reporting and the inevitable integration with speech recognition. *J. Digit Imaging*, 19, 98–104.

Lowe, H. J., Antipov, I., Hersh, W., and Smith, C. A. 1998. Towards knowledge-based retrieval of medical images. The role of semantic indexing, image content representation and knowledge-based retrieval. *Proc. AMIA Symp.*, pp. 882–6.

Maclin, P. S. and Dempsey, J. 1992. Using an artificial neural network to diagnose hepatic masses. *J. Med. Syst.*, 16, 215–25.

Marwede, D., Fielding, M., and Kahn, T. 2007. Radio: A prototype application ontology for radiology reporting tasks. *AMIA Ann. Symp. Proc.*, pp. 513–7.

Marwede, D., Schulz, T., and Kahn, T. 2008. Indexing thoracic CT reports using a preliminary version of a standardized radiological lexicon (RadLex). *J. Digit Imaging*, 21, 363–70.

Masolo, C., Borgo, S., Gangemi, A., Guarino, N., and Oltramari, A. 2003. WonderWeb Deliverable D18. http://wonderweb.semanticweb.org.

Mejino, J. L., Rubin, D. L., and Brinkley, J. F. 2008. FMA-RadLex: An application ontology of radiological anatomy derived from the foundational model of anatomy reference ontology. *AMIA Ann. Symp. Proc.*, pp. 465–9.

Minsky, M. 1975. A framework for representing knowlegde. In Winston, P. (Ed.), *The Psychology of Computer Vision.* McGraw-Hill.

Muller, H., Michoux, N., Bandon, D., and Geissbuhler, A. 2004. A review of content-based image retrieval systems in medical applications: Clinical benefits and future directions. *Int. J. Med. Inform.*, 73, 1–23.

Musen, M. A., Tu, S. W., Das, A. K., and Shahar, Y. 1996. EON: A component-based approach to automation of protocol-directed therapy. *J. Am. Med. Inform. Assoc.*, 3, 367–88.

Noy, N. F., Crubezy, M., Fergerson, R. W., Knublauch, H., Tu, S. W., Vendetti, J., and Musen, M. A. 2003. Protege-2000: An open-source ontology-development and knowledge-acquisition environment. *AMIA Ann. Symp. Proc.*, pp. 953.

Piraino, D. W., Amartur, S. C., Richmond, B. J., Schils, J. P., Thome, J. M., Belhobek, G. H., and Schlucter, M. D. 1991. Application of an artificial neural network in radiographic diagnosis. *J. Digit Imaging*, 4, 226–32.

Purves, I. N. 1998. PRODIGY: Implementing clinical guidance using computers. *Br. J. Gen. Pract.*, 48, 1552–3.

Rector, A. L., Bechhofer, S., Goble, C. A., Horrocks, I., Nowlan, W. A., and Solomon, W. D. 1997. The GRAIL concept modelling language for medical terminology. *Artif. Intell. Med.*, 9, 139–71.

Rector, A. L. and Nowlan, W. A. 1994. The GALEN project. *Comput. Methods Programs Biomed.*, 45, 75–8.

Rosenthal, D. I., Weilburg, J. B., Schultz, T., Miller, J. C., Nixon, V., Dreyer, K. J., and Thrall, J. H. 2006. Radiology order entry with decision support: Initial clinical experience. *J. Am. Coll. Radiol.*, 3, 799–806.

Rosse, C. and Mejino, J. L., Jr. 2003. A reference ontology for biomedical informatics: The Foundational Model of Anatomy. *J. Biomed. Inform.*, 36, 478–500.

Rubin, D. L. 2009. Informatics methods to enable patient-centered radiology. *Acad. Radiol.*, 16, 524–34.

Rubin, D. L., Dameron, O., Bashir, Y., Grossman, D., Dev, P., and Musen, M. A. 2006. Using ontologies linked with geometric models to reason about penetrating injuries. *Artif. Intell. Med.*, 37, 167–76.

Rubin, D. L., Rodriguez, C., Shah, P., and Beaulieu, C. 2008. iPad: Semantic annotation and markup of radiological images. *AMIA Ann. Symp. Proc.*, pp. 626–30.

Scarsbrook, A. F., Graham, R. N., and Perriss, R. W. 2005. The scope of educational resources for radiologists on the internet. *Clin. Radiol.*, 60, 524–30.

Seifert, S., Kelm, M., Moeller, M., Huber, M., Cavallaro, A., and Comaniciu, D. 2010. Semantic annotation of medical images. *Proc. SPIE Medical Imaging*, San Diego.

Siegel, E. and Reiner, B. 2001. The Radiological Society of North America's Medical Image Resource Center: An update. *J. Digit Imaging*, 14, 77–9.

Sistrom, C. L. 2005. Conceptual approach for the design of radiology reporting interfaces: The talking template. *J. Digit Imaging*, 18, 176–87.

Sistrom, C. L. 2008. In support of the ACR Appropriateness Criteria. *J. Am. Coll. Radiol.*, 5, 630–5; discussion 636–7.

Sistrom, C. L., Dang, P. A., Weilburg, J. B., Dreyer, K. J., Rosenthal, D. I., and Thrall, J. H. 2009. Effect of computerized order entry with integrated decision support on the growth of outpatient procedure volumes: Seven-year time series analysis. *Radiology*, 251, 147–55.

Sittig, D. F. and Stead, W. W. 1994. Computer-based physician order entry: The state of the art. *J. Am. Med. Inform. Assoc.*, 1, 108–23.

Smith, B., Ceusters, W., Klagges, B., Kohler, J., Kumar, A., Lomax, J., Mungall, C., Neuhaus, F., Rector, A. L., and Rosse, C. 2005. Relations in biomedical ontologies. *Genome Biol.*, 6, R46.

Smith, B. and Rosse, C. 2004. The role of foundational relations in the alignment of biomedical ontologies. *Stud. Health Technol. Inform.*, 107, 444–8.

Sowa, J. 1987. Semantic networks. In Shapiro, S. (Ed.) *Encyclopedia of Artificial Intelligence* (2nd Ed.). Wiley.

Spackman, K. A., Campbell, K. E., and Cote, R. A. 1997. Snomed RT: A reference terminology for healthcare. *Proc. AMIA Ann. Fall Symp.*, pp. 640–4.

Sparacia, G., Cannizzaro, F., D'Alessandro, D. M., D'Alessandro, M. P., Caruso, G., and Lagalla, R. 2007. Initial experiences in radiology e-learning. *Radiographics*, 27, 573–81.

Staab, S. and Studer, R. (Eds.) 2009. *Handbook on Ontologies.* Berlin: Springer.

Stoufflet, P., Ohno-Machado, L., Deibel, S., Lee, D., and Greenes, R. 1996. Geode-CM: A state transition framework for clinical management. In *Proc. 20th Ann. Symp. Comput. Appl. Med. Care.*

Tagare, H. D., Jaffe, C. C., and Duncan, J. 1997. Medical image databases: A content-based retrieval approach. *J. Am. Med. Inform. Assoc.*, 4, 184–98.

Tjahjono, D. and Kahn, C. E., Jr. 1999. Promoting the online use of radiology appropriateness criteria. *Radiographics*, 19, 1673–81.

Tolk, A. and Muguira, J. 2003. The levels of conceptual interoperability model (LCIM). *Proc. IEEE Fall Sim. Interoperability Workshop.* IEEE CS Press.

Tombropoulus, R., Shiffman, S., and Davidson, C. 1993. A decision aid for diagnosis of liver lesions on MRI. *Proc. Ann. Symp. Comput. Appl. Med. Care.*

Tuddenham, W. J. 1984. Glossary of terms for thoracic radiology: Recommendations of the Nomenclature Committee of the Fleischner Society. *AJR Am. J. Roentgenol.*, 143, 509–17.

Turnitsa, C. 2005. Extending the levels of conceptual interoperability model. *Proc. IEEE Summer Comp. Sim. Conf.* IEEE CD Press.

Wu, Y. C., Freedman, M. T., Hasegawa, A., Zuurbier, R. A., Lo, S. C., and Mun, S. K. 1995. Classification of microcalcifications in radiographs of pathologic specimens for the diagnosis of breast cancer. *Acad. Radiol.*, 2, 199–204.

Xu, X. W., Doi, K., Kobayashi, T., Macmahon, H., and Giger, M. L. 1997. Development of an improved CAD scheme for automated detection of lung nodules in digital chest images. *Med. Phys.*, 24, 1395–403.

Yoshida, H. and Dachman, A. H. 2004. Computer-aided diagnosis for CT colonography. *Semin Ultrasound CT MR*, 25, 419–31.

2

Informatics Constructs

Steve G. Langer
Mayo Clinic

2.1 Background

2.1.1 Terms and Definitions

Actor: In a particular Use Case, Actors are the agents that exchange data via Transactions, and perform operations on that data, to accomplish the Use Case goal (Alhir, 2003).

Class: In programming and design, the class defines an Actor's data elements, and the operations it can perform on those data (Alhir, 2003).

Constructs: Constructs are conceptual aids (often graphical) that visually express the relationships among Actors, Transactions, transactional data, and how they inter-relate in solving Use Cases.

Informatics: Medical Informatics has been defined as "that area that concerns itself with the cognitive, information processing, and communication tasks of medical practice, education, and research, including the information science and the technology to support these tasks" (Greenes and Shortliffe, 1990). More broadly, informatics is a given branch of knowledge and how it is acquired, represented, stored, transmitted, and mined for meaning (Langer and Bartholmai, 2010).

Object: An Object is the real world instantiation of a Class with specific data.

Ontology: A specification of a representational vocabulary for a shared domain of discourse—definitions of classes, relations, functions, and other objects (Gruber, 1993). Another way to consider ontology is the collection of content terms and their relationships that are agreed to represent concepts in a specific branch of knowledge. A common example is HTTP (Hypertext Transfer Protocol), which is the grammar/protocol used to express HTML (Hypertext Markup Language) content on the World Wide Web.

Protocol: Protocols define the transactional format for transmission of information via a standard Ontology among Actors (Holzmann, 1991).

Transactions: Messages that are passed among Actors using standard Protocols that encapsulate the standard terms of an Ontology. The instance of a communication pairing between two Actors is known as an association.

Use Case: A formal statement of a specific workflow, the inputs and outputs, and the Actors that accomplish the goal via the exchange of Transactions (Bittner and Spence, 2002).

2.1.2 Acquired, Stored, Transmitted, and Mined for Meaning

As defined above, the term "Informatics" can be applied to many areas; bioinformatics concerns the study of the various scales of living systems. Medical Imaging Informatics, the focus of this book, is concerned with the methods by which medical images are acquired, stored, viewed, shared, and mined for meaning. The purpose of this chapter is to provide the background to understand the constituents of Medical Imaging Informatics that will be covered in more detail elsewhere in this book. After reading it, the reader should have sufficient background to place the material in Chapters 1 (Ontology), 3 (HL7), and 4 (DICOM) in a cohesive context and be in a comfortable position to understand the spirit and details of Chapter 5 (IHE, Integration of the Healthcare Enterprise).

As will ultimately become clear, the goal of patient care is accomplished via the exchange of Transactions among various Actors; such exchanges are illustrated by a variety of constructs, consisting of various diagram types. These diagrams are ultimately tied rendered with the content Transactions, Protocols, and Actors that enable the solution of Use Case scenarios.

2.2 Acquired and Stored

When either humans or machines make measurements or acquire data in the physical world, there are several tasks that must be accomplished:

a. The item must be measured in a standard, reproducible way or it has no benefit.
b. The value's magnitude and other features must be represented in some persistent symbolic format (i.e., writing on paper, or bits in a computer) that has universally agreed meanings.
c. If the data is to be shared, there must be a protocol that can encapsulate the symbols and transmit them among humans (as in speech or writing) or machines (electromagnetic waves or computer networks) in transactions that have a standard, universally understood, structure.

2.2.1 Data Structure and Grammar

2.2.1.1 HL7

The Health Level 7 (HL7) standard is the primary grammar used to encapsulate symbolic representations of healthcare data among computers dealing in nonimaging applications (Henderson, 2007). It will be covered in detail in Chapter 3, but for the purposes of the current discussion it is sufficient to know just a few basic concepts. First, that HL7 specifies both events and the message content that can accompany those events. Second, some aspects of HL7 have strictly defined allowed terms, while other message "payloads" can have either free text (i.e., radiology reports) or other variable content. Consider Figure 2.1.

Finally, HL7 transactions can be expressed in two different protocols: the classical HL7 format (versions V2.x), which relies on a low-level networking protocol called TCP/IP (see Section 2.3), is exemplified in Figures 2.1 and the new XML format (for HL7 V3.x) is shown in Figure 2.2.

(a)

```
MSH|^~\&|RIMS|MCR|IHE-ESB|MCR|20101116103737||ORM^O01|1362708283|P|2.3.1|||||||||
PID||2372497|03303925^^^^MC~033039256^^^^CYCARE~AU0003434^^^^AU|03-303-925^^^^MC~03-303-925-
6^^^^CYCARE~AU0003434^^^^AU|TESTING^ANN^M.^^^||19350415|F|||||||||||||||||||||
PV1||O|^^^^ROMAYO||||||||||||||||||||||||||||||||||||||||||||||||| ORC|SC|429578441-1^MSS|429578441-
1^RIMS||NW||^^^201011161100^^NORM|||10181741^CLEMENTS^IAN^P||10181741^CLEMENTS^IAN^P|E2X-
REC||||||^^^| OBR|0001|429578441-1^MSS|429578441-1|07398^Chest-- PA \T\ Lateral^RIMS|NORM||||||||testing
interface to PCIL||^^^^N Chest-- PA \T\ Lateral|10181741^CLEMENTS^IAN^P||429578441-1|429578441-
1|07398||201011161037||CR||||||| |||&&&||||||||||07398^Chest-- PA \T\ Lateral^RIMS^^^|
ZDS|1.2.840.113717.2.429578441.1^RIMS^Application^DICOM|
Z01|NW|201011161037|0055|||MCRE3|201011161037|201011161100|N||
```

(b)

```
MSH|^~'&|RADIOLO|ROCHESTER|ESB||20101110072148||ORU^R01|1362696376|P|2.3.1|||||
PID|||06004163||Fall^Autumn^E.^^^||19720916|F||||||||||||||||||
PV1||O|RADIOLOGY^||||||||||||||||||||||||||||||||||||||||||||||
OBR|||429578288-2|07201^CT Head
wo^RRIMS|||201011100720|201011100721||||||201011100721||10247131^BRAUN^COLLEEN^M^^^^PERSONI
D||SMH|SMHMMB|429578288-3|N|201011100721||CT|F||^^^^^^R||||testing|99999990^RADIOLOGY
STAFF^BRAUN^^^^^PERSONID|||10247131|
OBX|1|TX|07201^CT Head wo^RRIMS|429511111|{\rtf1\ansi \deff1\deflang1033\ {\fonttbl{\f1\fmodern\fcharset0
Courier;}{\f2\fmodern\fcharset0 Courier;}} \pard\plain \f1\fs18\tx0604\par ||||||P|
OBX|2|TX|07201^CT Head wo^RRIMS|429511111|10-Nov-2010 07:20:00  Exam: CT Head wo\par ||||||P|
OBX|3|TX|07201^CT Head wo^RRIMS|429511111|Indications: testing\par ||||||P|
OBX|4|TX|07201^CT Head wo^RRIMS|429511111|ORIGINAL REPORT - 10-Nov-2010 07:21:00   SMH\par
||||||P|
OBX|5|TX|07201^CT Head wo^RRIMS|429511111|test\par ||||||P|
OBX|6|TX|07201^CT Head wo^RRIMS|429511111|Electronically signed by: \par ||||||P|
OBX|7|TX|07201^CT Head wo^RRIMS|429511111|Radiology Staff, Braun 10-Nov-2010 07:21 \par }||||||P|_
```

FIGURE 2.1 (a) Health Level 7 consists of messages, whose transfer is initiated by messages and events. This figure shows an Order. (b) This is the resulting OBX message that contains the content (a radiology report in this case from a CT).

2.2.1.2 DICOM

While HL7 has found wide acceptance in most medical specialties, it was found insufficient for medical imaging. Hence in 1993, the American College of Radiology (ACR) and National Electrical Manufacturers Association (NEMA) collaborated to debut DICOM (Digital Imaging Communications in Medicine) at the Radiological Society of North America annual meeting. DICOM introduced the concept of Service–Object Pairs, which relates for certain object types what services can be applied to them (i.e., store, get, print, display). DICOM is also much stronger "typed" then HL7, meaning that specific data elements not only have fixed data type that can be used, but fixed sizes as well.

2.2.1.3 XML

The eXtensible Markup Language (XML) is an extension to the original HTML (Hypertext Markup Language) that was invented by Tim Berners-Lee in the early 1990s (Berners-Lee and Fischetti, 1999). It differs from HTML (Figure 2.3) in that in addition to simply formatting the page's presentation state, it also enables defining what the content of page elements are. In other words, if a postal code appeared on the Web page, the XML page itself could wrap that element with the tag "postal-code." By self-documenting the page content, it enables computer programs to scan XML pages in a manner similar to a database, if the defined terms are agreed upon.

2.2.2 Content

While a protocol grammar defines the structure of transactions, the permitted terms (and the relationships among them) are defined by specific ontologies. It is the purpose of a specific ontology to define the taxonomy (or class hierarchies) of specific classes, the objects within them, and how they are related. The following examples address different needs, consistent with the areas they are tailored to address.

```
<Labrs3P00 T=" Labrs3P00">
    < Labrs3P00.PTP T="PTP">
        <PTP.primePrsnm T="NM">
            <fmn T="ST"> Jones </fmn>
            <gvn T="ST"> Tim </gvn>
            <mdn T="ST"> H </mdn>
        </PTP.primePrsnm>
    < /Labrs3P00.PTP>
    <Labrs3P00.SI00_L T="SI00_L">
        <SI00_L.item T="SI00">
            <SI00.filrOrdId T="IID">LABGL110802< /SI00.filrOrdId >
            <SI00.placrOrdID T="IID">DMCRES387209373</SI00.placrOrdID>
            <SI00.InsncOf T="MSRV">
                <MSRV.unvSvcId T="CE">18768-1<.MSRV.unvSvcId>
                < MSRV.svcDesc T="TX">Cell Counts< /MSRV.svcDesc>
            </SI00.InsncOf>
            <SI00.SRVE_L T="SRVE_L">
                <SRVE_L.item T="SRVE">
                    <SRVE.name T="CE">4544-3</ SRVE.name>
                    <SRVE.svcEventDesc T="ST">Hematocrit</SRVE.svcEventDesc>
                    <SRVE.CLOB T="CLOB">
                        <CLOB.obsvnValu T="NM">45< /CLOB.obsvnValu >
                        <CLOB.refsRng T="ST">39-49< /CLOB.refsRng >
                        <CLOB.clnRvlnBgmDtm T="DTM">199812292128</CLOB.clnRvlnBgmDtm >
                    </SRVE.CLOB>
                    <SRVE.spcmRcvdDtm T="DTM">199812292135</SRVE.spcmRcvdDtm >
                </SRVE_L.item>
            </SI00.SRVE_L>
    < Labrs3P00.SI00_L>
</Labrs3P00>
```


FIGURE 2.2 HL7 is available in two formats; the version 2.x in wide use today is expressed in the format shown in Figures 2.1. The HL7 V3.0 is encoded in XML as seen here; note this sample explicitly states it contains laboratory values.

2.2.2.1 SNOMED

Developed in 1973, SNOMED (Systemized Nomenclature of Medicine) was developed by pathologists working with the College of American Pathologists. Its purpose is to be a standard nomenclature of clinical medicine and findings (Cote, 1986). By 1993, SNOMED V3.0 achieved international status. It has 11 top level classes (referred to as "axis") that define: anatomic terms, morphology, bacteria/viruses, drugs, symptoms, occupations, diagnoses, procedures, disease agents, social contexts and relations, and syntactical qualifiers. Any disease or finding may descend from one or more of those axes, for example, lung (anatomy), fibrosis (diagnosis), and coal miner (occupation).

2.2.2.2 RadLex

While SNOMED addressed the need for a standard way to define illness and findings with respect to anatomy, morphology, and other factors, RadLex seeks to address the specific subspecialty needs of radiology. Beginning in 2005, the effort started with six organ-based committees in coordination with 30 standards organizations and professional societies (Langlotz, 2006). In 2007, six additional committees were formed to align the lexicon along the lines of six modalities; the result is now referred to as the RadLex Playbook.

2.2.2.3 ICD9

The International Statistical Classification of Diseases and Related Health Problems, better known as ICD, was created in 1992 and is now in its 10th version, although many electronic systems may still be using V9.0 (Buck, 2011). Its purpose is to classify diseases and a wide variety of signs, symptoms, abnormal findings, complaints, social circumstances, and external causes of injury or disease. It is used by the World Health Organization (WHO) and used worldwide for morbidity and mortality statistics. It is also often used to encode the diagnosis from medical reports into a machine-readable format that is used by Electronic Medical Record (EMR) and billing systems. The lexicon is structured using the following example: A00-B99 encodes infections and parasites, C00-D48 encodes neoplasms and cancers, and so on through U00-U99 (special codes).

2.3 Transmission Protocols

The previous section described two of the basic components of informatics constructs: symbols to encode concepts (ontologies) and grammars to assemble those symbols into standard messages. An analogy is helpful. Verbs, nouns, and adjectives form the ontology in speech. Subjects, predicates, and objects of the verb form the basis of spoken grammar. What is missing in both our healthcare messaging and speech example is a method to transmit the message to a remote "listener." The human speech solution to this challenge is writing and the printing press. The electronic analogs are computer transmission protocols.

2.3.1 TCP/IP

Transmission Control Protocol/Internet Protocol (TCP/IP) is a layering of concepts to enable the transmission of messages

```
<html>
<head>
<meta content="text/html;charset=ISO-8859-1 " http-equiv="Content-Type">
<title>html example</title>
<head>
<body>
<h1 style="text-align: center;">This is an Example of HTML formatting tags</h1>
<br>
The above part is bold and centered. This part is left-justified and
normal font size and weight<br>
<br>
This next. pan. is a table<br>
<br>
<table style="text-align: left; width: 100%;" border="1" cellpadding="2"
cellspacing="2">
<tbody>
<tr>
<td style="vertical-align: top;">1<br>
</td>
<td style="vertical-align: top;">3<br>
</td>
</tr>
<tr>
<td style="vertical-align: top;">2<br>
</td>
<td style="vertical-align: top;">4<br>
</td>
</tr>
</tbody>
</table>
<br>
And this is the end of this document.<br>
<br>
</body>
</html>
```

FIGURE 2.3 HTML (Hypertext Markup Language) is a text markup language that informs the appropriate Web browsers (e.g., Firefox) how to render a page, but has no provision for encoding the content meaning of the page. By contrast, XML (as seen in Figure 2.2) adds the capability to express the meaning of the page content through the use of agreed upon "tags."

consisting of bits from one computer to another. The rules of the protocol guarantee that all the bits arrive, uncorrupted, in the correct order. The layers referred to are a result of the original formulations by the Internet Engineering Task Force (IETF) of what has come to be known as TCP/IP. Basically, if one starts at the physical layer (the network interface card), the naming convention is physical or link layer (layer one), Internet layer (layer two), transport layer (layer three), and the application layer (layer four) Request for Comment, RFC 1122–1123). Several years later, the International Standards Organization created the seven layer Open Systems Interconnect (OSI) model, which can lead to confusion if one does not know which system is being referenced (Zimmermann, 1980). For our purposes, it is sufficient to know that the further protocols discussed below ride on top of TCP/IP and rely on its guarantees of uncorrupted packet delivery in the correct order.

2.3.2 DICOM

Yes, DICOM again. This can be a point of some confusion, but DICOM is both an ontology and a protocol. Recall from Section 2.2.1.2 the concept of Service–Object Pairs. The objects are the message content (i.e., images, structured reports, etc.). The services are the actions that can be applied to the objects, and this includes transmitting them. The transactions that are responsible for network transmission of DICOM objects have names like C-MOVE and C-STORE. To facilitate the network associations among two computers to perform the transfer, the DICOM standard defines the process of *transfer syntax negotiation*. This process, between the server (service class provider or SCP in DICOM) and client (service class user or SCU), makes sure that the SCP can provide the required service, with the same kind of image compression, and in the right format for the computer processor on the SCU.

2.3.3 HTTP

Recall from Section 2.2.1.3 that Tim Berners-Lee invented HTML, the first widely used markup language to render Web pages in a Web reader. However, there remained the need to transfer such pages from server computers to the users that possessed the Web-reading clients (i.e., Internet Explorer or Firefox). The Hyper Text Transfer Protocol was invented to fill that role (RFC 2616). As alluded to earlier, HTTP is an application level protocol that rides on the back of the underlying TCP/IP protocol. Since its beginning, HTTP has been expanded to carry not just HTML-encoded patients, but XML content and other encapsulated arbitrary data payloads as well (i.e., images, executable files, binary files, etc.). Another enhancement, HTTPS (S is for secure), provides encryption between the endpoints of the communication and is the basis for trusted Internet-shopping stores (i.e., Amazon) to online (RFC 2818).

2.4 Diagrams

2.4.1 Classes and Objects

We have defined a step at a time the components which shall now come together in the informatics constructs generally referred to as diagrams. When one begins to read actual informatics system documentation (i.e., DICOM or IHE conformance statements), a typical point of departure is the Use Case. We will see examples of those in the next section, but for now it is useful to know that Use Cases leverage Actors, and Actors can be considered to be the equivalent of the Class as defined in computer science.

Recall from Section 2.1.1 that a Class defines an Actor's data elements, and the operations it can perform on those data. A simple real world example might be the class of temperature sensors. A temperature sensor may actually consist of a variety of complex electronics, but to the outside world, the Class "Temperature Sensor" only needs to *expose* a few items: temperature value, unit, and possess an address to a remote computer can access and read it. Optionally, it may also permit the remote reader to program the update interval.

Explicitly, the definition of the Temperature Session Class would look like this:

Listing 2.1: A Textual Rendition of How One May Represent a Class in a Computer Language

```
Class "Temperature Sensor" {
Value temperature
Value unit
Value update-interval
Value sensor-address
Function read-temp (address, temp)
Function set-interval (address, interval)
Function set-unit (address, unit)
}
```

The Class definition above specifies the potential information of a "Temperature Sensor"; a specific instantiation of a Class is referred to as an Object. The following shows this distinction.

Listing 2.2: The Instantiation of a Class Results in an Object, Which Has Specific Values

```
Object Sensor-1 is_class "Temperature Sensor" {
temperature 32
unit F
update_interval 5
address sensor1.site1.com
read-temp (address, temp)
set-interval (address, interval)
set-unit (address, unit)
}
```

One way to think of Actors in IHE (which will be discussed in detail in Chapter 5) is that the IHE documentation defines the Actor's Class behavior and a real-world device is an object level instantiation.

2.4.2 Use Cases

In Section 2.1, Use Cases were defined as a formal statement of a specific workflow, the inputs and outputs, and the Actors that accomplish the goal via the exchange of Transactions. A goal of this section is to begin to prepare the reader to interpret the IHE Technical Frameworks, which will be covered in Chapter 5. IHE specifies real world use cases (called Integration Profiles) encountered in the healthcare environment, and then offers implementation guidelines to implement those workflows that leverage existing informatics standards (DICOM, HL7, XML, etc.). As such, Sections 2.4.2.1 through 2.4.2.2 will delve into the specifics of a single Integration Profile, Scheduled Workflow. [*Note*: The concept may have presaged the term, but the first formal mention of Integration Profiles occurs in IHE Version 5.0, which curiously was the third anniversary of the IHE founding.]

2.4.2.1 Actors

A key strategy in IHE is that it defines Actors to have very low-level and limited functionality. Rather than describing the behavior of large and complex systems such as an RIS (Radiology Information Systems), the IHE model looks at all tasks that an RIS performs and then breaks out those "atomic" functions to specific Actors. To take a rather simple example, a Picture Archive and Communication System (PACS) is broken out into the following series of Actors: image archive/manager, image display, and optionally report creator/repository/manager. To begin to understand this process, we start with a diagram that depicts just the Actors involved in the Scheduled Workflow Integration Profile (SWF).

For reference, the actors are

a. *ADT*: The patient registration admission/discharge/transfer system.
b. *Order Placer*: The medical center wide system used to assign exam orders to a patient, and fulfills those orders from departmental systems.
c. *Order Filler*: The departmental system that knows the schedule for departmental assets, and schedules exam times for those assets.
d. *Acquisition Modality or Image Creator*: A DICOM imaging modality (or Workstation) that creates exam images.
e. *Performed Procedure Step Manager*: A central broker that accepts exam status updates from (d) and forwards them to the departmental Order Filler or Image Archive.
f. *Image Display*: The system that supports looking up patient exams and viewing the contained images.
g. *Image Manager/Archive*: The departmental system that stores exam status information, the images, and supports the move requests.

2.4.2.2 Associations and Transaction Diagrams

Figure 2.4a shows what Actors are involved in the Use Case for SWF, but gives no insight into what data flows among the Actors, the ordering of those Transactions, or the content. For that we add the following information shown below. For reference, the transactions are

a. *Rad-1 Patient Registration*: This message contains the patient's name, Identifier number assigned by the medical centers, and other demographics.
b. *Rad-2 Placer Order Management*: The Order Placer (often part of a Hospital Information System) creates an HL7 order request of the department-scheduling system.
c. *Rad-3 Filler Order Management*: The department system responds with a location and time for the required resources.

Rad-4 Procedure Scheduled:

a. *Rad-5 Modality Worklist Provided*: The required resource is reserved and the exam assigned an ID number.
b. *Rad-6 Modality Performed Procedure Step (PPS) in Progress*: The modality informs downstream systems that an exam/series is under way.

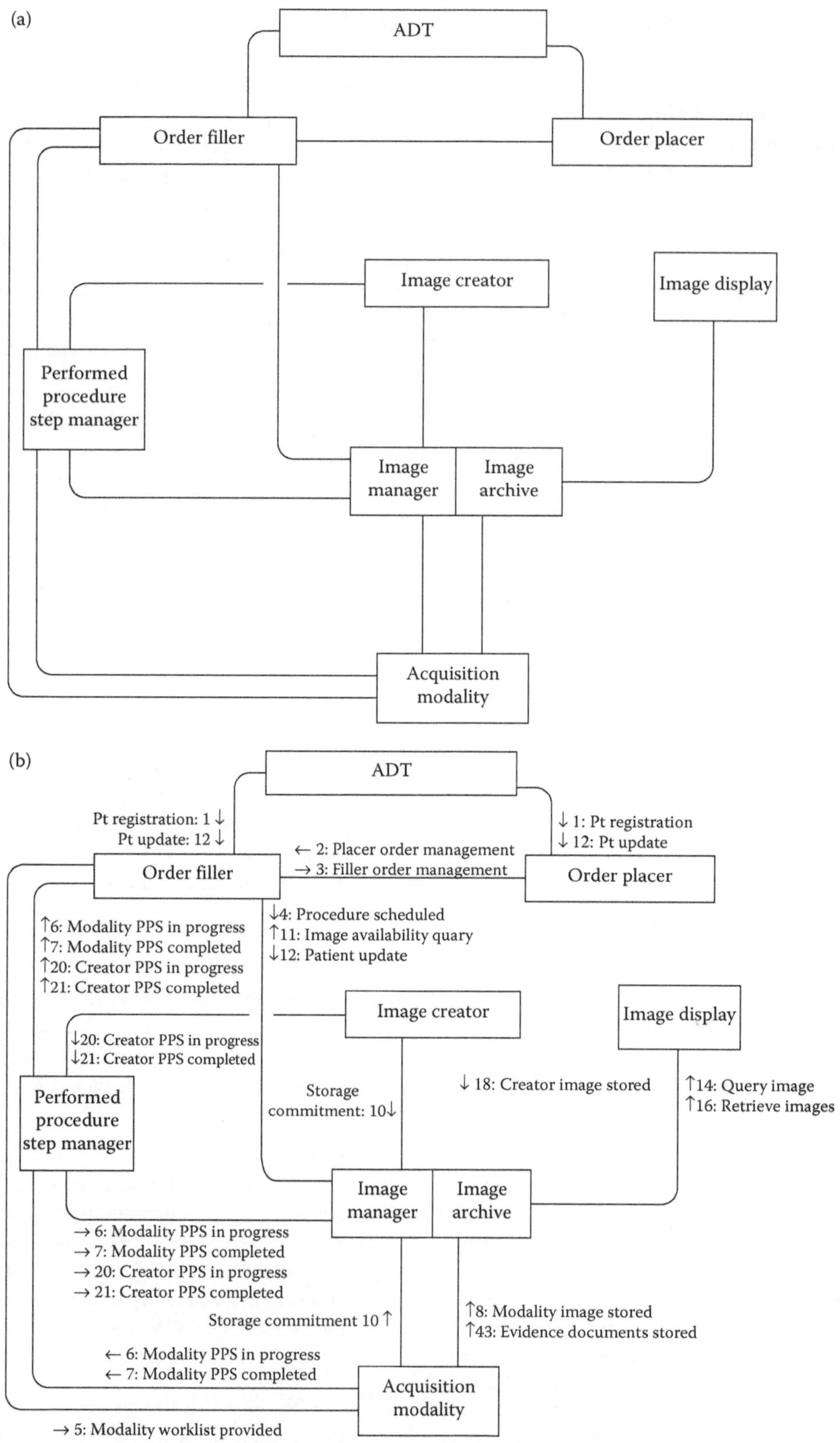

FIGURE 2.4 (a) The component Actors involved in the Scheduled Workflow Integration profile. (Adapted from *IHE Technical Framework* Vol. 1, V5.3, Figure 2.1.) (b) The same figure with the IHE Transactions included. The figure can be somewhat overwhelming because all the transactions are shown that are needed by the SWF Profile. (Adapted from *IHE Technical Framework* Vol. 1, V5.3, Figure 2.1.)

c. *Rad-7 Modality PPS Complete*: The modality informs downstream systems that an exam/series is complete.
d. *Rad-8 Modality Image Stored*: The image archive signals it has new images.
e. *Rad-10 Storage Commitment*: The archive signals the modality it has the entire exam and the modality can purge it.
f. *Rad-11 Image Availability Query*: An image consumer or medical record queries for the status of an imaging exam.
g. *Rad-12 Patient Update*: Updating the patient record with knowledge of the new exam.
h. *Rad-14 Query Images*: An image consumer queries for images in a known complete exam.
i. *Rad-16 Retrieve Images*: The image consumer pulls the images to itself.
j. *Rad-18 Creator Image Stored*: These transactions (18, 20, 21) are workstation-based replications of the Modality transactions (6–8).
k. Rad-20 Creator PPS in Progress.
l. Rad-21 Creator PPS Complete.
m. *Rad-43 Evidence Documents Stored*: the archive announces the storage of any other nonimage objects.

While complete, the information in Figure 2.4b can be overwhelming to take in all at once. For that reason, the diagrams discussed in the next section are used.

2.4.3 Interaction Diagrams

To simplify the understanding of all the data contained in the Transaction Diagram (Figure 2.4b), Interaction Diagrams (Figure 2.5) were created that show the same Actors, but isolate and group the Transactions based on their specific purpose in the overall SWF workflow (Booch et al., 1998). For instance, one can consider the functional groupings in the SWF workflow depicted in Figure 2.4b to be composed of the following:

a. Administrative processes
b. Procedure step processes
c. Patient update before order entry processes
d. Patient update after order entry processes
e. Patient update after procedure scheduling
f. Order replacement by the order placer
g. Order replacement by the order filler
h. And several exception scenarios

2.5 Mined for Meaning

Thus far this chapter has been largely a dry recitation of the methods and concepts behind medical imaging informatics. But we would be remiss if we did not point out what all this technology enables. Because of the standards and implementations outlined here, it is possible to create systems that can mine medical images for real world useful data; patient radiation history, scanner duty factors, the health of the imaging system components and the usage (or nonusage) of the PACS workstations at a site and whether there truly is a need for additional workstations.

2.5.1 DICOM Index Tracker

DICOM images contain a wealth of data that is largely unminable by most medical imaging practitioners. Of topical interest is radiation exposure; the national press has brought to full public discussion the use of medical radiation in diagnosis especially with the use of x-ray computed tomography (Brenner and Hall, 2007; Opreanu and Kepros, 2009). A group at the author's institution commenced to develop a flexible approach to store, harvest, and mine this source—not only for radiation dosimetry but also for other uses as well (Wang et al., 2010). This solution diverts a copy of the images at a site to the DICOM Index Tracker (DIT) and the image headers are harvested without the need for the image also. This makes storage needs relatively slight. Also, the system has a knowledge base of known modality software versions, and hence "knows" how to locate DICOM "shadow" tags, which encode information in nonstandard areas. This enables users to create single queries that can mine data across the myriad of modality implementations: the radiation dose record for a given patient, class of patients, class of exam, or performing sites as well as other query possibilities. It also enables time–motion studies of MRI and CT suite usage, throughput of dedicated chest rooms, and so on. This latter information has been used to a great extent by efficiency teams in developing both room scheduling and staffing models.

2.5.2 PACS Usage Tracker

The author's institution also found a need to validate usage patterns of PACS workstations to reduce hardware and licensing fees for underutilized workstations. Initially, simple user surveys were tried, but random spot checks on specific workstations when compared to user recollections were found to be widely divergent. The audit-logging requirements of HIPAA (Health Informatics Portability and Accountability Act) within the United States make it possible to track in our PACS the numbers of exams that were opened on specific workstations (what exams were opened is also possible, but this detail is ignored for our purpose). A central repository queries all the PACS workstations on a daily basis, and stores the exam-opened count to a database. The results are plotted on a Web form (Figure 2.6), which shows exam volume by workstation over a user-selectable period (French and Langer, 2010). This tool has been a great help to administrators seeking to assign PACS resources to areas where they are most needed, and reduce needless procurement.

2.5.3 PACS Pulse

It is also useful to be able to chart the performance metrics in a PACS, locate sources of latency, and troubleshoot areas when

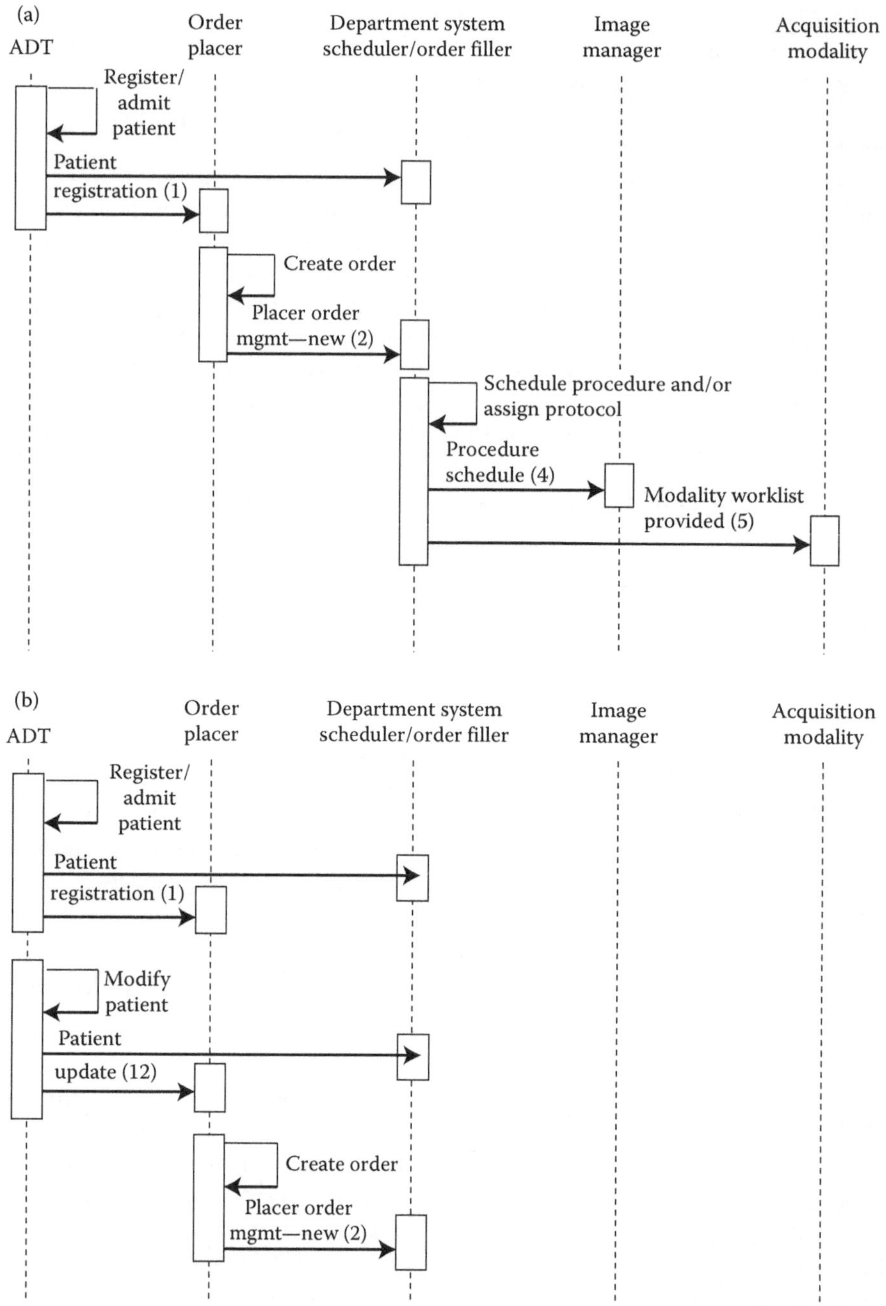

FIGURE 2.5 (a) A Process Flow diagram renders all the Actors as in the Actor–Transactions diagrams, but isolates the Transactions according to what phase they represent in the overall workflow. This happens to be the Administrative Transaction summary. (Adapted from *IHE Technical Framework* Vol. 1, V5.3, Figure 2.2–1.) (b) Another process flow diagram, summarizing Patient Update. (Adapted from *IHE Technical Framework* Vol. 1, V5.3, Figure 2.2–3.)

subsystems fail or are slow enough to be harming the practice. The PACS Pulse project uses log parsing from the PACS DICOM operations to accomplish this objective and enable real-time proactive management of PACS resources (Nagy et al., 2003). The same group has also developed a more sophisticated tool that leverages DICOM, HTML, and HL7 data feeds to monitor: patient wait times, order backlog times, exam performance to report turnaround times, delivery of critical finding times, reasons for exam repeat/rejects, and other metrics (Nagy et al., 2009). The brilliant assemblage of these data in a single Web-reporting tool offers Radiology managers the ability to make informed business decisions on staffing, equipment purchases, and scheduling, thus enabling the improved productivity, performance, and quality of service in the department.

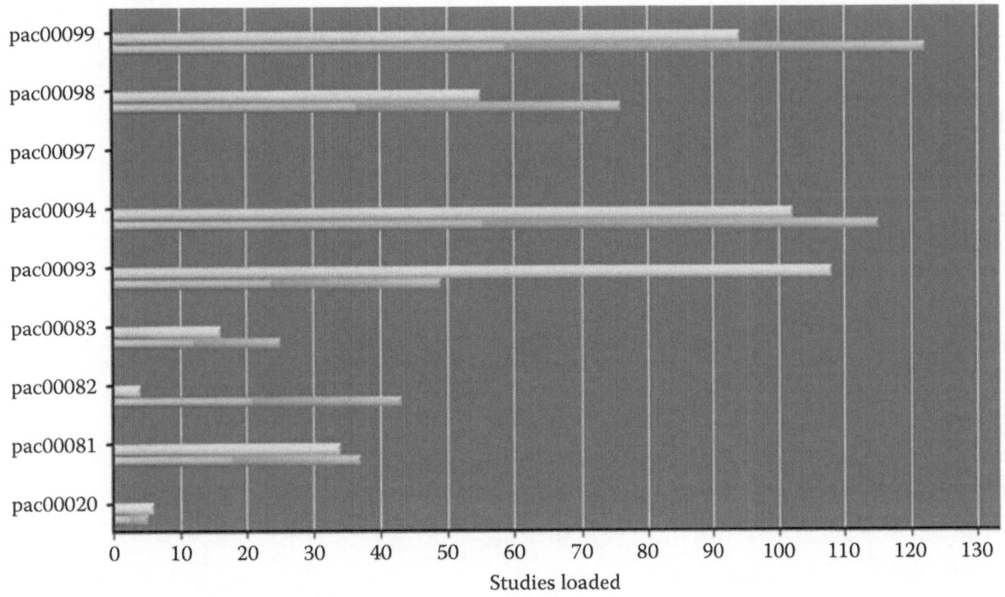

FIGURE 2.6 A snapshot of a Web page report on the study volumes opened on the PACS workstations in a department. For example, in the 2-week period shown here, PAC099 opened 95 studies in the first week and 123 in the second.

References

Alhir, S. S. 2003. *Learning UML*. Sabasrolpol, CA: O'Reilly and Associates.

Berners-Lee, T. and Fischetti, M. 1999. *Weaving the Web: The Original Design and Ultimate Destiny of the World Wide Web*. New York, NY: Harper Collins.

Bittner, K. and Spence, I. 2002. *Use Case Modeling*. Boston, MA: Addison-Wesley.

Booch, G., Rumbaugh, J., and Jacobson, I. 1998. *The Unified Modeling Language User Guide*. Toronto, CA: Addison-Wesley.

Brenner, D. J. and Hall, E. J. 2007. Computed tomography—An increasing source of radiation exposure. *N. Engl. J. Med.*, 357, 2277–84.

Buck, C. J. 2011. *ICD-9 for Physicians*. Philadelphia, PA: Saunders, a division of Elsevier Publishing.

Cote, R. A. 1986. *The Architecture of SNOMED*. Boston, MA: IEEE.

French, T. L. and Langer, S. G. 2010. Tracking PACS usage with open source tools. *J. Digit Imaging*. DOI:10.1007/s10278-010-9337-y.

Greenes R. A. and Shortliffe E. H. 1990. Medical informatics. An emerging academic discipline and institutional priority. *JAMA*. 263(8):1114–20.

Gruber, T. 1993. A translational approach to portable ontology specifications. *Knowledge Acquisition*, 5, 199–220.

Henderson, M. 2007. *HL7 Messaging*. Aubrey, TX: O'tech Healthcare Technology Solutions.

Holzmann, G. J. 1991. *Design and Validation of Computer Protocols*. Englewood Cliffs, NJ: Prentice-Hall.

IHE Technical Framework Vol. 1, V 5.5. Available at ftp://ftp.ihe.net/Radiology/TF_Final_Text_Versions/v5.5/. Last accessed December 2010.

Langer, S. and Bartholmai, B. 2010. Imaging informatics: Challenges in multi-site imaging trials. *J. Digit Imaging*, 24(1), 151–159.

Langlotz, C. P. 2006. RadLex: A new method for indexing online educational materials. *Radiographics*, 26, 1595–97.

Nagy, P. G., Daly, M., Warnock, M., Ehlers, K. C., and Rehm, J. 2003. PACSPulse: A web-based DICOM network traffic monitor and analysis tool. *Radiographics*, 23, 795–801.

Nagy, P. G., Warnock, M. J., Daly, M., Toland, C., Meenan, C. D., and Mezrich, R. S. 2009. Informatics in radiology: Automated Web-based graphical dashboard for radiology operational business intelligence. *Radiographics*, 29, 1897–906.

Opreanu, R. C. and Kepros, J. P. 2009. Radiation doses associated with cardiac computed tomography angiography. *JAMA*, 301, 2324–5; author reply 2325.

RFC. *RFC 1122* [Online]. Available at http://tools.ietf.org/html/rfc1122. Accessed September 1, 2010.

RFC. *RFC 1123* [Online]. Available at http://tools.ietf.org/html/rfc1123. Accessed September 1, 2010.

RFC. *RFC 2616* [Online]. Available at http://tools.ietf.org/html/rfc2616. Accessed September 1, 2010.

RFC. *RFC 2818* [Online]. Available at http://tools.ietf.org/html/rfc2818. Accessed September 1, 2010.

Wang, S., Pavlicek, W., Roberts, C. C., Langer, S. G., Zhang, M., Hu, M., Morin, R. L., Schueler, B. A., Wellnitz, C. V., and Wu, T. 2010. An automated DICOM database capable of arbitrary data mining (Including Radiation Dose Indicators) for quality monitoring. *J. Digit Imaging*, 24(2), 223–233.

Zimmermann, H. 1980. OSI reference model. *IEEE Trans. Commun.*, 28, 425.

II

Standard Protocols in Imaging Informatics

3

Health Level 7 Imaging Integration

Helmut König
Siemens AG Healthcare Sector

3.1 HL7 Basics

3.1.1 Brief History and Overview on the HL7 Standards Development Organization

Early development of the Health Level 7 (HL7) standard started in 1987 focusing on the communication of clinical data between hospital information systems. The name HL7 refers to the International Organization for Standardization (ISO) Open Systems Interconnection (OSI) reference model (ISO/IEC, 1994) for computer network protocols and its seventh level termed the application layer. HL7 is an American National Standards Institute (ANSI) accredited nonprofit standards developing organization. The standard has gained acceptance internationally with a growing number of HL7 international affiliate member organizations promoting the standard and working on national adaptation strategies.

HL7 is cooperating with numerous external healthcare standards developing organizations based on individual memoranda of agreement defining the scope and terms of the formal relationships. The Digital Imaging and Communications in Medicine (DICOM) Standards Committee and HL7 created a common working group in 1999: DICOM Working Group 20 and the HL7 Imaging Integration Work Group have common membership and focus on topics that address the integration of imaging and information systems. For standardization efforts in the intersection of their domains, DICOM and HL7 harmonize concepts to promote interoperation between the imaging and healthcare enterprise domains. HL7 and Integrating the Healthcare Enterprise (IHE) signed their initial associate charter agreement in 2005 to promote the coordinated use of HL7 standards in IHE integration profiles. IHE defines, in its published Technical Frameworks, a set of implementation profiles specifying standards-based solutions for common interface and integration problems such as the Scheduled Workflow integration profile for tracking of scheduled and performed imaging procedure steps.

3.1.2 HL7's Main Interoperability Goals

HL7 defines messages, document formats, and a variety of other standards to support care provision and communicate healthcare data across and between healthcare enterprises for the delivery and evaluation of healthcare services. While HL7 Version 2.x message specifications focus on the exchange of structured messages to achieve syntactic interoperability, Version 3 messages and the Clinical Document Architecture (CDA) format for structured documents strive toward semantic interoperability by using a reference information model and common vocabulary (coded concepts) as their foundation. HL7 Version 3 standards use XML instead of delimiter-based formatting known as the default encoding from HL7 Version 2.x messages.

3.1.3 Focus of HL7 Communication Standards

Since its inception, HL7 has specified standards for a large number of application areas. HL7 standards cover generic application fields such as patient administration, patient care, order entry, results reporting, document, and financial management. In addition to that, HL7 addresses the departmental information system communication needs of clinical specialties like laboratory medicine and diagnostic imaging. HL7 has entered into new clinical domains (e.g., clinical genomics and clinical pathology), cross-institutional and regional communication of healthcare data, health quality measure reporting, as well as clinical trials in the context of V3 standards development efforts.

3.2 HL7 Version 2.x Messages

HL7 has specified multiple Version 2 message standards collectively known as Version 2.x standards. In those standards, messages are the atomic units of data that are transferred between information systems. Real world events are associated with trigger events that initiate the exchange of messages between sending and receiving systems. Trigger events are labeled with an upper case letter and two digits, for example, "A01" for Admit/Visit Notification or "O23" for Imaging Order Messages. Message types such as Admission/Discharge/Transfer Message (ADT) or Imaging Order Message (OMI) define the purpose of the message. The transfer of messages largely follows a push model. In this case, the transaction is termed an unsolicited update.

Messages are hierarchically structured, essentially consisting of segments and data fields. A segment is a logical grouping of data fields specifying the order in which the fields appear. Segments of a message may be required or optional and are allowed to repeat (Figure 3.1).

Each segment is identified by a unique three-character code known as the Segment ID (e.g., "MSH" for the Message Header Segment or "PID" for the Patient Identification Segment). Data types are the basic building blocks that define the contents of a field. HL7 specifies the optionality and repetition of fields as well as data types that may comprise components and subcomponents. For instance, the Hierarchic Designator (HD) data type that is used to determine the issuer of an identifier has three components: Namespace ID, Universal ID, and Universal ID Type. The first component "Namespace ID" identifies an entity within the local namespace or domain, the second component "Universal ID" is a universal or unique identifier for an entity, and the third "Universal ID Type" specifies the standard format of the Universal ID.

For the representation of DICOM Unique Identifiers (UIDs), which are ISO object identifiers (ISO, 2005), the second and third component of the HD data type would be used, for example, |^1.2.345.67.8.9.1^ISO|. Please note that the vertical bar "|" is the default field separator and carets "^" are used as default component separators in HL7 V2.x messages.

Reference pointers (RP Data Type) uniquely identify data that is located on remote systems. This data type can be used to reference relevant DICOM objects (e.g., images and Structured Reporting [SR] documents) within HL7 2.x messages. The referenced DICOM objects would typically be located in image archives that provide a DICOM Web Access to DICOM Persistent Objects (WADO) (DICOM, 2009b) interface. As an alternative to reference pointers, the ED data type (Encapsulated Data) may be used, for example, for sending CDA Release 2 documents (ANSI/HL7, 2005) as an HL7 V2.x message payload between imaging and clinical information systems.

Controlled vocabularies represented in coding schemes such as the "Systematized Nomenclature of Medicine—Clinical Terms" (SNOMED CT) (IHTSDO) or "Logical Observation Identifier Names and Codes" (LOINC) (RIHC, 2000) are the basis for conveying coded elements with defined semantics. The basic components comprise the code value (code identifier), coding scheme designator (name of coding system), and code meaning (displayed text that explains the semantics of the coded concept or element). Influenced by HL7 Version 3 vocabulary standardization efforts, HL7 Version 2.6 (ANSI/HL7, 2007) has introduced the coded with no exceptions (CNE) and coded with exceptions (CWE) data types for coded elements replacing the coded element (CE) data type that is used in earlier versions of HL7 V2.x standards. While CNE mandates using codes drawn from HL7 defined tables, imported code tables, or tables that contain codes from external coding schemes, CWE also allows for locally defined codes and text replacing coded values.

For the purpose of conveying imaging-related data, further important HL7 data types are Date (DT), Time (TM), Date/Time (DTM), String Data (ST), Numeric (NM) and its use for quantities, Extended Person Name (XPN), Extended Address (XAD), and Extended Composite Name and Identification Number for Organizations (XON).

3.2.1 Representation of Messages

The Abstract Message Syntax is a special notation that describes the order, repetition, and optionality of HL7 V2.x control and data

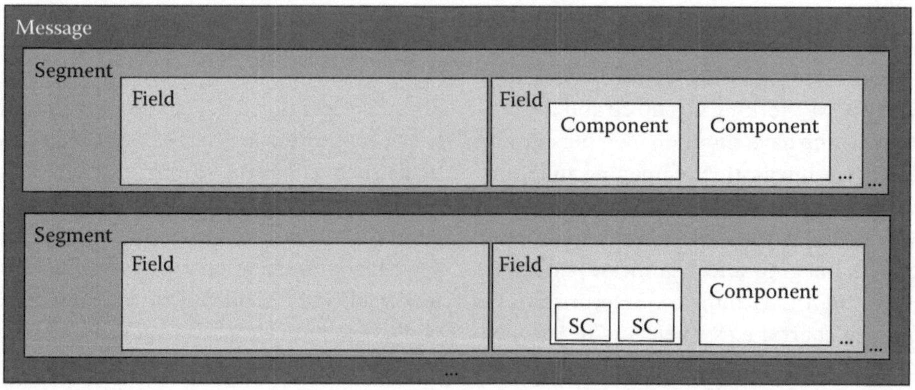

FIGURE 3.1 HL7 Version 2.x message structure.

ADT^A01^ADT_A01	ADT Message
MSH	Message Header
[{ SFT }]	Software Segment
[UAC]	User Authentication Credential
EVN	Event Type
PID	Patient Identification
...	...
ACK^A01^ACK	General Acknowledgment
MSH	Message Header
[{ SFT }]	Software Segment
[UAC]	User Authentication Credential
MSA	Message Acknowledgment
[{ ERR }]	Error

FIGURE 3.2 Abstract message syntax example.

segments. Segments are listed in the order they would appear in the message and are identified by their Segment ID (e.g., EVN for Event Type). One or more repetitions of a group of segments are enclosed in braces {. . .}. Brackets [. . .] indicating that the enclosed group of segments is optional. If segments are both optional and may repeat, they are enclosed in brackets and braces.

The A01 "Admit/Visit Notification" message starts with the message header segment, followed by one or more optional software segments, the optional user authentication credential segment, event type, and patient identification segments. The standard also includes detailed explanations on the trigger events, the specific use of segments in the context of individual trigger events and acknowledgment messages that are expected to be sent back by the receiving system. For the A01 "Admit/Visit Notification" trigger event a relatively simple general acknowledgment message is specified (Figure 3.2).

The sample ADT A01 message below comprises the MSH, EVN, and PID segment (Figure 3.3).

Patient John F. Doe was admitted on August 23, 2009 at 11:24 a.m. The message (HL7 Version 2.6) was sent from system ADTREG1 at the Good Health Hospital site to system RADADT, also at the Good Health Hospital site, 2 min after admission.

Further message types that are of interest in the context of order entry, results reporting, and document management are

- *ORM:* pharmacy and treatment order messages, also used for imaging orders in HL7 2.x versions prior to 2.5.

```
MSH|^~\&|ADTREG1|GOOD HEALTH HOSPITAL|RADADT|
GOOD HEALTH HOSPITAL |200908231126|SECURITY|
ADT^A01^ADT_A01|MSG00001|P|2.6<CR>
EVN|A01|200908231124||<CR>
PID|||1234567||DOE^JOHN^F|...<CR>
```

FIGURE 3.3 HL7 Version 2 ADT A01 sample message.

- *OMI:* imaging order messages specified in HL7 version 2.5, for example, digital x-ray procedures.
- *ORU:* unsolicited observation message used to transmit results, for example, laboratory measurements and results.
- *MDM:* medical document management, for example, notification on the creation and amendments of a CDA document.

3.2.2 Acknowledgment Messages

In HL7, the receiving application is expected to send back an acknowledgment message in response to the message sent as an unsolicited update by the sending application. The standard specifies two acknowledgment modes. The original acknowledgment mode that is most widely used is an application acknowledgment. After receiving a message an acknowledgment (ACK) message is sent back to the sending system to indicate the message was received. The HL7 standard makes no assumptions that the receiving system commits the data to save storage before acknowledging it. All that is required is that the receiving system accepts responsibility for the data. In a typical HL7 environment, a sender will assume the message was not received until it receives an acknowledgment message.

Enhanced mode includes an additional accept acknowledgment, which indicates that the data has been committed to save storage by the receiving system. It thereby releases the sending system from the need to resend the message. After processing the message, the receiving system may use an application acknowledgment to indicate the status of the message processing.

3.2.3 Message Encoding

Special characters are used to construct messages. The segment terminator is always a carriage return (hex 0d in ASCII). Other delimiters are specified in the MSH. The field delimiter that separates two adjacent fields within a segment is defined in the fourth character position of the message header segment (suggested value: "|"). Other delimiters are listed in the encoding characters field, which is the first field after the segment ID. The component separator (suggested value: "^") separates adjacent components of data fields. The repetition separator (suggested value: "~") is used to separate multiple occurrences of a field. The escape character (suggested value: "\") and sequences within textual data escape special characters, character set, and formatting information. The subcomponent separator (suggested value: "&") is a delimiter for adjacent subcomponents of data fields.

A large portion of HL7 messaging is transported by using the Minimal Lower Layer Protocol (MLLP) in combination with the Transmission Control Protocol and Internet Protocol (TCP/IP). MLLP specifies a minimal message wrapper that includes a start block (SB) character and an end block (EB) character immediately followed by a carriage return <CR>. The header block character is a vertical tab character <VT> (hex value: 0b). The end block character is a field separator character <FS> (hex value: 1c) (Figure 3.4).

<VT>	HL7 | Message Payload	<FS>	<CR>
(hex 0x0b)		(hex 0x1c)	(hex 0x0d)

FIGURE 3.4 Minimal lower layer protocol.

In addition to sending messages as unsolicited updates, HL7 also specifies a query/response model for HL7 queries and special protocols (e.g., for batch processing).

3.3 HL7 Version 3

HL7 V3 (ANSI/HL7, 2009) strives for consistency by basing the family of V3 standards on Unified Modeling Language (UML)-models. In order to improve the quality of standards, a framework for the HL7 standards development process has been defined. The HL7 Development Framework documents the entire lifecycle, tools, actors, rules, and artifacts of that process. It addresses project initiation, specification of domain-specific use cases and domain analysis models, as well as the mapping of those artifacts to the HL7 Reference Information Model (RIM) and refined models used to develop HL7 V3 standards such as messages, CDA documents, and services (Figure 3.5).

Use cases and storyboards describe the tasks and actors that are involved in interactions. Interaction models focus on the trigger events, abstract messages, and application roles of the sender and receiver of messages. Information models such as the RIM and the associated refined models specify the classes, attributes, and relations that form the content of V3 messages, CDA documents, and services. The RIM provides a static view of the information needed for the development of HL7 standards. It is used to derive domain-specific models that constrain and refine the RIM in a series of transformations. Domain Message Information Models (D-MIM) contain all the classes and relationship needed to specify the contents of domains such as laboratory and clinical genomics. Refined message information models (R-MIMs) constrain domain information models to specify the information content for specific message and document content within that domain (e.g., domain-specific order and results messages or CDA document types). Subsequently, R-MIMs are serialized to create Hierarchical Message Definitions (HMDs). An HMD is a tabular representation of the sequence of classes, attributes, and associations that defines the message content without reference to the implementation technology. The HMD defines the base message structure that is used as a template from which the specific message types are drawn. HL7 V3 supports the model-based development of standards by cloning artifacts from the RIM to represent concepts in the refined and constrained domain-specific models that are consistent with the RIM. Refined models may restrict vocabulary, cardinality, and relationships. XML Implementation Technology Specifications describe how V3 message and CDA document XML schemas are derived from the HMD and its associated Message Types (MT). A message type represents a unique set of constraints applied against the HMD common base message structure.

3.3.1 Reference Information Model

The RIM is a UML-based object information model that is the common source for the information content of V3 standards. It contains the core classes and relationships for covering information in the healthcare domain. In essence, this object information model defines a healthcare framework on entities associated with roles that define their kind of participation in acts (Figure 3.6).

Acts represent actions (i.e., events) in the context of healthcare provision and management (e.g., order placement, substance administration, performing procedures, and documentation of acts). Participations express the context of an act related to the involved roles and entities (e.g., who performed the act and where

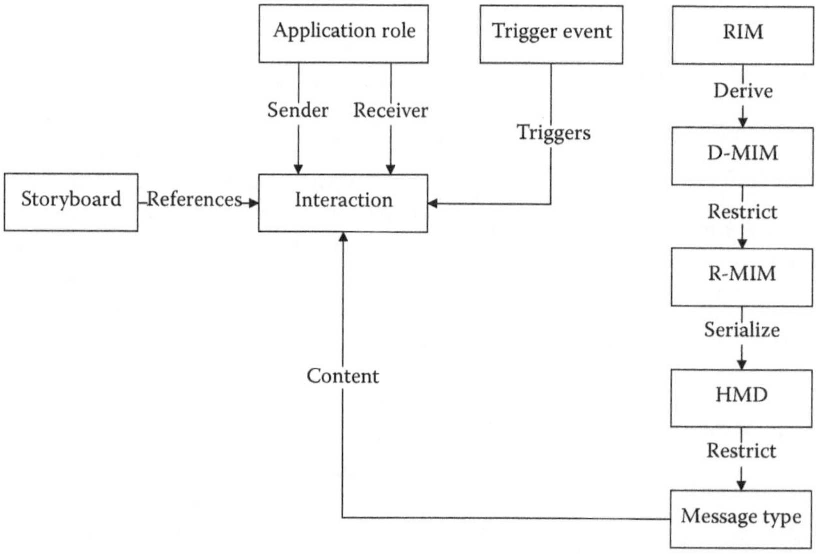

FIGURE 3.5 Overview on HL7 V3 methodology.

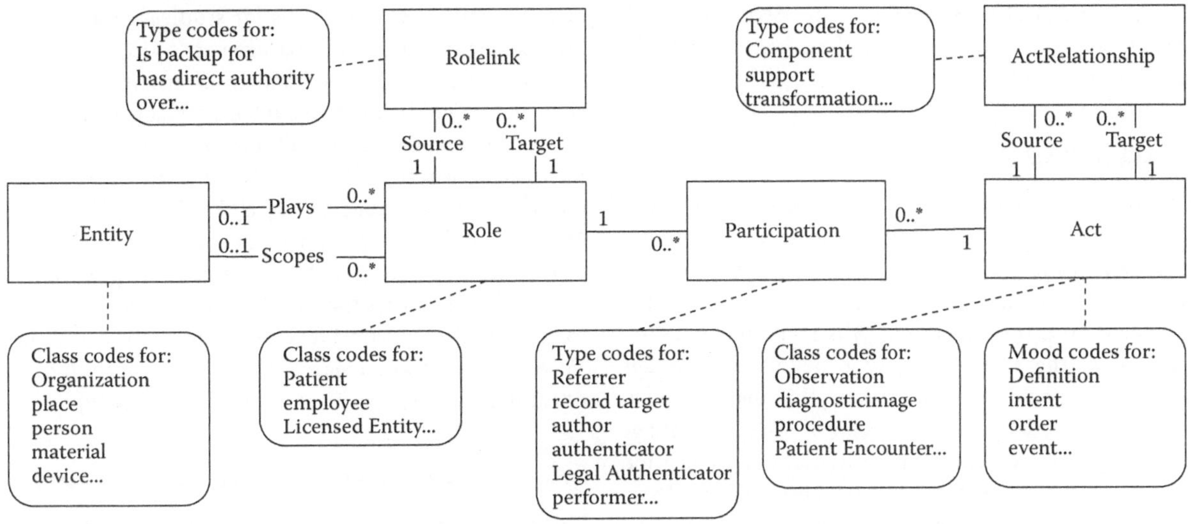

FIGURE 3.6 V3 RIM core classes.

it was performed). Entities are physical things and beings such as persons and organizations that play roles (e.g., patient, healthcare provider) as they participate in acts or are scoped by roles. Act relationships represent the binding of one act to another (e.g., for conclusions and diagnoses that are based on image data observations). Role links represent relationships between individual roles (e.g., to express that an employee of an organization is a backup for another employee). Three of the RIM core classes—Act, Role, and Entity—are represented by generalization–specialization hierarchies. Subtypes for these classes are added if one or more additional attributes or associations are needed that cannot be inherited from its parent classes. The Observation class is, for instance, a specialization of the Act class and inherits all attributes of Act and adds a value attribute. Observation itself generalizes the DiagnosticImage class that adds the subjectOrientationCode attribute. In order to distinguish concepts represented by classes that share a common set of attributes, the following coded attributes are used

- *classCode*: available in Act, Entity, and Role; defines concepts such as Observation and Clinical Document.
- *moodCode*: available in Act; describes activities as they progress from intended and requested to the event that has occurred.
- *determinerCode*: available in Entity; distinguishes whether the class represents an instance or a kind of Act or Entity.
- *code*: available in Act, Entity, and Role; an optional-coded attribute that allows for defining specific subtypes of a given act determined by the classCode attribute, for example, procedure codes.

No subtypes exist for the RIM core classes Participation, ActRelationship, and RoleLink. Distinct concepts are primarily represented by using the typeCode attribute (e.g., for defining referrer and authenticator participations for CDA document standards and V3 messages).

3.3.2 Vocabulary

Standardized vocabulary allows for the unambiguous interpretation of coded concepts conveyed between information systems. In contrast to arbitrary symbols, coded concepts have defined semantics (formal meaning). The use of coded concepts is an important building block in achieving semantic interoperability. HL7 V3 uses coded concepts for its core classes and associations in the RIM. The HL7 vocabulary specification defines the set of all concepts that can be used for coded attributes. The concepts are organized hierarchically as follows:

- Concept domains are named categories of concepts that are independent from code systems or specific vocabularies. Concepts domains are bound to one or more coded elements and may contain subdomains that can be used to further constrain the values. (E.g., ActRelationshipType is the concept domain for codes specifying the meaning and purpose of every ActRelationship instance. It is bound to the RIM attribute ActRelationship.typeCode and includes subdomains such as ActRelationshipEntry.)
- Value sets consist of one or more coded concepts constituting the intended values for a domain or subdomain. Value sets are similar to DICOM context groups (DICOM, 2009a) as they comprise the intended values for a given context and purpose. (E.g., the ActCode concept domain includes the ActCodeProcessStep subdomain that is associated with the ActCodeProcessStep value set. The latter value set includes concepts like filtration and defibrination for laboratory process steps.)
- Code systems define concepts that are used to populate coded attributes and their associated data type values. There are HL7-maintained systems (e.g., for mood codes and other HL7-specific concepts) and external systems (e.g., SNOMED-CT and LOINC) referenced by HL7. Code systems and value sets are assigned unique identifiers

(e.g., "2.16.840.1.113883.19.6.962" for SNOMED-CT). Concepts are guaranteed to be unique only within the context of their code system.

3.3.3 Data Types

The data type abstract specification defines the semantics of HL7 V3 data types independent of their technical representation that is based on specific implementation technologies. RIM attributes are data elements having a data type that defines the set of valid values and their meaning. Abstract data types are specified based on the formal data type definition language (DTDL) that uses a specific abstract syntax. This abstract specification is accompanied by the Extensible Markup Language (XML) Data Type Implementation Technology Specification (ITS) that defines the representation and encoding in XML. Compared to HL7 V2.x data type specifications, HL7 V3 takes a different approach with regard to the theoretical foundation and representation of data types. The goal is to harmonize data type specifications with ISO healthcare data types.

HL7 V3 uses globally unique instance identifiers (II) to identify a wide variety of objects. Instance identifiers consist of a root (unique identifier) and an optional extension (string). ISO Object Identifiers (OIDs), Distributed Computing Environment (DCE), Universal Unique Identifiers (UUIDs), and HL7-reserved unique identifiers (RUIDs) may be used for the II root. The use of OIDs with a maximum length of 64 characters is recommended for CDA documents and information intended to be exchanged between DICOM and HL7-speaking systems because HL7 V3 OIDs are based on the same identification scheme (ISO 8824). DICOM uses for unique identifiers (DICOM UID data type).

HL7 V3 data types for coded data elements typically comprise the following components:

- *code*: code value, for example, SNOMED-CT concept id "439932008" for "Length of structure."
- *codeSystem*: unique identifier of code system, for example, "2.16.840.1.113883.19.6.96" for SNOMED-CT.
- *codeSystemName*: common name of the coding system.
- *displayName*: name or title for the code, under which the sending system shows the code value to its users.

Coded values are extensively used in structured documents. The basic attributes of HL7 V3 code data types can easily be mapped to the basic-coded entry attributes of DICOM code sequences. The most generic HL7 V3 data type for coded values is the Concept Descriptor (CD) that contains the original code values, optional translations of those values into different coding systems, and qualifiers (used for postcoordinated terms, e.g., anatomic code for "hand" with qualifier "left" to specify laterality for paired anatomic structures). Other V3 code data types such as CE (Coded with Equivalents; may contain translation codes but no qualifiers) specialize the concept descriptor data type.

DICOM WADO (DICOM, 2009b) references may be used in V3 uniform resource locators to access DICOM images and documents through the HTTP/HTTPS protocol (GET request and response). DICOM UIDs for studies, series, and instances are used as HTTP request parameters to identify the persistent objects. Data may be retrieved in a presentation-ready format such as JPEG or in a native DICOM format (Figure 3.7).

The Physical Quantity (PQ) data type may be used to express measurement results. It comprises a value of type REAL and a coded unit as specified by UCUM (Unified Code for Units of Measure) and optional translations to different units.

Further important data types for imaging purposes are within the categories that have been listed for Version 2.x (refer to Section 1.2). Version 3 takes an approach that is different from Version 2.x in many regards like the specification of the details of timing, names, addresses, and generic data types.

3.3.4 Clinical Document Architecture

The CDA standard is an XML-based document standard created and maintained by HL7 Structured Document Work Group. CDA Release 2 (ANSI/HL7, 2005) has been published in 2005 following CDA Release 1 (ANSI/HL7, 2000) that is available since the year 2000. While Release 1 started with the unconstrained CDA specification and section content, CDA Release 2 redefined part of that content and added content focusing on the structured part of CDA documents (structured entries that represent the computer-processing components within document sections). CDA documents are persistent objects that are maintained by an organization as the custodian entrusted with its care (stewardship). Clinical documents are intended to be legally authenticated as opposed to messages that may be used to send unauthenticated results. Authentication applies to the whole document not just portions of the document (principle of wholeness). CDA documents are human-readable (i.e., the attested document content is required to be rendered for human readability).

The standard specifies the structure and semantics of clinical documents by leveraging the use of XML, the HL7 RIM, version 3 data types, and coded vocabularies. CDA documents are intended to be both human-readable and machine computable. Human readability implies that a single generic stylesheet renders the authenticated clinical content of any CDA document. Like the header and sections, the structured, machine computable part of CDA documents is primarily based on the HL7 Reference Information Model (RIM) and the use of the HL7 Version 3 Data Types. The CDA refined model (R-MIM) is derived from the RIM. It specifies the general constraints for CDA documents. CDA adheres to the HL7 development framework for

```
http://www.example.org/wado?requestType=WADO
&studyUID=1.2.840.113619.2.62.994044785528.11428
  9542805
&seriesUID=1.2.840.113619.2.62.994044785528.200
  60823223142485051
&objectUID=1.2.840.113619.2.62.994044785528.2006
  0823.200608232232322.3
&contentType=application%2Fdicom
```

FIGURE 3.7 WADO request for native DICOM object.

V3 standards described in Section 3.3. A CDA document is a defined and complete information object that can exist outside of a message, and can include text, images, sounds, and other multimedia content. A CDA document can be conveyed as an Multipurpose Internet MailExtensions (MIME)-encoded payload within an HL7 message. In that sense, CDA complements HL7 Version 2.x and V3 messaging specifications. Conformant CDA documents validate against the CDA schema and restrict its use of coded vocabulary to values allowable within the specified vocabulary domains. Additional constraints are introduced by CDA Implementation Guides (IGs) that specify templates to constrain CDA for defining report types such as Continuity of Care Documents (CCD), Public Health Case Reporting (PHCR), and Diagnostic Imaging Reports (DIR).

3.3.4.1 Basic Document Structure

A CDA document contains a clinical document header and a body. The <ClinicalDocument> XML element is the root element of a CDA document. The header includes the metadata describing the document (e.g., unique document id, document type code, and document version), information on participants (e.g., the patient as the record target, author, and authenticators), and relationships to other acts (e.g., parent documents, service events, orders, encompassing encounter). The CDA header is linked to the body through a component relationship. The CDA body that contains the clinical report is either represented as an unstructured blob or structured markup. Every CDA document has exactly one body. The nonXMLBody element is used for non-XML content (e.g., JPEG images or PDF documents) that is referenced if the non-XML data is stored externally or is encoded inline as part of the CDA document. The structuredBody element is used for XML-structured content and contains one or more document sections. Sections may nest and can contain a single narrative block. The narrative text in combination with its rendered multimedia content (the <renderMultiMedia> element references external multimedia that is integral to a document) comprises the attested content of the section. Sections may also include optional entries and external references representing structured content intended for machine processing. The CDA narrative block is wrapped by the <text> element within the <Section> element. XHTML-like components are used to reference in and out of the narrative block, to label contents such as lists and tables, and to suggest rendering characteristics. The narrative block contains the human-readable content to be rendered by CDA applications. CDA entries are associated with CDA sections through the entry relationship. Each section may contain zero to many structured entries (e.g., to represent observations, procedures, regions of interest, and substance administrations). CDA entries associated with sections and their narrative block represent structured content that is consistent with the narrative. Entry relationships (act relationships with specific type codes) between CDA entries allow for building the content tree of structured entries. CDA external references to acts, observations, procedures, and documents always occur within the context of CDA entries. Entries and external references are associated through a reference act relationship (the <reference> element wraps the external references). CDA external references itself are represented by single classes thus allowing for simple references to external objects that are not part of the attested document content.

CDA level one comprises the unconstrained CDA specification, while CDA levels two and three constrain CDA based on section-level and/or entry-level templates (Figure 3.8).

3.3.4.2 Context and Context Conduction

The context of a CDA document is set in the document header and applies to the entire document (e.g., human language and confidentiality) unless explicitly overridden in the document body, at section and/or entry level. The document context in the header includes the author and legal authenticator of the document as well as the record target. Among other context data, author information may be overridden for sections and entries. The subject of the document may also change for different sections and entries. The RIM concentrates on context as the participations associated with acts that can be propagated to nested acts based on act relationships between parent and child acts. Whether or not the context on an act can propagate to nested acts depends on the values of context control indicator (ActRelationship.contextConductionInd) and context control code (Participation.contextControlCode) RIM attributes. CDA constrains the general RIM context mechanism such that context always overrides and propagates.

DICOM SR documents contain observation context information that comprises the observer context (human or device observer and the organization to which they belong), subject context (patient, specimen, or fetus), and the procedure context (diagnostic imaging and interventional procedures). The initial or default context is set in the document header. The Patient, Specimen Identification, General Study, and General Equipment modules contain context information on the subject and procedure. The explicit rule for context propagation states that child nodes within the SR content inherit the accumulated observation context from its parent. Context information may be overridden by specifying observation context for content items at different levels of the content tree. Thus, DICOM context rules are matching the CDA default context mechanisms.

3.3.4.3 Document Versioning

In addition to the CDA key characteristics described above, document versioning is another important CDA aspect. A new CDA document may replace a former document version and/or append an addendum to a parent document. Those relations are represented by the relatedDocument act relationship between ClinicalDocument and the ParentDocument act classes. Every CDA document must have a unique clinical document identifier, which is used to identify the different CDA objects including replacement and addendum documents. CDA documents may also have a ClinicalDocument.setId and ClinicalDocument.versionNumber which support the CDA-versioning scheme. If a document is replaced, the new document will use the same set identifier value as its parent and will increase the version number by 1. An addendum document will typically use a set

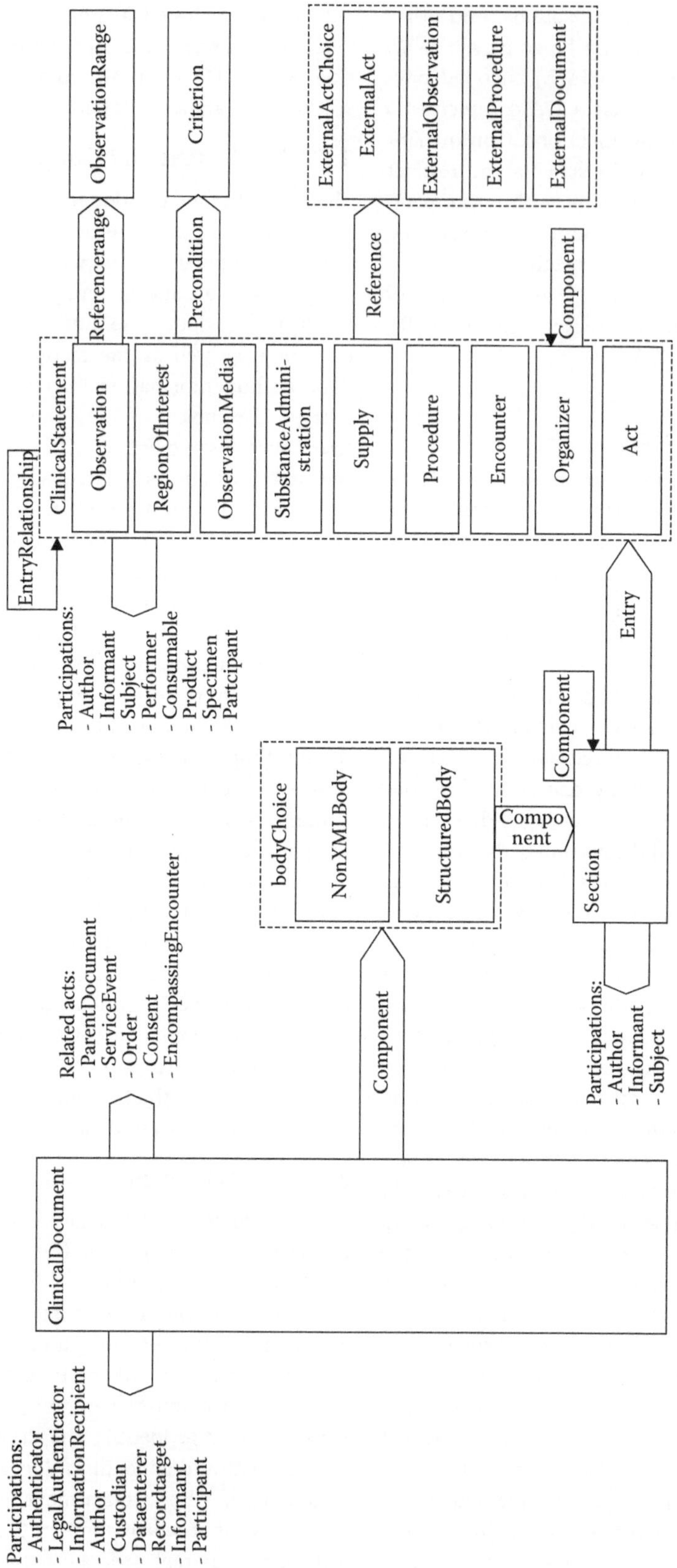

FIGURE 3.8 Overview on CDA structure.

identifier value that is different from the parent document and the version number will not be incremented.

DICOM SR instances are uniquely identified by their Service–Object Pair (SOP) Instance UID. DICOM does not specify relationships of SR documents to their parent documents or document version numbers. If DICOM SR documents are transformed to CDA R2, the original SR document will be the parent of the CDA document that is generated. In this case, the relatedDocument act relationship is used to express that a transformation has been performed.

3.3.4.4 CDA Document Exchange

HL7 Version 2.x and V3 messages can be used to convey CDA documents as a Multipurpose Internet Mail Extension (IETF, 1996) package. It is encoded as encapsulated data adhering to the Internet Engineering Task Force (IETF) recommendation "MIME Encapsulation of Aggregate Documents, such as HTML (MHTML)" (IETF, 1999).

CDA documents intended to be exchanged as the payload of HL7 version 2.x messages (e.g., medical records messages) are placed in the Observation (OBX segment as observation value encoded as V2.x-encapsulated data. The value of OBX value type should be set to "ED." The value of the coded OBX observation identifier should be the same as ClinicalDocument.code.

In order to support the exchange of CDA documents in imaging environments, DICOM has specified a way to encapsulate those documents in DICOM messages (DICOM Supplement 114 "DICOM Encapsulation of CDA Documents" [DICOM, 2007]).The goal is to make clinical information contained in CDA document available as input in the context of imaging procedures or imaging reports. CDA documents can thus be conveyed as DICOM objects using the DICOM Storage and Query/Retrieve Services. CDA documents are encoded in XML based on the normative HL7 v3 XML Implementation Technology Specification and wrapped in a DICOM container. Supplement 114 also defines the header data for encapsulated CDA objects and generally takes an approach that is similar to DICOM encapsulation of Portable Document Format (PDF) documents, which is a further option for document exchange if reports are generated or scanned into a PDF format.

DICOM Supplement 101 "HL7 Structured Document Object References" (DICOM, 2005) specifies how CDA documents can be referenced within DICOM SR documents and DICOM worklists. It also provides the basis to include HL7 CDA documents on DICOM interchange media. Supplement 101 specifies attributes for the type of the referenced CDA document (e.g., Release 2) and its Instance Identifier. Since Instance Identifiers that exceed 64 characters or use an extension cannot be encoded as a DICOM UID, DICOM specifies the mapping of HL7 Instance Identifiers to the local Referenced SOP Instance UID. Optional Retrieve Uniform Resource Identifiers (URIs) (IETF, 1998) can be used to access the referenced CDA document.

3.3.4.5 CDA Implementation Guides

Implementation guides (IGs) specify a set of constraints on documents that may be created on the base CDA standard. Document types that are specified based on IGs include the Continuity of Care Document (CCD) (HL7, 2007), Public Health Case Reporting (HL7, 2009b), and Diagnostic Imaging Report (DIR) (HL7, 2009a) that are based on CDA Release 2. An implementation guide includes the requirements and templates for specific report types. Templates are collections of constraints detailed in the IG requirements that are assigned a unique template identifier (templateId). Templates may specify constraints for the CDA header and the clinical non structured content of the document (CDA level one), sections within the structured body of the clinical document (CDA level two), and/or structured entries (CDA level three). The creator of a document, for instance, may use a templateId to assert conformance with certain constraints. In that sense, CDA IGs constitute conformance profiles. Requirements that do not add further constraints to the base standard are typically not included in the IGs. Recipients may choose to reject document instances that do not contain a particular templateId or may process those instances despite the absence of an expected templateId.

The CDA Implementation Guide for Diagnostic Imaging Reports (DIR) contains a consulting specialist's interpretation of a non invasive imaging procedure. It is intended to convey the interpretation results and diagnoses to the referring (ordering) physician and become part of the patient's medical record. The purpose of this Implementation Guide (IG) is to describe constraints on the CDA Header and Body elements. The DIR IG has been developed jointly by DICOM, and HL7 is consistent with the "SR Diagnostic Imaging Report Transformation Guide" (DICOM, 2010). The report may contain both narrative and coded data (Figure 3.9).

The ClinicalDocument element indicates the start and the end of the document. The namespace for CDA R2 is urn:hl7-org:v3. The ClinicalDocument/typeId element identifies the constraints imposed by CDA R2 on the content, while ClinicalDocument/templateId identifies the template that defines constraints on the content of CDA Diagnostic Imaging Reports. ClinicalDocument/code specifies the type of the clinical document (e.g., the LOINC code for "Diagnostic Imaging Report").

The DIR IG specifies constraints on header participants such as the author of the document and the referring physician who ordered the imaging procedure. If a referrer exists, he should

```
<ClinicalDocumentxmlns="urn:hl7-org:v3"xmlns:
voc="urn:hl7-org:v3/voc" xmlns:xsi="http://www.
w3.org/2001/XMLSchema-instance">
        <typeId root="2.16.840.1.113883.1.3"
extension="POCD_HD000040"/>
        <templateId root="2.16. 840.1.113883.
10.20.6"/>
        <id root="2.16.840.1.113883.19.4.27"/>
        <code code="18748-4"
codeSystem="2.16.840.1.113883.6.1"
codeSystemName="LOINC" displayName="Diagnostic
Imaging Report"/>
        ...
</ClinicalDocument>
```

FIGURE 3.9. Clinical document templateID sample.

```
<!-- transformation of a DICOM SR -->
<relatedDocument typeCode="XFRM">
   <parentDocument>
      <id root="1.2.840.113619.2.62.994044785528.20060823.200608232232322.9"/>
      <!-- SOP Instance UID (0008,0018) of SR sample document-->
   </parentDocument>
</relatedDocument>
```

FIGURE 3.10 CDA-related document transformation sample.

also be recorded as the information recipient. Information on the legal authenticator (typically a supervising physician who signs reports attested by residents) must be present if the document has been legally authenticated.

Acts that are related to the clinical document comprise orders (i.e., the Placer Order that was fulfilled by the imaging procedure), service events (the imaging procedure(s) that the provider describes and interprets), and parent documents (e.g., prior versions of the current CDA document or the original DICOM SR document that has been transformed to the current CDA document) (Figure 3.10).

Constraints on the CDA body comprise document section and structured entry constraints. Subject and observer context information may be overridden at the section level to record information on fetuses and section authors. The DICOM Object Catalog section is a special section that lists all referenced DICOM objects and their associated series and studies. It does not contain narrative text since it is machine-readable content used to look up the DICOM UIDs and information required to retrieve the referenced objects such as images that illustrate specific findings. The contents of this section are not intended to be rendered as part of the CDA document. The findings section contains the direct observations that were made in interpreting image data acquired in the context of the current imaging procedure. Further sections such as "Reason for Study," "History," and "Impressions" are optional. Reference values pointing to structured entries in the structured part of the document and WADO references may be used within section text (Figure 3.11).

```
<section>
 <templateId root="2.16.840.1.113883.10.20.6.1.2"/>
   <code code="121070" codeSystem="1.2.840.10008.2.16.4" codeSystemName="DCM"
        displayName="Findings"/>
   <title>Findings</title>
   <text>
   <paragraph>
    <caption>Finding</caption>
    <content ID="Fndng2">The cardiomediastinum is within normal limits. The
       trachea is midline. The previously described opacity at the medial right lung base
       has cleared. There are no new infiltrates. There is a new round density at the left
       hilus, superiorly (diameter about 45mm). A CT scan is recommended for further
       evaluation. The pleural spaces are clear. The visualized musculoskeletal
       structures and the upper abdomen are stable and unremarkable.</content>
   </paragraph>
   </text>
   <entry>
    <observation classCode="OBS" moodCode="EVN">
    <!-- Text Observation -->
    <templateId root="2.16.840.1.113883.10.20.6.2.12"/>
    <code code="121071" codeSystem="1.2.840.10008.2.16.4"
       codeSystemName="DCM" displayName="Finding"/>
    <value xsi:type="ED">
     <reference value="#Fndng2"/>
    </value>
    ...
    <!-- entryRelationships to Quantity Text, Code and Measurement Observations
         may go here -->
   </observation>
  </entry>
</section>
```

FIGURE 3.11 Findings section example with section text and observation entry.

```
<observation classCode="DGIMG" moodCode="EVN">
  <templateId root="2.16.840.1.113883.10.20.6.2.8"/>
  <!-- (0008,1155) Referenced SOP Instance UID-->
  <id root="1.2.840.113619.2.62.994044785528.20060823.200608232232322.3"/>
  <!-- (0008,1150) Referenced SOP Class UID -->
  <code code="1.2.840.10008.5.1.4.1.1.1" codeSystem="1.2.840.10008.2.6.1"
 codeSystemName="DCMUID" displayName="Computed Radiography Image Storage">
  </code>
  <text mediaType="application/dicom">
    <!--reference to CR DICOM image (PA view) -->
    <reference        value="http://www.example.org/wado?requestType=WADO&amp;
studyUID=1.2.840.113619.2.62.994044785528.114289542805&amp;
seriesUID=1.2.840.113619.2.62.994044785528.20060823223142485051&amp;
objectUID=1.2.840.113619.2.62.994044785528.20060823.200608232232322.3&amp;
contentType=application/dicom"/>
  </text>
  <effectiveTime value="20060823223232"/>
  <!-- entryRelationship elements containing Purpose of Reference or Referenced
       Frames observations may go here -->
</observation>
```

FIGURE 3.12 SOP instance observation example.

CDA DIR documents may contain structured CDA entries that are based on the DICOM Basic Diagnostic Imaging Report (Template 2000) and Transcribed Diagnostic Imaging Report (Template 2005). Most of the constraints have been inherited from those templates and their transformation specified in the draft "SR Diagnostic Imaging Report Transformation Guide" (DICOM Supplement 135). Each section may contain

- *Text Observations*: optionally inferred from Quantity Measurement Observation or Image.
- *Code Observations*: optionally inferred from Quantity Measurement Observation or Image references.
- *Quantity Measurement Observations*: optionally inferred from an image reference.
- *SOP Instance Observations (Figure 3.12)*: containing references (e.g., to images).

Spatial Coordinates (SCOORD) for regions of interest associated with linear, area, and volume measurements based on image data are not encoded in the CDA document. If it is desired to show images with graphical annotations, they should be based on DICOM Softcopy Presentation State objects that reference the relevant images. The procedure context may be overridden in the document body by using act or procedure classes associated with the individual sections.

A future Procedure Note CDA Implementation Guide is intended for the documentation of image-guided interventions and other interventional procedures.

3.3.5 V3 Messages

HL7 V3 messaging standards specify interactions that comprise the trigger event and application roles in the exchange of data. Application roles are abstractions of healthcare information system components, which send or receive HL7 V3 messages and define their responsibilities. Trigger events cause information to be exchanged between systems in a healthcare domain. Trigger events can be interaction-based, associated with state transitions (e.g., state transition of a focal class such as an observation class in a result message), or occur at the request of a human user.

V3 messages (Figure 3.13) are developed based on the HL7 development framework and derive their domain content (message contents) from the RIM (e.g., observations and associated interpretation ranges as part of laboratory result messages). Common Message Element Types (CMETs) are standardized model fragments intended to be reused in the message information models of individual content domains. The transmission wrapper of the message includes information on the interaction and the message itself, such as

- Message Identifier
- Creation time of the message
- Interaction Identifier
- Processing controls on the message

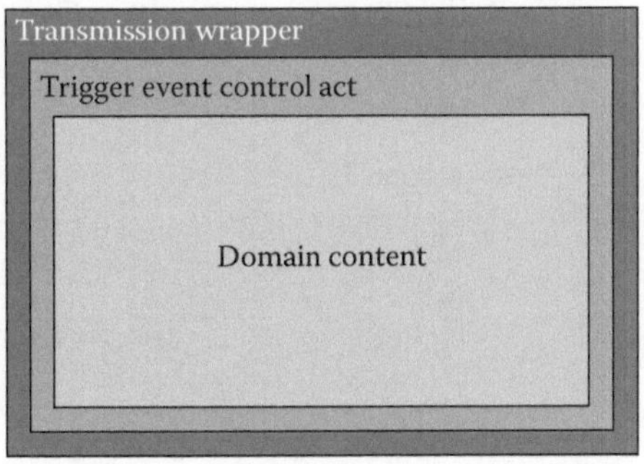

FIGURE 3.13 HL7 V3 message structure.

- Message Acknowledgment
- Identity of the sending and receiving systems

A "Trigger Event Control Act" is required for all messages except accept level acknowledgments, for which it is not permitted. It contains the control information for the subject of the message (the payload), that is, the trigger event and related information such as

- Trigger Event Code
- Date and time of the event that triggered the interaction
- Control Act Process Participations (e.g., information on the author or the originating device)

Although HL7 V3 messaging standards build on a number of concepts that are known from HL7 Version 2.x (e.g., accept level and application level message acknowledgment), the standards are not backward compatible to Version 2.x. V3 messaging standards are being developed in administrative (e.g., accounting and billing, scheduling of appointments) and clinical domains (e.g., order entry and results reporting). Those emerging standards will provide the basis for imaging-specific V3 message development.

3.4 Conclusions

Access to relevant information is one of the core elements in supporting the medical treatment process (Figure 3.14). From an information technology perspective, imaging processes start with order communications (Order Entry), followed by internal scheduling of the procedure, image acquisition, quality control, postprocessing, and interpretation of image data. Finally, results are communicated (Results Reporting) by messages and documents. Access to relevant information (i.e., clinical information in the Electronic Patient Record (EPR), prior imaging studies, and reports) is an essential requirement for selecting the appropriate imaging procedure and ensuring the completeness and accuracy of the interpretation and report. Relevant information is determined by the study context of the imaging procedure (e.g., current and prior imaging studies and reports) and in extension to that by the patient and clinical context (e.g., indication-based access to clinical information).

HL7-based communication of data plays an important role in the following areas:

- *Patient identity*: Management and cross-referencing of patient identifiers, exchange of demographic data, patient information reconciliation.
- *Patient registration and location tracking*: Admission, discharge, and transfer of the patient.
- *Order entry*: Imaging order placer and filler management, imaging order workflow management.
- *Results reporting*: Results and document management messages, DICOM SR evidence documents, and CDA Imaging Reports.

HL7 message and CDA document constraints are specified in many Integrating the Healthcare Enterprise (IHE) profiles addressing those areas for radiology, cardiology, and other medical disciplines. HL7 standards are widely implemented and successfully support the integration of imaging and clinical information systems.

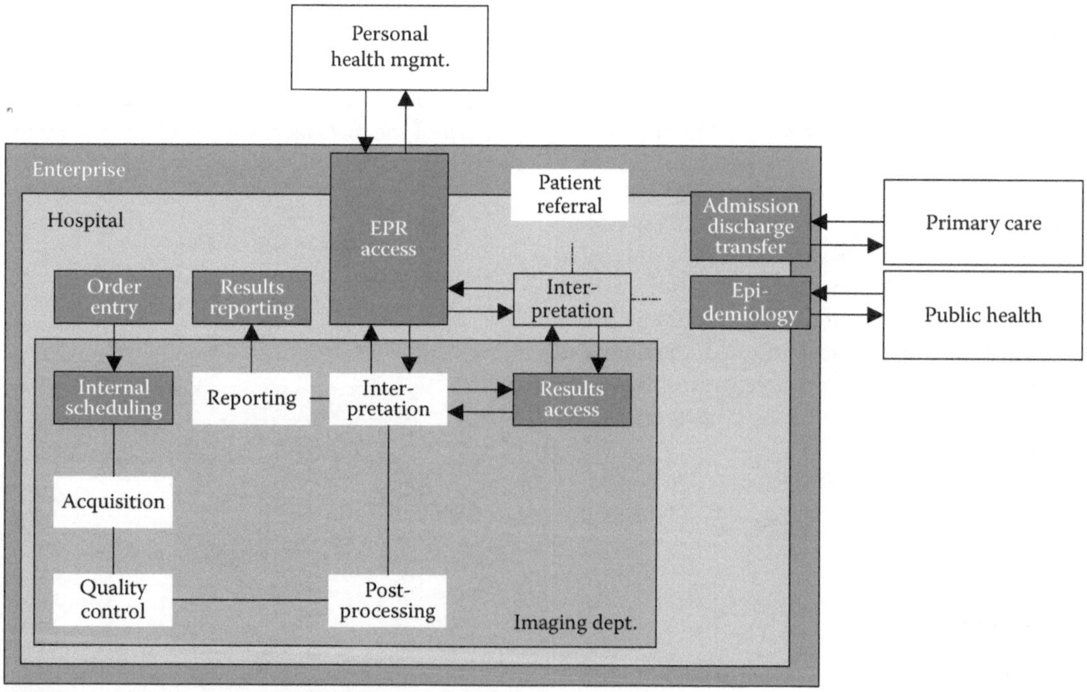

FIGURE 3.14 Integration of imaging and clinical information systems.

References

ANSI/HL7 2000. *ANSI/HL7 CDA, R1-2000. HL7 Version 3 Standard: Clinical Document Architecture, Release 1.*

ANSI/HL7 2005. *CDA, R2-2005. HL7 Version 3 Standard: Clinical Document Architecture (CDA), Release 2.*

ANSI/HL7 2007. *ANSI/HL7 2.6-2007. Health Level Seven Standard Version 2.6—An Application Protocol for Electronic Data Exchange in Healthcare Environments.*

ANSI/HL7 2009. *ANSI/HL7 V3 2009. HL7 Version 3 Standard: Normative Edition.*

DICOM 2005. *Digital Imaging and Communications in Medicine (DICOM). Supplement 101: HL7 Structured Document Object References.*

DICOM 2007. *Digital Imaging and Communications in Medicine (DICOM). Supplement 114: DICOM Encapsulation of CDA Documents.*

DICOM 2009a. *Digital Imaging and Communications in Medicine (DICOM) PS 3.16-2009. Part 16: Content Mapping Resource.*

DICOM 2009b. *Digital Imaging and Communications in Medicine (DICOM) PS 3.18. Web Access to Dicom Persistent Objects (WADO).*

DICOM 2010. *Digital Imaging and Communications in Medicine (DICOM), Supplement 135: SR Diagnostic Imaging Report Transformation Guide.*

HL7 2007. *HL7 Implementation Guide: CDA Release 2—Continuity of Care Document (CCD).*

HL7 2009a. *HL7 Implementation Guide for CDA Release 2: Diagnostic Imaging Reports (DIR)—Universal Realm, Release 1.0.*

HL7 2009b. *HL7 Implementation Guide for CDA Release 2: Public Health Case Reporting, Release 1.*

IETF 1996. *IETF RFC 2046. Multipurpose Internet Mail Extensions (MIME) Part Two: Media Types.*

IETF 1998. *IETF RFC 2046. Uniform Resource Identifiers (URI). Generic Syntax.*

IETF 1999. *IETF RFC 2557. MIME Encapsulation of Aggregate Documents, such as HTML (MHTML).*

IHTSDO *International Health Terminology Standards Development Organisation (IHTSDO). Systematized Nomenclature of Medicine—Clinical Terms.*

ISO 2005. *9834-1: Information Technology—Open Systems Interconnection—Procedures for Operation of OSI Registration Authorities: General Procedures and Top Arcs of the ASN.1 Object Identifier Tree.*

ISO/IEC 1994. *7498-1: Information Technology—Open Systems Interconnection—Basic Reference Model: The Basic Model.*

RIHC 2000. *Regenstrief Institute for Healthcare. Logical Observation Identifier Names and Codes.* Indianapolis.

4

DICOM

Steven C. Horii
University of Pennsylvania Medical Center

4.1 Introduction

4.1.1 A Brief History of ACR-NEMA

Most readers of this chapter likely have at least some knowledge of Digital Imaging and Communications in Medicine (DICOM). Some may know that what is presently DICOM began as the American College of Radiology-National Electrical Manufacturers Association (ACR-NEMA) Standard. A few will remember something about how and why the Standard was developed and why the particular organizations responsible for it got involved. The development of DICOM represents one of the major enabling factors for picture archiving and communications systems (PACS). The history of the effort stands as an example of how dedicated engineers, radiologists, physicists, and corporate administrators managed to achieve consensus on what were highly contentious issues.

Why the American College of Radiology (ACR) and the National Electrical Manufacturers Association (NEMA) were the organizations initially involved is straightforward. The ACR represents the professional interests of radiologists and radiological physicists. The College, through its Commissions and Committees, has a long history of establishing practice guidelines for the various aspects of radiology. The ACR also serves as the lobbying body for the specialty of radiology. NEMA is a large trade association representing a very diverse group of manufacturers of electrical and electronic equipment. NEMA is well known for standards, electrical plugs, and sockets ("wiring devices") serving as a familiar example. NEMA represents the manufacturers of medical imaging equipment and NEMA is a lobbying organization. The community that would use a standard for communicating digital images would be well represented with the ACR and NEMA, two organizations with a history of developing standards and responsive to the interests and needs of their members.

In 1982, the first major meeting on PACS was held. Organized by Sam Dwyer and Andre Duerinckx, the meeting was held in Newport Beach, California and most of the United States, European, and Asian groups who were working on, or planning, PACS were in attendance. One of the sessions of the meeting was devoted to standards. That standards for digital images would be needed if PACS were to be implemented was a consistent theme of the papers in that session (Baxter et al., 1982; Haney et al., 1982; Schneider, 1982; Wendler and Meyer-Ebrecht, 1982).

Radiologists and radiological physicists who were doing research using digital images were, at about the same time as the 1982 PACS meeting, complaining to the ACR about difficulties accessing these images. Since this involved medical imaging equipment that was regulated by the Food and Drug Administration's (FDA) Center for Devices and Radiological Health (at the time in transition from the Bureau of Radiological Health), the ACR made inquiries to the FDA about the problems the radiologists were having with digital images. The result was a meeting of representatives of equipment manufacturers (through NEMA), radiologists (through the ACR), and the FDA. The vendors agreed that a voluntary standard would be preferable to a regulatory one (SIIM, 2008). Shortly thereafter, in November 1983, the ACR and NEMA met to form the Digital Imaging and Communications Standards Committee (Horii, 2005).

The governance of the ACR-NEMA Digital Imaging and Communications Standards Committee (hereafter "ACR-NEMA Committee") was determined at the organizing meeting. So that both users and manufacturers would have an equal voice, the ACR-NEMA Committee would have co-chairmen; one elected from the NEMA medical imaging equipment manufacturers and one elected from the ACR. The first two Co-Chairs were Allan Edwin, then at General Electric Medical Systems, and Gwilym S. Lodwick, MD, then a radiologist at the Massachusetts General Hospital and representative of the ACR.

An important decision made early by the ACR-NEMA Committee was how to structure the group that would do the work on developing the Standard. It was recognized that a single large group would be unwieldy; rather, the developers would be divided into smaller working groups, each with its own governance. The first three working groups were

- Working Group I (WG I): Hardware Interface; Chairman: Owen Nelson of 3M Corporation;
- Working Group II (WG II): Data Format; Chairman: David Hill, PhD of Siemens Medical Systems; and
- Working Group III (WG III): Systems and Performance Specifications; Chairman: John Moore, PhD of Bio-Imaging Research and an ACR representative.

These groups would meet separately at regular intervals and could also meet together when necessary. The Chairmen would report on their activity to the ACR-NEMA Committee which acted in a coordinating and overall governing role. The author of this chapter was a member of WG II and III.

The first work of the three Working Groups was to survey existing standards for ideas. The members canvassed both the literature and colleagues and found a large number of potentially applicable standards. One of the original organizers of the meeting at the FDA was Sam Dwyer, III. Sam was a member of the Institute of Electrical and Electronics Engineers (IEEE), an organization that was well known in the world of standards developers. IEEE publications were useful to WG I in looking at the hardware that would drive the data across the interface. Perhaps, the widest searches were for potentially useful data formats. The members of WG II searched a large number of standards, including National Aeronautics and Space Administration (NASA), American Federation of Information Processing Societies (AFIPS), American National Standards Institute (ANSI), International Standards Organization (ISO), and various European and Asian standards organizations. NASA Landsat attracted interest because it addressed digital images produced by the Landsat series of satellites (Billingsley, 1982). However, the format included many more data elements that were thought unnecessary for medical images. The most influential standard was one published by the American Association of Physicists in Medicine (AAPM) as their Report Number 10 (Maguire et al., 1982). (Note: this is *not* the report of AAPM Task Group 10.) This standard proposed a data structure for digital medical images recorded on magnetic tape. The significant part of this standard was the way the data structure was organized. It defined key-value pairs. The "key" would be a field such as "Patient Name" and the "value" would, in the standard definition, describe how the patient name was to be encoded (in the notation of the Report: Patient name: = <character string>). When put into practice, the value field would contain the actual patient name, encoded as specified in the standard. There were major advantages to this idea: it is self-documenting in that the keys name the data elements, fields were not restricted in length, and data elements could be added easily. As should be evident from the structure of the ACR-NEMA and DICOM data elements, the idea for the structure of these was drawn directly from the AAPM Report 10 standard.

4.1.2 Why ACR-NEMA?

The introduction of imaging methods that produced digital images began in the mid-1970s to late 1970s with nuclear medicine followed shortly by computed tomography (CT) and ultrasound. In particular, the rapid growth of CT resulted in radiology departments and practices having to determine how best to display and store the images. The earliest CT systems used Polaroid images taken of the console display as a way to record the images for viewing. These did not fit the existing radiology film library logistics very well, though mounting multiple

Polaroid prints on larger sheets of paper that fit into film jackets was one popular method. Manufacturers responded with film recorders that could expose film to the CT image displayed on a cathode ray tube (CRT); the film could be processed in conventional x-ray rapid processors and handled like any other radiographic films. This also provided the advantage of viewing on existing light boxes or multiviewers. Researchers wanted access to the images and the idea of retaining the digital image data was attractive from an archival standpoint as well. Since CT machines used minicomputer hardware, the various peripherals used with those computers were readily available and supported by the operating systems of those machines. Magnetic tape was a natural choice as a result. Researchers could read the tapes physically, but the manufacturers made their storage formats proprietary, fearing that other vendors might be able to read their images.

Some institutions had CT machines from several vendors. This meant that researchers wanting access to the images had to be able to read multiple tape formats. Having to sign multiple nondisclosure agreements was the norm. Other research groups resorted to "hacking" the tape format, so they could read the images without the restrictions of nondisclosure. The AAPM Report 10 proposed standard for magnetic tape storage of digital images was an early user community response to this problem of multiple incompatible formats. It was also applied to nuclear medicine as many of the digital processing systems used there also used magnetic tape for storage.

For the radiologist, proprietary formats meant that the images could be viewed either on analog film or on a dedicated "console" provided by the CT vendor. Using film meant giving up the range of attenuation values that CT could represent, since film typically could reproduce about 256 visually distinct gray levels, but CT images had 4096. At the dedicated console, in actuality an early imaging workstation, the radiologist could adjust the window width and level controls of the images, could do distance and area measurements on them, and examine the attenuation value of the pixels. Images could be displayed so as to optimize the display of bone, soft tissues, or lung. To do this on film meant recording the same image with different settings of window width and level—a practice fairly widely used. The independent viewing consoles were necessary not only because of the ability to manipulate the digital images, but because the machine console was needed by the technologists to run the CT system. The proprietary formats also meant that a vendor-specific independent viewing console was needed for each different vendor's CT machine.

Early visionaries realized that having digital images also meant that they could be communicated using the then nascent digital communications networks. In particular, teleradiology was an early application for medical digital images. The growth of local area networks, and the appearance of standards for them, also suggested to some that such networks could be used to move digital images over hospital-scale facilities, avoiding the movement, and potential loss or misplacement of carts full of radiographic film (Ausherman et al., 1970; Lemke et al., 1979). As described in the former section, it was this interest in taking the advantage of digital images and digital communications that developed the pressure needed to start the ACR-NEMA work.

Another important aspect of the ACR-NEMA collaboration was the protection from antitrust litigation that NEMA offered vendors. The US antitrust laws prohibited manufacturers from trying to fix prices or engage in other anticompetitive practices. NEMA, as a trade association, had a legal staff that developed rules for member vendors that would protect them from antitrust allegations. The rules forbid discussion of prices, market share, or product plans at any meeting held under NEMA auspices. The minutes of NEMA meetings were reviewed by the lawyers and were made publicly available. For the imaging equipment manufacturers, it meant that fierce competitors in the marketplace could meet and discuss standards and other topics (e.g., safety requirements) that could be beneficial to the industry and their consumers without fearing that one of their competitors could accuse them of anticompetitive practices.

4.1.3 ACR-NEMA to DICOM

Work on the ACR-NEMA Standard began shortly after the formation of the Committee and the initial Working Groups. Progress was rapid, and by 1985, the first version of the ACR-NEMA Standard was published. It was given the designation of ACR-NEMA 300-1985 and the title was that of the Committee, "Digital Imaging and Communications." This first version of the ACR-NEMA Standard defined a high-speed, parallel digital interface, the connector and cabling to be used, the hardware to drive the interface, and a data format for the information to be moved across the interface. The interface employed a widely available 50-pin connector, but did not define a network interface.

Though Robert Mecalfe and David Boggs published the first paper on the Ethernet network in 1976 (Metcalfe and Boggs, 1976), it was not until 1985 that a slightly modified version of Ethernet was published as a standard by the IEEE as IEEE 802.3-1985 (IEEE, 1985). While those more familiar with DICOM than the ACR-NEMA Standard have asked why a network interface was not initially specified, the fact is that a standardized version of Ethernet was not available during the time of the ACR-NEMA Standard development. The developers of the ACR-NEMA 300–1985 standard certainly knew that operation over a network would be desirable. If the 50-pin connector was used to transfer information to a computer, the computer could act as a network interface and would manage the protocol needed to operate on a network.

Once the ACR-NEMA 300-1985 standard was released, engineers from both manufacturers and institutions scrutinized the standard and found some inconsistencies. Suggestions from users for additions to the data elements were also received by the ACR-NEMA Committee. As a result, work on version 2 of the standard began almost immediately and ACR-NEMA 300 Version 2 (officially designated ACR-NEMA 300-1988) was published in

1988. This version of the Standard was actually implemented by a number of vendors and a test of the implementations was carried out at Georgetown University in 1990 (Horii et al., 1990). All the implementations were successful at communicating and, though it was anticipated that the testing might take two or more days, the successful "round robin" testing was completed in a single day. Much of the historical information in what follows was presented at the DICOM Workshop held on the tenth anniversary of the publication of DICOM in 2003 (NEMA, 2003a,b).

Though the ACR-NEMA effort was initiated by radiologists and manufacturers, by the time of the introduction of the ACR-NEMA Version 2 standard, expansion beyond the original goal of being able to connect imaging equipment to another device was evident. Working Group IV, chaired by Hartwig Blume, PhD, of Philips Medical Systems, was formed to examine data compression methods that might be employed by users of the Standard. Working Group V was formed to address how the Standard could be applied for exchange of images using media. The Working Group (WG) Chairman was John Hoffman of General Electric Medical Systems and the first media to be addressed was 9-track magnetic tape, at the time the primary archival storage media for CT.

Beyond additional capabilities of the Standard, other specialties, notably cardiology, pathology, dentistry, and gastroenterology, were all users of imaging and with an increasing proportion of the images in digital form. Since an initial requirement to be a voting member of the ACR-NEMA Committee was to be either a member of the ACR or NEMA, other specialties were left out. Besides the two WGs noted above, three other Working Groups had been formed. WG VI was tasked with evaluating reports of errors or inconsistencies in the Standard and was chaired by Bill Bennett of Kodak. WG VII had taken on the task of representing multidimensional data and began its work with surveys of existing multidimensional image data representations. WG VIII was formed to examine how the Standard would need to be expanded to interface with radiology and hospital information systems. WG VIII, chaired by Bob Thompson of the University of North Carolina, turned out to have a pivotal role in the transformation of ACR-NEMA.

In looking at the task of how PACS fit in healthcare operation with radiology and hospital information systems (RIS and HIS, respectively), it was rapidly determined that a model of the information structure used in radiology was necessary so that descriptions of the information that would have to move between RIS, HIS, and PACS could be made as unambiguous as possible. In what would turn out to be a significant event, Fred Prior (then at Siemens Medical Systems) submitted an information model that he and his colleagues had been working on. It represented radiology information in an entity–relationship form. The "things" in radiology, such as patients, images, and reports, were shown in diagrammatic form with connectors representing the relationships between them. Each "thing," or entity, also had attributes attached to it that characterized it. These, it was clear, could be the ACR-NEMA data elements. Some readers may recognize this as a form of object-oriented information modeling. Most of the members of WG VIII promptly purchased textbooks on object-oriented modeling and design so as to better understand this then new approach to representing information.

Two other major influences were operating about the same time. Siemens and Philips had, in Europe, jointly developed a standard that put ACR-NEMA onto a network. This was known as the "Standard Product Interconnect," or SPI. Siemens and Philips decided to submit this to the ACR-NEMA Committee for consideration in moving ACR-NEMA from a point-to-point to a network standard. An energetic radiologist named W. Dean Bidgood, Jr., had joined WG VIII. Though a radiologist, Dr. Bidgood was very interested in getting other users of digital imaging in medicine to sign on to the standards effort. He foresaw that it would be important, as hospitals shifted to digital healthcare records, that imaging use a common standard. Much of his time was spent meeting with committees and attending conferences of other medical imaging specialties and explaining what the ACR-NEMA Standard was and how it could be useful to them.

Working Group VI, though initially charged with maintenance and correction of the Standard, also became the group that would examine how the ACR-NEMA Standard could be expanded to an implementation that could run over standard networks. With the submission from Siemens and Philips of the SPI proposal, the group began an intensive examination of how the ACR-NEMA protocol could be run over a network interconnection. What the computer between a piece of imaging equipment and a network would have to do was fundamentally what the WG determined. Networks by this time had a number of standards, both open (such as the ISO/IEEE 802.3 Ethernet standard) and vendor-specific (such as DECnet). A detailed model of how network communication worked was embodied in a standard known as the International Standards Organization-Open Systems Interconnection (ISO-OSI). This standard specified digital communication is taking place in seven layers. The user interacts with the top layer and the physical interconnection is at the bottom. The ACR-NEMA WG and Committee agreed that this was a useful model to follow. Additionally, some parts of ISO-OSI had been implemented and supported some of the functions that a networked version of ACR-NEMA would need. With the wide availability of Ethernet and its increasing use as a local area network in radiology departments and hospitals, it was clear that a way to operate the ACR-NEMA Standard over Ethernet would be important. Between the 1990 demonstration of ACR-NEMA Version 2 and 1992, a major undertaking of the ACR-NEMA Working Groups resulted in a preliminary 9-part proposed standard. Informally, this was known as ACR-NEMA Version 3. The radiology community and the MedPACS Section of NEMA very much wanted the vendors to adopt this new standard and a demonstration of it at the Radiological Society of North America (RSNA) meeting in 1992 was chosen as a way to show the radiology community what ACR-NEMA was doing. With coordination by the RSNA Electronic Communications Committee, chaired by Laurens Ackerman, MD, of Henry Ford Hospital, funding from NEMA and the RSNA was secured for the development of software for the demonstration. A competitive bidding process resulted in the Electronic Radiology Laboratory of the Mallinckrodt Institute of Radiology being selected as the software developer. By July 1992, a workshop was held in St. Louis to explain how the software would work. The code (about 28,000 lines) and 300 pages of documentation were distributed to the participating vendors. In September 1992, the first "connect-a-thon" was held in Chicago. Though 5 days were allocated for testing, all of the 25 vendors who had applied to participate were able to connect and transmit messages successfully within 2 days. On November 29, 1992, the ACR-NEMA Prototype Demonstration opened in the *info*RAD area of the RSNA meeting. It was the largest exhibit that had ever been held in *info*RAD. Attendees were quite surprised to see the names and logos of all the vendors in one exhibit as these manufacturers were competitors and had never previously exhibited together.

With the transition to networked operation and the increasing participation of specialties in addition to radiology, expansion of the standards effort was inevitable. As participants from outside the United States also increased, the need to change the name of the Committee to reflect this much-expanded constituency and to draft a new set of bylaws so that voting members could come from organizations other than the ACR and NEMA was evident. The name of the ACR-NEMA Digital Imaging and Communications Standards Committee was changed to Digital Imaging and Communications in Medicine, which also provided the acronym for the name of the Standard, DICOM.

4.2 Communication Fundamentals

4.2.1 An Example of Communication Failures

Suppose that on a well-earned day off, you are playing golf with some friends. You have your pager with you since you know that your cell phone may not work in all areas that you frequent. Your pager goes off and you check the message. It is from your administrative assistant. You take out your cell phone only to discover that the golf course is one of the places the phone does not work. You borrow a phone that does work from one of your friends and call your administrative assistant. You are a bit irritated at the interruption, but once your administrative assistant explains the situation, you relent; he received a call from Japan from Dr. Sato inviting you to speak at an upcoming meeting there. Now, it is an international call you have to make, though you were told you would have to return the call in the evening anyway because of the time difference.

The inability of your cellular telephone to connect to the network in the area is an example of a low-level communications failure. In this example, you might have a Global System for Mobile (GSM) phone while the area you happen to be in is a code division multiple access (CDMA) only area. These are two of the major communication protocols used by cellular telephones and they are not compatible. A similar problem with a land line telephone might be trying to connect an older telephone that uses a four-prong plug to a newer modular jack.

Later in the evening, you call the number you were given. You have determined that it is now 9:00 a.m. in Japan, and the message was that any time after 8:00 a.m. there would be fine. You also realize that it is Saturday morning in Japan, though Friday evening where you are. When the call is answered, the person answering is speaking a language you do not understand. He does, however, realize that you are also not understandable to him, so he puts his wife on the phone. You reached Dr. Sato's home and her husband, who speaks only Japanese, answered the telephone. Dr. Sato, fortunately, does speak English and you talk about the meeting invitation.

The failure in this instance is a level up from hardware. This time, your phone call goes through with no problems. However, you are initially confronted with an unfamiliar language. Even though the low-level communication worked, at a higher level—that of language—it initially failed.

In discussing the meeting, since you have never been to Japan before, you realize you should ask about the weather. Dr. Sato tells you that in Hiroshima when the meeting will take place, the average temperature is 34°C. You figure you had better take warm clothing. She then gives you the number for the audiovisual support group so you can discuss presentation requirements. You travel to the meeting successfully, but get off the plane in Hiroshima in sweltering heat.

In this instance, the failure is one of definitions of units. You spoke English with Dr. Sato and she answered your question about the weather, but she replied in degrees Celsius and your assumption was degrees Fahrenheit. Admittedly, a bit of research on your part would have allowed you to avoid this problem—but, you would have had to do some unit conversion to do so.

When you spoke with the audiovisual support group, you told them that you wanted single projection. They seemed a bit puzzled and asked if you were bringing your presentation or sending it ahead. You said you would be bringing it. When you show up at the lecture hall, you find that a single 35-mm slide projector has been set up with a projection screen. The audiovisual person is quite pleased that they were able to find a 35-mm slide projector. You explain hastily that you are sorry, but you meant an electronic projector. Fortunately, other speakers will be using a computer and electronic projector, so changing the setup is fairly simple. They had asked if you were going to send your presentation (they were thinking 35-mm slides) ahead so that they could load them into the projector tray for you. You were thinking you would pack your USB drive with the presentation files on it.

In this instance, the communication failure is a bit more subtle. Hardware and language were correct and units were not involved. What differed here were the models that you used and the one the audiovisual group used. Your model of "single projection" is an electronic projector connected to a computer or with a connection for a laptop. The audiovisual group has this model, but they also have a different one for "single projection." They thought of projection with a modifier (single or dual) as being for 35-mm film slides. For electronic projection, their terminology is "computer projection" and the assumption is that a single electronic projector is used.

4.2.2 The Layered Model of Communication

As was mentioned in Section 5.1.2, the model of communication that ACR-NEMA adopted was the layered model proposed by the ISO. In the example above of multiple communication failures, a simple-layered model is implied. At the lowest level, the physical connection (in the example, cellular telephone links) has to be compatible. Once a physical connection is established, languages and rules of speaking (e.g., each person waits for the other to be done speaking before speaking himself or herself) are needed. This is a protocol of sorts. At the next higher level, definitions of terms and units in the chosen language have to agree. Finally, at the uppermost level, conceptual models have to be consistent. The ISO-OSI-layered model formalized this sort of construct (see Figure 4.1) (ISO, 1994). At the bottom, or physical, layer, devices are connected by hardware. Communication is typically bidirectional, so the layers are designed to move information both down and up. The various layers in between each perform functions on the data moving from the top layer down or the bottom layer up. A major advantage of a layered model is that a layer can be exchanged for another if technology improves. As an example, the physical layer of Ethernet has changed many times. The original physical layer used coaxial cable. When twisted-pair cabling and electronics to drive it at high speed were developed, the coaxial cable layer could be replaced by the twisted-pair copper wire layer without having to change the layers above. Now, fiber optic and wireless physical layers exist, but the applications that run above these layers can remain the same.

In the ISO-OSI-layered model, there are seven layers: physical, datalink, network, transport, session, presentation, and application. Each of these layers is defined by a standard or series of standards that describe what the layer does and how it interfaces with the adjacent layers. Some concepts in the ISO-OSI series of standards have proven very useful, particularly to DICOM. However, it is difficult to find full implementations of the ISO-OSI standard. The most widely used model (and standard) is the one employed by the Internet: the Transmission Control Protocol/Internet Protocol or TCP/IP. TCP/IP uses layers analogous to, but not identical with, those in the ISO-OSI standard.

As an example of how layered communication works, let us examine the process of mailing a check to pay a bill. You have the invoice in hand, you write a check for the amount needed, place the check and copy of the invoice in an envelope, add postage, and drop it in the mail. The mail room of the company that receives your payment sends the envelope to their accounts payable department. A person there opens the envelope, removes your check and invoice copy, enters the payment information into the company computer system, sends the check on for deposit, and routes the invoice copy for filing.

Though most of these steps have been replaced by on-line bill paying, the steps and what happens in them illustrate how a layered communication system works. At each step, what started out as a check for payment has various pieces of additional information added, or removed, depending on what happens in that step. On your side, for example, you attached the check to a copy of the invoice as instructed. This adds information to the check indicating for what the payment is intended. At the corresponding accounts receivable department of the company to whom you sent the check, the invoice is removed and the information on it used to credit your payment to the appropriate account. In putting your check in the mail, you are performing the equivalent of putting information onto a network. The Postal Service (acting like the physical layer of a communications network) goes through a number of steps to deliver the envelope to the proper destination. It does not, however, have to open the envelope to

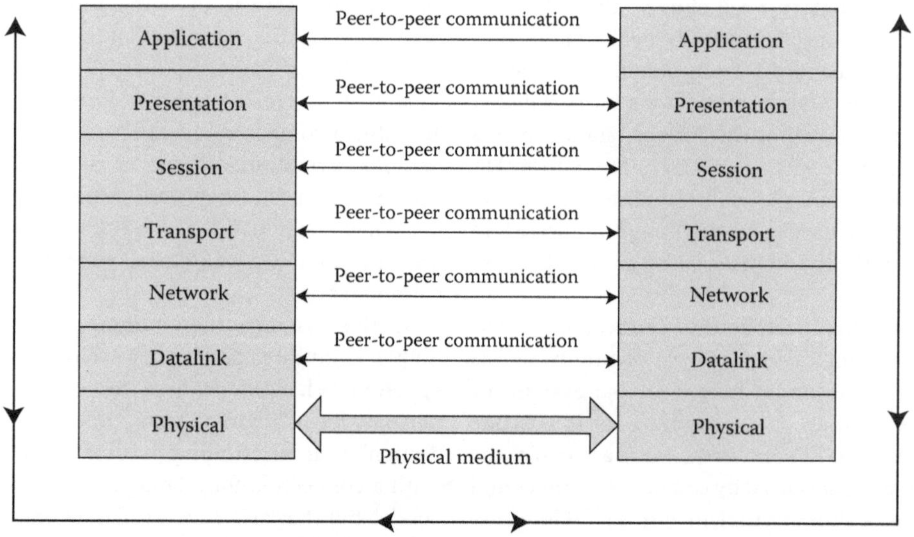

FIGURE 4.1 The OSI model.

determine where to deliver it; that is information you put on the outside of the envelope (or, as is typical, the company sent an envelope with a window to show the address printed on the invoice when folded and enclosed). Similarly, the mail room person at the company you paid does not have to open the envelope to deliver it internally. When it gets to accounts payable, the envelope is opened and the information on the invoice is used. What has happened is the equivalent of your handing the check and invoice directly to the accounts payable person who received it. For the purposes of you as payer and the accounts receivable person as payee, you have performed peer-to-peer communication. This peer-to-peer communication is another aspect of layered communication. Each layer, to the software that runs that layer, can be thought of as communicating directly with its peer on the other side of a network connection, even though the communication goes down through the "stack" of layers and back up through the stack on the other side. Most of us, when using the Internet, do not think about our data getting put into packets. The packets have various amounts of information added to them and removed from them as they move through the Ethernet TCP/IP communications layer. The sender and recipient of the data are not, and generally need not be, aware of what the various layers are doing.

4.2.3 DICOM and the Layered Model

As DICOM adopted a network, rather than point-to-point interface, the maturity of the Ethernet and TCP/IP standards made for a simple choice for one protocol that would be supported. Because so much of the early thinking about a network interface involved the ISO-OSI model, the DICOM Standard also included the ISO-OSI protocol. Some parts of the ISO-OSI standard were adopted directly by DICOM, specifically the Association Control Service Element (ACSE) (ISO, 1996). When two DICOM-conformant devices communicate, they begin the process by opening an Association. Over this Association, very basic information is exchanged so that subsequent communication can proceed smoothly. When the layers of TCP/IP were mapped against the ISO-OSI layers, it was clear to the DICOM Standards Committee that there was a gap between the TCP/IP layers and the ISO-OSI Application layer. To close this gap and allow applications to interact with a layer at the same level as the ISO-OSI Application layer, DICOM uses an Upper Layer (UL) protocol that was created by the DICOM Standards Committee. This UL protocol allows the ACSE to interface to TCP/IP.

Figure 4.2 shows the DICOM communication protocol layers. Since the initial publication of the Standard in 1993, it was determined that virtually no implementations of the ISO-OSI Standard existed. Also, the vendors were surveyed and none offered the original point-to-point interface in products by 2003. Accordingly, both the ISO-OSI protocol and the original ACR-NEMA point-to-point hardware and protocol were retired from the Standard. It is important to note that the ISO-OSI communications model and the use of the ACSE are retained by the DICOM Standard as they are necessary for definitions and implementation.

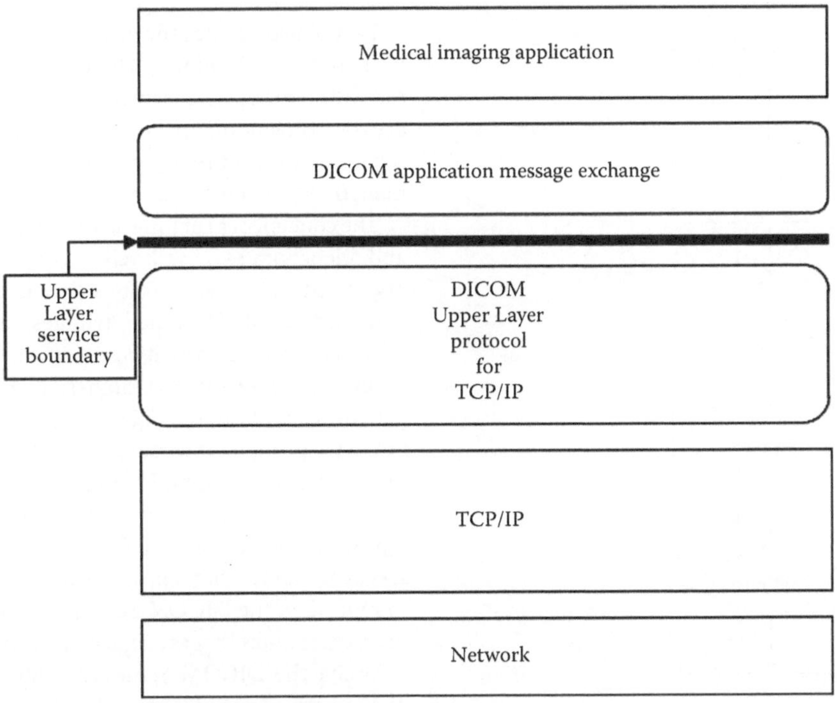

FIGURE 4.2 The DICOM communication protocol. (NEMA. 2009h. *National Electrical Manufacturers Association: Digital Imaging and Communications in Medicine (DICOM) Part 8: Network Communication Support for Message Exchange. PS 3.8-2009.* Rosslyn, VA: NEMA. Copyright NEMA. With permission.)

4.3 The "What" in Medical Imaging: The DICOM Information Object

4.3.1 The DICOM Information Model

The original entity–relationship diagram of radiology information developed by WG VIII was examined very closely. Aspects of it were reviewed by various DICOM WG members and features of the diagram were clarified and a simple core information model evolved. This is shown in Figure 4.3. This simple model says that a patient has a study, the study consists of series, and the series have images. The patient, study, series, and images are each entities in the model. The relationships are shown by the arrows, noting that the arrows do not show information flow, but the hierarchy of the relationship. That is, it would not be correct to say that a study has a patient (though they are related). There are also descriptors of how many entities are related. For example, a patient (implying 1 patient) can have multiple (*n*) studies. Each study can have multiple (m) series, and each series can have multiple (m, again) images. The "patient" entity (as do the others) also has descriptors, or attributes, attached to it that describe the patient. This core information model is central to the basic DICOM Information Model, shown in Figure 4.4.

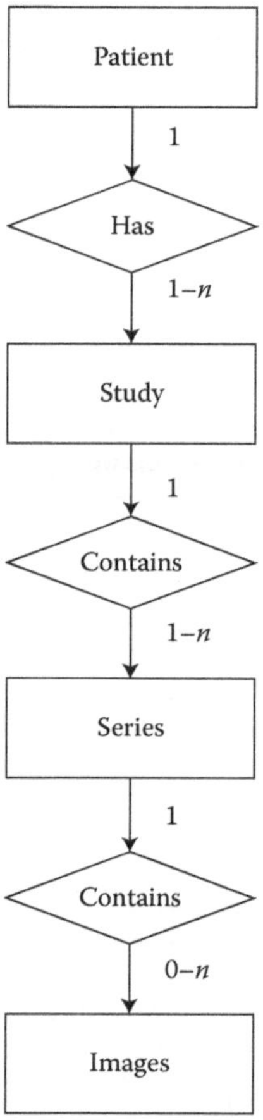

FIGURE 4.3 The DICOM core info model.

4.3.2 Information Model to Information Object Definition

The DICOM Information Model consists of a number of entities (the rectangular boxes in Figure 4.4). The DICOM Standard defines each of those entities by the attributes needed to identify and describe it. Each of these definitions is known as an Information Object Definition, or IOD (NEMA, 2009b).

The DICOM IOD is one of the fundamental parts of the Standard. It is the IOD that defines the various types of images that DICOM supports. With the addition of images from nonradiology specialties, the DICOM Standard now includes a multitude of IODs. Part 3 of the DICOM Standard contains these Information Object Definitions. The IODs are made up of the attributes that describe the object. For example, a CT image would have attributes describing the number of pixels in the rows and columns of the image, the date and time the image was generated, the equipment that generated it, and so on. Each of these attributes is represented in DICOM with a particular structure. They consist of a tag, which is formed from two hexadecimal numbers; the first representing the group to which the attribute belongs (a carryover from the ACR-NEMA Standard) and the second, a sequential number within that group. The groups were originally used by the ACR-NEMA Standard to associate related attributes together. For example, patient information could be put in one group whereas how an image was acquired (equipment, date, time, institution, etc.) would be in another. The attributes are named in DICOM Part 3, for example:

(0010, 0010) Patient's Name

Part 3 also defines the attributes. (In this case, the definition is "Patient's full name.") The two numbers in parentheses represent the group and element numbers of "Patient's Name" and are, by convention, an ordered pair shown in parentheses and separated by a comma (e.g., (0010, 0020) means group 0010, element 0020, which is "Patient ID").

The concept of grouping attributes together persists in DICOM and incorporates new levels of organization. Rather than be organized into related groups as was done in the ACR-NEMA Standard, in DICOM, the attributes are organized based on the information models that result in the IOD. Since many image types will share some attributes, such as those identifying the patient with whom the image is associated, those common attributes are grouped into modules. Patient's Name, for example, is included in the "Patient Identification Module." This module can then be incorporated into the IOD for images without having to repeat all the attributes. Within modules, there are sometimes sets of attributes that repeat. To make the modules more compact to represent, the DICOM Standard defines "macros." These represent attributes that are repeated within a module.

When the DICOM Standard evolved from the ACR-NEMA one, it carried with it the organization of attributes that defined various images. However, DICOM also adopted some object-oriented modeling principles, one of which is that the entities in an information model should not contain parts of other entities

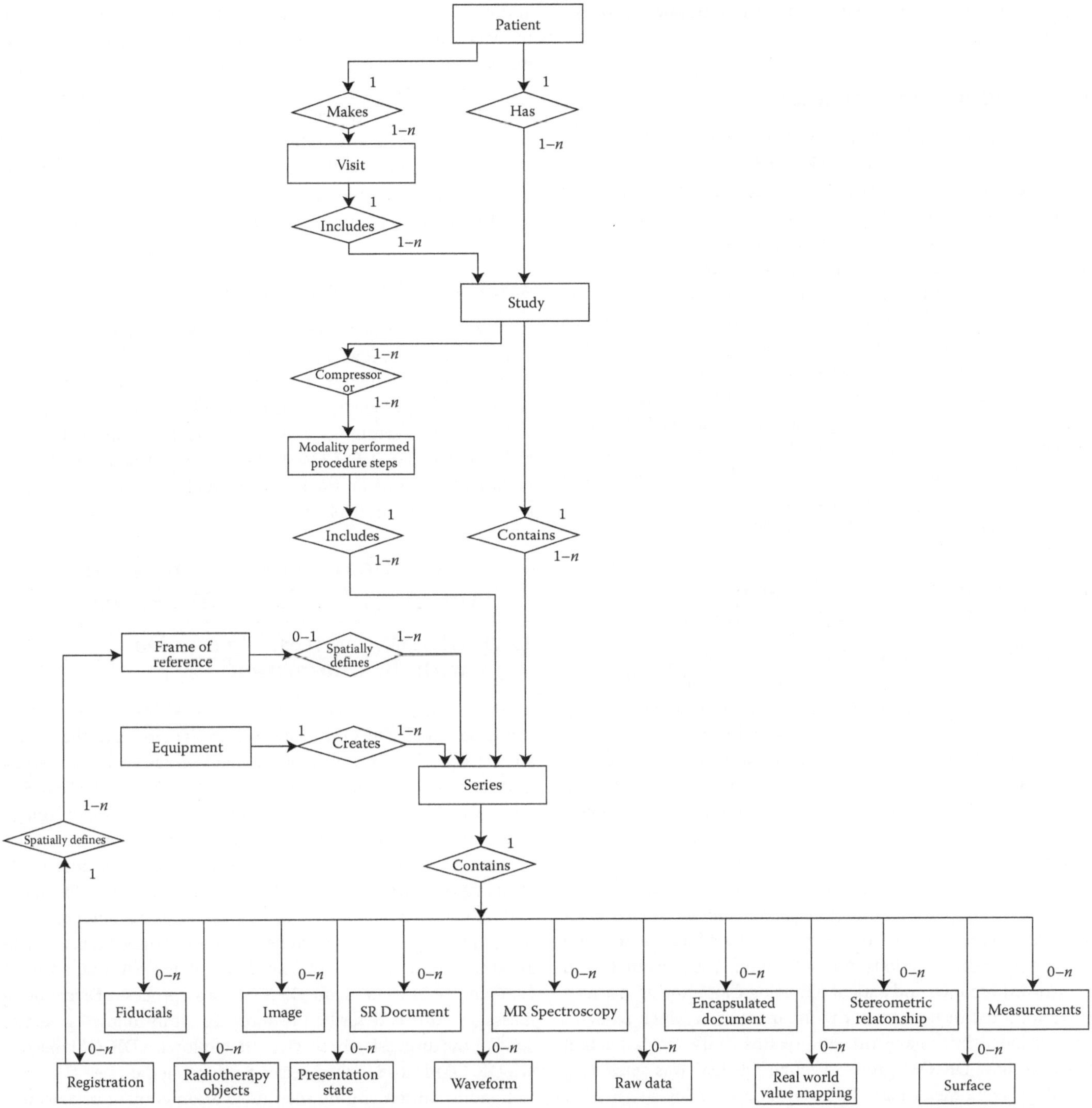

FIGURE 4.4 The DICOM basic info model. (From NEMA. 2009b. *National Electrical Manufacturers Association: Digital Imaging and Communications in Medicine (DICOM) Part 3: Information Object Definitions. PS 3.3-2009*. Rosslyn, VA: NEMA. Copyright NEMA. With permission.)

in the model. This raised a conflict because some of the ACR-NEMA structure for images (that is, the lists of attributes that make up the definition of the image) contains parts of more than one entity. The CT Image IOD, for example, contains the "Patient Module," which has attributes describing the patient, not the CT image. Newer IODs, originally those developed for supporting interfaces to other information systems (chiefly radiology and hospital information systems), followed object-oriented modeling and each IOD referred to one entity in the information model. Rather than re-cast all of the existing image IODs (CT, MR, US, etc.), DICOM has two types of IODs: composite and normalized. Composite information object definitions, as the name implies, may contain parts of several entities in the information model. Most of the image IODs developed from the ACR-NEMA Standard fall into this category. Normalized IODs are those that refer to a single entity in the information model.

These have coexisted peacefully since the introduction of the DICOM Standard.

4.3.3 Definition to Instance

One way to think of an IOD is as a blank form. It has fields to be filled in. Though the fields are blank until information is entered in them, the form nonetheless has a structure and differs from other forms because of the fields it contains. The DICOM IOD defines the fields on the "form." If those fields get "filled in," that is, if the attributes are assigned a value, the attributes now represent a particular instance of the IOD. DICOM uses the phrase, "real-world model" and the models (and resulting IODs) represent the structure of the data that will transform the model into an actual object. Instead of, "this is how a CT image is defined," by assigning values to the attributes in the CT Image IOD, an object representing a real CT image (e.g., "this is a CT image on a patient named John Doe") results. When an attribute is assigned a value, for example, by the CT machine scanning a patient, it is known in DICOM terms as a data element. Creating an instance of an IOD, or instantiation, is a fundamental process that the DICOM defines.

4.3.4 Information Modules to Suit Particular Imaging Techniques

One of the major values of DICOM and the way objects are defined is that the definitions are readily extensible. Where the original DICOM standard (the 1993 version) defined almost exclusively information objects encountered in radiology, the current version of the standard defines many images outside the radiology domain as well as imaging techniques in radiology that were at best in development at the time the original standard was published. Endoscopy, radiation oncology, and ophthalmology are examples of specialties now having extensive IOD representation in DICOM. For radiology, imagings such as MR spectroscopy, digital mammography, and computer-aided diagnosis (CAD) are examples of the many IODs added. In addition to imaging, representations of graphical data, such as that from electrocardiography and patient physiologic monitoring, are now also supported in the Standard. The manner in which DICOM has expanded into many imaging realms is also an important aspect of how DICOM grows. Though DICOM was started by radiology equipment manufacturers and radiologists, neither group has particular expertise in other imaging areas. What they have developed is an understanding of how DICOM is structured and the principles to which the Standard adheres (e.g., avoiding multiple ways of doing the same thing). Other medical specialties that use imaging have the domain knowledge to build the information models they need to represent their imaging. Once the information models are constructed, they can then be turned into a series of IODs to support the various images. The module and macrostructure of the DICOM Part 3 also allow new IODs to take advantage of what is already in the Standard. For example, almost all images will need to be identified with the patient from whom they originate. So, a new IOD can use the Patient Module and not have to develop new attributes if the ones needed are in the Module. Similarly, all digital images need a description of the image matrix size (e.g., 1024 × 1024 pixels), so rather than defining attributes that describe the number of pixel rows and columns in the image, a new IOD can reference the existing General Image Module (which incorporates the Image Pixel Macro—which includes attributes for the number of pixel rows and columns). This development process is one reason why DICOM has been able to expand to include nonradiology imaging very rapidly.

4.3.5 The DICOM Data Set

As described in Section 4.3.3, when DICOM Attributes are assigned a value, they are known as Data Elements. The collection of Data Elements that are required by the DICOM Information Object Definition forms a particular instance of the Information Object.

In DICOM terminology, when an IOD is instantiated, it is called a Data Set. A DICOM Data Set is the fundamental information for a particular image that an application communicates, stores, retrieves, or displays.

4.4 The "What to Do" in Medical Imaging: The DICOM Service Class

4.4.1 DICOM Services: What to Do with the Information

In Section 4.3, the concepts of the DICOM Information Object Definition and DICOM Data Set explained that these constructs describe the information that defines types of medical images, both in the abstract (the IOD) and specific (the Data Set). However, this is not yet sufficient for communication of images since various actions are required as part of the communication process. In addition to sending information, the sender of the information expects some action to be taken on the information and the receiver is presumed to be able to carry it out. For example, is an image being sent to be stored or printed? Once images are stored, how are they retrieved? To accomplish what an application needs DICOM to do, DICOM uses a process of exchanging messages. This is described in more detail in following Section 4.6. For the discussion of services that follows, a DICOM message is a DICOM Data Set plus the service for that data set.

In addition to defining the information content of objects to be communicated, DICOM also defines the "what to do" with that information. These actions are called services and include a series of basic functions that can operate over a DICOM network (NEMA, 2009c). Section 4.3.2 noted that there are two types of DICOM IODs: Composite and Normalized. Primarily because the Composite IODs describe images and Normalized IODs are used with reports and other nonimage information, these two types of IOD need different services. Services used with Composite IODs have a "C-" prefix and those used with Normalized IODs, an "N-" prefix. These services are communicated in parts of a DICOM message intended for them. Collectively, the services are known as DICOM Message Service

TABLE 4.1 DICOM DIMSE Service Primitives

Name	Group	Type
C-STORE	DIMSE-C	Operation
C-GET	DIMSE-C	Operation
C-MOVE	DIMSE-C	Operation
C-FIND	DIMSE-C	Operation
C-ECHO	DIMSE-C	Operation
N-EVENT-REPORT	DIMSE-N	Notification
N-GET	DIMSE-N	Operation
N-SET	DIMSE-N	Operation
N-ACTION	DIMSE-N	Operation
N-CREATE	DIMSE-N	Operation
N-DELETE	DIMSE-N	Operation

Elements, or DIMSEs. Table 4.1 is a listing of the DIMSEs, whether composite or normalized, and service type.

The DICOM C-STORE service is the basic operation used to transmit an image (more properly a DICOM Data Set or Information Object Instance, but "image" is used for simplicity in this section) across the DICOM interface. Note that there is no "display" service; to display an image, it is "stored" to a workstation. To store an image in a PACS storage system, it is "stored" to the system. The C-GET service is used to request an image from a device that has it. C-MOVE is used when one device wants to send an image to another. The C-FIND service is employed to locate a particular image. Finally, the C-ECHO service is used to verify that a DICOM connection is functioning. It is analogous to a "ping" used to verify that a network connection is working. Note that all the DIMSE C services are operations; that is, they cause something to happen to, or with, the information.

The DIMSE N services are somewhat different. The N-EVENT-REPORT is a notification service. It is intended to notify users of a service about an event regarding a DICOM message. The N-GET service is used to retrieve information. N-SET provides a means for requesting that information be modified. N-ACTION requests that a particular action be performed. N-CREATE is used to request that an instance of an IOD plus services be created. N-DELETE requests that an instance of an IOD plus services be deleted. Like the DIMSE C services, these DIMSEs are operations.

4.4.2 Building Services from Service Primitives

DIMSEs are known as service primitives. A medical imaging application may need one or more of these DIMSEs to provide the more complex services an application needs. At the application level of a DICOM–conformant system, storage is performed by the Storage Service Class. This makes use of the C-STORE DIMSE. However, to find and retrieve a Data Set, an application uses the Query/Retrieve Service Class. This requires the C-FIND, C-GET, and C-MOVE DIMSEs. DICOM Service Classes are built from one or more of the DIMSEs.

Service Classes are the application-level services that DICOM implementations provide. These invoke the DIMSE service primitives that are needed to carry out the functions of the Service Class. The DICOM Service Classes are listed in Table 4.2. The Storage Service Class is the basic operation to store images. The

TABLE 4.2 DICOM Service Classes

Service Class	Description
Verification	Verifies Application level communication between two DICOM AEs.
Storage	Used for transfer of DICOM Information Object Instances between AEs.
Query/Retrieve	Allows an AE to query another to determine if it has a particular Information Object Instance. The query uses a very limited set of keys and is not intended as a general database search tool. The retrieve service allows an AE to request a transfer of a remotely located Information Object Instance to itself or another AE.
Study Content Notification	*This service class has been retired.*
Patient Management	*This service class has been retired.*
Results Management	*This service class has been retired.*
Print Management	Supports the printing of images and image-related data on hardcopy media.
Media Storage	Supports the transfer of images and associated information between DICOM AEs using Storage Media.
Storage Commitment	Allows an AE to request that a Storage Service Class Provider commit to storing transferred images for an implementation-specific period.
Basic Worklist Management	Facilitates AE access to worklists.
Queue Management	*This service class has been retired.*
Application Event Logging	Supports the transfer of Log Event Records for centralized logging or storage.
Instance Availability Notification	Allows one DICOM AE to notify another AE of SOP Instances that may be retrieved.
Media Creation Management	Supports the creation of Interchange Media containing composite SOP Instances once the SOP Instances have been transferred to the media creation device (i.e., initiates the writing process on the media).
Hanging Protocol Storage	Allows one DICOM AE to send a Hanging Protocol SOP Instance to another DICOM AE.
Hanging Protocol Query/Retrieve	Facilitates access to Hanging Protocol composite objects.
Substance Administration Query	Facilitates obtaining detailed information about substances or devices used in imaging. It also supports obtaining approval to administer a contrast agent or pharmaceutical to a specific patient.
Color Palette Storage	Allows one DICOM AE to send a Color Palette SOP Instance to another AE.
Color Palette Query/Retrieve	Facilitates access to Color Palette composite objects.

Query/Retrieve Service Class is used to locate an image and retrieve it to a specified location. For printing images, the Print Management Service Class is used. The Media Storage Service Class is employed for storing images on various media (e.g., Compact Disc). Storage Commitment is used to verify that a device to which an image is sent has received it and is taking responsibility for it. Basic Worklist Management facilitates access to various worklists. Application Event Logging is used to facilitate the network transfer of various event log records for central logging or storage. The Relevant Patient Information Query Service Class facilitates access to relevant patient information (as distinct from Query/Retrieve which finds and retrieves an image). The Instance Availability Notification Service Class is used to let DICOM applications notify each other of the presence and availability of Information Object Instances and associated services. The Media Creation Management Service Class is used for management of a media creation service by instructing an appropriate device to create interchange media to record images already sent to the device. Because the arrangement of images when displayed is often important, the Hanging Protocol Storage Service Class allows for the storage of the information necessary to enable such image arrangement. A companion to this is the Hanging Protocol Query/Retrieve Service Class, which facilitates access to stored hanging protocol information object instances. The query regarding administration of various substances or devices used as part of an imaging study is supported by the Substance Administration Query Service Class. The Color Palette Storage Service Class is used for the exchange of Information Object Instances describing the color palette to be used with an image. The companion to this is the Color Palette Query/Retrieve Service Class. Other Service Classes are tied to specific types of IODs, so this is not a complete listing of all services provided by DICOM.

In addition to defining services, it is necessary in a PACS environment to know what a particular DICOM conformant device does. Devices may use a service, such as a piece of image equipment using the C-STORE service to store an image. They may also provide a service, such as an archive storing the image that it is sent to. These two roles are described in DICOM as "Service Class User (SCU)" and "Service Class Provider (SCP)." A device such as an archive can be both a provider and user of a service. An archive is an SCP for other devices that want to send images to it. It is an SCU when it uses the Storage Service Class provided by workstations to receive the images it sends.

4.5 The Fundamental Functional Unit in DICOM: The Service–Object Pair

4.5.1 Construction of a Service–Object Pair

With IODs, DICOM defines the manner in which images and other information objects are described and what information needs to be communicated with them so that they may be displayed and interpreted. The DICOM Services are the fundamental operations and notifications needed to communicate what is intended for the information objects. Together, the IODs and Services are the fundamental functional units of DICOM. In DICOM terms, the combination of an Information Object and associated Services is called a Service–Object Pair, or SOP. Any discussion of DICOM among those who use it is rife with descriptions of SOPs as it is possible to understand what a device does from a description of the SOPs it generates or receives. Because the constituents of SOPs are considered objects and methods in object-oriented terminology, the various DICOM SOPs are SOP Classes. The SOP Class used for verifying connectivity, the Verification SOP Class, is unusual in that it has no IOD. Since the intent is not to transmit data, but to verify that DICOM connectivity is operating, it does not need an IOD with it. The DIMSE it invokes is C-ECHO. This is done because SOP Classes are handled by software at the application level whereas DIMSEs operate at a lower level of the DICOM communication protocol. Various devices that communicate using DICOM are known in DICOM terms as Application Entities, or AEs. The process of DICOM communication involves communicating SOP Classes between AEs.

As should be apparent, the DICOM Standard defines as many SOP Classes as are needed for the communication of the various information objects that DICOM covers. As a way of unambiguously identifying the SOP Classes, each has a unique identifier, or UID. The DICOM SOP Class UID is itself constructed according to a standard. To keep these identifiers unique among many other data structures using identifiers, the DICOM Standards Committee applied for, and received, a "root" to be used as the basis for the UIDs it generates. The UID root is provided by the International Standards Organization and is unique; no other organization or user may use that root in the construction of a UID. The DICOM UID root is "1.2.840.10008." Because this root UID is assigned for DICOM use, it is present in all of the SOP Class UIDs. As an example, the UID for the CT Image Storage SOP Class is 1.2.840.10008.5.1.4.1.1.2. The ISO standard defines UIDs (ISO, 2008) as having all numeric components with each component separated by a period ("."). It is very important to understand that UIDs should be used, as they are in DICOM, for only identification purposes. Since organizations may define UIDs for instances of a SOP Class (an SOP Instance UID), there may be a temptation on the part of programmers to attach meanings (semantics) to various UID components (e.g., the first component after the root being used to indicate that the IOD instance is of a contrast-enhanced image). Doing so can have negative impacts on operability with other facilities that do not use such a scheme as their software may not parse the UID to find the facility (NEMA, 2009d).

As mentioned above, when the particulars of the component DIMSEs of a Service and the attributes of an Information Object are instantiated, the resulting SOP Instance is assigned a UID of its own. There may be several UIDs within a DICOM message since, for an examination that has series within the study, both the DICOM Study and Series are assigned instance UIDs that must be different. In theory, institutions should apply for their own UID root (or roots) to use with such instance UIDs. However, in practice, many institutions use the instance UIDs that the vendor of the PACS system sets up.

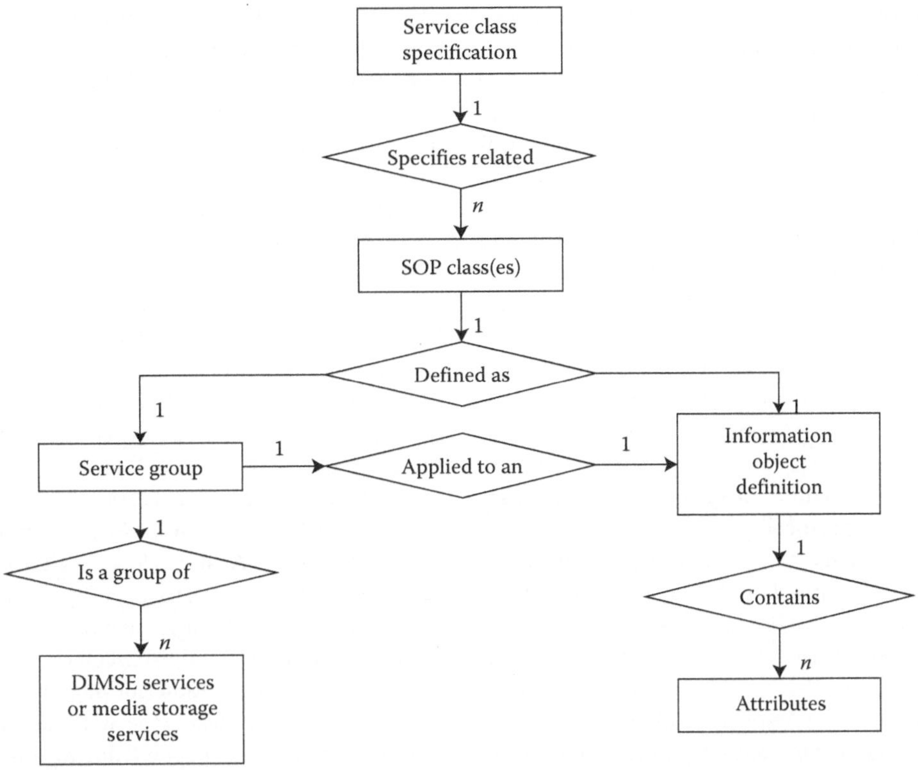

FIGURE 4.5 Relationship DICOM information structures. (From NEMA. 2009c. *National Electrical Manufacturers Association: Digital Imaging and Communications in Medicine (DICOM) Part 4: Service Class Specifications. PS 3.4-2009.* Rosslyn, VA: NEMA. Copyright NEMA. With permission.)

4.5.2 Deconstructing an SOP Class: Making Things Work

The various pieces described up to this point can be put into their own information model. Figure 4.5 serves as a summary of the various structures that form the high-level information model of DICOM.

The DICOM SOP Classes are too numerous to list in their entirety in this chapter. Part 4 of DICOM. "Service Class Specifications" (see NEMA, 2009c) provides the descriptions of DIMSEs, DICOM Service Classes, and DICOM SOP Classes.

As an example of what an SOP Instance does, this section will examine what happens when a CT Image Storage SOP Instance is used to communicate a CT image to a PACS storage system. After the initial Association Establishment (see Section 4.2.3 for a brief description; a more detailed one follows in Section 4.6) between two AEs, the Storage Service Class SCU is the CT machine. It invokes the CT Image Storage SOP Class which in turn uses the DIMSE C-STORE Service. The parameters needed to use that service are provided by the CT machine. The attributes defined for the CT Image IOD are instantiated by the CT machine as well. The combined CT Image Instance (a DICOM Data Set) and the Storage Service Class make up the CT Image Storage Service–Object Pair instance. Using DICOM Message Exchange (see Section 4.6), the CT machine communicates this message with the Storage Service Class SCP, in this case a PACS storage system. The Storage Service Class can be thought of as a command which the storage system understands to be applied to the CT Image Information Instance it just received. The PACS storage system proceeds to store the CT image. What may not be apparent is that storage systems need not store the DICOM Data Set as a unit, in fact, it is atypical to do so. Instead, for reasons of improved performance, many PACS store the DICOM Attribute values in their own database schema, typically in a series of relational database tables. This is not prohibited by the DICOM Standard, though if a PACS is DICOM conformant and claims to be able to return a DICOM Data Set when requested, it must reassemble the DICOM Data Set it received. If DICOM Data Sets are stored on media, the situation is different. DICOM specifies a file format that is a Data Set with the file structure used by the media employed.

4.6 DICOM Message Exchange: The Fundamental Function of DICOM

4.6.1 Exchanging Messages

In any conversation, there are preliminaries that are done so quickly that most people participating in a conversation do not even think of them. For example the questions:

- Do you have the attention of the person you want to speak with?
- Are you going to ask a question or make a statement?

are usually answered without any spoken exchange or may be the subject of visual cues. For example, if the person you want to speak with is looking at you, or turns to look at you, you have answered the first question. You probably already have formulated the answer to the second question. Once these initial steps are done, the conversation begins and follows a protocol defined by the situation and culture. The first words you speak will usually identify the language in which you intend to communicate. How loudly and rapidly you speak are often cued by whether or not the person with whom you are speaking appears to hear and understand you. The cues are typically facial expressions before any spoken response. Cultural factors will determine how close you stand or sit next to each other or if you continue after the initial opening or wait until asked. Communication among people is more complicated if there are multiple speakers as then a manner of speaking in some order needs to be included in the protocol.

Conversational conventions have a parallel in electronic communication. For AEs using DICOM, communication is carried out through the exchange of messages (NEMA, 2009g). Association Establishment was discussed briefly in Section 4.2.3. This is the first step that DICOM AEs take in communicating information. Establishing an Association allows the two AEs to determine the basics under which subsequent communication will take place. To make it simpler to follow what happens during Association Establishment, the DICOM Standard uses the term "requesting AE" for the AE that initially requests an Association and "accepting AE" for the AE with which the Association is requested. Within Association Establishment, the first step is that the requesting AE proposes an Application Context. This is primarily to allow for future extension as, at present, there is only one Application Context defined by DICOM. Private Application Contexts may be defined by organizations using DICOM, but private Application Contexts are not registered by the DICOM Standards Committee and may pose a problem for communication if not recognized by the accepting AE. Application Entities have identifying titles. These are often provided by the manufacturer or local organization and are up to 16 characters in length (e.g., "CT1_Main_Hosp," "PACS_Archive_01"). In addition, devices on a network must have a network address, typically an Internet Protocol (IP) address. Because DICOM-conformant devices may support multiple functions, a single AE Title may have more than one IP address. During Association Establishment, an AE states what its AE Title is, what AE Title it is calling, what its IP address is, and what IP address it is calling. A parameter called "User Information" is included and is used to negotiate the maximum length of data values that AEs can accept. In addition, there are parameters exchanged that allow AEs to report whether or not the Association is accepted and if there is any diagnostic information should the Association be rejected. There are two levels of rejecting an Association in DICOM; transient (if a device is, say, temporarily too busy to accept another Association) and permanent (if a device has failed above the Association level). The value of this is that a requesting AE receiving a transient rejection may try again later whereas receipt of a permanent rejection tells the requesting AE that it should not try again.

A major function of establishing an Association is negotiation. It is during Association Establishment that two AEs negotiate the basics of their communication and what they can do within DICOM. These key aspects of exchanging messages are established through the DICOM Presentation Context. The Presentation Context includes a list of definitions and it is these definitions that are important for the subsequent message exchange. The Presentation Context definition list includes the Presentation Context Identifier, an Abstract Syntax Name, and one or more Transfer Syntax Names (see table on "Proposed Presentation Contexts…"). The Presentation Context Identifier is simply a number used by the AEs during the Association and is unique only for that Association. Abstract Syntaxes correspond to DICOM SOP Classes. The Abstract Syntax Name is the SOP Class Name used in the Standard (e.g., "CT Image Storage") and the Abstract Syntax UID is the SOP Class UID (e.g., 1.2.840.10008.5.1.4.1.1.2). Why not just call these the SOP Class Name and UID? This is in part done to differentiate the role during Association Establishment (negotiation) from that for subsequent communication (message exchange). What is used in the Presentation Context is the SOP Class UID.

The negotiation that happens during Association establishment can be extended. DICOM includes an "extended negotiation" feature that allows AEs to negotiate some of the specializations that apply to SOP Classes. DICOM IODs have general Modules as well as modality-specific Modules. There are attributes in the modality-specific Module that may be used to replace the similar attribute in the general Module when constructing the DICOM message. This process is known as specialization and needs to be negotiated so that both AEs recognize that the specialized attribute carries some information that differs from the nonspecialized version.

The Transfer Syntax describes how AEs can communicate data. That there is more than one Transfer Syntax is a result of the ways in which DICOM devices can encode the data to be communicated. The basic DICOM Data Element consists of a Tag (the "group, element" pair; e.g., 0010, 0010 "Patient name"), the Value Length, and then the Value (the data itself). However, the DICOM Data Element may also contain a field called the Value Representation (VR) between the Tag and Value Length. If present, it describes how the data is represented. It is a two-character abbreviation. For the "Patient name" example, the VR is "PN" (for "person name"). Other names, such as 0008, 0090 "Referring Physician's Name," are also represented as PN. The Value Representation for each data element is listed in DICOM Part 6, Data Dictionary (NEMA, 2009f). The utility of including the VR is that a device does not have to support its own implementation of the DICOM Data Dictionary to determine the VR of a Data Element it receives. However, some devices may not include the VR as it is not required. To allow AEs to let each other know if they include the VR in their data elements, the Transfer Syntax includes a parameter that is either "Explicit VR," meaning that VR is included in Data Elements, or "Implicit VR," meaning that it is not.

In the evolution of computers, the number of bits that constitute the internal representation of data varied. This collection of bits is usually called a computer "word." This is the size of the "chunk" of data that a computer manipulates with its low-level instructions. Communication interfaces, however, are typically based on transferring eight-bit bytes; convenient since alphanumeric characters are usually represented by a single byte. (The Asian character sets are an exception to this.) For binary numbers, a computer's word usually contains more than one byte. When encoding a multibyte binary number for communication, some computers encode the least significant (low-order) byte first, then the following bytes in order of increasing significance. This is known as "little endian" encoding. The opposite encoding is used by other computer types; that is, the most significant (high-order) byte is encoded first, followed by the other bytes in order of decreasing significance. This is called "big endian" encoding. For devices to interpret binary values properly, the method of encoding must be known. This is another component of the Transfer Syntax. DICOM Transfer Syntaxes also have UIDs. To make possible communication between the widest variety of devices, the DICOM Standard defines a default Transfer Syntax that all devices must support. They may optionally support others, but they must be able to support the default if the device requesting the Association proposes only the default Transfer Syntax in its Presentation Context. The default DICOM Transfer Syntax is "Implicit VR Little Endian" and has the UID 1.2.840.10008.1.2. This Transfer Syntax means that the AE uses Data Elements that do not contain the VR field and encodes binary values with the least significant byte first. Specification of this default means that a computer that normally encodes binary values with the most significant byte first must be able to change this byte order for DICOM communication, though its internal representation will not change. Supporting this default Transfer Syntax also means that a device must be able to interpret the Value Representation of a Data Element. However, if it can also support an Explicit VR Transfer Syntax and the device it is communicating with can as well, then both the devices may be able to skip the process of determining the VR from the Data Dictionary.

Application Entities typically act as Service Class Users, Service Class Providers, or both. This role description is also included in the Presentation Syntax description. Finally, whether or not extended negotiation is supported is indicated.

DICOM supports the use of some data compression methods. If compression is used, it can be negotiated at Association Establishment through the AE proposing a Transfer Syntax that includes one of the compression methods supported by the Standard. The basic Transfer Syntax for all compressed data is Explicit VR Little Endian.

An example Presentation Context is

- Presentation Context ID: 1
- Abstract Syntax: CT Image Storage
- Transfer Syntax: Explicit VR Little Endian
- Role: SCU
- Extended Negotiation: not supported.

The numeric values would be

- Presentation Context ID: 00000001 (binary)
- Abstract Syntax: 1.2.840.10008.5.1.4.1.1.2
- Transfer Syntax: 1.2.840.10008.1.2.1
- Role: SCU
- Extended negotiation: not supported.

The Association Negotiation phase involves the requesting AE including a list of the Presentation Contexts it proposes for the Association. The accepting AE would respond with a list of which of the proposed Presentation Contexts it accepts. Once this is accomplished, the exchange of the accepted SOP Instances (proposed and accepted Abstract Syntaxes) can ensue.

4.6.2 Successes and Failures: Reporting Errors in DICOM

There are two ways to terminate an Association. The orderly way is for the requesting AE to request that the Association be closed. The accepting AE replies to this and the Association closes. An accepting AE may not request that an Association be closed using this orderly close mechanism. If something goes wrong with either of the two AEs during message exchange, either device may abort the Association. Unlike the orderly close, the aborted Association is not acknowledged by either AE. Whether or not an aborted Association is re-tried is dependent on the applications involved. There are diagnostics provided by the DICOM Upper Layer Service, so AEs may use this information in an attempt to recover. Since problems in layers below the DICOM Upper Layer may also result in an Association abort (e.g., TCP/IP errors, network hardware errors), network services outside the scope of the DICOM Standard may, and typically do, provide additional information about the causes of the communication failure in these instances. Unfortunately, most errors in DICOM message exchange result in fairly terse error descriptions. These are meaningful to a DICOM expert, but may mean little to a user. However, the errors reported can be of help in troubleshooting problems with DICOM communication. If there is an error, there are various parameters set that can help determine where the problem occurred. Most applications do not report these errors in detail to the user, though they are usually logged in error logs. Unfortunately, the user may only see a message such as "DICOM error occurred."

Since DICOM defines an Upper Layer (UL) protocol for TCP/IP, DICOM communication that uses TCP/IP can use the error-reporting mechanisms of the TCP/IP services. Establishing an Association in the DICOM UL protocol uses the Transport Connect request mechanism of the TCP. The interaction between the DICOM UL and TCP/IP is very strictly defined. It uses a structure called a "state machine" (NEMA, 2009h) that defines each event possible and the allowable state transitions for those events. The DICOM state machine is translated into computer software to implement the protocol. Undefined or incorrect state transitions are reportable as errors for troubleshooting purposes.

4.7 DICOM Conformance

4.7.1 Why Conformance?

For a standard to provide benefits to users and manufacturers, the various implementations of it must actually meet the standard. In some instances, for example, purity of pharmaceuticals or pharmaceutical ingredients, there are organizations that test products to make sure they conform to the set standards. For pharmaceutical products, in the United States, this is typically the US Pharmacopeia. If a product meets the standards set for it as demonstrated by testing, it can add "USP" to its label. Standards whose conformance is set by statute or law is a *de jure* standard. In other instances, standards may be set by widespread use because of popularity. Many such standards may start out as a feature of a product and other vendors follow it (and may have to license it if patented) so as to compete in the marketplace. An example is the QWERTY layout of keyboards. There are numerous stories of the reason for this particular layout of keys, but that multiple manufacturers followed, and continue to follow, the layout is an example of a *de facto* standard. In some instances, *de facto* standards become *de jure* ones if a standards body takes it up and passes it through its various processes. An example of this is the hypertext markup language (HTML). HTML began with Tim Berners-Lee (Wikipedia) who posted a guide to the "tags" used to describe text and image formatting. Though HTML has some overlap with the Standard Generalized Markup Language (SGML), an ISO Standard (ISO, 1986), the use of HTML primarily for Web descriptions over the print ones of the original SGML and the increasing use of HTML by multiple organizations led the Internet Engineering Task Force (IETF) to adopt it as a standard in 1995 (Wikipedia). Since the IETF establishes standards for the Internet that implementers must follow, what started as a *de facto* standard became a *de jure* one.

DICOM is a voluntary standard. That is, conformance to it is not enforced by any regulatory body. There is no "DICOM police" to impose penalties on vendors who implement DICOM improperly. Rather, DICOM is market regulated. If a vendor's DICOM implementation is flawed, as users attempt to interface it to PACS or other information systems, the flaws will become apparent (if not detected by vendor testing prior to production and sales) and various testing will determine the nature of the flaws. Large trade shows, the Radiological Society of North America and European Congress of Radiology in particular, were initially used as "connect-a-thons" where vendors had to demonstrate conformance to DICOM so they could participate in the demonstrations.

Within DICOM, there are many options that vendors may elect. For a given IOD, there are attributes that must be present, those that must be present, but can be described as being "value unknown," those that must be present if certain conditions are met, and those that may be present but are not required. Most vendors adhere to these definitions and include the attributes that are required by the Standard. However, DICOM also provides for "private" attributes. These have tags that are readily identifiable as being nonstandard (odd numbered group number). Vendors are supposed to use these for information that their systems may use for internal or proprietary processes. In some instances, private attributes are used by vendors if attributes they need are not in DICOM but might be generally useful. In these cases, vendors may subsequently propose that these private attributes be made standard. A problem for users and other vendors is the use of private attributes to carry information that does have standard attributes or that is required in conjunction with the standard attributes for proper interpretation of the vendor's information. This can be done in an attempt to force users to employ the vendor's implementation, though often argued by the vendor that this operation outside DICOM is necessary. This behavior is not prohibited by DICOM, but certainly violates the spirit of the Standard.

Another major aspect of flexibility in DICOM includes the Transfer Syntaxes. Section 4.6.1 describes how Transfer Syntaxes are specified. They describe whether or not the data elements contain the Value Representation and if multibyte data is sent as least or most significant byte first. If the Transfer Syntaxes of two DICOM implementations cannot be negotiated (unlikely since there is a default Transfer Syntax that all implementations must support), the two implementations will not be able to exchange DICOM messages.

4.7.2 Specifying DICOM Conformance

To make it simpler for vendors of DICOM to describe what their implementation does and for users to determine if a DICOM device will not work with their other DICOM equipment, the DICOM Standard specifies how to describe conformance. This is done through the DICOM Conformance Statement, the structure of which is specified in Part 2 of the Standard (NEMA, 2009a). As a result of this, DICOM Conformance Statements all contain the same sections, making it simpler, though not easy, to evaluate compatibility. The Appendix to this chapter includes a short description of how to "read" a DICOM Conformance Statement. It is important to understand that devices whose DICOM Conformance Statements "match" are not guaranteed to work together, though it means they are likely to. However, a mismatch in DICOM Conformance Statements does guarantee that two devices will not communicate.

A DICOM Conformance Statement usually has a title page that describes the device (and any options) to which the Conformance Statement applies, the version of the Conformance Statement, and the document date. Following this is a short text description of the network and media capabilities supported by the device. This text description is supposed to be written using lay terminology and is specifically not to contain DICOM acronyms. This is so that a reader unfamiliar with the details of DICOM may nonetheless have some idea of what the Conformance Statement covers. What network services are supported by the device in terms of transfer (i.e., sending and receiving images), query/retrieve (finding and moving information), workflow management (e.g., transfer of notes, reports, and measurements), and print management are described in a table. If a service is not supported, it is still listed in the table, but the table indicates that that service is not supported. If a device supports the storage of DICOM Data Sets on media (a Data Set is a File Set on media—it has some media-specific additions), a separate table is included that describes the Standard media supported and the role (write a DICOM File Set, read a DICOM File Set, or both). This much of the Conformance Statement is introductory and is supposed to serve as a quick overview of what the device does. This section of the Conformance Statement is followed by a detailed table of contents for the remainder of the document.

The balance of the Conformance Statement describes the "real-world model" of the device's function (as a diagram) followed by sections that list the SOP Class names supported along with their DICOM UIDs. How Associations are established is covered next, then the list of Presentation Contexts (Abstract Syntaxes, Transfer Syntax, role, and extended negotiation support) that the device proposes at Association Establishment is included as another table. If media are supported, there are sections that describe the particulars of the media usage, including a separate real-world model. The media-particular Application Profiles are specified and any augmentations and Private Application Profiles used are described. In an Annex, the Conformance Statement includes a detailed list of the Attributes in the IODs created by the device (both Standard and Private). This is a very useful part of the Conformance Statement as an experienced DICOM troubleshooter may turn to this section first if Associations are established correctly, but the applications are having problems interpreting the information received.

4.7.3 What DICOM Conformance Means

As complicated as it sounds, a Conformance Statement fundamentally says of a device, "this is what I can do, this is how I propose to do it, these are my basic communications specifics, and this is what my DICOM Information Objects contain."

4.8 Beyond Radiology: The Growth of DICOM

4.8.1 Internationalization of DICOM

Part of the reason for the change from ACR-NEMA to DICOM was to enable others besides NEMA and ACR members to become voting members of the DICOM Standards Committee. Many of the manufacturers of imaging equipment were based either in Europe or Asia and radiology professional societies also have worldwide membership. The ACR-NEMA bylaws were changed along with the name, enabling other interested parties to become full participants in the DICOM standards process. An early active group was the Comité Européen de Normalisation (CEN). CEN had a Technical Committee 251 that was tasked with standards for medical imaging. Once a formal liaison was developed, joint work with CEN led to the issuing of the MEDICOM Standard under the CEN auspices. This standard was DICOM endorsed by CEN and enabled many European countries to specify DICOM in their national healthcare standardization efforts. A most important facet of this was the development of an implementation of DICOM independent of the one done in the United States by Mallinckrodt Institute of Radiology for the RSNA. The CEN development effort was done by Peter Jensch (OFFIS at the University of Oldenburg), Andrew Hewett (Oldenburg), Emmanuel Cordonnier (University of Rennes), and Rudy Mattheus (VUB Brussels). The software was demonstrated along with the US reference implementation by Mallinckrodt Institute and the two implementations were fully interoperable.

In Asia, the Japanese Medical Image Processing Standards group was an early adopter of the ACR-NEMA Standard, issuing their MIPS-89 Standard in 1989. This was a Japanese language version of the ACR-NEMA 300-1985 (version 1) Standard. With the release of DICOM, the Japanese became much more active in DICOM and were tasked with, among other things, a method for supporting the Asian character sets that needed more than one byte. The Japanese also had developed a unique media standard, for the 5.25″ magneto-optical disk. It included hardware security features so that the resulting disks could only be read by a drive that had the security firmware in it. Such drives were to be regulated and sold only to the medical market. There was no DICOM provision for such media, so a solution that allowed the Japanese to use this standard was worked out. This did not prevent the uses of DICOM, without the special firmware security features, for nonmedia message exchange.

Eventually, with increasing participation by European and Asian standards bodies and professional societies, DICOM was put forward to ISO Technical Committee 215 for adoption as a standard. In 2006, ISO issued ISO 12052:2006 "Healthcare informatics—Digital imaging and communication in medicine (DICOM) including workflow and data management." DICOM has become a full international standard.

4.8.2 Nonradiological Imaging

Challenges for DICOM: Shortly before the change from ACR-NEMA to DICOM, the idea of expanding to other specialties that used digital imaging came under serious consideration by the Standards Committee. Dr. W. Dean Bidgood, Jr., a radiologist and member of the DICOM Standards Committee, began the first explorations of working with nonradiology specialty and professional societies. Members of cardiology, dentistry, gastroenterology, ophthalmology, and pathology professional societies were invited to attend the DICOM Standards Committee meetings and educated about the Standard and how it could be expanded to serve their specialties. Some of the organizations, notably cardiology (through the American College of Cardiology—ACC), dentistry (American Dental Association—ADA), ophthalmology (American Academy of Ophthalmology—AAO), pathology (College of American Pathologists—CAP) began participation in earnest, proposing new Working Groups and establishing secretariats for them.

At the time, pathology was largely involved through terminology. The College of American Pathologists had SNOMED—the Systematized Nomenclature for Medicine—that defined many of the terms used in anatomic descriptions. The DICOM Standards Committee also had a Working Group developing Structured Reporting and standardized terminology was of great interest. The early collaboration between the CAP and DICOM was in the area of nomenclature and controlled vocabularies. There was digital imaging in pathology, typically used for telepathology applications that tended to use video camera-equipped microscopes. The spatial resolution that pathologists needed was largely achieved through magnification. With the development of whole slide digital scanning systems, very high resolution images could be created from glass slides. These scanning systems yield very large image sizes—ranging from a typical 60,000 × 80,000 pixels to an extreme of 250,000 × 500,000 pixels. With the bit depth needed for color, these pixel matrices translate into file sizes of 15–375 gigabytes. Since the pathologist is interested in more than one focal depth, scanning at 10 focal planes would increase these file sizes per slide by an order of magnitude. For one slide scanned at the highest possible resolution and at 10 focal planes, the resulting 3.75 terabyte file is more data than many radiology departments generate in a year. The automated scanners can scan 1000 slides per day. If all the digital data is stored, it is beyond the capacity of any PACS archive. Pathologists, however, have to retain the glass slides and tissue blocks from which they work much as radiology has to keep patient images for some length of time. Digital whole slide images are not needed for archiving, but for teaching and some telepathology applications, they probably would be stored. Pathologists have also examined the use of lossy compression and determined that it could be used on digital whole slide images without a loss of clinical diagnostic quality, potentially reducing the file sizes by a factor of 20–50 (Ho et al., 2006).

The large pixel arrays resulting from whole slide imaging are another problem for DICOM. The maximum pixel array size for DICOM is fixed by the size of the attribute (16 bits) used to carry the row and pixel dimensions. This means that images cannot be larger than 64 × 64 K, a limit exceeded by whole slide images. Working Group 26 is for pathology imaging and a proposal that uses a "tiled" mechanism has been proposed. The idea is to cut the very large images into (maximum) 64 × 64 K tiles. The use of these tiled images would support a viewing model that is very much like using a microscope and glass slide. At "low power," a digital viewer would see a low-resolution, downsampled version of the whole image. As "magnification" is increased, progressively less and less downsampling is done until, at "high power," the full-resolution tile is viewed. At high magnification, corresponding to the resolution of the full image, the field of view of a microscope is such that only a small portion of the whole slide is seen.

DICOM has also been used in dentistry for digital intraoral radiographs and panoramic mandibular/maxillary imaging. In ophthalmology, DICOM is used for fundoscopic imaging, fluorescein angiography, and tomographic imaging. In addition, the various quantities used to describe ophthalmic lensometry and refractive measurements are supported in DICOM. Cardiology uses many of the existing DICOM imaging IODs, but has added support for electrocardiographic and physiologic waveforms and quantities. A general class of "visible light imaging" crosses several subspecialties including digital photography for dermatology, pathology, dentistry, and ophthalmology. It is also used to describe endoscopic images in DICOM form.

While the original intent of DICOM was to support diagnostic radiology, it took participation of experts in radiation oncology to develop support for the various aspects of radiation treatment including planning, imaging, dosimetry, and the treatment itself. Besides the support for nonradiological imaging, in many instances, the reporting needs of these specialties are also supported in DICOM through the structured reporting IODs. This resulted from many nonradiological imaging devices producing reports from measurements made. However, these tend to be in proprietary, or at least non-DICOM, formats. Through the translation or mapping of these reports into DICOM SR, they can be managed using the DICOM Services.

The surgical-operating room, with the many devices present, bears some similarity to pre-DICOM radiology. The various pieces of equipment in an operating room often capture or generate digital information, but if it is made available as digital output, it is in a proprietary or nonstandard format. Single-vendor equipment suites allow communication between the devices, but generally not with other hospital information systems without custom interfacing. DICOM WG 24 is working at addressing the issue of DICOM in surgery. Initial work has been on coordinate systems and patient physical modeling.

The advantages of DICOM have been noted by nonmedical imaging domains. An early adopter of the DICOM structure—information object definitions and services—was the nondestructive testing industry. Nondestructive evaluation (NDE) uses many of the imaging methods used in medical imaging: radiography, ultrasound, and computed tomography among them. Examples include the inspection of aircraft parts for cracks and solid rocket motor propellant for voids or cracks. The industry uses the structure of DICOM with application-specific image objects and services. The standard, called digital imaging and communication in nondestructive evaluation (DICONDE), is managed by the American Society for Testing and Materials (ASTM).

Security screening, most obvious at airports but used increasingly in many public thoroughfares, also uses imaging. Radiographic and computed tomographic imaging are most common, but development and testing of microwave and backscatter radiography are ongoing (TSA). The aggregation of resulting images and the communication of them for screening purposes have many elements in common with medical imaging including person identification and privacy. NEMA established a Division of

Security Imaging and Communications (NEMA). The Division is working on a derivative of DICOM for security imaging applications called digital imaging and communications in security (DICOS). NEMA notes that the proposed standard structure will be based on DICOM.

4.9 How DICOM Grows and Expands

4.9.1 The DICOM Working Groups

Given the pervasive nature of DICOM in medical imaging and the increasing uses for which it is employed, a natural question might be how growth of DICOM is achieved and managed. From the outset, a goal of the ACR-NEMA and subsequently the DICOM Standards Committee has been to incorporate new features into the standard as rapidly as possible. Typically, a new revision of DICOM is issued on a yearly basis. The revisions incorporate new material as well as clarifications and corrections where needed. Such a goal means a large amount of work. When the ACR-NEMA Committee was first established, it was recognized that work could often be done faster if the groups doing the work were small. As a check against a small group having too much of a parochial view, work would be reviewed by other groups and by the main Committee. Groups could also request additional membership if they needed help from experts or those with particular experience. The structure developed was a series of working groups, each with a chairman who reported to the full Committee. From the original three WGs, the DICOM Standards effort currently has 27. This includes a replacement of the original three WGs by new ones. Table 4.3 is a list of the current working groups and their names (which describes their domains). Readers should check the NEMA Web Site (NEMA) as Working Groups are added fairly frequently.

TABLE 4.3 DICOM Working Groups

WG-01: Cardiac and Vascular Information	WG-15: Digital Mammography and CAD
WG-02: Projection Radiography	WG-16: Magnetic Resonance
WG-03: Nuclear Medicine	WG-17: 3D
WG-04: Compression	WG-18: Clinical Trials and Education
WG-05: Exchange Media	WG-19: Dermatologic Standards
WG-06: Base Standard	WG-20: Integration of Imaging and Information Systems
WG-07: Radiotherapy	WG-21: Computed Tomography
WG-08: Structured Reporting	WG-22: Dentistry
WG-09: Ophthalmology	WG-23: Application Hosting
WG-10: Strategic Advisory	WG-24: Surgery
WG-11: Display Function Standard	WG-25: Veterinary Medicine
WG-12: Ultrasound	WG:26: Pathology
WG-13: Visible Light	WG-27: Web Technology for DICOM
WG-14: Security	

Working Groups (WGs) are formed in response to requests for new applications of DICOM. A professional society, manufacturer member, agency, or an individual or group with sufficient interest may propose the formation of a Working Group. Those who are not members of the DICOM Standards Committee may submit requests for new projects through member organizations or manufacturers. The request must be accompanied by a description of the proposed application including deliverables, an estimate of the number of meetings per year, and what organization or institution will serve as the secretariat. New applications and projects for DICOM are called work items. If the work item is approved by the DICOM Standards Committee, the Working Group may start its efforts. A list of approved work items, the WG to which they are assigned, a description, and the date of approval is available from NEMA. If the WG members are not expert in DICOM, they may request that those who (typically, members of another WG) attend some of the early meetings, or on an *ad hoc* basis, to provide guidance. Work on DICOM is entirely voluntary. Members of WGs and the Committee are funded by their organization, agency, or company. The secretariat must also contribute its work of supporting the WG.

4.9.2 Supplements and Change Proposals

The work item done by a WG, if approved by the DICOM Standards Committee when completed, becomes a DICOM Supplement. Supplements are made available for public comment—those who do choose to comment need not be DICOM Committee or WG members. It is through supplements that the DICOM standard grows. Once the public comment period has passed and comments addressed, the Supplement may be circulated to DICOM Committee members for ballot. Members vote on the supplement and may also comment on it, especially if their vote is negative. The WG that created the Supplement is required to address all negative comments. If the Supplement passes ballot, it is scheduled for incorporation in the next revision of the Standard. Should a Supplement not pass ballot (a rare occurrence) the reasons for its rejection are evaluated and, once corrected, it may be re-submitted for ballot. A list of the Supplements is available on-line. For each Supplement on the list, the parts of the Standard to which it applies, the title, status, and the version of the Standard in which the Supplement content appears are provided (Clunie, 2001). The text of the Supplement is also available for download.

In addition to Supplements, users of the DICOM Standard may find errors, omissions, conflicts, and parts that need clarification. Those who are finding these may submit them to Working Group 6 which considers the submission and, if valid, a Change Proposal is issued. Change Proposals must also be available for public comment and balloted before incorporation in the Standard. The Web site referenced above for Supplements also has a similarly structured list for Change Proposals. Change Proposals were called "Correction Proposals," but "Change" is the current terminology for these.

The basic model for expansion of DICOM is to have those with domain expertise propose work items. The Working Group mechanism means that the developers can concentrate on the work item and other WGs will assist with turning the result into "DICOMese." A major advantage of using DICOM is that once the Information Objects needed for a new imaging technique are developed, the existing DICOM Services may be used with them. If additional services are needed, they can also be added to the DICOM repertoire through the WG developmental process.

4.10 DICOM and Integrating the Healthcare Enterprise

4.10.1 The Relationship of DICOM and IHE

As readers are no doubt aware, a PACS cannot function without interfaces to other information systems in a healthcare facility. Should this not be the case, the PACS would have to duplicate many of the nonimaging functions that RISs and HISs perform. For example, the PACS would have to serve as a registration authority for patients and either find their existing medical record numbers in its database, or assign a new one. As it stands, the HIS performs this function and patient demographic information is usually passed to the RIS and then to the PACS.

As noted in Section 4.1.3, ACR-NEMA WG VIII examined the information flows between a PACS, RIS, and HIS. Radiology and hospital information systems use a standard for communication of information known as HL7. While DICOM includes management information that is intended to be communicated to and from RIS and HIS, these domains were thought outside of the DICOM purview. The necessity of communication between PACS, RIS, and HIS meant that custom interfaces had been developed. The problem is analogous to the early situation with imaging equipment, though both DICOM and HL7 standards addressed the imaging and information system domains, respectively. To help improve this situation, the RSNA and the Health Information Management Systems Society (HIMSS) formed a joint group called Integrating the Healthcare Enterprise (IHE) in 1997. A number of other societies and agencies have joined the IHE and it is now an international organization.

4.10.2 Profiling Standards

The goal of the IHE is to determine how HL7 and DICOM can be used to support the communication necessary between information and imaging systems. Rather than create new standards, the approach taken by IHE has been to determine what information needs to be exchanged and what services are needed to do so. The features and options in the HL7 and DICOM standards that can be used to carry out the services and communication are described in profiles. The profiles serve as a way of narrowing the often multiple possible ways of performing a function using DICOM and HL7 to a smaller (or even single) set of methods. IHE also goes beyond both DICOM and HL7 in developing models of operations in a healthcare facility. An illustrative example is how patient information is reconciled. An unconscious patient brought into an emergency department may need imaging studies before the patient can be identified. Such studies are usually done with a temporary name and identifier, which is then changed to the correct values once known. IHE developed a profile for this; Patient Information Reconciliation (PIR). This profile describes how this can be done using existing HL7 and DICOM. The IHE Technical Framework (IHE, 2008) is a series of documents that collects the profiles and descriptions and describes them. Since IHE extends across a number of healthcare domains in addition to radiology, there are Technical Frameworks for other specialties. Most of the IHE documents are available on-line (IHE). A most helpful document is the IHE User's Handbook for Radiology (IHE, 2005) as it describes how users may request conformance to IHE profiles in purchasing documents.

4.11 Conclusion

Since its beginning as the ACR-NEMA Standard, the DICOM Standard has grown both in extent and influence. It is ubiquitous in radiology and is rapidly becoming so in other specialties that use imaging. The methods used by the DICOM Standards Committee to expand, refine, and (when necessary) correct the Standard have resulted in its rapid building and maintenance. Both the Standard's structure and its development methods serve as a model for other nonmedical imaging standards. All of the work on DICOM was done through the voluntary efforts of many individuals from industry, healthcare professions, governments, and other standards organizations. Those who are engaged in almost any aspect of medical imaging owe a debt of gratitude to these dedicated individuals.

4.12 Appendix to Chapter 4

4.12.1 Reading a DICOM Conformance Statement

This Appendix assumes a basic understanding of DICOM (as provided, for example, by Chapter 4) and knowledge of the terminology used to describe computer networks (e.g., IP Address, port number, DHCP, DNS).

Although the DICOM Standard has no body that enforces it (It is a voluntary standard.), there is a mechanism by which vendors can describe how they conform to the specifics of the Standard. The DICOM conformance process involves the vendor describing, in a standard document, what they can and cannot do and what options are employed (where choices are available to implementers).

Part 2 of the DICOM Standard (NEMA, 2009a) describes the manner in which a Conformance Statement must be structured. For this reason, DICOM Conformance Statements all contain the same sections. Part 2 also contains some examples for both implementers and users to make how descriptions of conformance are translated into actual documents. A template is also provided so that implementers should find it simpler to produce their Conformance Statements.

It is important to understand what a Conformance Statement means to a user who is attempting to determine what a device does using DICOM. Chiefly, DICOM Conformance does not guarantee that two devices will operate together properly if their DICOM Conformance Statements "match." Though the probability that they will communicate correctly is high, DICOM Conformance cannot guarantee this. On the other hand, if the DICOM Conformance Statements of two devices do not "match," it is very nearly certain that the devices will not be able to communicate properly. As an example, a two-prong 110 V AC power plug (which is, in the United States, defined by a NEMA Standard) will fit into a three-prong 110 V AC receptacle, but there is no guarantee that the equipment so powered will operate correctly. (The two-prong system is not grounded when plugged into a three-prong receptacle.) The same two-prong power plug will not plug into a 220 V AC receptacle. The standards for the plug and receptacle in this case do not match. The design is purposely made so the unsafe situation of plugging 110 V equipment into a 220 V outlet cannot easily occur.

This Appendix will use the template provided in DICOM Part 2. Readers are encouraged to use the information in this Appendix to review some of the DICOM Conformance Statements of equipment they use to broaden their understanding of how such equipment communicates using DICOM.

4.12.1.1 The Cover Page

The cover page of a DICOM Conformance Statement is, despite its usual brevity, very important. First, it identifies the equipment manufacturer and equipment to which the Conformance Statement applies. Since a single vendor may often make a number of different DICOM conformant devices, there may be many Conformance Statements. Examining the cover page will help the reader determine if he or she has the appropriate Conformance Statement. In addition, the cover page will also describe the version of the product described. Most current imaging equipment has much software that runs on it. Since software may be updated to fix problems or provide new features, the product (or software) version is important to describe. Vendors must make sure the DICOM Conformance Statement they have produced reflects the version of the product described. The cover page also includes the date the document was created.

Some vendors add a section following the cover page that describes trademarks used, copyright, and any legal disclaimers.

4.12.2 Conformance Statement Overview

The overview begins with a nontechnical description of the network services and media storage capabilities of the product. This paragraph is then followed by a table that summarizes the network services used by the SOP Classes that the device supports. The table is divided into sections for Transfer, Query/Retrieve, Workflow Management, and Print Management. For each of these sections, the supported SOP Classes are listed if the SOP Class is a Service Class User or Service Class Provider.

This table is followed by one that describes the DICOM Media Services (i.e., CD, magneto-optical disk, DVD) that the device supports. If a device does not provide any DICOM Media Services, some vendors will omit this table while others will include it but show in the columns that they do not support the Media Services.

4.12.3 Table of Contents

Vendors usually place the table of contents following the Overview section. Note that in some DICOM Conformance Statements created using earlier Conformance Statement templates, there may be a different order of these first sections. The cover page may be followed by the revision history with a table of contents immediately after that.

4.12.4 Introduction

The introduction not only serves as a summary of the Conformance Statement, but also contains important subsections. The first is the revision history. It describes the various versions of the document with dates of their creation and the versions of the product to which they apply. The revision history also usually includes a brief description of the reason for the revision.

Following the revision history is a short section that describes the audience for which the document is intended. Not all manufacturers include this subsection. If they do include it, it is supposed to indicate to the reader what knowledge of DICOM is assumed. The template has suggested language for this paragraph.

A remarks section is intended to provide further guidance to the reader about the scope of the Conformance Statement, any disclaimers, and (if applicable) a direction to supported IHE Integration Statement(s).

The terms and definitions section provides a ready reference for the DICOM terms that appear in the subsequent sections of the document. The terms also have the DICOM definitions included, so the section actually serves as a glossary. The Conformance Statement template provides an example that vendors are starting to incorporate.

The basics of DICOM communication section is included by some vendors (recommended in the Conformance Statement template) and provides a short narrative description of how DICOM communication works.

A definition of abbreviations used in the Conformance Statement is next. The final section in this part of the Conformance Statement is for references. Readers should note that the references may include the operation, service, and reference manuals for the product that is the subject of the Conformance Statement. References to the DICOM Standard itself are included in this section as appropriate.

4.12.5 Networking

This section of the DICOM Conformance Statement serves to set out the network-related services to which the equipment

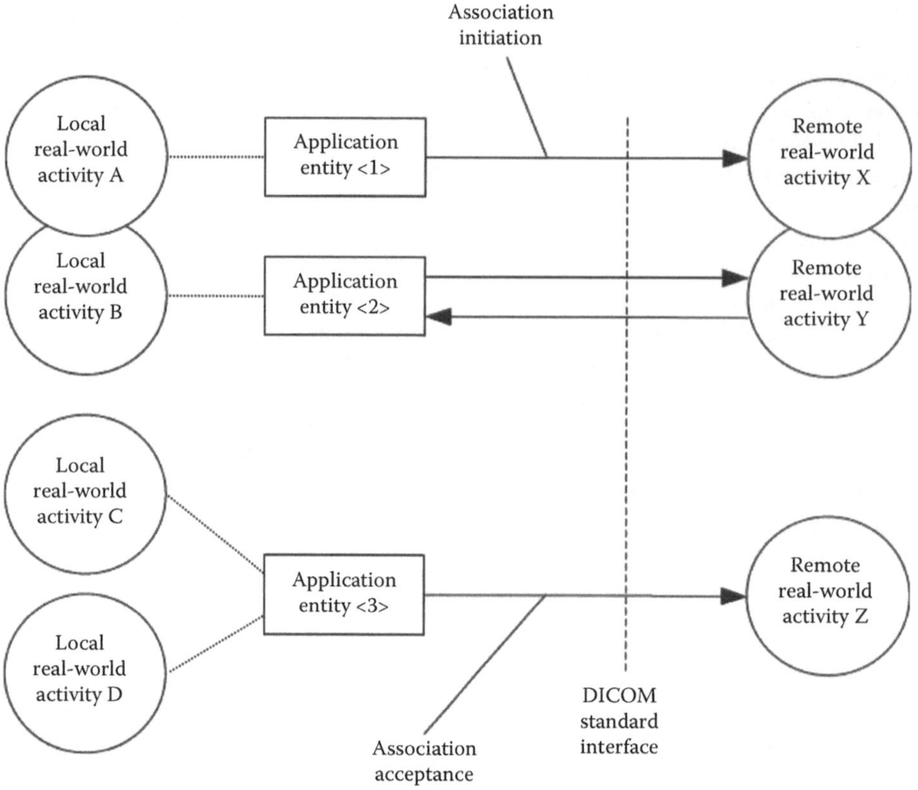

FIGURE 4.6 Template for Network Application Data Flow Diagram. (From NEMA. 2009a. *National Electrical Manufacturers Association: Digital Imaging and Communications in Medicine (DICOM) Part 2: Conformance. PS 3.2-2009*. Rosslyn, VA: NEMA. Copyright NEMA. With permission.)

conforms. If the equipment also conforms to media services, those are covered in a separate section.

Most often this section begins with a brief Introduction that sets out in short sentences or a list the various network services that the DICOM implementation supports. Note that an introduction is not required by DICOM Part 2, but some manufacturers will include it to make the parts that follow more readily understandable. The first part of this section required by the Standard is the Implementation Model. The implementation model itself is presented in three subsections. The first is the Application Data Flow Diagram (Figure 4.6). The diagram provides an overview of the real-world activities that the equipment supports and the DICOM Application Entities (AE). Real-world activities include such things as "Store an exam." The relationships of the AE to the real-world activities on both sides of the DICOM interface are shown. The diagram shows data flow, so the arrows do mean directionality. For example, a real-world activity such as the ECHO Service is shown with a bidirectional arrow since the service can be invoked by equipment on either side of the DICOM interface. Relationships of the real-world activities are also shown. If two real-world activities are interactively related, their real-world activity symbols on the diagram are shown as overlapping.

A simple example of a Data Flow Diagram for a hypothetical ultrasound machine is shown in Figure 4.7.

In Figure 4.7, the local real-world activity on the ultrasound machine side is storing the instance of an ultrasound image. The Application Entity is called "US_System" and the real-world activity across the DICOM Interface is a storage system that provides the DICOM Storage service as the SCP. The arrows show the direction of data flow. The arrow between the Application Entity and the remote activity represents a DICOM Association. This example would typically be a part of a diagram that shows all of the real-world activities in the conformant equipment that uses DICOM. Some manufacturers show the Application Entities that handle the real-world activities as separate boxes; others group all DICOM functions in a single Application Entity. The Standard does not dictate how vendors

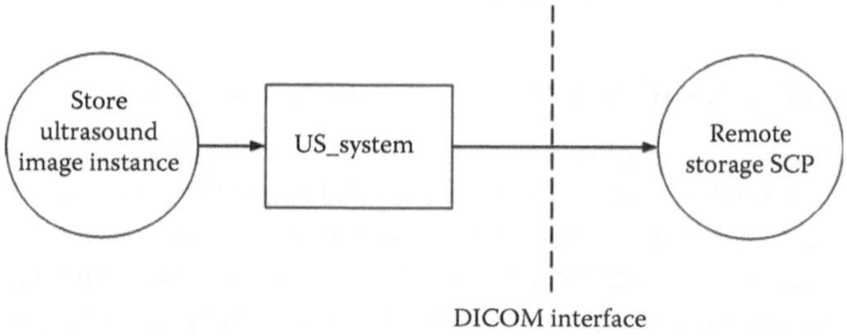

FIGURE 4.7 An example Application Data Flow Diagram for a hypothetical ultrasound machine.

should represent their implementations, only that the various local and remote real-world activities be shown.

What follows the Application Data Flow Diagram is a brief description of the real-world activities in the diagram. This can be thought of as the "caption" for the diagram. The description of the real-world activities is usually followed by a bulleted list of the Application Entities, or (when the AE encompasses all DICOM Network functions) a bulleted list of the DICOM functional components of the AE. Some manufacturers provide more detail with a series of paragraphs that provide the functional definition of each Application Entity shown in the diagram. For the example, only a single paragraph would be needed. For the template (Figure 4.6), three paragraphs are required. Each paragraph contains a general description of the functions performed by the Application Entity and the DICOM services used to accomplish these functions. Manufacturers are required to describe not only the DICOM Service Classes used, but also the lower level DICOM services, such as Association Services.

Some real-world activities may require that they be done in sequence. If this is the case, a subsection describing the necessary sequencing and any constraints on such sequencing is included following the AE functional definitions.

The major substance of the DICOM Conformance Statement is provided after these first sections. Each Application Entity for which DICOM Conformance is claimed is described in detail in subsections that follow the Application Data Flow section. These are known as the AE Specifications. Each AE Specification includes a series of subsections (usually numbered).

The first of these subsections is a listing (in the form of a table) of the SOP Classes to which the AE is conformant. This subsection begins with a specific, required statement: "This Application Entity provides Standard Conformance to the following SOP Class(es)." The table that follows has a first column that lists the SOP Class UID Name, a second column that lists the particular SOP Class UID, a third column that indicates if the AE is a user of the SOP Class (SCU), and the fourth column that indicates if the AE is a provider of the SOP Class (SCP). In some instances, manufacturers have chosen to create two tables: one for all of the SOP Classes for which the AE is an SCU and another table for all the SOP Classes for which the AE is an SCP. These tables are very important for understanding the conformance of a piece of equipment. For example, if one piece of equipment claims conformance to a particular SOP Class as an SCU, for it to communicate the use of that service successfully with another device, that device would have to conform to that same SOP Class as an SCP. If the equipment makes use of the specializations of an SOP Class to which it claims conformance that is noted in the table and detailed subsequently. Specializations are usually standard SOP Classes from which a manufacturer has created a superset through the addition of private attributes. DICOM provides specific methods for devices to negotiate whether or not they will accept such specialized SOP Classes.

The tables of SOP Classes (recall that such a table is required for each AE) are followed by a subsection that details the Association Policies of the AE. The policies that the AE uses for establishing or accepting an Association are described in the subsections. The first is a table containing the Application Context Name that an AE proposes at the start of Association establishment. For the current and previous versions of DICOM, there is only one Application Context Name, so the table has only a single entry.

The next subsection describes the Number of Associations that an AE may initiate. These are also presented as tables. The first table is titled "Number of Associations as an Association Initiator for 'Application Entity n'" (where "Application Entity n" is the name of the AE being described) and lists the maximum number of Associations that the AE may initiate. The second table is titled, "Number of Associations as an Association Acceptor for 'Application Entity n'" and is a similar list of the maximum number of Associations the AE may accept. The number of Associations is the number that an AE may initiate or accept simultaneously. However, AEs may have some restrictions on how these Associations are distributed. For example, an AE may support 10 simultaneous Associations, but will support only two with any particular remote AE. So, if each of the remote AEs with which it communicates uses two simultaneous Associations, the example AE would be able to support simultaneous Associations with five such remote AEs.

DICOM supports the ability for an AE to have asynchronous Associations. That is, on any single Association, there may be multiple transactions that are outstanding. If the AE supports an Asynchronous Nature, the next subsection is a table that lists the number of outstanding transactions that an AE supports. In general, at the time of this writing, very few (if any) devices support an Asynchronous Nature of transactions on an Association. If an AE does not support an Asynchronous Nature, this subsection usually states that though some vendors will include the table with the number of outstanding asynchronous transactions equal to one which has the same functional meaning as not supporting an Asynchronous Nature.

Following these subsections is the Implementation Identifying Information. This is provided in a short table (sometimes two) with one entry for the Implementation Class UID (this is where a vendor's DICOM UID root can be found) and another table (or table entry) for the Implementation Version Name. The UID follows the DICOM Standard for UIDs and includes the UID that a vendor uses for their products. Vendors apply for these UID roots and are guaranteed that they are unique. The Implementation Version Name is often the name given to the software that supports the AE.

The next major section of the Conformance Statement describes the Association Initiation Policy of the AE. This includes a number of subsections that provide the details of how a particular activity of an AE will initiate an Association. The subsections of this section are repeated for each activity of the AE. As an example, if an AE has two real-world activities that it supports, this section would contain descriptions of the Association Initiation Policy for each of the two activities.

For some AEs, the SOP Classes have activities that need to be sequenced properly if they are to work as expected. If an AE requires sequencing of activities, a subsection detailing the required sequencing is included as the Description and Sequencing of Activities. If this subsection is included, it usually has a narrative description of activity sequencing. The DICOM Conformance specification also recommends that manufacturers include an illustrative diagram.

When an Association is initiated, the AE activity that initiates the Association proposes the DICOM Presentation Contexts it would like to use. The next subsection is a table of the Proposed Presentation Contexts. This Presentation Context Table has four major divisions: Abstract Syntax, Transfer Syntax, Role, and Extended Negotiation. A DICOM Presentation Context consists of a Name and a UID, so the first division has a column for each. Similarly, the Transfer Syntax has both a name and UID, resulting in two columns as well. Role and Extended negotiation have no components, so each has a single column. For each Abstract Syntax proposed, an activity typically also proposes more than one Transfer Syntax. Figure 4.8 shows the Presentation Context Table for the hypothetical ultrasound AE shown in Figure 4.7.

For the AE activity of storing an ultrasound image, the hypothetical ultrasound machine AE will propose three Transfer Syntaxes at Association initiation: Implicit VR Little Endian, Explicit VR Little Endian, and JPEG Lossless Baseline (a lossless compressed Transfer Syntax). A real ultrasound system would continue this table. (The row with the ellipses illustrates that the table may have additional entries; an actual table would have any additional rows filled-in.) If an AE proposes Extended Negotiation, an additional table (Extended Negotiation as an SCU) is required to define the Extended Negotiation.

During DICOM communication, SOP Classes have specific behaviors. These are detailed in DICOM Part 4. For purposes of a Conformance Statement, these behaviors (error codes, error and exception handling, time-outs, etc.) are described in the subsection that follows the Presentation Contexts. This is called the SOP Specific Conformance for "SOP Class(es)" subsection and typically includes a table that describes the SOP Classes supported behave under different statuses. A second table is used to define communication failure behavior. Note that the components of this subsection repeat for each of the SOP Classes proposed in the Presentation Context Table.

If an AE activity acts as an Association acceptor, the SOP Specific Conformance subsection is followed by a section that details the Association Acceptance Policies of the AE activity. This section begins with a subsection with the name (and usually a short description) of the AE activity. What follows are subsection tables that are the parallel of those in the previous section that describes behavior as an Association initiator. Instead of a table of proposed Presentation Contexts, this section has a table of Acceptable Presentation Contexts. This would be followed, if necessary, by a table detailing the Extended Negotiation as an SCP. Also to match the Association initiation section, this section includes tables of SOP Specific Conformance for the SOP Classes accepted. The Association Acceptance Policy section also repeats for each of the AEs that can act as an Association acceptor. Because there is a default Verification SOP Class that must be accepted by DICOM conformant devices, all AEs will have this section of the Conformance Statement at least completed for the Verification SOP Class.

The networking portion of the Conformance Statement concludes with a section that describes how the equipment conforms to the network interface and DICOM-supported protocols. The section begins with a description of the Physical Network Interface used by the equipment. This is almost always Ethernet 10/100/1000BaseT using an RJ-45 connector. Some manufacturers will indicate in this subsection whether or not the network interface autonegotiates speed (if not, what the permissible fixed speed settings are) and if full- or half-duplex communication is supported. Since most equipment includes network interface electronics supplied by a specialty manufacturer of such interfaces (rather than design and build their own interfaces), the specifications in this subsection are the same as those found in other commercial computer systems.

On occasion, manufacturers may specifically state which of the DICOM network communication "stacks" they support. Since the ISO-OSI stack was retired from DICOM, the DICOM UL/TCP/IP protocol stack is the only one currently supported by the Standard.

A subsection describing any Additional Protocols, such as those for system management, used by the equipment follows the specification of the physical network interface. DICOM Part 15 (NEMA, 2009j) defines support for System Management and Security Profiles. If the equipment supports these, a table of System Management Profiles is included. This subsection is also used for descriptions (if employed or supported) of

- DHCP behavior for configuration of the local IP address.

Proposed Presentation Contexts for "US_System"

Abstract Syntax		Transfer Syntax		Role	Extended Negotiation
Name	UID	Name List	UID List		
US Image Storage	1.2.840.10008.5.1.4.1.1.6.1	Implicit VR Little Endian Explicit VR Little Endian JPEG Lossless Baseline	1.2.840.10008.1.2 1.2.840.10008.1.2.1 1.2.840.10008.1.2.4.70	SCU	None
...	...	...	...	...	...

FIGURE 4.8 Example presentation context table.

- DNS operations to obtain an IP address based on the hostname information.
- Use of NTP or SNTP for time synchronization and the available NTP configuration alternatives used.
- What options and restrictions are used if the equipment supports DICOM Web Access to DICOM Objects (WADO).

If IPv4 and IPv6 are supported, a subsection detailing the specific features of IPv4 and IPv6 is included. In addition, if the security and configuration details of IPv6 are used, that information is also described.

Note that the entire section on Additional Protocols may be absent if the equipment does not support them.

Various devices in an electronic imaging network usually need to be configured for the particular network being used and the other devices on the network. Following the section on protocols is one devoted to Configuration. One configuration issue is the AE Title/Presentation Address Mapping. This mapping translates AE Titles into the Presentation Addresses used on the network. Some equipment may have more than one AE operating under the single AE Title of the device. DICOM permits this, but if this is the case, the tables in this subsection allow for this to be described. Local AE titles are listed in an Application Entity Configuration Table which has a column for the AE, the default AE Title, and the default TCP/IP port. The situation of multiple AEs under a single AE Title is shown by listing the AEs with the same default AE Title in the table. Some manufacturers provide a table that lists the AEs (usually by the name used in the equipment) along with the SOPs supported by that AE. The table then tells the user that these AEs are configurable on the equipment.

If applicable, the next subsection describes configuration of Remote AEs in a manner similar to that for the local AEs.

The final part of this section, and of the network portion of the Conformance Statement, is a description of the operational parameters used. DICOM has a number of configurable parameters, including various time-outs, maximum object size constraints, maximum PDU size for the network, and configurable Transfer Syntaxes. The default values and the acceptable range for these are described in a Configuration Parameters Table. This table lists both the general (such as time-outs) and AE specific (such as maximum object size) that the equipment allows to be configured. Manufacturers are also encouraged to describe other device configuration details in this section.

4.12.6 Media Interchange

If a device supports the exchange of DICOM Objects through the use of removable media (CD, DVD, etc.), the Conformance Statement continues with a series of sections that parallel those for networking, but with media-specific items. This part of the Conformance Statement begins with an Application Data Flow Diagram, much as the networking part does. Instead of SCU and SCP roles, media have File-Set Reader (FSR) and File-Set Creator (FSC). For media, the equivalent of a device that is both SCU and SCP is the File-Set Updater (FSU). In place of a description of the network interface particulars, this part describes the DICOM Media Storage Application Profile(s) supported. Because this part of the Conformance Statement is so similar to that for networking, it is not described in detail here.

4.12.7 Support of Character Sets

Since DICOM is an international standard, supporting character sets other than US ASCII are necessary. However, for conformance, it is important to describe what character sets are supported by equipment. It is also important for a manufacturer to describe

- What a device does if it receives a DICOM Object that uses a character set it does not support.
- What character set configuration (e.g., options) a device supports, if it does.
- Mapping and/or conversion of character sets across DICOM Services and Instances.
- Query capabilities for attributes that include nondefault character sets.
- How characters are presented to the user, including capabilities, font substitutions, and limitations.

4.12.8 Security

If a device claims conformance to any of the Security Profiles that DICOM supports, the details of the conformance to the Security Profiles are described in this section. Security Profiles may provide Association level and Application level security. The particulars of the security measures supported (e.g., allowing only certain IP addresses to open an Association at the Association level and biometrics at the Application level) are described.

4.12.9 Annexes

A great deal of information about how an implementation conforms to DICOM is described in an Annex. The first part of the Annex describes the IOD Contents. This includes both Standard and Private IODs. This part is usually provided as tables by the manufacturer and is essentially an equipment-specific data dictionary. The first table is usually a list of the DICOM Information Entities (IE, e.g., Patient, Study, Equipment, etc.) and either the local name for the entity or the DICOM Modules included in the IE. Subsequent tables then typically expand the description of the module contents. These detail tables are very important as they are lists of the Attributes that a manufacturer includes in their DICOM Dataset. These tables typically use abbreviations to describe the expected state of an attribute (e.g., VNAP—value not always present). When conformance problems are suspected, a very useful troubleshooting technique is to compare the DICOM Data Set produced by a piece of equipment with what the Conformance Statement says it should contain.

If an Application depends on certain Attributes in an IOD it receives, a Usage of Attributes from Received IODs subsection following the tables of Attributes is used to describe the Attributes the Application needs if it is to function correctly.

Some Attributes are used in multiple SOP Classes. The Attribute Mapping subsection allows manufacturers to include a table showing the mapping of particular Attributes across the SOP Classes in which they are used. Though not required, if an Attribute is used in a field of another protocol (e.g., HL7), manufacturers are strongly encouraged to include a description of such mapping in this subsection.

DICOM has a concept of Attribute coercion. This is not intended to be pejorative, but to describe how the value of an Attribute might be changed by an Application Entity. For example, a storage SCP that takes its database values from a master patient index might change the value of a received Patient Name Attribute. Some changes to DICOM SOP Instances require a change of the Instance UID. Such a change is another example of DICOM coercion. For conformance, it is important to know what Attributes an AE may coerce or modify and under what circumstances. The subsection on Coerced/Modified fields provides a place in which a manufacturer can describe such Attribute coercion or modification.

Though private Attributes are those a manufacturer may create and use for their own purposes, they should not be secret. A DICOM-conformant device receiving a Dataset containing private Attributes needs to know that they are present. Equipment is not expected to need private Attributes from a different manufacturer to operate correctly, but such Attributes are expected to be retained if a Dataset is stored and returned if the Dataset is requested. The Data Dictionary of Private Attributes subsection is for the listing of private Attributes. The format is intended to be the same as that used in DICOM Part 6 (NEMA, 2009f). If a manufacturer uses private SOP Classes and Transfer Syntaxes, they are supposed to be listed in this subsection. (Separate subsections are for detailed descriptions of private SOP Classes and Transfer Syntaxes.)

Primarily for DICOM Structured Reports (SR), Coded Terminology and Templates may be used. This subsection is intended for the description of the support and content of Coded Terminology and Templates when they are used by an AE.

The use of Coded Terminology usually leads to the definition of Context Groups. These detail how the Coded Terminology is used in a specific context. A table in this subsection describes the Context Groups, their default values, if they are configurable, and how they are used. If private Context Groups are used, they are described in a table that follows the same structure as for the standard Context Groups.

Two subsections that follow Context Groups are for Template Specifications and Private Code definitions. These are self-explanatory.

DICOM provides a method for standardizing the grayscale display of various display devices. This is described in the DICOM Grayscale Standard Display Function (NEMA, 2009i). If a device supports this part of the Standard, it is described in the Grayscale Image Consistency section.

The final two sections of the DICOM Conformance Statement are nonetheless essential to conformance if the equipment uses the features described. The first is the description of any Standard Extended/Specialized/Private SOP Classes. For each SOP Class that falls into one of these categories, there is a subsection that describes it in detail. Such description is intended to follow the structure for the description of Standard SOP Classes. Finally, any use of Private Transfer Syntaxes is described. For each such Private Transfer Syntax used, there is a following subsection that describes it in detail. The description is required to be the same as the description of Standard Transfer Syntaxes in DICOM Part 5 (NEMA, 2009e).

References

ASTM. *ASTM* [Online]. Available at http://www.astm.org/COMMIT/SUBCOMMIT/E07.htm. Accessed September 7, 2010.

Ausherman, D. A., Dwyer III, S. J., and Lodwick, G. S. 1970. A system for the digitization, storage and display of images. *American Association of Physicists in Medicine Conference*. Kansas City, Kansas.

Baxter, B., Hitchner, L., and Maguire Jr., G. 1982. Characteristics of a protocol for exchanging digital image information. In *PACS for Medical Applications*, pp. 273–277. Bellingham, WA: SPIE.

Billingsley, F. 1982. LANDSAT computer-compatible tape family. In *PACS for Medical Applications*, pp. 278–283. Bellingham, WA: SPIE.

Clunie, D. 2001. *DICOM Standard Status* [Online]. Available at http://dclunie.com/dicom-status/status.html#BaseStandard 2001. Accessed September 7, 2010.

Haney, M., Johnston, R., and O'Brien Jr., W. 1982. On Standards for the storage of images. In: *PACS for Medical Applications*, Bellingham, WA: SPIE.

HL7. *V2 messages* [Online]. Available at http://www.hl7.org/implement/standards/v2messages.cfm. Accessed September 7, 2010.

Ho, J., Parwani, A. V., Jukic, D. M., Yagi, Y., Anthony, L., and Gilbertson, J. R. 2006. Use of whole slide imaging in surgical pathology quality assurance: Design and pilot validation studies. *Hum. Pathol.*, 37, 322–31.

Horii, S. C. 2005. Introduction to "Minutes: NEMA Ad hoc Technical Committee and American College of Radiology's Subcommittee on Computer Standards". *J. Digit Imaging*, 18, 5–22.

Horii, S. C., Hill, D. G., Blume, H. R., Best, D. E., Thompson, B., Fuscoe, C., and Snavely, D. 1990. An update on American College of Radiology-National Electrical Manufacturers Association standards activity. *J. Digit Imaging*, 3, 146–51.

IEEE. 1985. Institute of Electrical and Electronics Engineers: 802.3-1985 *IEEE Standards for Local Area Networks: Carrier Sense Multiple Access with Collision Detection.*

IHE. *Integrating the Healthcare Enterprise* [Online]. Available at http:www.ihe.net. Accessed September 7, 2010.

IHE. 2005. *IHE Radiology User's Handbook* [Online]. Available at http://www.ihe.net/Resources/upload/ihe_radiology_users_handbook_2005edition.pdf. Accessed September 7, 2010.

IHE. 2008. *Integrating the Healthcare Enterprise: IHE Technical Framework: Vol. 1 Integration Profiles, Rev 9* [Online]. Available at http://www.ihe.net/Technical_Framework/upload/ihe_tf_rev9-0ft_vol1_2008-06-27.pdf. Accessed September 7, 2010.

ISO. 1986. International Standards Organizations: ISO 8879: 1986: Information processing—Text and Office Systems—Standard Generalized Markup Language (SGML).

ISO. 1994. International Standards Organization: ISO/IEC 7491-1: 1994: Information technology—Open Systems Interconnection—Basic Reference Model.

ISO. 1996. International Standards Organization: ISO/IEC 8649: 1996: Information technology—Open Systems Interconnection—Service definition for the Association Control Service Element.

ISO. 2008. International Standards Organization: ISO/IEC 9834-3: 2008: Information technology—Open Systems Interconnection—Procedures for the operation of OSI Registration Authorities: Registration of Object Identifier arcs beneath the top-level arc jointly administered by ISO and ITU-T.

Lemke, H., Stiehl, H., Scharnweber, H., and Jackel, D. 1979. Application of picture processing, image analysis and computer graphics techniques to cranial CT scans. *Sixth Conf. on Computer Applications in Radiology and Computer Aided Analysis of Radiological Images.* Newport Beach, CA: IEEE Computer Society Press, June 18–21.

Maguire, Jr., G., Baxter, B., and Hitchner, L. 1982. An AAPM standard magnetic tape formt for digital image exchange. In *PACS for Medical Applications*, pp. 284–293. Bellingham, WA: SPIE.

Metcalfe, R. and Boggs, D. 1976. Ethernet distributed packet switching for local computer networks. *Commun. ACM*, 19, 395–404.

NEMA. *NEMA* [Online]. Available at http://medical.nema.org. Accessed September 7, 2010

NEMA. *NEMA SICD* [Online]. Available at http://www.nema.org/prod/security. Last accessed May 18, 2011.

NEMA. *NEMA Work Items* [Online]. Available at http://medical.nema.org/dicom/getinfo/Work-Items-DSC.pdf. Accessed September 7, 2010.

NEMA. 2003a. *DICOM Anniversary Conference and Workshop* [Online]. Baltimore, MD: NEMA. Available at http://medical.nema.org/dicom/workshop-03. Accessed September 7, 2010.

NEMA. 2003b. *The DICOM Story* [Online]. NEMA. Available at http://nema.medical.org/dicom/workshop-03/pres/mildenberger.ppt. Accessed September 7, 2010.

NEMA. 2009a. *National Electrical Manufacturers Association: Digital Imaging and Communications in Medicine (DICOM) Part 2: Conformance. PS 3.2-2009.* Rosslyn, VA: NEMA.

NEMA. 2009b. *National Electrical Manufacturers Association: Digital Imaging and Communications in Medicine (DICOM) Part 3: Information Object Definitions. PS 3.3-2009.* Rosslyn, VA: NEMA.

NEMA. 2009c. *National Electrical Manufacturers Association: Digital Imaging and Communications in Medicine (DICOM) Part 4: Service Class Specifications. PS 3.4-2009.* Rosslyn, VA: NEMA.

NEMA. 2009d. *National Electrical Manufacturers Association: Digital Imaging and Communications in Medicine (DICOM) Part 5: Data Structures and Encoding . PS 3.5-2009.* Rosslyn, VA: NEMA.

NEMA. 2009e. *National Electrical Manufacturers Association: Digital Imaging and Communications in Medicine (DICOM) Part 5: Data Structures and Encoding . PS 3.5-2009: p. 59.* Rosslyn, VA: NEMA.

NEMA. 2009f. *National Electrical Manufacturers Association: Digital Imaging and Communications in Medicine (DICOM) Part 6: Data Dictionary. PS 3.6-2009.* Rosslyn, VA: NEMA.

NEMA. 2009g. *National Electrical Manufacturers Association: Digital Imaging and Communications in Medicine (DICOM) Part 7: Message Exchange. PS 3.7-2009.* Rosslyn, VA: NEMA.

NEMA. 2009h. *National Electrical Manufacturers Association: Digital Imaging and Communications in Medicine (DICOM) Part 8: Network Communication Support for Message Exchange. PS 3.8-2009.* Rosslyn, VA: NEMA.

NEMA. 2009i. *National Electrical Manufacturers Association: Digital Imaging and Communications in Medicine (DICOM) Part 14: Grayscale Display Function. PS 3.14-2009.* Rosslyn, VA: NEMA.

NEMA. 2009j. *National Electrical Manufacturers Association: Digital Imaging and Communications in Medicine (DICOM) Part 15: Security and System Management Profiles. PS 3.15-2009.* Rosslyn, VA: NEMA.

Schneider, R. 1982. The role of standards in the development of systems for communicating and archiving medical images. In *PACS for Medical Applications*, pp. 270–271. Bellingham, WA: SPIE.

SIIM. 2008. *Society for Imaging Informatics in Medicine History Subcommittee: Transcript of an interview with Joseph N. Gitlin* [Online]. Available at http://www.siiweb.org/WorkArea/showcontent.aspx?id=5932. Accessed September 7, 2010.

TSA. *TSA* [Online]. Available at http://www.tsa.gov/approach/tech/ait/index.shtm. Accessed September 7, 2010.

Wendler, T. and Meyer-Ebrecht, D. 1982. Proposed standard for variable format picture processing and a codec approach to match diverse imaging devices. In: *PACS for Medical Applications*, pp. 198–305. Bellingham, WA: SPIE.

Wikipedia. *HTML* [Online]. Available at http://en.wikipedia.org/wiki/HTML. Accessed Septemeber 7, 2010.

5

Integrating the Healthcare Enterprise IHE

Steve G. Langer
Mayo Clinic

5.1 Background

5.1.1 Terms and Definitions

Affinity Domains: A group of healthcare enterprises that have agreed to work together using a common set of policies and share a common infrastructure.

Connectathon: Global annual meetings where neutral referees gather to test vendor claims of IHE Compliance. Currently, there are Connectathons in North America, Europe, and Asia (Connectathon, 2011).

Data Model: As defined in the IHE TF-V4.0, a collection of informatics constructs (scope definition, Use Case roles, referenced standards, Interaction Diagrams, and Transaction definitions) required to implement an Integration Profile. The analogy to computer science is that the Data Model defines the attributes of a Class (see Chapter 3).

DICOM: Digital Imaging and Communications in Medicine are both a Protocol and Ontological set for communicating image information among medical devices and applications. For a thorough treatment, the reader is directed to Chapter 4.

HL7: Health Level 7 is Ontology for communicating medical information among nonimaging medical applications and devices; it is usually expressed over the network in one of two Protocols, the classic method is a text stream. More recently, HL7 has come to be expressed over XML. For a more thorough treatment, the reader is directed to Chapter 3.

IHE: Integration of the Healthcare Enterprise is the result of a multisociety and industry collaboration designed to clarify and formalize the implementation of Uses Cases in medicine.

Integration Profiles: An Integration Profile may be considered to be an instance of the IHE Data Model applied to a specific Use Case; the analogy to an Object instantiation of a software Class (see Chapter 2) is obvious. The term was introduced in IHE V5.0 Part A.

Technical Frameworks: IHE addresses the Use Cases in various medical specialties in separate volumes of documents. Hence, there are volumes that address Radiology; other volumes are concerned with Radiation Oncology, and so on.

Transaction Model: Formally required to consist of a scope definition, model of the real world, and an information model.

5.1.2 Ontologies, Use Cases, and Protocols: Oh My!

To fully understand the concepts in this chapter, it may be helpful if the reader has covered the chapters on DICOM, HL7, and Informatics Constructs. The last describes general constructs used in imaging informatics and in particular how they are used in IHE documents, and the first two describe the main standards used in imaging informatics. Capitalized nouns that are not defined in this chapter (such as Ontology, Use Case, Protocol, etc.) have been defined in "Chapter 2: Informatics Constructs," otherwise they are defined here.

5.1.3 Vendor and Market Acceptance

There are several compelling reasons for medical imaging practitioners to be supportive of IHE. IHE provides what computer scientists call an abstraction layer; that is it is a high-level abstraction of many details below. For example, it is easier for a CT scanner purchaser to say to a vendor, "I would like to have a scanner that supports IHE Consistent Presentation of Images," rather than,

> "I would like a scanner that supports the DICOM Gray-Scale Standard Display Function, Grayscale Softcopy Presentation State, Color Softcopy Presentation State, DICOM Store-Commit, Presentation Lookup Tables, Value of Interest Lookup Tables ..." and over 12 other requirements.

A customer seeking to validate the DICOM behavior of a scanner would have to request and read the scanner's DICOM Conformance Statement; a document that is often over one hundred pages long (see Chapter 4 for how to interpret these). In contrast, IHE Conformance Statements are typically very brief, often less than five pages (GE PACS IHE Conformance, 2006). As a result, it is vastly simpler for a purchaser to have some confidence in their purchasing choices by simply comparing the IHE Conformance statements of the equipment that must interact.

As one would expect, vendor acceptance of IHE began with a few luminaries that were engaged with the initial society consortium. Today that number has grown and vendors participate in Connectathons that occur around the world in North America, Europe, and Asia. The North American Connectathon was the first and is now in its 11th year. In 2010, the Chicago event hosted 500 engineers testing 150 imaging informatics systems from 105 different companies (Connectathon, 2010). The tests are set up by volunteer informatics professionals from academia and sponsoring societies; the results are interpreted and reported by volunteers trained by the Society of Imaging Informatics in Medicine (SIIM, 2010).

5.1.4 DICOM and HL7 versus IHE

An often repeated misunderstanding regarding IHE is that it is a communication standard like DICOM and HL7. Actually this is a second level misunderstanding because at least in DICOM's case, it is not only a Protocol (as defined in Chapter 2 but also a communication Ontology that defines the allowed terms and relationships among defined objects.) However, IHE is unlike either of those other efforts; it is neither a Protocol that defines Transaction grammar, nor an Ontology that describes the standard terms and their relationships to each other for defined objects. Rather, IHE *leverages* other informatics standards such as DICOM, HL7, and XML *to resolve Use Cases* via *implementation guidelines* called Integration Profiles. If existing standards are insufficient to accomplish the Use Case goals, IHE committees work with other standards committees to augment their standards.

5.2 Early Efforts

5.2.1 Years 1–3

5.2.1.1 History of the Early IHE

In 1997 RSNA (Radiological Society of North America), HIMSS (Healthcare Information and Management Systems Society), several academic centers, and a number of medical imaging vendors embarked on a program to solve integration issues across the breadth of healthcare informatics (Channin, 2000; Channin et al., 2001). It was christened Integration of the Healthcare Enterprise (IHE). Originally, it was confined to Use Cases found in Radiology. Since that time, the scope has increased to include many so-called Technical Frameworks: anatomic pathology, cardiology, eye care, information technology infrastructure, labs, patient care coordination, patient care devices, quality, radiation oncology, and the original radiology (IHE, 1997).

IHE is based on the concepts of imaging informatics and informatics constructs (as covered in Chapter 2). To briefly summarize those points, those constructs include

a. *Use Cases:* What is the task to be done?
b. *Actors:* What agents will accomplish that task?
c. *Transactions:* What messages and contents will be flow among the actors in the course of completing the task?
d. *Integration Profiles:* the synthesis of the above.

The relationships among the objects defined above are detailed in various diagrams which are the principle forms of documentation in the IHE Technical Framework volumes. For the remainder of this chapter, we will follow the historical evolution of the IHE Integration Profiles within Radiology only.

5.2.1.2 The First Frameworks: Radiology Years 1 and 2

The definitive source for IHE Radiology documentation, including the earliest documents, is hosted by the IHE Web site (IHE FTP). The historical record is a bit sketchy in the first 2 years, and in fact the mapping between document *versions* and *IHE Year* is not obvious for the first several revisions as Table 5.1 shows.

As one could expect, the first efforts in Year 1 required a bit of iteration to settle on a standard nomenclature. It may also be said that some early hopes the initiative had may have been overzealous (i.e., its adoption would eliminate the need for third-party devices to translate between HL7 and DICOM) (Dreyer, 2000). However, the initial concepts were being defined and a path created to formalize them. In Year 1, the primary mission was to develop plans to integrate the Radiology department within the context of the hospital at large. In classical terms, that meant integrating the RIS and PACS with the Hospital Information System.

Since IHE began in the Radiology domain, many actors and profiles were defined that have since been recognized to have more general application. These have been moved over time to the Information Technology Framework. Some examples of

TABLE 5.1 The Relationship between IHE Document Versions and IHE Years

IHE Year	Document Version	Date Ratified
Year 2	V1.5	March 2000
Year 2	V4.0	March 2000
Year 3	V5.0	October 2001
Year 4	V5.3	April 2002
Year 4	V5.4	December 2002
Year 5	V5.5	November 2003
Year 6	V6.0	May 2005
Year 7	V7.0	May 2006
Year 8	V8.0	June 2007
Year 9	V9.0	June 2008
Year 10	V10.0	Pending 2010

Actors that have moved are Auditing, Secure Node, Authenticator (Kerberos), Consistent Time, and so on. However, in Year 1, there had not yet been invented such a granular taxonomy, and the effort defined the following Actors: ADT/Patient Registration (functions classically assigned to the HIS), Order-Placer (again usually found in the HIS), Departmental Scheduler/Database (RIS), Image Manager (PACS), Acquisition Modality, and Image Archive (PACS) (Smedema, 2000). Even at this early stage, one can see the revolutionary change in thinking that was beginning to occur; that is the separation of the *task* from the concept of the *task performer.* An early concept in IHE was to subdivide tasks to their most fundamental (aka atomic) unit. Hence, the very familiar concept of a PACS is exploded in IHE to several atomic Actors: Image Manager, Image Archive, Image Viewer, and possibly other components as well. This concept will be seen to pervade IHE as it evolves.

By Year 2, the concepts and nomenclature in IHE began to be refined. The Technical Framework formalized the constructs for representing both the Transaction Model and the Data Model (IHE Radiology Technical Framework V4.0). The former was defined to consist of scope definition, Use Case roles (defining Actors and their tasks), the standards that are referenced (e.g., DICOM), Interaction Diagrams, and message definitions. The Data Model was formally required to consist of a scope definition, model of the real world, and an information model (consisting of entity–relationship diagrams, described more fully in Chapter 4).

Chapter 3 of V4.0 provided a useful mini-index (Table 5.2) to the framework and summarized the Actors and the Transactions they were required to support. The following Actors were formally defined

a. *Acquisition Modality:* A system that creates and acquires medical images, for example, a Computed Tomography scanner or Nuclear Medicine camera.
b. *ADT (Admission/Discharge/Transfer) Patient Registration:* Responsible for adding and/or updating patient demographic and encounter information.

TABLE 5.2 The Relation between Actors and the Integration Profiles They Must Support

Integration Profile Actor	SWF	PIR	CPI	PGP	ARI	KIN	SINR	SEC	CHG	PWF
Acq. Modality	X	X	X	X		X		X	X	
ADT Patient Reg.	X	X						X	X	
Audit Record Rep.								X		
Charge Processor								X	X	
DSS/OF	X	X		X				X	X	X
Enterprise Rep. Repository							X	X		
Ext. Rep. Access					X		X	X		
Image Archive	X	X	X	X	X	X		X		X
Image Creator	X		X			X		X		X
Image Display	X		X		X	X		X		X
Image Manager	X	X	X	X	X	X		X		X
Order Placer	X	X						X		
Postprocessing Manager								X		X
PPS Manager	X	X	X	X		X		X		
Print Composer			X					X		
Print Server			X					X		
Report Creator							X	X		
Report Manager							X	X		
Report Reader					X		X	X		
Report Repository					X		X	X		
Secure Node								X		
Time Server								X		

Source: Reprinted from IHE *Radiology Technical Framework* V9.0 Vol. 1, Table 2.3-1. With permission.

c. *Order Placer:* An enterprise-wide system that generates orders for various departments and distributes those orders to the correct department.
d. *Department System Database:* A departmental database-based information system (for instance, Radiology) which stores all relevant data about patients, orders, and their results.
e. *Department System Scheduler/Order Filler:* Department-based information system that schedules resources to perform procedures according to orders it receives from external systems or through the user interface.
f. *External Report Repository Access:* Performs retrieval of clinical reports containing information generated from outside the Radiology department and presented as DICOM Structured Reporting Objects.
g. *Image Archive:* A system that provides long-term storage of images and presentation data.
h. *Image Creator:* A system that creates additional images and/or Grayscale Softcopy Presentation States and transmits the data to an Image Archive. It also makes requests for storage commitment to the Image Manager for the images and/or Presentation States previously transmitted and generates Performed Procedure Steps.
i. *Image Display:* A system that offers browsing of patients' studies with a series of images. In addition, it supports the retrieval of selected set of images and their presentation characteristics specified by modality (size, color, annotations, layout, etc.).
j. *Image Manager:* A system that provides functions related to safe data storage and image data handling. It supplies image availability information to the Department System Scheduler.
k. *Master Patient Index:* A system that maintains unique enterprise-wide identifier for a patient.
l. *Performed Procedure Step Manager:* A system that provides a service of redistributing the Modality Performed Procedure Step information from the Acquisition Modality or Image Creator to the Department System Scheduler/Order Filler and Image Manager actors.
m. *Print Composer:* A system that generates DICOM print requests to the Print Server Actor. Print requests include presentation state information in the form of Presentation Look-Up Tables (Presentation LUTs).
n. *Print Server:* A system that accepts and processes DICOM print requests as a DICOM Print SCP and performs image rendering on hardcopy media. The system must support pixel rendering according to the DICOM Grayscale Standard Display Function.
o. *Report Creator:* A system that generates and transmits draft (and optionally, final) reports, presenting them as DICOM Structured Reporting Objects.
p. *Report Manager:* A system that provides functions related to report management. This involves the ability to handle content and state changes to reports and to create new DICOM Structured Reporting Objects based on these changes.
q. *Report Reader:* A system that can query/retrieve and view reports presented as DICOM Structured Reporting Objects.
r. *Report Repository:* A system that provides long-term storage of reports and their retrieval as DICOM Structured Reporting Objects.

Moreover, Chapter 3 also lists the following Transactions:

1. *Patient Registration:* The patient is registered/admitted. This will generate a visit or encounter event as well as a registration event if the patient is not preexisting.
2. *New Order:* An order is created via an order entry system (Order Placer); an order may contain procedures that cross multiple departments. Department-specific orders/ procedures are forwarded to the appropriate department. The Order Filler informs an Order Placer about the order's status changes. An order may also be generated by the Order Filler in a department and submitted to the Order Placer.
3. *Order Cancel:* A previously placed order is terminated or changed. Either the Order Placer or the Departmental System Scheduler/Order Filler may need to change order information or cancel/discontinue an order. When order information change is necessary, the IHE Technical Framework: Year 2 requires that initiator cancel the order and generate the new one using new information. All systems that are aware of the order are informed of the change, including the Image Manager if the order has been scheduled as one or more procedures.
4. *Procedure Scheduled:* Schedule information is sent from the DSS/Order Filler to the Image Manager.
5. *Modality Worklist Provided:* Based on a query entered at the Acquisition Modality, a modality worklist is generated listing all the items that satisfy the query. This list of Scheduled Procedure Steps with selected demographic information is returned to the Acquisition Modality.
6. *Modality Procedure Step In Progress:* An Acquisition Modality notifies the Performed Procedure Step Manager of a new Procedure Step.
7. *Modality Procedure Step Completed:* An Acquisition Modality notifies the Performed Procedure Step Manager of the completion of a Procedure Step.
8. *Modality Images Stored:* An Acquisition Modality requests that the Image Archive store acquired or generated images.
9. *Modality Presentation State Stored:* An Acquisition Modality requests that the Image Archive store the Grayscale Softcopy Presentation State (GSPS) for the acquired or generated images.
10. *Modality Storage Commitment:* An Acquisition Modality requests that the Image Manager take responsibility for the specified images and/or GSPS objects the Acquisition Modality stored.
11. *Images Availability Query:* The Department System Scheduler/Order Filler asks the Image Manager if a particular image or image series is available.

12. *Patient Update:* The ADT Patient Registration System informs the Order Placer and the Department System Scheduler/Order Filler of new information for a particular patient. The Department System Scheduler may then further inform the Image Manager.
13. *Procedure Update:* The Department System Scheduler/Order Filler sends the Image Manager updated order or procedure information.
14. *Query Image:* An Image Display provides a set of criteria to select the list of entries representing images by patient, study, series, or instance known by the Image Archive.
15. *Query Presentation State:* An Image Display provides a set of criteria to select the list of entries representing image Grayscale Softcopy Presentation States (GSPS) by patient, study, series, or instance known by the Image Archive.
16. *Retrieve Images:* An Image Display requests and retrieves a particular image or set of images from the Image Archive.
17. *Retrieve Presentation States:* An Image Display requests and retrieves the Grayscale Softcopy Presentation State (GSPS) information for a particular image or image set.
18. *Creator Images Stored:* An Image Creator requests that the Image Archive store new images.
19. *Creator Presentation State Stored:* An Image Creator requests that the Image Archive store the created Grayscale Softcopy Presentation State objects.
20. *Creator Procedure Step In Progress:* An Image Creator notifies the Performed Procedure Step Manager of a new Procedure Step.
21. *Creator Procedure Step Completed:* An Image Creator notifies the Performed Procedure Step Manager of the completion of a Procedure Step.
22. *Creator Storage Commitment:* An Image Creator requests that the Image Manager take responsibility for the specified images and/or GSPS objects that the Creator recently stored.
23. *Print Request with Presentation LUT:* A Printer Composer sends a print request to the Print Server specifying Presentation LUT information.
24. *Report Submission:* A Report Creator sends a draft or final report to the Report Manager.
25. *Report Issuing:* A Report Manager sends a draft or final Report to the Report Repository.
26. *Query Report:* A Report Reader provides a set of criteria to select the list of entries representing reports by patient, study, series, or report known by the Report Repository or External Report Repository Access.
27. *Retrieve Report:* A Report Reader requests and retrieves a report from the Report Repository or External Report Repository Access.

5.2.1.3 Year 3

In its third year, IHE had grown to the point where the documentation was split into two documents: Part A and B (IHE Radiology Technical Framework V5.0). The concepts (and in fact language) of the Transaction and Data Models were largely unchanged from the prior year. However, Chapter 3 in this version not only introduces the new Actors and Transactions invented in this version, but also a new term, Integration Profiles. The first seven Integration Profiles were

a. *Scheduled Workflow (SWF):* Specifies transactions that maintain the consistency of patient and ordering information as well as providing the scheduling and imaging acquisition procedure steps.
b. *Patient Information Reconciliation (PIR):* Extends the *Scheduled Workflow Integration Profile* by offering the means to match images acquired of an unidentified patient (e.g., during a trauma case) or misidentified patient.
c. *Consistent Presentation of Images (CPI):* Specifies a number of transactions that maintain the consistency of presentation for grayscale images and their presentation state information (including user annotations, shutters, flip/rotate, display area, and zoom). It also defines a standard contrast curve, the Grayscale Standard Display Function.
d. *Access to Radiology Information (ARI):* Specifies a number of query transactions providing access to radiology information, including images and related reports, in a DICOM format as they were acquired or created.
e. *Key Image Notes (KIN):* Specifies transactions that allow a user to mark one or more images in a study as significant by attaching to them a note managed together with the study.
f. *Simple Image and Numeric Reports (SINR):* Facilitates the growing use of digital dictation, voice recognition, and specialized reporting packages, by separating the functions of reporting into discrete actors for creation, management, storage, and viewing.
g. *Presentation of Grouped Procedure (PGP):* Provides a mechanism for facilitating workflow when viewing images and reporting on individual requested procedures that an operator has grouped (often for the sake of acquisition efficiency and patient comfort) into a single acquisition.

The incremental Actors are

a. *Department System Database:* Was deprecated and is no longer in use.
b. *Enterprise Report Repository:* A system that stores Structured Report Export Transactions from the Report Manager.

And the Transactions incurred the following changes and additions:

1. *Placer Order Management:* Replaces the old New Order transaction.
2. *Filler Order Management:* Replaces the old Order Cancel.
3. The old Creator Storage Commitment was deprecated and replaced with nothing.
4. *Structured Report Export (New):* A Report Manager composes an HL7 Result transaction by mapping from DICOM SR and transmits it to the Enterprise Report Repository for storage.

5. *Key Image Note Stored (New):* An Acquisition Modality or an Image Creator sends a Key Image Note to the Image Archive.
6. *Query Key Image Notes (New):* An Image Display queries the Image Archive for a list of entries representing Key Image Notes by patient, study, series, or instance.
7. *Retrieve Key Image Note (New):* An Image Display requests and retrieves a Key Image Note from the Image Archive.

5.2.2 Year 4

5.2.2.1 New Radiology Integration Profiles, Actors, and Transactions

In Year 4, the final version of the Radiology Technical Framework was published as V5.4 (December 2002). Both, V5.4 and the V5.3, drafts were published in four-volume sets; furthermore, both were standardized on a format that has continued in the Radiology Technical Frameworks up to the present: Volume 1 lays out the Integration Profiles, Volumes 2–3 lay out the required Transactions in detail, and Volume Four details extensions to the Framework for international audiences (starting with European nations). Also, the mini-index of Actors, Transactions, and Integration Profiles (formerly found in Chapter 3) has been moved to Volume 1, Chapter 2. Another innovation in Year 4 is the use of summary tables that track the relationship between Actors and the Integration Profiles they must support, and another table that tracks Integration Profiles and the Transactions they implement (see Tables 5.2 and 5.3). The remaining chapters in Volume 1 detail each Integration Profile.

The following new Integration Profiles were defined

a. *Basic Security (SEC):* Provides institutions the ability to consolidate audit trail events on user activity across several systems interconnected in a secured manner.
b. *Charge Posting (CHG):* Describes standardized messages sent from the Order Filler to describe charges for procedures.
c. *Postprocessing Workflow (PWF):* Describes mechanisms to automate the distributed postprocessing of images, such as 3-D Reconstruction and Computer Aided Detection (CAD).

The following Actors were added

a. *Audit Record Repository:* A system unit that receives and collects audit records from multiple systems.
b. *Charge Processor:* Receives the posted charges and serves component of the financial system. Further definition of this actor is beyond the current IHE scope.
c. *Postprocessing Manager:* A system that provides functions related to postprocessing worklist management. This involves the ability to schedule postprocessing worklist items (scheduled procedure steps), provide worklist items to postprocessing worklist clients, and update the status of scheduled and performed procedure steps as received from postprocessing worklist clients.
d. *Secure Node:* A system unit that validates the identity of any user and of any other node, and determines whether or not access to the system for this user and information exchange with the other node is allowed. Maintains the correct time.
e. *Time Server:* A server that maintains and distributes the correct time in the enterprise.

And the following incremental Transactions were defined

1. *Authenticate Node:* Any two actors exchange certificates in order to validate the identity of another node.
2. *Maintain Time:* Synchronize the local time with the time maintained by the Time Server.
3. *Record Audit Event:* Create and transmit an Audit Record.
4. *Charge Posted:* The Department System Scheduler/Order Filler sends descriptions of potential procedure and material changes.
5. *Account Management:* The ADT Patient Registration Actor informs Charge Processor about creation, modification, and ending of patient's account.
6. *Worklist Provided:* Based on a query from a worklist client (Image Creator), a worklist is generated by the worklist manager (Postprocessing Manager) containing either Postprocessing or CAD workitems that satisfy the query. Workitems are returned in the form of a list of General Purpose Scheduled Procedure Steps.
7. *Workitem Claimed:* A worklist client (Image Creator) notifies the worklist provider (Postprocessing Manager) that it has claimed the workitem (i.e., started a General Purpose Scheduled Procedure Step).
8. *Workitem PPS In Progress:* A worklist client (Image Creator) notifies the worklist provider (Postprocessing Manager) that it has started work (i.e., created a General Purpose Performed Procedure Step).
9. *Workitem PPS Completed:* A worklist client (Image Creator) notifies the worklist provider (Postprocessing Manager) of the completion of a General Purpose Performed Procedure Step.
10. *Workitem Completed:* A worklist client (Image Creator) notifies the worklist provider (Postprocessing Manager) that it has finished the workitem (i.e., completed a General Purpose Scheduled Procedure Step).
11. *Performed Work Status Update:* The worklist provider informs other interested actors of the on-going status and completion of the performed work.
12. *Evidence Document Stored:* An Acquisition Modality or Image Creator sends measured or derived diagnostic evidence in the form of a DICOM Structured Report to the Image Archive.
13. *Query Evidence Documents:* An Image Display queries the Image Archive for a list of entries representing Evidence Documents.
14. *Retrieve Evidence Documents:* An Image Display requests and retrieves an Evidence Document from the Image Archive.

TABLE 5.3 The Relation between Integration Profiles and the Transactions They Must Implement

Integration Profile Transaction	SWF	PIR	CPI	PGP	ARI	KIN	SINR	SEC	CHG	PWF
Patient Registration	X	X								
Placer Order	X	X								
Filler Order	X	X								
Procedure Scheduled	X	X		X						
MWL Provided	X	X		X						
MPS In Progress	X	X	X	X		X				
MPS Completed	X	X	X	X		X				
Mod. Images Stored	X	X	X	X						
Mod. Pres. Stored				X						
Storage Commitment	X	X	X	X		X				X
Images Avail. Query	X	X		X						X
Patient Update	X	X								
Procedure Update										
Query Images	X		X	X	X	X				X
Query Pres. States			X	X	X	X				
Retrieve Images	X		X	X	X	X				X
Retrieve Pres. States			X	X	X	X				
Creator Img. Stored	X		X	X						X
Creator Pres. Stored			X	X						
CPS In Progress	X		X	X		X				
CPS Complete	X		X	X		X				
Print Request, LUT			X							
Report Submission							X			
Report Issuing							X			
Query Reports					X		X			
Retrieve Reports					X		X			
Struct. Report Export							X			
KIN Stored						X				
Query KIN						X				
Retrieve KIN						X				
Authenticate Node								X		
Maintain Time								X		
Record Audit Event								X		
Charge Posted									X	
Account Mgmt									X	
Worklist Provided										X
Workitem Claimed										X
Workitem Completed										X
Workitem PPS In-Progress										X
Workitem PPS Completed										X
Performed Work Status Update	X									X
Evidence Documents Stored										X
Query Evidence Documents										X
Retrieve Evidence Documents										X

Source: Reprinted from IHE *Radiology Technical Framework* V9.0 Vol. 1, Table 2.4-1. With permission.

5.2.3 Year 5

5.2.3.1 New Radiology Integration Profiles, Actors, and Transactions

Version V5.5 (ratified in November 2003) follows the formatting and structure of the prior year (IHE Radiology Technical Framework V5.5). There are again four volumes with the same overall naming and content conventions, and Volume 1, Chapter 2 once again has the index of all content in this version.

Integration Profiles have been grouped in a new hierarchy: Workflow, Content, and Infrastructure. This action foreshadows the eventual segregation of Integration Profiles into their own separate Technical Frameworks, as we shall see. The only new Integration Profiles in this version are

a. *Reporting Workflow (RWF):* Addresses the need to schedule, distribute, and track the status of the reporting workflow tasks such as interpretation, transcription, and verification.
b. *Evidence Documents (ED):* Defines interoperable ways for observations, measurements, results, and other procedure details recorded in the course of carrying out a procedure step to be output by devices.

Actor changes in this version include

a. *Evidence Creator:* the new name for the Actor formally known as Image Creator to clarify that it is used to create more than just images.

Transactions were also updated, and their growing number led to them being given numeric aliases in addition to their verbal name. The following itemizes the new Transactions and their assigned numeric alias:

1. *Query Postprocessing Worklist (37):* A worklist is generated by the worklist manager (Postprocessing Manager) containing either Postprocessing or CAD workitems that satisfy the query. Workitems are returned in the form of a list of General Purpose Scheduled Procedure Steps.
2. *Query Reporting Worklist (46):* A query from a Report Creator worklist client, a worklist is generated by the Report Manager containing reporting task workitems that satisfy the query.

5.2.4 Year 6

5.2.4.1 Creation of New Domains

The documentation for Year 6 was not ratified until May 2005, about 18 months after the previous version (IHE Radiology Technical Framework V6.0). In the interim, many new concepts had been considered and a major conceptual advance that was occurring in this period was the introduction of *domains* other than Radiology. Coinciding with the adoption surge of IHE in many other professional societies (e.g., American College of Cardiology and Laboratory Healthcare Partnership), IHE was also gaining ground in Europe and Asia. The result is that multiple Domains began to be defined: Radiology (RAD) was the first, but was joined by others including Information Technology Infrastructure (ITI). Also added from 2003 to 2005:

- IHE Cardiology—2003.
- IHE Eye Care—2005.
- IHE Technical Infrastructure (ITI)—2004 While inaugurated in late 2004, ITI came too late to be reflected in the Year 5 Radiology Technical Framework. Hence, the first mention of it in the Radiology document is in V6.0 in 2005.
- IHE Laboratory—2003.
- IHE Patient Care Coordination (PCC)—2005.
- IHE Patient Care Devices (PCD)—2005.

The upshot of this increased specificity drove the adoption of more specific naming conventions for IHE Actors and Transactions. For instance, until Year 5 one could only refer to a Transaction by its verbal name; for example, "Procedure Scheduled." Since only one Technical Framework (Radiology) existed for one domain (radiology), there was no ambiguity and it was universally understood that the speaker was referring to the "Radiology Technical Framework Procedure Scheduled" transaction. After Year 5, Transactions began to be numbered, hence one could say Transaction 4 (the number assigned to Procedure Scheduled) and informed listeners would be aware of the intent. However, there is now a more structured reference style: *Domain_code* TF-*volume*: *section* (where TF mean Technical Framework). An explicit example may help:

ITI TF-1: 3.1 represents "Information Technology Infrastructure Technical Framework Volume 1: Section 3.1. Similarly, a precise reference to the aforementioned "Procedure Scheduled" transaction would be encoded with "RAD-4"; a succinct (if cryptic) notation alluding to the Transaction being described in the Radiology Technical Framework.

5.2.4.2 Radiology Profiles Migrated to ITI Domain

The following Integration Profiles were moved into the ITI domain in V1.1 of the ITI Technical Framework:

a. *Basic Security (SEC):* was moved and renamed the Radiology Audit Trail Option on ITI-Audit Trail and Node Authentication (ATNA). The Time and Secure Actors accompanied this move to the ITI Technical Framework. Also, Transactions RAD 32–34 were deprecated.

5.2.4.3 New Radiology Integration Profiles, Actors, and Transactions

Two new Integration Profiles were introduced

a. *Nuclear Medicine Image (NM):* Specifies how Acquisition Modalities and workstations should store NM Images and how Image Displays should retrieve and make use of them.

b. *Portable Data for Imaging (PDI):* Specifies actors and transactions that allow users to distribute imaging-related information on interchange media.

Two new Actors were defined to accompany the PDI Profile:

a. *Portable Media Creator:* Assembles the content of the media and writes it to the physical medium.
b. *Portable Media Importer:* Reads the DICOM information contained on the media, and allows the user to select DICOM instances, reconcile key patient and study attributes, and store these instances.

The new Transactions include

1. *Distribute Imaging Information on Media (RAD-47):* A standard format for representing images and reports on portable media such as compact disks.
2. *Appointment Notification (RAD-48):* The DSS/OF sends the Order Placer the date and time of for one or more Scheduled Procedure Steps.
3. *Instance Availability Notification (RAD-49):* The Image Manager informs the DSS/OF and others of the availability status of instances at specific storage locations.

5.2.5 Year 7

5.2.5.1 Creation of Other Domains

In May 2006, Year 7 of IHE documentation was ratified in the form of V7.0. The documentation set consisted of Volumes 1–3; hence, the full accurate reference to a given volume would be "IHE Year 7 RAD TF-1": (References to a specific section would follow the colon.) No new domains were created.

5.2.5.2 Radiology Profiles Migrated to ITI domain

Scope changes to existing Radiology Integration Profiles were highlighted in RAD TF-1: 1.7. No further Integration Profiles were migrated to the ITI domain (IHE Radiology Technical Framework V7.0).

5.2.5.3 New Radiology Integration Profiles, Actors, and Transactions

As in the prior several releases, Chapter 2 (RAD TF-1: 2) served as an index to the remaining volumes. No new Integration Profiles or Transactions were defined.

5.2.6 Year 8

5.2.6.1 Creation of Other Domains

In Year 8, V8.0 was ratified in June 2007. The documentation expanded to four volumes in this release (IHE Radiology Technical Framework V8.0). The following new domains were created in this issue:

- IHE Quality, Research and Public Health Domain (QRPH)—2007.
- IHE Radiation Oncology (RO)—2007.

5.2.6.2 Radiology Profiles Migrated to ITI domain

None.

5.2.6.3 New Radiology Integration Profiles, Actors, and Transactions

The new Integration Profiles created in this release included

a. *Cross-Enterprise Document Sharing for Imaging (XDS-I):* Specifies actors and transactions allowing users to share across enterprises sets of DICOM instances.
b. *Mammography Image (MAMMO):* Specifies how DICOM Mammography images and evidence objects are created, exchanged, and used.
c. *Import Reconciliation Workflow (IRWF):* Specifies actors and transactions that allow users to share imaging information across enterprises.

New Actors:

a. Importer (for the Charge Posting Profile):

And the new Transactions introduced are

1. *Provide and Register Imaging Document Set (RAD-54):* For each document in the Submission Set, the Imaging Document Source actor provides both the documents as an opaque octet stream and the corresponding metadata to the Document Repository.
2. *WADO Retrieve (RAD-55):* Issued by an Imaging Document Consumer to an Imaging Document Source to retrieve DICOM objects over HTTP/HTTPS protocol.
3. *Import Procedure Step In Progress (RAD-59):* The Performed Procedure Step Manager receives progress notification of an importation Procedure Step and in turn notifies the Order Filler, Image Manager, and the Report Manager.
4. *Import Procedure Step Completed (RAD-60):* The Performed Procedure Step Manager receives completion notification of an importation Procedure Step and in turn notifies the Order Filler, Image Manager, and the Report Manager.
5. *Imported Objects Stored (RAD-61):* A system importing DICOM Objects or digitized hardcopy sends imported DICOM Composite Objects to the Image Archive.

5.3 Current Status

5.3.1 Year 9

5.3.1.1 Creation of Other Domains

In Year 9, Version 9.0 of the Radiology Technical Framework was ratified in June 2008 (IHE Radiology Technical Framework V8.0). The Anatomic Pathology domain was formed in 2008 with the ratification of V1.2 of its Technical Framework and defines the Pathology Workflow Integration Profile.

5.3.1.2 Radiology Profiles Migrated to IT Domain

None.

5.3.1.3 New Radiology Integration Profiles, Actors, and Transactions

The following Integration Profile was added to the Technical Framework

a. *Teaching File and Clinical Trial Export (TCE):* Provides for selecting images and related information for the deidentification and export to systems that author and distribute teaching files or receive information for clinical trials.

In an attempt to become more responsive to needs that arise outside of the annual publication cycle of the large Technical Frameworks, the concept of IHE Supplements has been created. These introduce new Profiles, Actors, or Transactions in-between the annual publications of the Framework. The supplemental Profiles added to the 2008–2009 testing cycle included

a. *Cross-Enterprise Document Sharing for Imaging (XDS-I.b):* An enhancement to XDS-I that enables Web services.
b. *Basic Image Review (BIR):* Adds a basic image viewer to the *PDI* Profile.
c. *MR Diffusion Imaging (MDI):* A Profile that includes the workflow for MR diffusion imaging.
d. *CT/MR Perfusion Imaging with Contrast (PIC):* Enhances the RAD-8 and RAD-16 Transaction to handle enhanced DICOM MR/CT objects.
e. *Mammography Acquisition Workflow (MAWF).*
f. *Radiation Exposure Monitoring (REM):* Facilitates the collection and distribution of information about estimated patient radiation exposure resulting from imaging procedures.
g. *Image Fusion (FUS):* Addresses the ability to convey registered data from one system to another for further processing, storage, and display, and also the ability to present repeatable fused displays consisting of a grayscale underlying image and a pseudocolor overlay image.

5.3.2 Year 10

5.3.2.1 Creation of Other Domains

None at the time of this writing in spring 2010.

5.3.2.2 Radiology Profiles Migrated to IT Domain

As of this writing, V10.0 of the Framework is yet to be ratified. However, there seems to be some momentum toward deprecating the Radiology Portable Data for Imaging (PDI) Profile and migrating to the more general ITI Cross-Enterprise Document Media Interchange (XDM) Profile which provides document interchange using a common file and directory structure on common media formats. This permits the patient to use physical media, as well as e-mail, to transport medical documents.

5.3.2.3 New Radiology Integration Profiles, Actors, and Transactions

Nothing new has been added in the pending Framework version.

5.3.3 Aids to Learning

By now, the reader should be familiar with the pattern IHE takes: define an expertise domain, define the Use Cases/Actors/workflows in that domain, define the Transactions that enable those workflows, and finally summarize the result in an Integration Profile. A useful method to visualize this hierarchy is to imagine a tree whose trunk is the domain, the main branches are Actors, the second-order branches are the Integration Profiles those Actors must implement, and the leaves are the Transactions. Figure 5.1 demonstrates this.

The author prepared a Web site based upon the forgoing concepts to aid programmers in learning the data requirements for developing the software to implement IHE Actors at his site. The result was shown at several international informatics meetings (Langer and Persons, 2005, 2008). It is also on-line for those interested (IHE Web V2.0).

5.3.4 Enterprise Issues

5.3.4.1 Point-to-Point Scaling Issues: DICOM and HL7

At this point, the reader has amassed the historical of view of how the IHE Radiology Technical Framework has evolved. A key point to bear in mind is that the underlying protocols used in most of the Profiles to date are HL7 and DICOM. These are point-to-point protocols that require effort at both ends of a connection to create it and maintain it. The simplified diagram (Figure 5.2) shows how tangled such an approach can become with relatively few Actors.

As can be seen, it does not take long before the complexity becomes costly, error prone, and virtually unmanageable to maintain. A better approach is needed.

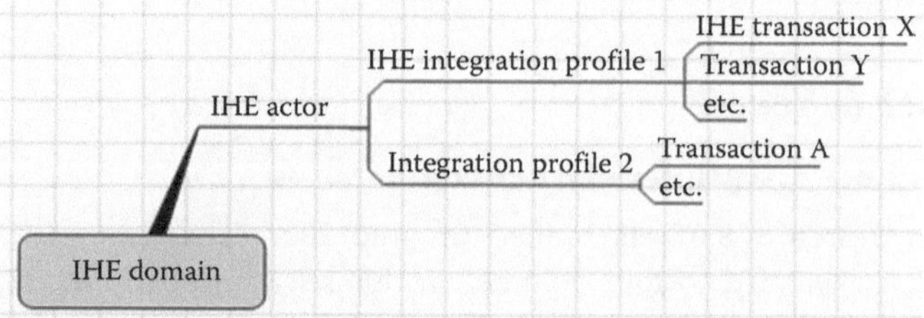

FIGURE 5.1 "IHE for the Impatient" Web tutorial example. The live site for V2.0 traces the Actors, Implementation Profiles and required Transaction for two domains: Radiology (RAD) and Information Technology Infrastructure (ITI).

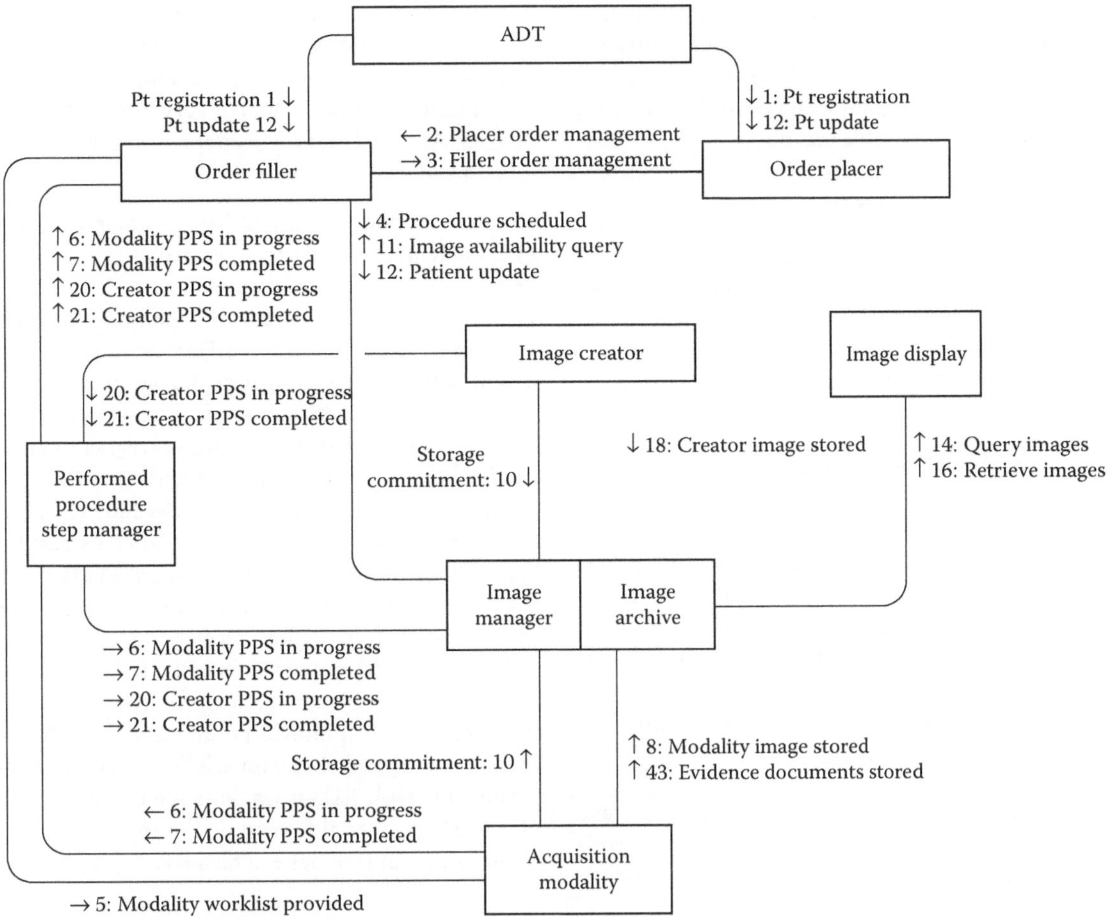

FIGURE 5.2 The nature of point-to-point interfaces for the simple case of Scheduled Workflow Integration Profile. One can see the complexity of the mesh increases geometrically with the number of Actors. (Reprinted from IHE *Technical Framework V9.0* Vol. 1, Figure 3.1-1, With permission.)

5.3.4.2 Moving to a Subscription Model: SOA and XML

The "scaling issue" was noted by informaticists about 20 years ago; a technology named CORBA (Common Object Request Broker Architecture) was invented in 1991 to solve it (CORBA, 2004). It was also applied to medical imaging via the extension known as CORBAmed (2007). With CORBA, application software programmers learned a new paradigm—that of publishing results to an ORB (Object Request Broker). Interested parties would "subscribe" to the ORB for messages of interest to them. CORBA continues to be used today, but largely through an accident of historical timing, it has been largely eclipsed.

In 1993, Tim Berners-Lee was working for the Center for European Research Nuclear (CERN) and was faced with a problem; how could he share high-energy physics data, figures, and plots with physicists throughout the world? He could have simply continued to set up a file server and allowed users to connect to it via file-sharing tools like ftp (file transfer protocol); a slow and tedious process where users do not know what they are getting until they retrieve it and view it locally. Instead, he envisioned and created the World Wide Web; a client–server model that enabled users to see the text, images, and other content immediately in their Web browser (Berners-Lee, 1996). Initially, the WWW was based on two key technologies: HTTP (Hyper Text Transfer Protocol) was the protocol that transferred data among clients and their servers, and the content was rendered in the HTML (Hyper Text Markup Language) standards (see Chapter 2 for more detail). However, HTML could only specify "how" to render a page; it could not specify the "what" contained in the page. Enter XML (eXtensible Markup Language); it extended HTML with the ability to self-document the content of pages.

The invention of XML immediately stimulated the developer community to consider ways to use it to perform point-to-point, and later publisher–consumer models of communicating among Actors. A conceptual abstraction to the approach taken by CORBA for the publishing-consumer model called SOA (Service Oriented Architecture) was developed. It is not tied to any one transport service (and indeed can use CORBA), but it has largely come to be almost synonymous with XML and Web services for performing SOA applications over the WWW (SOA, 2008a,b). By *abstracting* services from the underlying programming languages and operating systems, it is trivial to have a client written in Java (Sun Microsystems, Santa Clara, CA) "consume" data from a C# application running on a .NET server (Microsoft Corporation, Redmond WA) and a legacy C program wrapped in Web services. Figure 5.3 shows conceptually how this is achieved.

5.3.4.3 Affinity Domains and XDS

Finally, we are in a position to see and appreciate the future directions of IHE. Recall that at its inception, IHE was created to enable workflow between departments in the same hospital. The bulk of this chapter has been dedicated to following the evolution of a single department—Radiology—in that journey. The new objectives are to enable:

1. Patient care between different healthcare centers, using the same patient identifier but perhaps different ordering systems (i.e., two different Radiology departments using different exam identifiers). This arrangement can be considered an Affinity Domain of the first kind.
2. Patient care between different healthcare centers even if those centers use different identifiers for the patient, different ordering systems, and different exam identifiers; this cooperative care arrangement can be named an Affinity Domain of the second kind.

To see how these scenarios will be accomplished, we must leave the familiar confines of the Radiology Technical Framework and look toward several ITI Integration Profiles. The primary Profiles for this goal are

a. *EUA (End User Authentication):* This Profile fulfills the goals of the authenticator shown in Figure 5.3.
b. *PIX (Patient Identifier Cross-referencing):* Creates a Master Patient Index (MPI) that maps the various identifiers a patient is known by across a collection of healthcare centers to a single, common identifier.
c. *XDS (Cross-Enterprise Document Sharing):* Enables a number of healthcare delivery organizations belonging to an Affinity Domain to cooperate in the care of a patient by sharing clinical documents.

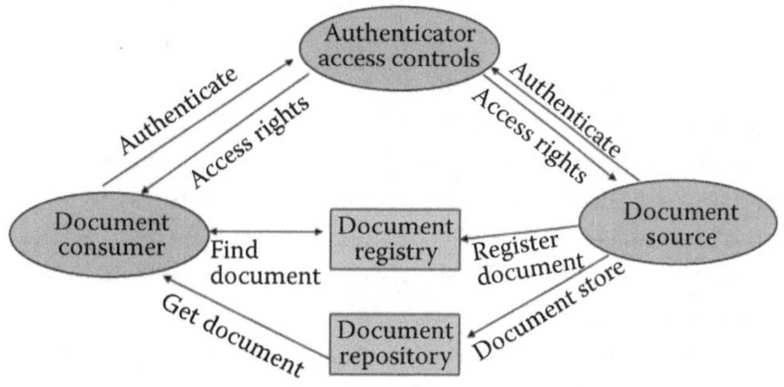

FIGURE 5.3 Web services and SOA architectures allow interested parties (consumers) to subscribe to well-understood sources (repositories). Central controls enforce proving a consumer's identity (authentication) and once proven regulating what the consumer can see (authorization). It is obvious that this model can be expanded to an arbitrary number of sources and consumers, each built on an arbitrary operating system, language, and Web server.

EUA is necessary for a healthcare worker to log in with a single set of credentials that can be checked against an enterprise-wide identity and set of access permissions. With XDS (which also requires PIX), a healthcare worker can pull the medical records from any source in the Affinity Domain, and the records for that patient will have the identifiers the worker expects for that site in the site's informatics system. An analogous Profile, XDS-I (where the "I" means imaging) was created to enable the same functionality for imaging departments.

XDS was initially introduced in the 2005 ITI TF-1, and at that point it could not fully realize all the goals outlined above. In 2006, ITI TF-2 portended the creation of an extension that would be called the "XDS Domain Federation Integration Profile" that was hinted to use further services to enable order tracking, problem lists, and enhanced security. In brief, the designers had begun to realize that the scaling problem would be insurmountable with the old point-to-point standbys: DICOM and HL7. There was a brief version named XDS.a that was quickly replaced in late 2007 with XDS.b which is the current designation (IHE XDS.b, 2010). With this version, the embrasure of modern SOA methods within IHE has established a solid precedent.

5.4 Summary

We have followed the development of IHE from its inception as a Radiology-based endeavor to communicate more broadly within a single medical center, to the expansion of multiple domains (cardiology, lab, pathology, etc.) among loosely federated medical centers scattered across large distances (Affinity Domains). In this evolution, some things have changed: some Profiles have been moved from the Radiology domain to the IT domain, old standards have been augmented (DICOM and HL7), and new ones embraced (SOA with XML and Web services). But some founding concepts remain:

a. Pick a domain.
b. Identify the workflows within it.
c. Model the workflows with Use Cases.
d. Identify the atomic Actors within the Use Case.
e. Model the Transaction data that needs to be exchanged to enable the workflow.
f. And finally look for standards that can be leveraged to accomplish (e).

With the scaling issue addressed, the limits of the IHE approach are bound only by its adoption—which is largely up to the customer base.

References

Berners-Lee, T. WWW: Past, present and future. *IEEE Computer Magazine*, 29(10), 1996.

Channin, D.S. 2000. M:I-2 and IHE: Integrating the Healthcare Enterprise, year 2. *Radiographics*, 20(5), 1261–62.

Channin, D.S., Pariso, C., Wanchoo, V., Leontiev, A., and Siegel, E.L. 2001. Integrating the Healthcare Enterprise, a primer: Part 3, What does IHE do for me? *Radiographics,* 21(5), 1351.

Connectathon. 2010. Available at http://www.ihe.net/News/ihe-na-connectathon_pr_2010-01-25.cfm. Accessed May 2011.

Connectathon. 2011. Available at http://www.ihe.net/Connectathon/#. Accessed May 2011.

CORBA. 2004. Available at http://en.wikipedia.org/wiki/Common_Object_Request_Broker_Architecture#Overview. Accessed May 2011.

CORBAmed 2007. Available at http://www.i2-health.org/wiki-container/CORBAmed. Accessed May 2011.

Dreyer, K. J. 2000. Why IHE. *Radiographics*, 20, 1583–84.

GE IHE Conformance Statement. 2006. Available at http://www.gehealthcare.com/usen/interoperability/docs/IHE_IS_PACS_3_2017753_204r1_3.pdf. Accessed May 2011.

IHE. 1997. Available at http://ihe.net. Accessed May 2011.

IHE FTP site. Available at ftp://ftp.ihe.net/Radiology/TF_Final_Text_Versions. Accessed May 2011.

IHE *Radiology Technical Framework V4.0*. Available at ftp://ftp.ihe.net/Radiology/TF_Final_Text_Versions/v4.0/. Accessed May 2011.

IHE *Radiology Technical Framework V5.0*. Available at ftp://ftp.ihe.net/Radiology/TF_Final_Text_Versions/v5.0/. Accessed May 2011.

IHE *Radiology Technical Framework V5.5*. Available at ftp://ftp.ihe.net/Radiology/TF_Final_Text_Versions/v5.5/. Accessed May 2011.

IHE *Radiology Technical Framework V6.0*. Available at ftp://ftp.ihe.net/Radiology/TF_Final_Text_Versions/v6.0/. Accessed May 2011.

IHE *Radiology Technical Framework V7.0*. Available at ftp://ftp.ihe.net/Radiology/TF_Final_Text_Versions/v7.0/. Accessed May 2011.

IHE *Radiology Technical Framework V8.0*. Available at ftp://ftp.ihe.net/Radiology/TF_Final_Text_Versions/v8.0/final/. Accessed May 2011.

IHE *Radiology Technical Framework V9 Vol.1*. Available at ftp://ftp.ihe.net/Radiology/TF_Final_Text_Versions/v9.0/final/ Accessed May 2011.

IHE XDS.b. 2010. *ITI V6.0 TF-1 Section 10.7*. Available at http://www.ihe.net/Technical_Framework/index.cfm#IT. Accessed May 2011.

Langer, S.G. and Persons, K. 2005. *IHE for the Impatient.* Orlando, FL: SCAR.

Langer, S.G. and Persons, K. 2008. *IHE for the Impatient V2.* Washington: SIIM, Seattle.

Langer, S.G. *IHE Web V2.0*. Available at http://iheweb.nfinite-horizons.org/. Accessed May 2011.

SIIM. Available at http://siimweb.org/. Accessed May 2011.

Smedema, K. 2000. Integrating the Healthcare Enterprise (IHE): The radiological perspective. *Medica Mundi*, 44(1), 39–47.

SOA. 2008a. Available at http://en.wikipedia.org/wiki/Service-oriented_architecture. Accessed May 2011.

SOA. 2008b. Available at http://en.wikipedia.org/wiki/Web_Services. Accessed May 2011.

III

Key Technologies

6

Operating Systems

Christos Alexakos
University of Patras

George C. Kagadis
University of Patras

6.1 Introduction

Current computer systems—workstations, servers, and mobile devices—are built upon a variety of hardware parts and external devices. Each computer system may consist of multiple processors, memory, hard disk drives, network cards or be connected with input/output (I/O) devices such as monitor, keyboard, printers, and scanners. From the early years of the worldwide adoption of computers, the complex hardware infrastructure raised the necessity of a mediation layer that will allow common users to interact with any computer system in a simple way providing a layer of transparency to hardware parts and connected devices. This was the motivation for the creation of *operating systems* (OSs), which provide users a better and simpler model of a computer system and are simultaneously responsible for managing automatically and without user interference the various resources (hardware parts and devices).

A definition of operating system (OS) is provided by Silberschatz and Galvin (1994, p. 54): "An *operating system* is a program that acts as an intermediary between a user of a computer and the computer hardware." In simple words, the OS is a piece of software which is executed directly to the computers' hardware and acts as a mediator for other software applications used by users including the Graphical User Interface (GUI). OS functionality is the factor which permits a software application (i.e., Microsoft Word, Excel) to be able to be executed in computer systems with different hardware setup.

OSs evolution is closely tied with the computers' architecture on which they run. Tanenbaum (2001) presents five generations of OSs that are historically associated with the evolution of computers:

- The First Generation (1945–55) called "Vacuum Tubes and Plugboards" describes an era when computers consisted of mechanical relays or vacuum tubes and their programming included wiring up plugboards.
- The Second Generation (1955–1960) "Transistors and Batch" introduces transistors in the computers' architecture. The key fact of this time period is the appearance of programming languages as Assembly or FORTRAN whose programs were "written" on cards or magnetic tapes in order to be executed. In 1959, IBM presented the IBM 7090 computer system with the ancestors of modern OSs, Fortran Monitor System (FMS), and IBSYS, the last one based on SHARE Operating System (SOS).
- The Third Generation (1965–1980) "ICs and Multiprogramming" starts with the aspect of creating an OS which runs in computer systems with different hardware. The first attempt was IBM's OS/360 which resulted in a complex OS. At that time, new computer techniques appeared, with *multiprogramming* being the most significant. Multiprogramming allows a computer to run more than one program "simultaneously" by time-scheduling the usage of the resources (processor, memory, I/O devices). In this concept, Compatible Time Sharing System (CTSS) and MULTiplexed Information and Computing Services (MULTICS) were the first systems supporting multiple users and provided the basis for the development of next-generation OSs. Moreover, in 1969, the well-known UNIX

OS appeared, which can be considered to be the most closely related one to the modern OSs.

- The Fourth Generation (1980–Present) "Personal Computers" introduces microcomputers (today known as personal computers) whose architecture is similar to the computers currently used. The modern OSs started to provide supporting, simpler programming codes imported by keyboard and saved in digital files. Furthermore, the first user-friendly interfaces are developed starting from simple character-based command line shells such as the well-known Microsoft Disk Operating System (MS-DOS), Berkeley Software Distribution (BSD), and UNIX. In recent years, GUIs with windows, icons, and mouse cursors are supported by the OSs including today's most common OSs such as Microsoft Windows family, Linux distributions, and the MacIntosh OS (Apple Corporation, Cupertino, CA). Moreover, the appearance of small computers (embedded systems and mobile devices) leads to the development of OSs such as Windows Mobile Family (Microsoft Corporation, Redmond, WA), iPhone (Apple Corporation, Cupertino, CA), Symbian (Symbian Foundation, London, UK), and Android (Google Corporation, Mountain View, CA). All the modern OSs are based on the same functional principles and are presented in this chapter.

The aim of this chapter is to provide a short—but descriptive—presentation of the main concepts and principles on which the modern OSs work. Section 6.2 describes the main architecture of a modern OS. Section 6.3 focuses on the main tasks and responsibilities of OSs. Moreover, a short presentation of the most common OSs is included in Section 6.4. The last section presents the conclusion of the main OSs technologies as well as future trends.

6.2 Operating Systems Architecture

OSs reside on a hardware layer and have two major roles. The first one is to provide the necessary abstraction layer to the users in order to easily work with the software applications and the peripheral devices. The second is to sufficiently manage the hardware resources of a computer systems aiming to achieve the best performance for users.

The abstraction layer which an OS provides is nothing but the way common users "see" a computer system. The underlying reality is a bit different; computers consist of chips, boards, disks, keyboards, monitors, drives, and maybe, printers or scanners. All these hardware processes command in primitive machine languages which consist of digital signals and arrays of 0s and 1s as byte code. The problem can become more complex if someone considers that each vendor uses a different command set for each hardware part or device. Thus, in a world without OSs, users must know how each hardware part in the computer system works at the lowest level and every time that they want to do a job, they must give the appropriate commands composed in each resource's machine language code. This situation is reminiscent of the first generation of computers where wiring up plug boards was an obligation in order to do a mathematical calculation. The OS is the basic part of a computer system which hides from common users all aforementioned knowledge requirements and presents them a computer system model with simple entities such as hard drives to store data, monitor to see results, keyboard to import data, printers to print on a piece of paper, and so on. The main concept is that users do not need to know how a specific hardware piece works but need to know what it can do. Thus, the OSs are responsible to keep a transparent layer between the user and the hardware. In a similar way, the OSs provide an abstraction layer to the programmers in order to develop software applications using the hardware resources. Thus, the majority of the software applications are developed based on the OS on which they will get executed (software for Microsoft Windows, for UNIX, for MAC OS) and not for a specific hardware combination.

The management of a computer system's resources is the task of trying to make all the pieces of a complex system to work together. A computer system consists of one or more processors, memory, buses, expansion cards (graphic, network cards), data storage (hard disks), and other external I/O devices (keyboard, disk drives, monitor, etc.). In order to execute a job, a computer system uses the Central Process Unit (CPU) of the processor to make the calculations, the memory in order to temporarily save data which is used in the calculations, and buses to transfer data between the resources and I/O devices such as keyboard for importing data from the user, network cards to transform and transmit data through a network or graphic cards to translate the data to signals for the monitor. The role of the OS is to manage and activate these resources in order to execute a specific job whose steps are described in terms of language machine programs—that is, calculate an addition, save data to memory, read data from memory, transfer data to hard disk, and other similar commands. The operation systems identify which is the next step that must be executed, check if the associated resource is free, and then supervise its execution. Moreover, modern OSs use time-efficient algorithms in order to permit the execution of more than one job concurrently in a computer system. They arrange the usage of resources in an efficient way ensuring that many jobs will use the available resources in a specific time slot. This technique is well known as multitasking.

Figure 6.1 denotes a layer-based drawing of the main parts of a computer system, where the placement of OSs in the computer system functionality is clearly defined.

As depicted in Figure 6.1, the OS is a software which exists between the applications used by users and the actual hardware which comprises a computer system. Users can interact with a computer system with two subcategories of software, user interfaces provided by OSs, and third-party software applications developed in order to assist them to execute specific tasks in the computer system. User interface functionalities are confined to allow users to execute simple computer procedures such as read, write, delete a file from a hard disk drive, view the folder

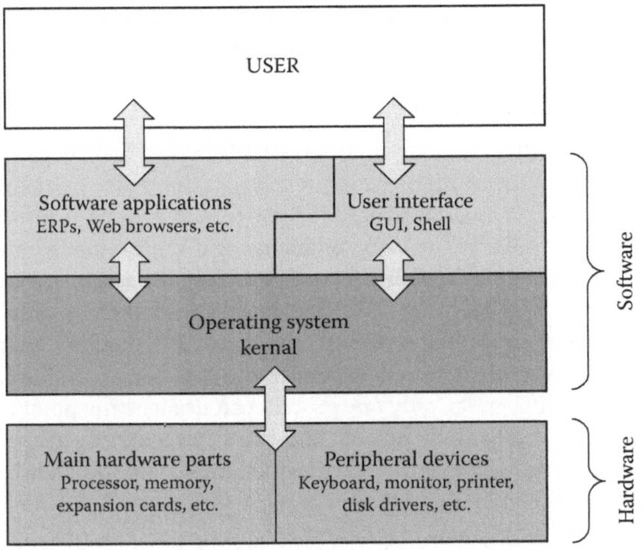

FIGURE 6.1 Role of operating systems in computer systems functionality.

structure, set user permissions, and start other programs. Common computer users will have some experience working with an OS and executing such tasks in their career. There are two types of user interfaces, the text-based shell where the users type the commands in a text form and the GUIs where the users use the mouse cursor, icons, and forms in order to give commands to the OS. The second way of user interaction consists of software applications that have been developed for executing specific tasks. Examples of such software applications are text editors, Web browsers, computer-aided design (CAD) applications, Enterprise Resource Planning (ERP) systems, CD-DVD ROM burners, and other commonly used software applications. These applications usually are composed in high-level programming languages (C++, C#, Visual Basic, Java) and consequently compiled in code which is "understandable" from a specific OS. Some software applications providing graphical user interface to users are partially developed utilizing the OS's GUI functionalities and components.

Computer users interact with the computer using the user interface of either the OS or the specific software application. This software consists of code which can be translated from the OS in order to be executed in the hardware. The basic component of an OS is the *Kernel* which is responsible for receiving a command from the software in the above layer (as depicted in Figure 6.1) and executing the appropriate procedure in the hardware (low layer). The Kernel's primary aim is to manage the hardware resources allowing other programs to run and use these resources. Wulf et al. (1974) identify as the key aspect for resource management the definition of an execution domain (memory address space) and the protection mechanism which mediates the accesses to the resources within this domain.

Software applications running in a computer system can be categorized in two major categories: (a) software running in *kernel mode* and (b) software running in *user mode*. Software running in kernel mode consists of the OS software. User mode includes software which is executed on top of the OS's kernel and it is implemented independently from the OS. OS software, running on kernel mode, provides services regarding the process management and scheduling, memory management, and resource allocation. Moreover, in some OSs, there are additional services provided by the software running in kernel mode such as device management, file management and access, multiple user support, and user-friendly interface. In other OSs, these services are provided by software running in user mode. These OSs are more flexible since these services can be developed from third-party organizations, usually for a specific purpose, and also they can be loaded dynamically during the system's operation. There are three major categories of OS kernels, referring to the mode where their services are executed

- *Monolithic Kernels* are developed in order to execute all the aforementioned services in kernel mode as one software. In the early years, some of the monolithic kernel's services were made modular in order to be developed separately from the OS, nevertheless, they are executed in the kernel mode. Known monolithic kernels are included in MS-DOS, Windows 9x series (95,98,Me), and some BSD and Solaris UNIX-like OSs.
- *Client Server Kernels (Microkernels)* separate the execution of operating system services such as device drivers, file systems, and user interface code from the kernel mode to the user mode. The rest of the operating systems services running in kernel mode are executed by the OS's *microkernel*. Representatives of microkernel-based OSs are MAC OS X, GNU Hurd, MINIX 3, and Symbian.
- *Hybrid Kernels* are a combination of monolithic kernels and microkernels. In hybrid kernels, most of the operating systems services run on kernel mode, but there is the ability for some of them to be implemented from third-party organizations and run in user mode. Microsoft Windows NT kernel, which is used in the earlier versions of Microsoft's OSs such as Windows XP, Vista, and 7, is developed on top of the Hybrid Kernel concept.

6.3 Usability and Features

As mentioned above, OSs are responsible for executing tasks assigned by the user to the computer's hardware. In order to do that, it must efficiently manage the process execution, the resource allocation, memory access, and file access. Moreover, it must provide users secure access to software applications and private data through user-friendly interfaces. The main issues in these tasks and the most common solutions adopted by the modern OSs are presented in the following sections.

6.3.1 Process Management

An OS must ensure the execution of a task which is assigned by a software application. In order to do that, it identifies each task as a process, allocates the appropriate resources, and monitors the execution. Furthermore, it is responsible for the time scheduling of the execution of the processes' steps in order to accomplish the parallel execution of the assigned tasks, thereby attaining better performance.

6.3.1.1 Processes and Threads

In OSs, a process is realized as an occurrence of a computer program which is being executed. It consists of program codes and the appropriate resources which must be used. In the modern Oss, processes made up of one or more threads are independent subprocesses executed in the CPU. An example of process realization by the user is through the Task Manager utility of Microsoft Windows Family operating systems or the *ps* command in Unix.

In each step of the process, the OS executes a process depending on the following characteristics (Silberschatz et al., 2004)

- A program with the machine code for execution.
- The memory space where the code and the data is stored.
- Descriptors of resources used by the process such as file descriptors (Unix terminology) or handles (Windows).
- The current process execution state.
- Internal security information such as process owner and process' allowable operations.

The most useful specification of the definition of a process is its state. Process states are used by OSs to identify what operations must be scheduled for continuing the execution of the process. The primary states of a process are (Stallings, 2005):

- *Created* is the initial state of a process when it is created. In this state, OSs identify the tasks and the resources needed for process execution, decide the appropriate schedule, and set the process to the "ready" state.
- *Ready or Waiting* state is when the data and code of the process is loaded in the main memory and waits its turn for execution by the CPU.
- *Blocked* state is when a process is paused waiting for some event such as reading a file, keyboard input, and so on.
- *Terminated* state denotes that a process has either completed its execution or is explicitly being killed by the OS.

OSs ensure that processes are kept separated and the needed resources are allocated with the intention that they are less likely to interfere with each other and cause system failures. Some OSs also support mechanisms for safe interprocess communication to enable processes to interact with each other.

Modern OSs, in order to improve efficiency in the process management, separate concurrently running tasks, which are called threads. A process may include multiple threads which share resources such as memory, while different processes do not share these resources. The fact that a thread is executed inside a process confuses the usage of the thread which is a different concept and can be treated separately. Threads use the same allocated memory and interact only with other processes. Furthermore, processes are executed in a group of resources (CPU, Memory, I/O devices, storage devices, etc.) but threads are considered as entities executed only on the CPU.

6.3.1.2 Multitasking

A main feature of modern computer systems is that they can execute a number of tasks simultaneously. Computer users are able to listen to music from an MP3 player running in a personal computer while, at the same moment, read their e-mails on a web browser. This capability is based on the *multitasking* feature supported by modern OSs. Multitasking is the method by which processes share common resources such as CPU, memory, hard disk drives, and I/O devices. In a traditional computer where only one CPU exists, only one process can use it at any point of time. This means that only one task can be executed by a computer system. Multitasking provides a solution to this problem by sufficient scheduling of which process can use a specific resource at any given time and when another waiting process consumes the resource. This switching for resource usage between two or more processes is frequently occurring resulting in the illusion of parallel execution of these processes.

6.3.1.3 Process Scheduling

The improvement achieved by multitasking creates the demand for efficient scheduling of resource usage by the processes. When a computer works in multitasking mode, there are a number of processes competing which one will first use a resource, mainly the CPU. The functional component of OSs which is responsible to decide which process will be assigned to a resource is called *scheduler*. The scheduler is executing a *scheduling algorithm* in order to decide the next process which will use a particular resource aiming to increase the computer performance and avoid conflicts which result in system failure. The existing scheduling algorithms try to follow the above criteria:

- *Fairness*: Each process must be treated fairly regarding its assignment to a resource.
- *Throughput*: As many as possible processes must be executed in a given time.
- *Response time*: Interactive users must have the minimal response time.
- *Efficiency*: CPU must work without stop at any time.
- *Turnaround*: Waiting time of batch user must be minimized.

The scheduler's behavior is different in every OS regarding the goals of the computer system where it is installed. Thus, scheduling algorithms can be distinguished in three main categories:

- *Algorithms for Batch Systems*. Batch systems are focused in the execution of processes which do not need to interact with users. These algorithms are not only simple but also more efficient regarding the faster performance of the CPU.
- *Algorithms for Interactive Systems*. Interactive systems must be always available for users' interaction. Thus, these algorithms cannot permit a process running for a long

time, leaving others to wait. The availability of resources is essential and the algorithms are more complex.

- *Algorithms for Real-Time Systems.* Real-time systems also require the availability of the resources in order to execute their processes as fast as possible. Although the requirements seem higher than interactive systems, in real-time systems, software applications are designed to execute small processes with reduced requirements of time and resources. Thus, the complexity of these algorithms is lower from those in interactive systems.

6.3.2 Memory Management

Each computer system has two vital hardware parts, the CPU and the physical memory. These two pieces of hardware characterize usually the performance of a computer. Thus, OSs include mechanisms for the best use of these two resources aiming for the best performance for a particular computer system. Multitasking methods orientate to efficient CPU usage. Regarding memory, OSs meet the challenge to allow the execution of one or more programs simultaneously when the required amount of memory is higher than the existing physical memory capacity. In this concept, all the modern OSs use the method of *virtual memory.*

6.3.2.1 Virtual Memory

Random Access Memory (RAM) is the physical memory of modern OSs. The RAM's read/write speed is higher than the respective one of hard disk drives, but its cost is also much higher from that of the hard disk drives. Nowadays, the price of 4GB of RAM is about the same as that with a hard disk drive of 2TB—about 500 times more capacity. Also, the programs today consist of graphical interfaces, images, videos, 3D objects which require huge amounts of memory. This, in conjunction with the adoption of multitasking, makes the existing amount of RAM inadequate for running software applications efficiently.

In order to solve this problem, OSs have an internal mechanism which uses a part of the hard disk drive's capacity in order to temporarily save data planned for RAM. The portion of the hard disk's storage used is called Virtual Memory. Virtual memory operation is based on the fact that when a process is paused or if it waits for an event, its data consumes memory space which reduces the available memory capacity at the specific time point. Thus, it is easier to copy the data saved in physical memory to the hard drive when a process is inactive and free the allocated memory space. When the process has to be activated again, the data will be transferred back to memory. The part of the hard disk drive where the memory data is stored is called *page file*. Page file holds portions of RAM, called pages, and the OS constantly swaps data back and forth between the page file and RAM. Figure 6.2 demonstrates how the data is swapped between RAM and virtual memory during a computer system operation.

Virtual memory is considered as an efficient solution in the problem of limited RAM. Nevertheless, there are some other factors which define the performance of a computer system. OSs allow the execution of programs simultaneously even if the total amount of the required memory is higher than that of the RAM's capacity. But if the requirements of memory for each running program are high, then a significant performance drop is noticed during swapping between tasks. Thus, users, while working simultaneously in applications with high memory, often notice a slight pause between the changing of the active programs. The best case achieving the highest performance of a computer system is when the RAM capacity is enough to store the data required by the software applications which the users run simultaneously. If this is not happening, the OS has to constantly swap information back and forth between RAM and the hard disk. This is called thrashing, and it can make your computer feel incredibly slow.

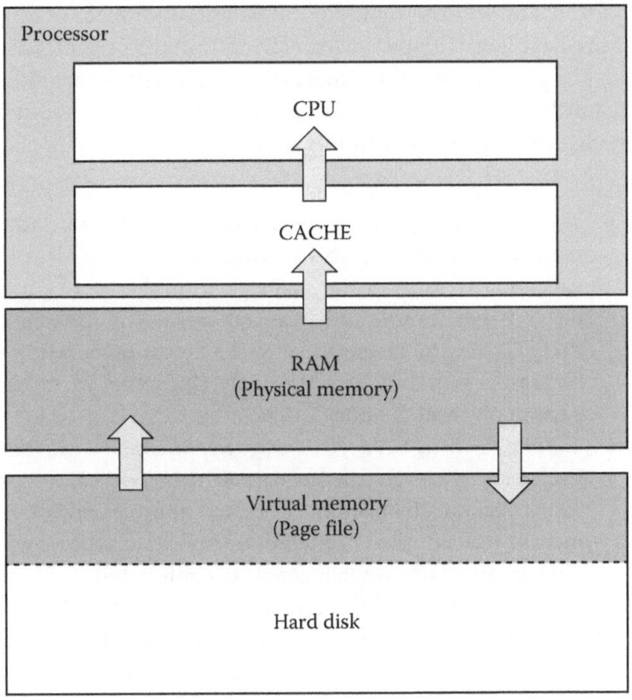

FIGURE 6.2 Virtual memory operation.

6.3.2.2 Memory Paging

Memory paging is the mechanism, supported by an OS, that addresses the temporary removal of data portions from the physical memory to virtual memory (page file) whenever it is necessary. The effectiveness of paging is based on the fact that programs running in a computer require a part of the assigned virtual memory at each time point of their execution.

The challenge in memory paging mechanism is the selection of the memory page (memory portion) in the physical memory that must be removed when a new process needs the physical memory in order to execute a task. OS uses algorithms in order to identify which page must be removed from the physical memory and saved in the page file of virtual memory. This decision is a complex task and usually modern OSs use a combination of algorithms to find the right page to be removed. A lot of algorithms have been proposed for solving the problem of selection. The problem is that the most efficient algorithms are either difficult to be implemented or require a respective amount of time which is not feasible in this situation. The most important algorithms are listed below:

- *The Optimal Page Replacement Algorithm*. This algorithm removes the page which is predicted to be used late for execution. This is the optimal algorithm that can be used but it is not feasible to be implemented. The problem is that OSs have no means to know the execution duration of each process in order to estimate the time point when each page in physical memory will be used.
- *The Not Recently Used (NRU) Page Replacement Algorithm*. Each page file is marked whenever it is used. The mark resets periodically from the OS. Each time a page replacement is needed, the OS chooses one of the marked pages. This is an easily implemented algorithm but does not guarantee the optimal solution.
- *The First-In, First-Out (FIFO) Page Replacement Algorithm*. In this case, the OS keeps the earliest memory page which is inserted in the physical memory and removes it.
- *The Second Chance Page Replacement Algorithm*. This algorithm is a combination of FIFO and NRU and chooses the memory page which is longer in the memory (by FIFO) and which is not recently used (NRU).
- *The Clock Page Replacement Algorithm*. This algorithm places virtually the pages in a cycle. The pages are marked when used in a similar way as that with the NRU. Moreover, there is a pointer that each time a point moves clockwise and when a page replacement is required, it removes the first nonrecently used page.
- *The Least Recently Used (LRU) Page Replacement Algorithm*. The basic assumption of LRU is that the pages that have been heavily used in the last few task executions are more likely to be used in the near future. Although, this algorithm is closer to the optimal solution, it is not considered efficient because it requires an additional portion of memory in order to keep the statistics of memory page usage.
- *The Working Set Page Replacement Algorithm*. The algorithm is based on the preloading of data used by a process (called *working set*) in the memory in order to avoid the requirement for page replacement during process' execution.
- *The WSClock Page Replacement Algorithm*. WSClock is a variant of Clock Page Replacement Algorithm and it uses working sets instead of pages. Its implementation is simple and its performance is sufficient. Thus, WSClock is widely used in the modern operation systems.

6.3.3 Resource Management

The main functionality of a computer system is based on the CPU, the memory, and the I/O devices. OSs are designed in order to manage efficiently these three main components. Although CPU and memory define the calculative performance of the computer systems, the I/O devices are the components which are usually used in order to provide services to the users; keyboard allows users to import data, monitor shows data, disk drives permanently save data, printers imprint data to a piece of paper, network cards transfer data to other computer systems, and a lot of other devices. Given the fact that modern computer systems can execute more than one task simultaneously, the appropriate management of the existing resources is essential in order to avoid system failures.

6.3.3.1 Input/Output Devices

The great number and variety of I/O devices that can be attached to a computer system lead to the necessity of some kind of prototyping their architecture, so that an OS can easily manage new imported or manufactured devices. Thus, an I/O device in modern computer systems consists of two parts (Tanenbaum and Woodhull, 2006):

1. *Device Controller*. Controller is an electric device consisting of chips which control the device. It provides an intermediate layer which accepts commands from the OSs, such as write, read, or more complex commands.
2. *Device*. The device provides a very simple interface in order to electronically interact with the controllers. On some occasions, devices have to follow some industrial standards for this interface in order to interact with third-party device controllers. For example, hard disk drives follow interface standards as Integrated Drive Electronics (IDE), Small Computer System Interface (SCSI), or Serial Advanced Technology Attachment (SATA), in order to be attached to the associated disk controller (i.e., IDE controller).

There are three ways of interaction with a device:

- *Busy waiting*. In this case, the OS activates the device and starts the execution of a task from it (i.e., copy files). While the task is executed, the OS periodically checks if the task is finished. When the device finishes its task, the resulting data is taken by the device driver and saves it to memory in order to be used for the next steps of a process.
- *Interrupts based*. In this case, the OS does not poll all the time the device controller to check if the task is finished. Instead, the device controller generates a special signal to the OS, called interrupt, in order to inform it about the task's competition. Interrupts are widely used in the modern computer systems.
- *Direct memory access (DMA)*. In this method, a chip is used—the DMA—in order to allow a device to read and write directly to and from the memory without using the CPU. The OS initializes the process for the direct transfer of data between device and memory. When this transfer completes, the controller sends an interrupt to the OS.

6.3.3.2 Device Drivers

Each device controller has its own command set, depending on its nature or the manufacturer. Thus there is different software, running as part of the OS that interacts with each controller. This software is known as the *device driver*. Device drivers are software created in order to be installed in the OS allowing the second to be empowered with the specific commands needed to operate the associated device. Thus the majority of hardware parts or peripheral devices are accompanied by installation CDs with device drivers for the different OSs (MS Windows, Linux, etc.). Moreover, nowadays, the OS installation software is delivered including a large number of device drivers, especially for older devices whose drivers are not easily found.

The device driver's code is implemented by the device manufacturer and not from the OS's development team. In order to simplify the implementation process of a driver, OS designers publish a well-defined model which describes in detail the methods of how a driver interacts with the OS. Figure 6.3 illustrates the placement of device drivers and how they are used for their communication with the device controllers in order to manage the I/O devices.

The common procedure to be executed by a device driver in order to do a specific job with the device includes the following steps:

1. The OS gives a command to the device driver. Device driver's first responsibility is to check if the command's arguments are valid.
2. Driver "translate" arguments' values in a format understandable from the device.

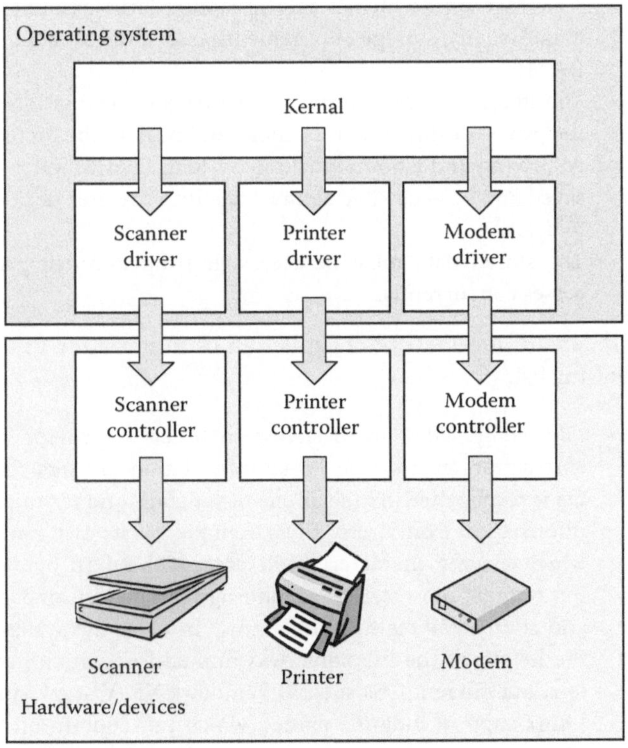

FIGURE 6.3 Operating system–devices interaction.

3. Consequently, it checks if the device is available. The majority of device drivers are keeping a queue of commands for execution.
4. When the device is available, the device driver initializes the execution process by interpreting the OS's command to a sequence of commands which must be executed on the device.
5. The device driver controls the execution of the command sequence by interacting with the device controller. During the execution, it checks if the device controller is ready to execute the next command, and sends it to the device controller.
6. Finally, when the last command is executed from the device controller, the device driver may react in two ways regarding the device and the assigned job. If the job requires some additional work by the controller (i.e., copy verification), the device driver waits until the final interrupt comes from the device controller. In the other situation, there is no work left and the device driver informs the OS that the assigned job is completed.

6.3.3.3 Deadlocks

The biggest challenge of OSs as far as resource management is concerned is to avoid *deadlocks*. Deadlock is the situation when each process in a group of processes is waiting for another process in the same group to release a specific resource. The fact that each process waits for another to do something in order to continue results on the fact that none of them will ever create the appropriate conditions that will activate any of the other group members, including indefinite waiting.

In the early 1970s, Coffman et al. (1971) determined four conditions that must be occurring in order to cause a deadlock situation:

1. *Mutual exclusion condition*. The processes need exclusive control of the required resource.
2. *Hold and wait condition*. The processes keep holding a resource granted earlier until another resource will be assigned to them.
3. *No preemption condition*. The OS cannot forcibly release a resource by a process. Only the process can release a resource.
4. *Circular wait condition*. Processes exist in a circular chain of two or more processes. Each process is waiting for a resource held by the next member of the chain.

The aforementioned four conditions must be met simultaneously in order to result in a deadlock situation. In case one of the four is not present, then a deadlock situation is impossible. OSs use different strategies in order to manage a deadlock situation. Regarding the possibility for a deadlock to occur or result in system failure, different policies are utilized which can be categorized in the following general strategies:

- *Ignorance of the problem*. The OS ignores the problem and waits for some external event to occur that will solve the deadlock situation (i.e., process cancellation by the user).

- *Detection and recovery*. The OS leaves the deadlock to occur, but when it detects them, it acts in order to solve them.
- *Dynamic avoidance*. The OS carefully allocates the resources preventing any possibility for the occurrence of a deadlock situation.
- *Prevention*. The OS system design contains the appropriate structural functionalities (i.e., resource force release) in order to negate one of the four conditions which result in a deadlock.

6.3.3.4 Multiprocessors

Although the majority of computer systems include only one CPU, recently, the manufacturing of computer systems with parallel CPUs is gaining a bigger part of the market, especially for target-specific systems aiming to calculate fast and accurate complex mathematics. These systems can support over 1000 CPUs and the majority of them are following the shared-memory multiprocessor model or just the multiprocessor model as is likely called. Multiprocessors are systems with more than one CPU having access to a common memory. Because of their different architectures, the multiprocessor's OSs have some special features which are not met in the classic OSs. In general, the multiprocessor's OSs are similar to the traditional ones, regarding the resource and memory management. Their behavior differs in the management of the process execution on the CPUs. There are four types of multiprocessors, each one managing the existence of more than one CPU in a dissimilar way. These types are listed below:

- *Separate Operating System for Each CPU*. Each CPU has its own copy of the OS and a preassigned memory portion. The processes are redirected randomly to each OS which undertakes the responsibility to get executed to its associated CPU. However, this approach may result in the CPU being overloaded while another one stays idle.
- *Master–Slave Multiprocessors*. In this model, a CPU undertakes the role of the Master and all the processes are assigned to it. There is an OS instance which is associated with the master CPU. The master CPU is responsible to keep a list of waiting processes and to reassign them to the first available CPU (slaves). This model works fine for small microprocessors, but in a large number of CPUs, the master CPU is likely to be overloaded from the large number of requests and assignments.
- *Symmetric Multiprocessors (SMP)*. In the third model, a copy of the OS is preloaded in the memory and any CPU can run it. Memory and processes are dynamically balanced since there is only one copy of the OS in the memory. The only restriction in this approach is that two CPUs cannot use simultaneously the OS copy.

During the last few years, the processors' technology evolution resulted in the manufacturing and promotion of processors with multiple CPU cores (Intel Multi-Core Technology, AMD Multi-Core Processing). Processors with multiple cores are not considered as multiprocessors. Multiprocessors support a lot of CPUs. Multi-Core processors include a single CPU with more than one core which means that they can execute processes faster than a CPU with a single core. Thus, the OSs treat the multicore processors as a single CPU and not as multiprocessors.

6.3.4 File System

The most important user requirement of a computer system is to be able to store and retrieve data. In modern computers, the unit which is used in order to save data to hard disk drives, floppy disks, or other storage device is the *file*. The OS is responsible to retrieve, create, delete, and save data to the files of the storage devices that are connected to the computer system on which it is installed. Moreover, it is responsible for managing a mechanism which allows the organization of the files in *directories* (folders). The way an OS can have access to the files and directories of a storage device is called *file system*. Well-known file systems for hard disk drives are the New Technology File System (NTFS), File Allocation Table (FAT) supported mainly by Microsoft Windows OSs, Unix file system (UFS), extended file system (EXT) for UNIX—like OSs and Hierarchical File System (HFS), and HFS Plus for Mac OS. In this section, the main characteristics of files and directories that are supported from the majority of the file systems are described.

6.3.4.1 Files

File is the storage unit of data in all the computer systems. Files were created in order to fulfill the following requirements:

- A storage unit which is able to store a large amount of data. Usually, storage requirements exceed the capacity of RAM.
- The stored data must be retained after the termination of the process using it. For example, an image in the Picture Archiving and Communication Systems (PACS) must be saved and be accessible at any time until the user decides to delete it.
- The stored data must be accessible from different processes concurrently.

The main characteristics of a file can be summarized in the following list:

- *File Name*. Files are designed in order to provide an abstraction layer of the stored information to users. The OS is responsible to hide all the file's details and technical information from users. Thus, each file has its own name which usually characterizes its content. Modern operating systems allow strings containing characters, numbers, and some symbols for file naming. In older file systems the length of the file name was limited to eight characters, but current OSs such as Windows XP, Vista, 7, and Linux support long file names, which can contain up to 255 characters. Furthermore, some OSs—mainly Unix-like—distinguish between upper and lower case letters.

Usually, file names are separated in two parts: the first contains the file name and the second identifies the format of data contained in it. The second part is named *file extension* and usually determines the software application which accesses the file.

- *File Type.* OSs support a variety of file types. In the majority of them files are distinguished between *regular* files, *directories*, and *system special files*. With regard to regular files containing user data, directories are system files with the structure of the file system and system special files are used by the OS to store data used by it in its procedures. Regarding regular files, there are two types: (a) ASCII files which contain data as a sequence of characters and (b) binary files where their content is represented in binary format. The content of an ASCII file is easily read by users using a simple text editor (like Notepad). Conversely, the binary files' content is not understandable by common users and a third-party software application is needed in order to access the file's information. Example of binary files are the documents created in the Microsoft Word application. If users try to open with Notepad, the content will be incomprehensible to them, but using the Microsoft Word application, they will be able to read the stored document.
- *File Access.* Another characteristic of files is the mechanism used by the OSs to access data stored in them. In older OSs, a file's content was parsed from the beginning of the file to the end which is defined as *sequential access*. Modern OSs can read the content of a file out of order or identify the file's areas rather than commencing the read process from the beginning of the file. Files allowing access in their content in any order are called *random access files*.
- *File Attributes.* Apart from their name and content data, files are also associated with some additional information by the OSs. This kind of information refers to their size, the user who created them, the date and time of their creation, date and time of last access to them, and user access permissions. OSs are responsible to associate this information to each file and to easily display it to the users.
- *File Operations.* Each OS uses its own operation to retrieve the data stored in a file. Nevertheless, the most common file operations are the creation, deletion, opening, closing, reading, writing, appending content, seeking data, attributes retrieval, attributes setting, and renaming.

6.3.4.2 Folders/Directories

The number of files in a modern computer system may sometimes reach billions. Therefore, most of the file systems use directories (or folders) in order to provide a structural mechanism for tracking files. File system directories usually include both files and other directories leading to tree-like structures of the file system. The directory organization mechanisms meet differences in various file systems. Below is a list of the representative forms of directory systems.

- *Single-Level Directory Streams.* In this form, only one directory exists—called root directory—and all the files are in this directory. The main disadvantage of this structure is the difficulty for a multiuser OS to set access permissions for the files.
- *Two-Level Directory Systems.* This structure allows the existence of directories only inside the root directory. This was arrived at as a solution to improve the single-level approach in the case of multiuser systems. Users could have one or more directories where they were able to create or delete files, but they could not see or modify the contents of other users' personal directories.
- *Hierarchical Directory Systems.* The two previous approaches fail when the number of files in computer systems increases. In order to keep the user privacy implemented by the two-level directory systems, this approach structures the directory system in a tree-like graph allowing directories to contain other directories. The possibility to create more than one directory allows users to logically organize their files in groups, making their tracking simpler.

The hierarchical directory systems form the most common structure used in modern OSs. One significant advantage is that the location of a file or directory can be defined by the route to it starting from the root directory and following the directory structure, as depicted in Figure 6.4. This route is called *absolute path* and is expressed by the names of the directories which compose the route separated by a special character—usually backslash (\) or forward slash (/). For example the file *img00.gif* which resides inside a directory called *images* which is under the root directory can be defined in Linux-like OSs by the absolute path/ *images/img00.gif.* In windows OSs, the starting point of the absolute path is the letter of the partition or drive where the file exists. In the previous example, if the file img00.*gif* is saved in the *D* partition the absolute name would be *D:\images\img00. gif.* The absolute path uniquely defines a file in the file system, no file can have two or more absolute paths. Besides absolute path, there is the relative path which denotes the route from a directory to a file or another directory following the links created by the tree-structure.

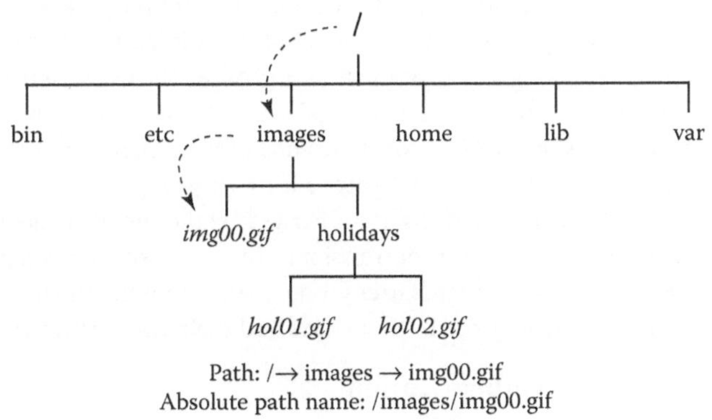

FIGURE 6.4 File system—absolute path.

6.3.5 Security and Privacy

All modern OSs meet the requirement of supporting multiuser usage, meaning that more than one user can use the same computer system simultaneously without the other user's interference. A user working in a multiuser environment demands from the OS to fulfill two major requirements:

1. To work independently from other users, meaning that his processes will not mix up with the ones of the other users.
2. To keep this data private which other users are not able to have access to.

In order to fulfill the above requirements, OSs utilize a mechanism for user authentication and identification. The distinction of the processes between users is accomplished by tagging each process with a number which is the unique identifier of the user and is called **USER ID**. For the second requirement, OS associates each file with each user's access permissions.

6.3.5.1 User Management and File Permissions

User authentication allows OSs to identify a person as the system's user. In order to manage security restrictions to files or directories, OSs are empowered with a protection system which defines the operation which a user can do on each file or directory. Ferraiolo et al. (2007) refer that a protection system is defined by the notions of *users, subjects, objects, operations*, and *permissions*. Subjects are usually the processes running on behalf of a user and objects, in file systems, considered the files and directories. Harrison et al. (1976, p. 462) gave a definition of the configuration of a protection system denoting that "A configuration of a protection system is a triple (S, O, P), where S is the set of current *subjects*, O is the set of current *objects*, and P is an access matrix, with a row for every subject in S and a column for every object in O." The access matrix is implemented in most modern OSs by using an Access control list (ACL) attached to each file. The ACL is a data structure (usually a table), which associates each operation that can be executed on a file (open, read, write, delete, etc.) with the operating system's users and defines their eligibility from them.

Although the implementation of ACLs differs between the various OSs, all of them share a common feature which is *user groups*. User group is an abstract notation defining a group of users that share common permissions on a set of files or directories. It is for each user to belong to one or more user groups. The usage of user groups comes from the fact that an OS is usually managing tens of users and must keep ACLs for thousands of files. Therefore, each file's ACL contains additional entries for user groups access permissions. These permissions are inherited by the member users. Thus, instead of defining permissions for each user, OS keeps access rights for the system's user groups. This simplifies the management of security from both the OSs and their administrators.

6.3.5.2 User Authentication

In computer systems, the term authentication refers to the process determining that a user's claim identification is legitimate (Ferraiolo et al., 2007). Authentication is an essential process of OSs security mechanisms as it ensures that the particular user can operate the computer system and access the data in the permitted files. Authentication is mainly based on the following factors:

1. Something the user knows.
2. Something the user has.
3. Something the user is.

These factors lead to the implementation of different authentication methods and procedures, each one with different complexity, cost, and security properties. The most common authentication methods are

- *Passwords*. This method is based on the first factor. For each user, there is a set of a user name and a password which only the user knows. Password is usually encrypted in the OS and no user can read it, not even the administrators. Users use their passwords in order to pass the authentication process. The password is private for each user and, in the majority of Oss, is not even shown in the login window while the user types it.
- *Physical Object*. In this case, users pass the authentication process by using a physical object that they possess. The most common example is the bank's Automated Teller Machine (ATM), where the user inserts a card in a specific slot on the computer system and by using a 4-digit Personal identification number (PIN) he is identified by it. Other examples are smart cards containing the user's authentication credentials and security certificates.
- *Biometrics*. Technology evolution allows a computer to identify users by measuring their physical characteristics that are hard to forge. Users are identified by their fingerprints, voice, eyes, or the shape of their face.

6.3.6 User Interface

OSs' role as a mediator between the computer's hardware and the user demands the adoption of the appropriate mechanism which allows users to give commands, to view data, and to execute operations with software applications. This interface to the user is called *shell*. The shell is a software which allows users to use the services of an OS kernel. Generally, there are two categories of shells regarding the way they interact with the user: command line interfaces (CLIs) and GUIs.

6.3.6.1 Command-Line Interface

In CLIs, users interact with an OS by typing commands which initiate the execution of specific tasks. CLI's command are text based, users type a sequence of words, symbols, and numbers based on a particular syntax and then presses the "Enter" key. Then, a command-line interpreter receives, analyzes, and executes the requested command.

The first use of CLIs was in the 1950s when teletypewriter machines (TTY) were connected to computer systems in order to replace the mechanical punched card input technology.

Afterward, they are fulfilled in the two well-known Oss, MS-DOS and Unix in the 1980s (Stephenson, 1999). Although, in early years, the CLIs were replaced by GUIs, they are still used in server administration and the remote operation of connected computers using Telnet protocol.

The structural features of a CLI are its commands' syntax and semantics. All the commands follow specific grammar and their operations are uniquely defined. Commands usually include their name, a set of properties, and a set of arguments. Properties are usually used to customize the operation's execution or results. Arguments (or parameters) are values used as input to the operation or optionally capture or redirect its output. Unlike GUI's menus and buttons, command line commands are stating what users want similar to the expression they could use in written language. For example, in a UNIX environment, the command *cp img00.gif /images* denotes the operation of *copying* the file *img00.gif* to the directory/*images*. Although CLIs differ among OSs, there are similarities in their syntax and semantics. Therefore, a user familiar with an OS CLI can easily learn to use another.

6.3.6.2 Graphical User Interfaces

Although CLIs cover all the operations that a user can do in an OS, they are considered obscure and difficult for nonexpertise users. Thus, from the beginning of computer's history, a new type of interaction with the user including icons, menus, windows, and mouse pointers is created, the well-known GUIs that are present in the majority of the modern OSs.

The history of GUIs dates back to the 1960s with the oN-Line System (NLS) which was developed in Stanford Research Institute by Douglas Engelbart (Engelbart and English, 1968). Lately, the concept of NLS was the basis for the Apple Macintosh interface which was the first GUI similar to the modern ones such as Microsoft Windows or Linux KDE.

GUIs use a combination of technologies and devices (mouse, digit pens) in order to provide users the ability to interact with the computer. The most common combination of elements satisfied in modern GUIs is WIMP that stands for Window, Icon, Menu, and Pointing device. The WIMP style of interaction allows a physical input device to control the position of a cursor, usually a mouse. The information presented is organized in windows or icons. The available commands are placed in menus. The structured elements of a modern WIMP are the windows manager which facilitates the interactions between windows, applications, and the windowing system and the desktop environment on which the WIMP's elements are displayed. Moreover, mobile devices (PDAs, SmartPhones) use WIMP elements in a different way than the classic WIMP due to constraints in display space and the input devices (phone keyboard). These GUIs are characterized as postWIMP user interfaces (van Dam, 1997).

6.4 Commonly Used OSs

The history of OSs, as mentioned before, starts about the same time as with the birth of computer systems, back to the 1950s. All these years, hundreds of OSs have been developed and used in all kinds of developed computer systems. Some of them became known worldwide because they are used in commonly used computer systems such as personal computers (workstations), servers (Web, application, and database), and mobile devices (PDAs, mobile phones).

The most usable category of computer systems today is personal computers (PCs); PCs exist in almost every company, public organization, or even at homes. Workstations and laptops are the main product categories of Personal Computers. Moreover, another product family similar to PCs are the Servers that are used for specific purposes—Web servers, database servers, file servers, and so on. Although servers have advanced performance requirements compared to PCs, their architecture is similar. Thus, their OSs have a lot of similarities, and even OS development companies and organizations publish similar OSs for both PCs and Servers.

In the following sections, there is a presentation of the most common OSs families; Microsoft Windows, Linux, and Mac OS. The statistics given by W3Shools (2010)—a global training organization for Web development—shows that 99.5% of their visitors use OSs which belong to one of these three families.

6.4.1 Windows Family

Microsoft Windows are OSs including GUIs produced by Microsoft (Microsoft, 2010). Today, Microsoft Windows holds the biggest piece of the personal computer market which is estimated at 90%. Microsoft success was based on the easy-to-use interface provided at low-cost personal computers in the mid-1980s. The first OS of the Microsoft Windows Family was produced in 1985 as an add-on the existing MS-DOS OS. Afterward, Microsoft developed some improved versions and at the start of the 1990s, announced its Windows NT OS technology which was more reliable and stable, but it was promoted for business use. Windows NT 2.5, 3.5, 4.0, and 2000 are based on this technology. Windows XP, released in 2001, brought the NT technology to home users providing a powerful and stable GUI-based OS. Simultaneously, Microsoft adjusted the NT technology to meet the needs of server systems and released Windows Server 2003. The most recently released OSs of Microsoft are Windows 7—as successor of Windows Vista—and Windows Server 2008 R2.

In the last few years, Microsoft concentrated on the development of OSs for supporting the newly released 64-bit processors. Although the first attempt was made in Windows XP, Windows Vista is the first Microsoft's OSs which was released in two versions, one for 32-bit processors and one for 64-bit processors.

6.4.2 Unix/Linux

Unix was initially designed in 1970 at Bell Labs. Unix evolution continued until 1991 when its successor Linux was developed by Linus Torvalds and posted free on the Internet. Linux was designed for the IBM PC (Intel 80386) architecture, which is the basis for today's personal computers (Linux, 2010). This fact, in addition with its free distribution, was the main reason why Linux became an early alternative to other UNIX workstations, such as the ones from Sun Microsystems and IBM.

Linux success is based on two reasons. The first one is that it was free and moreover released under the GNU Public License (GPL) of Free Software Foundation (FSF). The second was the publication of its source code. Linux was an open-source software which resulted in a lot of research institutions, programmers' communities, and software companies starting to develop improved versions of it adding features, improving functionalities, and creating stable software applications. Thus, Linux is not considered as a unique OS, but Linux's kernel is the basis of more third-party developed OSs. All these OSs constitute the Linux family OSs and they are called Linux distributions. Some well-known Linux distribution systems are Arch Linux, Ubuntu, Suse, Fedora, and Red Hat Enterprise Linux.

An innovation of Linux distributions is that the majority of them are offering a copy of the OS in a CD that is able to run on a computer system without the need for installation. The Live CD—as it used to be referred—boots the entire OS from CD without installing it on the hard disk. Live CDs appear to be for demonstration purposes, but in the next few years, they will be used in testing systems and in target-specific devices that do not have hard disk drives.

6.4.3 Mac OS

Mac OS is a family of OSs developed by Apple in order to support the Apple Macintosh computer systems. The first Mac OS released together with the first Macintosh computer in 1984. Afterward, nine more releases were produced by Apple concluding to Mac OS X at 2002. Although the first releases follow Apple's independently designed kernel, the Mac OS X is a Unix-based OS with a user-friendly GUI (Apple, 2010).

Although Mac OS is considered to provide stable performance and one of the best and most efficient graphical interfaces, it is not very popular in the home users' market. The main reason is that Apple restricts its use to their brand computer systems without aiming to distribute Mac OS versions for IBM-compatible Personal Computers. In the past there were some projects, such as Start Trek, initiated by Apple in order to run Mac OS in common PCs but without significant results. In 2006, Apple started to replace its traditional Apple processors with Intel x86, which resulted in the earlier version of Mac OS X 10.6 running in a x86 architecture. Nevertheless, Apple forbids the installation of Mac OS X in "non-Apple-labeled computers."

6.5 Conclusion

OSs play an essential role in the operation of a computer system. They are the software placed between the user's software applications and the hardware of a system, providing users an understandable abstraction model of the computer. Moreover, they are responsible for managing and monitoring the execution of the processes and coordinating the resources' usage aiming to avoid system shutdowns and providing the best performance in terms of reliability and efficiency. The OS is the biggest and the most complex software running on a computer. They consist of many subsystems that are responsible for managing resources, ensuring user privacy, and providing user-friendly interaction interfaces to the user.

Because of the significance of OSs, a lot of effort in their improvement has been made from the research and industry community for the last 30 years. This effort inncluded the principles of OS design that were presented in this chapter. These principles are used with some minor improvements to all the OSs released in the last 10 years. Although the basis of OSs is established, the research continues in more specific areas. As mentioned in the chapter's introduction, the evolution of OSs follows the evolution of computer systems. Therefore, today the research on operation system focuses on small computer systems such as PDAs and SmartPhones aiming to adjust traditional OSs functionalities to different hardware and operational requirements. Recently released OSs such as iPhone OS, Symbian OS, and Android are promising efficient performance and more functionalities in the upcoming mobile devices. Furthermore, the introduction of new interaction devices like smart cameras, microphones, and motion sensors creates a new research path for the development of advanced user–computer interfaces which will be much different from the traditional WIMP-based GUIs.

References

Apple. 2010. *Mac OS X*. Available at http://www.apple.com/macosx/.

Coffman, E. G., Elphick, M. J., and Shoshani, A. 1971. System deadlocks. *ACM Computing Surveys*, *3*(2), 67–78.

Engelbart, D. and English, W. N. 1968. A research center for augmenting human intellect. *Proceedings of the 1968 FJCC*, 33, 395–410.

Ferraiolo, D. F., Kuhn, R. D., and Chandramouli, R. 2007. *Role-Based Access Control*, 2nd edition. Norwood: Artech House.

Harrison, M. A., Ruzzo, W. L., and Ullman, J. D. 1976. Protection in operating systems. *Communications ACM*, 19, 461–471.

Linux. 2010. *Linux* [Online]. Available at http://www.linux.org/.

Microsoft. 2010. *Microsoft Windows Family Home Page*. Available at http://www.microsoft.com/windows/.

Silberschatz, A. and Galvin, P. B. 1994. *Operating Systems Concepts*. Reading: Addison-Wesley.

Silberschatz, A., Cagne, G., and Galvin, P. B. 2004. *Operating System Concepts with Java*, 6th edition. Hoboken: John Wiley & Sons.

Stallings, W. 2005. *Operating Systems: Internals and Design Principles*, 5th edition. Upper Saddle River: Prentice-Hall.

Stephenson, N. 1999. *In the Beginning... Was the Command Line*. New York: Avon Books.

Tanenbaum, A. 2001. *Modern Operating Systems*, 2nd edition. Upper Saddle River: Prentice-Hall.

Tanenbaum, A. and Woodhull, A. S. 2006. *Operating Systems Design and Implementation*, 3rd edition. Upper Saddle River: Prentice-Hall.

van Dam, A. 1997. Post-WIMP user interfaces. *Communications ACM*, 40, 63–67.

W3Schools. 2010. *OS Platform Statistics*. Available at http://www.w3schools.com/browsers/browsers_os.asp. Accessed March 2010.

Wulf, W., Cohen, E., Corwin, W. et al. 1974. HYDRA: The kernel of a multiprocessor operating system. *Communications ACM*, 17(6), 337–345.

7

Networks and Networking

Christos Alexakos
University of Patras

George C. Kagadis
University of Patras

7.1 Introduction

Through human history, many efforts were made toward efficient interpersonal communication, eliminating the problem of long distances. Traditional mail, phone, and FAX are some paradigms of tools used by people in order to discuss, exchange information, or entertain themselves. In the twentieth century, technology evolution brings computers to people's daily routine raising the need of communication through them. At the beginning, there was the need for exchanging research information and experiment results within the academia, later, the need for exchange of corporate information among the employees. Today, computer communications are commonly used and most people's communications can be executed using computers, such as chatting with a friend, watching movies, reading books, studying courses, playing video games, and so on. The technology area which deals with communication through computer systems is called *computer networking.*

Although the term computer networks is not standardized, a widely accepted definition is given by Tanenbaum (2003, p. 1) "a computer network is a collection of autonomous computers interconnected by a single technology." The interconnection technology of computer networks deals with all the aspects of computer networks, ranging from the interconnection medium (cables, radio frequency [RF]) to special hardware and software. The first computer networks consisted of wired computers located in a room and their evolution resulted in the worldwide network of Internet, which is the most known and used computer network. Thus, it is the subject of research around computer networks and the application field of networks' state-of-the-art technologies. Internet's technology is also applied to smaller networks such as enterprise and university networks.

Computer networks provide a variety of services and functionalities to computer users. In the business sector, private enterprise networks simplify the collaboration among employees allowing e-mail message exchange, resource sharing (diagrams, documents, images), or written conversation establishing (chat). Moreover, they facilitate the communication of an enterprise with third entities such as vendors and customers. Business-to-Business, Business-to-Customer applications, and Web site presentations help an enterprise to become more efficient and competitive in terms of modern economy. For home use, users can entertain themselves, communicate with friends, learn, or even order goods. Furthermore, Internet services offer new types of socializing such as forums, blogs, and social networks.

This chapter aims to present the main technology concepts and techniques which are applied to modern computer networks. Section 7.2 deals with the main computer network classifications based on their geographic coverage area and their connection topology. Section 7.3 deals with the physical infrastructure of computers' interconnection presenting the wired and wireless solutions that are commonly applied today. In Section 7.4, there is a short presentation of the major design models of a computer network. Section 7.5 explains the basic mechanism of Internet

Protocol (IP), which is the basis of data transmission in commonly used computer networks. Specifications and technologies regarding the available network software applications are presented in Section 7.6. Finally, Section 7.7 presents some of the future trends in computer networks.

7.2 Networks Categorization

Each computer network is characterized by a variety of factors which are associated with their physical connection medium, regional coverage, functional mode, and network topology. Computer networks are often classified by the scale of the region, which is covered by their nodes. Moreover, computer networks are also identified by their network topology, which describes the physical interconnection of their various elements.

7.2.1 Regional Coverage Categorization

Computer networks are classified into three categories in accordance with their covered geographical area: the Local Area Networks (LAN), Wide-Area Networks (WAN), and Metropolitan Area Networks (MAN) (Liu Sheng and Lee, 1992). Their main characteristics are presented in the following sections.

7.2.1.1 Local Area Networks

A LAN is a computer network that spans in a relatively small area. LANs are usually established in a building or a group of neighboring buildings. LANs are used mostly by companies and organizations in order to achieve the collaboration among their employees, interenterprise information exchange, and data security. Most LANs connect personal computers and office automation devices (printers, scanners, etc.). Each node in a LAN comes as an individual computer system, with its own Central Processing Unit (CPU) executing software programs, being able to use network share resources such as files stored in a different node, network printers, intranet applications, and so on. Moreover, LANs provide tools to users in order to communicate with each other.

The various LANs can be defined based on the following characteristics:

- *Network Topology:* The geometric arrangement of devices on the network is usually one of Ring, Bus, Tree, and Star topologies.
- *Communication Protocols:* The rules and encoding specification of data transfer are often following the peer-to-peer or client/server network architectures.
- *Physical Medium:* Nodes and network devices in LANs can be connected using wired infrastructure (twisted-pair wire, coaxial cables, or fiber optic cables) or wireless communication via radio waves.

The quality of data transfer on LANs' physical medium in a limited distance permits the data transmission at very fast rates (up to 1250 Mb/s in 10 Gb Ethernet), which is much faster than data transfer ratios over a telephone line. Besides the limited area coverage, another disadvantage of LANs is the limited number of computers that can be attached to a single LAN. Figure 7.1 depicts the way nodes are connected in a LAN network.

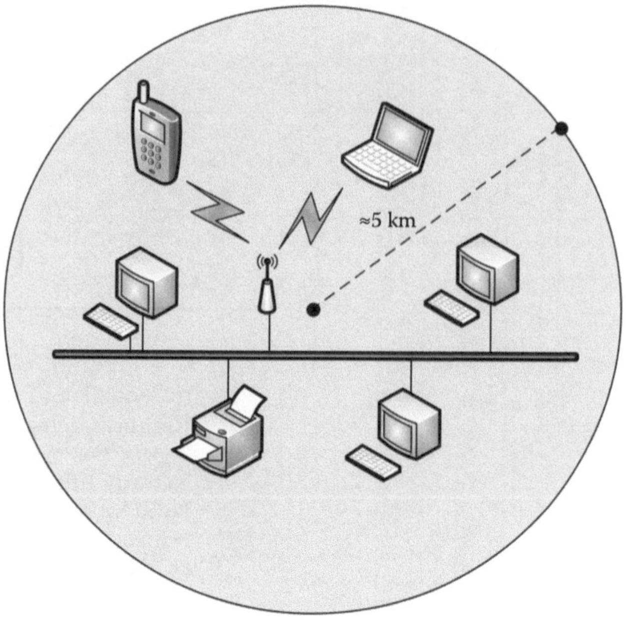

FIGURE 7.1 Local area network.

7.2.1.2 Metropolitan Area Networks

MANs are networks which cover larger geographical areas than LANS. The Institute of Electrical and Electronics Engineers (IEEE) define WANs as "A computer network, extending over a large geographical area such as an urban area and providing integrated communication services such as data, voice, and video" (IEEE, 2002, p. 6). WANs are usually established in a campus or a city and connect some LANs using high-speed connections, known as *backbones*.

MANs are characterized by three basic features, which discriminate them from the other computer networks:

1. *Geographical Area Coverage.* A MAN's range is between 5 and 50 km, which classifies it between LANs and WANs.
2. *Ownership.* Unlike LANs, WANs are not usually owned by a single organization or company. MAN's network equipment and links often belong to a service provider, which sells the service to the users or a consortium of organizations which need to collaborate with each other.
3. *Role.* MANs are mostly used as mediators to LANs in order to connect them with other networks in a metropolitan area. Also they provide services of sharing regional resources over a high-speed network.

MAN's connection technology is a combination of technologies used in LANs and WANs. For the interconnection between network nodes in a MAN, fiber optic cables and wireless technologies are commonly used. Figure 7.2 illustrates the usage of MANs as a mediator for shared connection between different LANs.

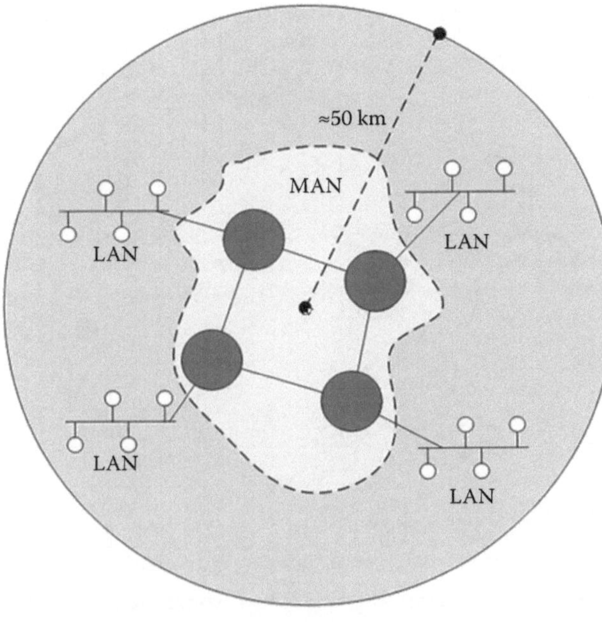

FIGURE 7.2 Metropolitan area network.

7.2.1.3 Wide Area Networks

WANs are networks that cover a large geographic area. WANs consist of connected nodes located in a state, province, country, and even worldwide. They are often comprised of connections of smaller networks such as LANs and MANs. The Internet is the most popular WAN. Furthermore, some subnetworks of Internet are considered as MANs such as Virtual Private Networks (VPN). On a smaller scale around the world, there are some corporate or research networks which use leased lines in order to create a MAN. Figure 7.3 depicts how a WAN interconnects various LANs and MANs in order to achieve data exchange between them.

WAN's nodes are usually connected over the telephone carriers. In order to achieve data transfer over telephone lines, WANs consist of a variety of devices which properly transform the signal traveling through the carriers. Devices, such as modems, WAN

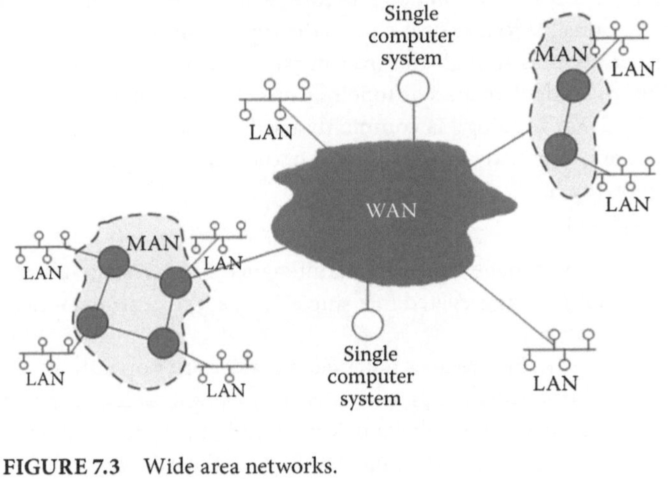

FIGURE 7.3 Wide area networks.

switches, access servers, channel service unit/digital service unit (CSU/DSU), Integrated Services Digital Network (ISDN) Terminal Adapters, comprise the basic infrastructure of a WAN. WANs are characterized by a variety of types of connections between their nodes (computers). The most common connection technologies are listed below (CISCO, 2003):

- *Point-to-Point Links.* A point-to-point link is a single preestablished WAN connection path from one node to another through a carrier network, such as a telephone company. On this occasion, these lines are leased from the provider which allocates pairs of wire and network equipment in order to establish a private line. Generally, the cost of leased lines is relative to the required bandwidth (data transfer rate) and distance between the two connected nodes.
- *Circuit Switching.* Switched circuits are data connections that are not permanently established. These connections are dynamically initiated when a request for data transfer occurs and they are terminated after the completion of the communication. This type of communication is very similar to a voice telephone call. A representative example of Circuit Switching is the ISDN connection.
- *Packet Switching.* In packet switching, the nodes share common carrier resources which decrease the cost of network usage. Utilizing the appropriate network infrastructure, the carrier creates virtual circuits (connections) between two nodes and transfers the data in stamped packages in order to identify their destination node.

7.2.2 Computer Network Topologies

A computer network's physical topology represents the physical design of a network including the devices, location, and cable installation. Bus, Ring, Star, Tree, and Mesh (Yeh and Siegmund, 1980), presented in the following sections, are the most common topologies in today's computer networks.

7.2.2.1 Bus Topology

In a bus topology computer network, all the network nodes are connected to a single data transmission medium, usually to a cable. This medium is called *bus* or *backbone* of the network and it has at least two endpoints. In the cases where there are exactly two endpoints the topology is characterized as "linear bus" whereas when there are more than two endpoints it is called a "distributed bus" topology.

In bus topology, data travels as a signal across the bus until it reaches the destination node which receives it. In order to avoid signal collision when two signals are traveling on the bus at the same time, a method called carrier sense multiple access with collision detection (CSMA/CD) is used to handle this type of situation. Furthermore, in each endpoint of the bus, there are terminators that are responsible to absorb the signal in order to avoid back reflection to the network. The nodes are just waiting for receiving a signal and they are not responsible for forwarding

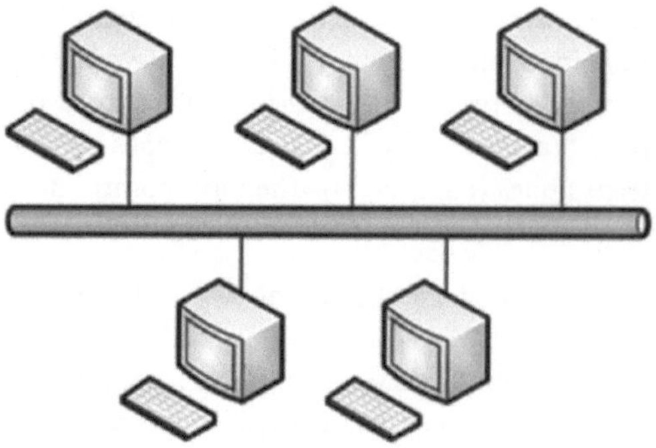

FIGURE 7.4 Bus topology.

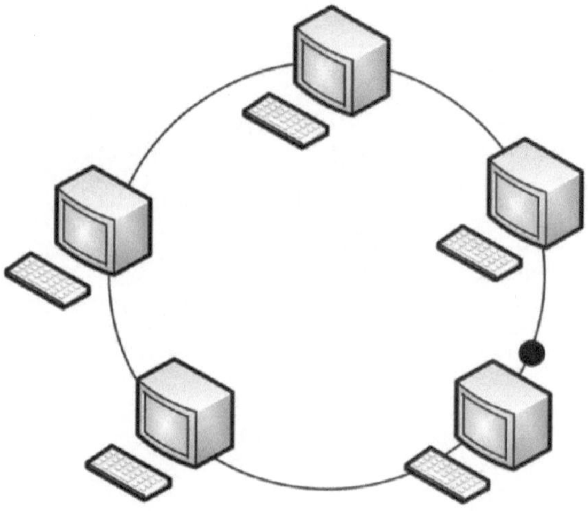

FIGURE 7.5 Ring topology.

the signal to the rest of the network. Thus, the bus topology is classified as *passive*. Figure 7.4 illustrates computer connections in bus topology.

Some of the advantages and disadvantages of the bus topology are listed below.

Advantages:

1. When a node fails (i.e., shut down), the network is still alive and the failure does not affect the communication between the other nodes, because data transmission takes place via the bus.
2. It has better performance regarding the data transfer rate compared to ring and star topologies.
3. Its installation is easy, fast, and low cost when there is a small number of nodes.
4. It is easily extended by expanding the bus medium.

Disadvantages:

1. The effort of management of two transmissions at the same mode confine network performance.
2. A failure in bus cable will result in network deactivation.
3. The number of nodes that can be supported is relative to the length of the bus cable.
4. It has decreased performance in heavy traffic and when additional nodes are added.
5. Although the installation is easy, maintenance and troubleshooting is difficult, thereby increasing its cost over time.

7.2.2.2 Ring Topology

In ring topology, the nodes are connected in such a way that they illustrate a logical ring, as depicted in Figure 7.5. Data is transmitted in circular fusion from one node to another. The most commonly used ring topology is the *token ring* which is standardized as IEEE 802.5 (IEEE, 1998). For the establishment of a computer network following the token ring topology, each node has an attached network device called the multistation access unit (MSAU) that is responsible for the transmission of the signal between the nodes. Each MSAU has two ports: a ring in (RI) port and ring out (RO) port. The RI of each node is connected to the RO of the neighbor node, and last node's RI is connected to the first node's RO in order to "close" the circle.

The usage of ring topology has some advantages and disadvantages:

Advantages:

1. There is no need of additional network devices apart from the connected computers (nodes).
2. It is easily installed and administrated.
3. A network failure is easily located decreasing thus the effort for troubleshooting.

Disadvantages:

1. A node's failure might cause the interruption of data signal transmission.
2. The expansion of the network can cause network disruption because at least one connection must be disabled until the completion of the new node's installation.

7.2.2.3 Star Topology

Star topology is most widely used in LANs and especially Ethernet networks. It is based on a central connection point to which all nodes are connected. The central connection point may be a computer *hub* or a simple *switch*. The hub (or switch) is responsible to transmit data to the right destination node and generally manages all the transmissions through the network. Figure 7.6 depicts the star topology of a computer network.

The star topology is commonly used because its advantages overtake its disadvantages in most of the cases as depicted in the following list.

Advantages:

1. The management and maintenance is easy since all the effort is aggregated in a single device, the central connection point.
2. Node's failure does not affect the network's operation.
3. The network's expansion is easy by using a cable to connect the new node to the hub. Moreover, it is easy to expand the hub's ports by attaching new hubs to the existing one.

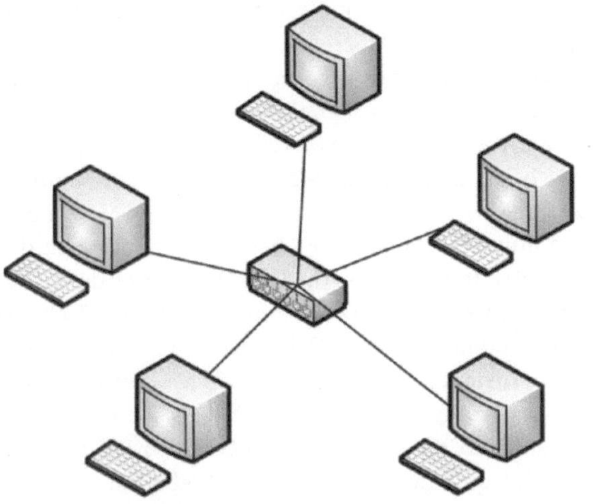

FIGURE 7.6 Star topology.

Disadvantages:

1. In case of central connection point's failure, network stops its operation.
2. The covered geographic area of the network is dependent on the length of the cable which connects a node to the central point. Because in most star topology computer networks, the cables have a maximum allowable length, the potential covered area of the network is confined.

7.2.2.4 Tree Topology

The tree (or expanded star) topology is a combination of bus and star topologies. It consists of a group of nodes assembled in star topologies which are connected to a backbone, as depicted in Figure 7.7. Tree topology is used in order to exploit the advantages of the two combined topologies. It can provide fast data transfer in long distances because of bus topology's backbone and also it can easily be expanded and maintained due to the independency of star-configured node groups.

Advantages:

1. It provides fast point-to-point communication between the star-configured node groups.
2. It is easily expanded when few nodes are added.

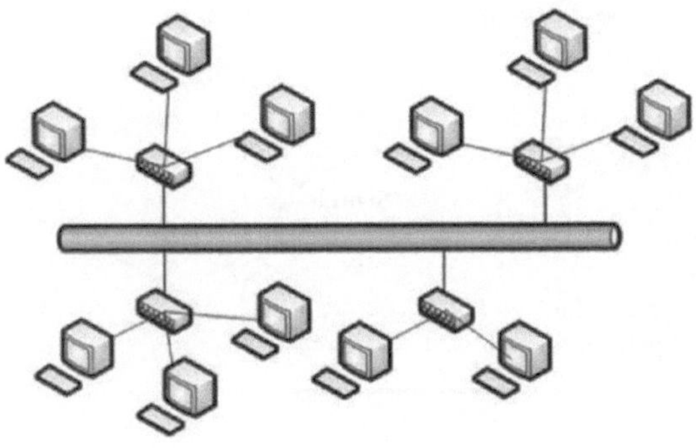

FIGURE 7.7 Tree topology.

3. It is supported by several network equipment vendors.
4. The network can cover a bigger geographic area.

Disadvantages:

1. Each star-configured node group has limited coverage geographic area.
2. Backbone's break will cause communications failure.
3. It is difficult to be installed and configured because of the high demands in wiring and equipment.

7.2.2.5 Mesh Topology

In this topology, all the nodes are connected to each other as illustrated in Figure 7.8. When data has to be transmitted from one node to another, the entire network automatically decides the shortest path to the destination node. The decision is made after negotiation of all nodes, which results in the path that the data must follow to reach its destination, even if the direct connection is broken down. Mesh topology is mainly adopted by wireless computer networks where it is easily to be installed and deals with the problem that a wireless connection works only if the two nodes can "see" each other. Mobile *ad hoc* networks (MANET) are wireless networks based on the mesh topology.

Advantages:

1. A node's failure does not have any impact to the network operation.
2. Multiple data can be transferred simultaneously through different routes.

Disadvantages:

1. A lot of connections may be inactive for a long time in case they are not used for routing the transmitted data.
2. Its installation is difficult for wired computer networks.

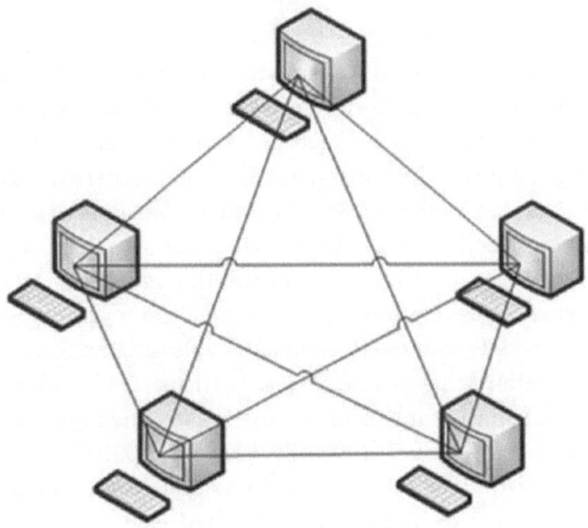

FIGURE 7.8 Mesh topology.

7.3 Network Infrastructures

Computer communication networks are composed of various elements in order to manage data transfer between two machines. Both software and hardware are used for aggregating data transmission through a network. Various network devices and equipment are used in order to install a computer network. Cables or antennas are responsible to transfer the data as signals in wired or wireless networks. Devices such as routers, hubs, and switches have mechanisms for connecting two separated networks or direct data transmission through the network nodes until data reaches its ultimate destination. A computer network installation demands the appropriate devices and equipments regarding the physical connection among its nodes.

7.3.1 Wired Networking

Wired networks are networks where the physical transfer medium is a cable. Their first form has been a small number of computers wired with a cable. The evolution of techniques for transmitting data over telephone and power cables in conjunction with the technology progress on cable design and materials available in today's networks can even spread over large geographical areas providing fast and reliable data communication. There are many types of wired networks with different infrastructure requirements. In the following sections, the most commonly used network infrastructures are presented such as Ethernet and Internet connections technologies using telephone lines.

7.3.1.1 Ethernet

Ethernet is a family of networking technologies used for the installation of LANs. In 1985, Ethernet was standardized by IEEE in the IEEE802.3 standard. The main characteristic of Ethernet is the support of carrier-sense multiple access with collision detection (CSMA/CD) (IEEE, 2009) technique in order to transmit data over a cable. Moreover, an Ethernet consists of network devices as routers, switches, and hubs in order to interconnect the participant computers.

CSMA/CD is a technique for detecting and resolving signal collisions on the physical transmission medium. When a device wants to send a signal to the network, it first checks the availability of the transmission medium and then sends the signal reducing the possibility of signal collision's occurrence. Nevertheless, it is very much possible for signal collisions to occur. In case of signal collision, CSMA/CD stops the signal, informs the network's node, and tries to retransmit the signal.

Ethernet's most common physical medium is either twisted-pair or fiber optic cables. Twisted pair cables are made of copper and are similar to telephone cables. Their major classification is made based on their maximum transfer rate. The most common categories are 10BASE-T, 100BASE-TX, and 1000BASE-T, which can transfer data at rates of 10 Mbit/s, 100, and 1000 Mbit/s, respectively. At the endpoints of a twisted-pair cable, there are connectors in order to plug them to network devices (network cards). This connector is an 8-position modular connector with 8 pins, usually called RJ45.

Fiber optics are cables with optical fibers which direct light pulses that carry the data signal. They permit data transmission over longer distances than twisted-pair cables in high transfer rates, with minimal signal. Moreover, their cost is high, thus they are mainly used in connections between neighbor buildings and building's floors. The fiber optic's bandwidth (transfer rate) is related to its length. The longer a fiber optic is, the lower the transfer rate gets. There are two major types of fiber optics: (a) Single Mode Fiber Optic Cables that can transfer data up to 10 Gb/s for distances up to 3 km and (b) Multimode Fiber Optic Cable which transfers data at 100 Mbit/s for distances up to 2 km, 1 Gbit/s to 220–550 m, and 10 Gbit/s to 300 m.

The main equipments used in the installation of an Ethernet network are the hubs, switches, and routers. Hubs and switches are devices that connect multiple twisted-pair or fiber optic Ethernet devices together and permit data exchange. The main difference between them is that hubs allow only one connected device to transmit to the network at the same time resulting in collisions and retransmissions when the others try to transmit data, in contrary to switches that manage multiple data transfers. Thus, there is a limited number of hubs that can be used in an Ethernet network.

Routers are devices used for the interconnection of two or more different computer networks. Routers are more advanced devices compared to hubs and switches. Their primary responsibility is to determine the path that an incoming data signal will follow to the network until its destination node. Routers translate the notation that defines the receiver node from an external network to the one defined in its network, the opposite translation is made in the cases of outgoing data to the external network. Furthermore, routers support the interconnection with other nodes of network by different types of physical medium such as twisted copper, fiber optics, or wireless. Routers are also used for the connection of office LANs to the Internet over broadband technologies. Figure 7.9 demonstrates the network infrastructure of an Ethernet network.

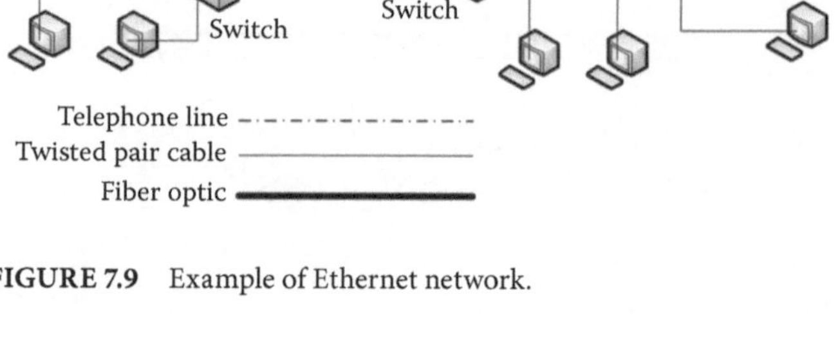

FIGURE 7.9 Example of Ethernet network.

7.3.1.2 Telephone Line-Based Networks

The Internet is the most well-known computer network. Although its main infrastructure is composed of cable connections, usually of fiber optics, in order to transmit data faster in long distances, the majority of the connected nodes (homes and organizations) use telephone lines for transmitting data to the Internet's backbone network. The challenge in these technologies is to deliver high transfer rates over the existing public switched telephone networks (PSTN).

Dial-up Internet Access was for years the only technology for connecting to the Internet. The only device needed in this case is a modulator–demodulator, commonly known as *modem*. A computer is attached with a modem which makes a call to another modem located in the Internet service provider (ISP) and establishes a modem-to-modem link, allowing the computer to communicate with other computers over the Internet. The modem's connection bandwidth is limited to 56 kbit/s which is considered the lowest compared to the other Internet connection technologies.

Integrated Service Digital Network (ISDN) is a telephone data service standard defining how data is transmitted over PSTN without interfering with the voice transfer of a phone call. ISDN is considered *broadband* technology's family member. Broadband is a term used to characterize Internet access technologies with high data transfer rate. The basic equipment used for ISDN connection is an ISDN modem which allows the connection to the PSTN of telephone, fax, and computer working simultaneously over the telephone line. ISDN connections usually deliver data on 64 kbit/s or 128 kbit/s (ISDN Basic Rate Interface—BRI), but there are some implementations with higher rates.

T-1/DS-1 are highly regulated services for voice and data transmission. T-1 connections are traditionally intended for organizations and enterprises. Thus, their use was in the connection of enterprise LANs to the Internet or interenterprise WANs. T-1 aims to deliver high quality services to enterprises consequently increasing their maintenance cost. T-1 connections can transmit data up to 1.544 Mbit/s.

Digital Subscriber Line (DSL) contains a set of technologies providing digital data transmission over PSTN at high transfer rates. DSL technologies are able to transmit data from 144 kbits/s to 200 Mbits/s, but new technologies are continuously developed increasing the delivered bandwidth. The most common DSL technology used worldwide is the Asymmetric Digital Subscriber Line (ADSL) and its improvements. ADSL delivers different transfer rates for downloading and uploading data in order to provide better services to the subscribed users, since the average users download more than they upload. Table 7.1 presents a list of major DSL technologies and their maximum download transfer rate. The basic equipment used for establishing a DSL connection is the digital subscriber line access multiplexer (DSLAM) and a DSL modem (also DSL Transceiver or ATU-R). DSLAM is located in the facilities of an ISP and is connected to the Internet's backbone network. DSL modem establishes a connection to the DSLAM through telephone lines.

TABLE 7.1 xDSL Common Technologies

xDSL Technology	Download Bandwidth
ISDN Digital Subscriber Line (IDSL)	128 kbit/s–144 kbit/s
High Data Rate Digital Subscriber Line (HDSL/HDSL2)	1.544 Mbit/s–2 Mbit/s
Symmetric Digital Subscriber Line (SDSL/SHDSL)	1.544 Mbit/s–2 Mbit/s
Symmetric High-Speed Digital Subscriber Line (G.SHDSL)	192 kbit/s–4 Mbit/s
Asymmetric Digital Subscriber Line (ADSL)	8 Mbit/s
Asymmetric Digital Subscriber Line 2 (ADSL2)	12 Mbit/s
Asymmetric Digital Subscriber Line 2 Plus (ADSL2+)	24 Mbit/s
Very High Speed Digital Subscriber Line (VDSL)	52 Mbit/s
Very High Speed Digital Subscriber Line 2 (VDSL2)	200 Mbit/s

7.3.2 Wireless Networking

The high cost and the installation difficulties of wired networks made the academic and industrial community to search for new ways of transmitting data, especially regarding the physical medium. These efforts concluded in the evolution of wireless networks which mainly use radio waves to transfer the data signal instead of cables. Today, wireless networks are widely used for personal communication through wireless LANs or mobile and cellular networks. Although there is a variety of wireless networks and technologies such as IEEE 802.16 (WiMAX) and Satellite Internet, in the following sections there is a presentation of today's most significant and widely used wireless technologies.

7.3.2.1 IEEE 802.11 (Wi-Fi)

Following the example of communication standardization in wired LANs, IEEE collected a set of standards for wireless LAN communications which is named IEEE 802.11 (IEEE, 2007). IEEE 802.11 is also known as Wi-Fi from Wi-Fi Alliance's trademark which is attached to the majority of wireless network devices.

Wi-Fi networks need no physical wired connection in order to operate. The data is transferred from sender to receiver by using RF technology. RF is a frequency within the electromagnetic spectrum associated with radio wave propagation. RFs are supplied to antennas which are responsible for propagating the signal through air. The basic equipment of a Wi-Fi is the Access Point (AP) and the wireless network adapter. APs are empowered with antennas and special hardware which are used for broadcasting wireless signals that computers can detect and retrieve through their attached wireless network adapter.

The major problem in the Wi-Fi networks, which follow all the RF-based networks, is that the area between the two participants in a communication, the AP and a computer, must be clear of physical barriers like buildings, hills, mountains, and so on.

Thus, the installation of APs requires a soil survey in order to find the appropriate location that will cover the majority of the neighboring areas.

Today, there are a number of Wi-Fi standards defined by IEEE with different characteristics based on the retained RF band, data transfer rate, the covered area, and their sensitiveness to physical barriers. The most used standards are

- *802.11a*-based networks operate using the relatively unused 5 GHz band and transfer data at 54 Mbit/s. Moreover, their signals are absorbed more easily by solid objects in their path due to their smaller wavelength resulting in increased bandwidth.
- *802.11b* standard defines wireless networks which are using the 2.4 GHz band and transmit data up to 11 Mbit/s rate. It is more resistant to solid barriers than 802.11a but suffers interference from other networks operating in the same band as microwave ovens, Bluetooth devices, baby monitors, and wireless telephones.
- *802.11g* is an improvement over 802.11b increasing the transfer rate up to 22 Mbit/s. 802.11g is the most used standard today.

In 2003, IEEE combined all its wireless standards and their improvements to one single standard named IEEE 802.11-2007.

7.3.2.2 Bluetooth

Bluetooth wireless technology is a short-range communication technology. Bluetooth is aiming to be the interconnection between a small number of portable devices, usually one to five, rather than creating massive wireless LANs (Haartsen et al., 1998). The key features in this technology are robustness, low power consumption, and low cost. Thus, Bluetooth is mostly used for data exchange between mobile phones, PDAs, and laptops. Furthermore, it is widely used for connecting computer peripherals such as mouse, keyboards, and printers without using wires.

Bluetooth operates at the 2.4 GHz band and uses the adaptive frequency hopping (AFH) capability to reduce interference between wireless technologies sharing this band. The only equipment used in Bluetooth networks is a Bluetooth network adapter. In the market, the majority of mobile phones and laptops have embedded Bluetooth adapters. The Bluetooth range is specified as 10 m and is able to transmit data up to 3 Mbit/s rates.

7.3.2.3 General Packet Radio Service

General Packet Radio Service (GPRS) is the most common mobile system for transmitting data over mobile networks such as for Global System for Mobile Communications (GSM) and Universal Mobile Telecommunications System (UMTS). GPRS provides mobility management, session management, and data transport based on the Internet Protocol (IP) (Hämäläinen, 1999). Also, it provides some additional functionalities such as billing and location tracking. Today's mobile phones can access the Internet through GPRS and their users are able to read e-mail and visit Web pages.

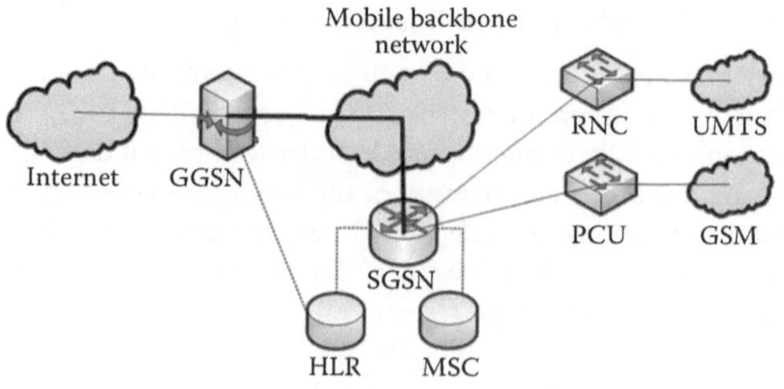

FIGURE 7.10 GPRS core network architecture.

Figure 7.10 defines GPRS core network infrastructure, which consists of the following components:

- *Gateway GPRS support node (GGSN)* is the main component of GPRS since it is responsible for the network's interconnection with other packet-switched networks like the Internet.
- *Serving GPRS support node (SGSN)* is responsible to deliver data segments between two mobile stations. Its functionality is similar to that of routers.
- *Home location register (HLR)* includes a database with all the required information regarding the subscribers of a GSM network.
- *Mobile switching center (MSC)* is responsible for or routing voice calls and SMS as well as other services (such as conference calls, FAX, and circuit-switched data) both to mobile network and classic PSTN.
- *Packet control unit (PCU)* is controlling the frequency channels for data transmission.
- *Radio Network Controller (RNC)* is an element of UTMS network for managing the RF used and the data transmitted in and out of the subnetwork.

7.4 Network Design and Reference Models

A necessary condition in order to establish a communication between two computer systems in a network is the common support of all the operations that take place during data exchange including signal transmission, destination identification, quality of service, data representation, and so on. The way these operations are implemented in a computer network depends on its design. In most cases, a network design consists of a stack of layers or levels. Each layer provides a specified function and offers certain services to higher layer hiding the details of their implementation. Following the layered approach, today's computer networks are designed based on two well-known reference models: the ISO Open Systems Interconnection (OSI) Reference Model (ISO/IEC, 1994) and the Transmission Control Protocol/Internet Protocol (TCP/IP) Model (Braden, 1989).

7.4.1 ISO Reference Model

The OSI model (as it is commonly known) is a collection of international standards for network communication protocols used in a stack of layers that is presented in Figure 7.11. The OSI model is proposed from the International Standards Organization (ISO) and International Electrotechnical Commission (IEC) based on the model presented by Zimmermann (1980) and its revision from Day (1995).

The OSI model has seven layers, and its design is based on the following principles:

1. Definition of a layer where a different abstraction is needed between communication operations.
2. Layers' functions are clearly and specifically defined.
3. Layers' functionalities must be able to define internationally standardized protocols.
4. Layers' design must achieve the minimal piece of information transferred between layers.
5. Layers' size must be large enough to include similar functionalities and small enough to achieve a simply described architecture.

OSI layers are separated into two groups: (a) the media layers consist of operations made in order to ensure the physical transfer of data in bits (signal) from one node to another and (b) the host layers support functionalities for managing the communication channel and the interpretation of the transmitted data. The functionalities of the seven layers of the OSI model are

The *Physical Layer* is responsible to ensure that the data's bits will be transmitted correctly over the physical transmission medium (cable, RF). This layer is attached to the network hardware and takes care that when a node sends 1 bit, the destination node will receive 1 bit and not 0 bits.

The *Data Link Layer* locates the most appropriate line to transmit the data to the next node in a computer network, eliminating the possibility of error occurrence during transmission. Moreover, regarding broadcast networks, the data link layer is responsible to manage access to the shared channel.

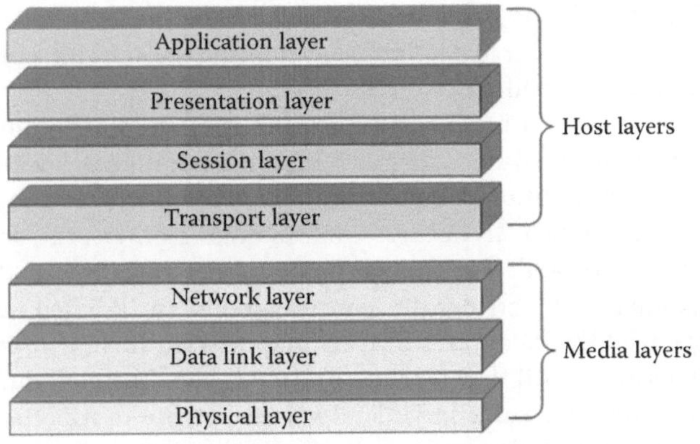

FIGURE 7.11 OSI reference model.

The *Network Layer* contains operations for determining the route which will be followed in a computer network in order that data reaches the destination node. The routing operation is critical for computer networks. Depending on each network, the followed routes can be predefined or established at the start of a communication or dynamically changed during data transmission. In the last case, data fragmentation may exist. Another responsibility of the network layer is to manage the different notations of the node's addresses on different subnetworks.

The *Transport Layer* manages and monitors the data transfer providing at the same time the appropriate abstraction to the upper layers regarding the hardware used. In transport layer, the data that must be transmitted is separated into small pieces that are forwarded to the network layer in order to be sent to their destination where they are joined back to one. This separation permits better management of data transmission through a computer network until it arrives at the destination node.

The *Session Layer* supports the establishment of communication sessions between users of different machines. Operations in this layer manage when and how data is sent from one node to another.

The *Presentation Layer* is concerned with the syntax and semantics of the information transmitted. This layer is responsible for defining an abstract form of transmitted data structures in order to be correctly interpreted from the receiver.

Finally, the *Application Layer* includes the data representation standards used in the user's applications. An example of such an application protocol is the HyperText Transfer Protocol (HTTP), which is widely used on the Internet and represents the data of a Web page that can be presented on a Web browser (i.e., Internet Explorer, Mozilla Firefox, etc.).

7.4.2 TCP/IP Reference Model

Although the OSI Model describes the main principles of the design of a computer network, today's most used reference model is the TCP/IP model, which is part of the Internet Protocol Suite. The reason is that it is the basis of the majority of LANs and WANs used today, as well as the Internet. The TCP/IP model is based on the TCP/IP protocol and was created by the Defense Advanced Research Projects Agency (DARPA) in order to implement the predecessor of Internet, the ARPANET. The model was first presented by Cerf and Kahn (1974), but there have been some improvements over the last few years. The related documents and protocols of the TCP/IP model are maintained by the Internet Engineering Task Force (IETF). The basic challenge behind the TCP/IP model is the aspect of keeping the communication alive between two network nodes, even if some of the network devices or transmission lines in between suddenly fail operation. Thus, data transmission routes can be easily changed during the transmission. Furthermore, the TCP/IP

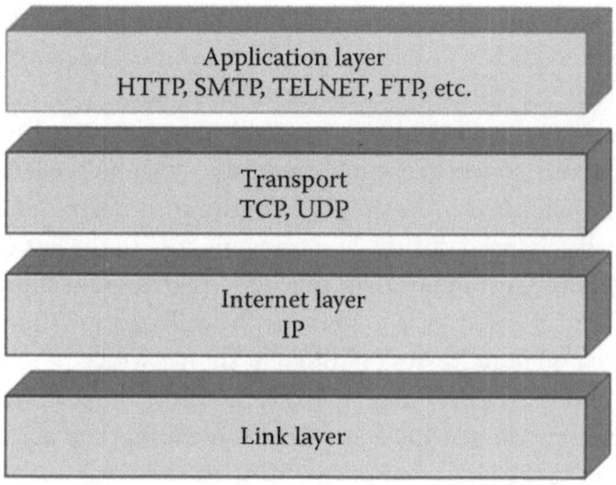

FIGURE 7.12 TCP/IP reference model.

model provides a flexible architecture-supporting applications with demanding requirements such as transfer of big-size files or real-time speech and video transmissions.

Figure 7.12 presents the four layers of the TCP/IP model. Unlike the OSI model, the TCP/IP model is focusing on the software implementation of computer communications and it is designed in order to work in a variety of hardware options. Thus, the physical communication layers are excluded from its definition. The main layers are described below:

Link Layer is responsible to transfer data packets (data fragments) from the sender node to the receiver. This layer includes specification of network addressing translation such as Media Access Control (MAC) used by the IP. Although some of these specifications are implemented by hardware, standards regarding physical data transmission are not explicitly defined.

Internet Layer is the core layer of the TCP/IP Model. Its role is to manage the transmission of data packets into any network without preestablished routes to the destination. Packets are traveling through network infrastructure, knowing which the destination is, but not exactly how to reach it. Packets are traveling through different networks, continuously changing their travel path until finding the target node. This ensures that the data will finally reach its destination apart from hardware failures or different network walkthroughs.

Transport Layer is similar to that of the OSI model. It allows two network nodes to establish a communication. The two defined protocols of this layer are TCP and User Datagram Protocol (UDP). TCP defines processes and mechanisms ensuring that the data will reach the receiver intact without time limits, contrary to the UDP, which has less check mechanisms aiming for faster data transmissions, thus it is used in applications needed for fast delivery of data such as real-time speech or video.

Application Layer contains protocols regarding specific network applications. Such protocols are the virtual terminal (TELNET), file transfer (FTP), and electronic mail (SMTP), Domain Name System (DNS), and HTTP.

OSI and TCP/IP models both follow the layered stack-based design methodology and they have a lot of similarities. But their basic principles are different and thus have some major differences. The OSI model is more detailed and defines standardized protocols for all the steps of data transfer through a computer network, from the physical transmission (signal bits) to the application-specific data aiming for the best performance of fast and reliable data communication. On the other hand, the TCP/IP model deals with large networks that most of them consist of smaller different ones from the perspective of equipments and physical transmission mediums. Its main purpose is to ensure that data will reach its destination, traveling dynamically through different networks. Thus, in the TCP/IP model, contrary to the OSI model, there are no media layers or presentation and session layer.

7.5 Communication over Modern Networks

Computer data communication, as mentioned above, is based on series of standardized protocols defining in detail all the steps of the transmission processes. Nowadays, the majority of the used networks around the world are built following the standards of the Internet Protocol Suite, which also contains the aforementioned TCP/IP model. Such networks are the Internet, enterprise and campus LANs/MANs, and small offices/home LANs, all supporting either wired or wireless connections. Thus, in this section, the most basic protocol of the Internet Protocol Suite, IP, will be presented as the basic concept of today's computer networks.

7.5.1 Internet Protocol

IP is used in order to transmit data across packet-switched computer communication networks (Postel, 1981). Its main functionality is to transfer blocks of data called *datagrams* between two computers. Moreover, IP provides datagrams fragmentation and reassembly in order to be able to operate in networks supporting small packet transfers. The key concept of IP's transmission mechanism is that each computer in the network is identified by a fixed-length address, known as *IP Address*. The IP address is structured in order to define the location and the network where a computer is residing.

There are two basic functionalities of IP: addressing and fragmentation. Each datagram contains additional headers which are additional information attached to the actually transmitted data. IP headers contain among others, the destination's IP address. The various modules installed in the computers and network devices (routers) in an IP-based network read the datagrams headers and forward them through the network until they reach their destination. These modules share common rules in order to interpret the destination's IP address and also are able to execute operations for making decisions regarding the path that a datagram will follow, which also is called routing. Furthermore, each of the aforementioned modules supports mechanisms for fragmenting and assembling datagrams. Information needed from a node in order to assembly a fragmented datagram is stored in the IP header and transmitted with the datagram fragment. IP uses four basic mechanisms in order to transmit data over a network:

1. *Type of Service* is a generalized set of parameters which characterize the choices provided in the various subnetworks consisting of the entire network. These parameters are used by the network devices, called gateways, in order to select how the datagram will be transmitted through a particular network.
2. *Time to Live* defines the duration that a datagram can travel through the network until it reaches its destination. It is like a countdown which when it reaches zero, the datagram is destroyed.
3. *Options* are parameters for specific functionalities in special communications. Information in options may include timestamps, security, and special routing.
4. *Header Checksum* is a code generated according to the transmitted data. It is used from the destination computer for verification of the received data.

7.5.2 IP Address

IP Address is the fundamental mechanism of IP-based networks by pointing the destination node in a data transmission. IP Address is a numerical number label assigned to each node of the network. The concept behind IP address is to serve two principal functions: computer or network identification and location addressing. Today, Internet Protocol Version 4 (IPv4) is the dominant protocol-defining address structure. Its successor, Internet Protocol Version 6 (IPv6), is also an active protocol worldwide and is supported from the majority of network equipments worldwide, but it is not as popular as IPv4.

IPv4 is commonly referred to as IP Address because of its wide acceptance. Thus in personal computers, the term IP Address is used in order to define the IPv4 address of a computer or device (printer, network hard disk drive, camera, etc.) in a network. IPv4 is defining an address label of a 32-bit number consisting of 4 bytes. Each byte can be represented in the decimal numbering system as a number from 0 to 255. Thus, IPv4 addresses are represented by four decimal numbers (0–255) separated by three dots. For example, an IPv4 address is "192.168.2.13."

The continuous growth of Internet over the last few years resulted in the foundation of a global organization for managing the assignment of IP addresses to the computers connected to the Internet, the Internet Assigned Numbers Authority (IANA). IANA, in order to manage address allocation worldwide, is cooperating with five Regional Internet Registries (RIRs), which are responsible to assign IP Addresses to Local Internet Registries (LIRs), which usually are associated with each country. IANA with RIRS and LIRs log the assigned address so it is easy to determine the geographic location (country or city) of a connected computer to the Internet. Moreover, IETF, in order to define Internet's architecture, classified the various networks in five classes A, B, C, D, and E. Practically, only classes A, B, and C are used. The classification is referred as *classful network* architectures. Each class can be determined by the first number of the IPv4 address. Each class defines the number of networks and connected addresses (computers) that can be represented using its notation. Table 7.2 lists the three classes and their characteristics.

TABLE 7.2 Classful Network Classes

Class	Range of First Number	Number of Networks	Number of Addresses
Class A	0–127	$2^7 = 128$	$2^{24} - 2 = 16.777.214$
Class B	128–191	$2^{14} = 16{,}384$	$2^{16} - 2 = 65.534$
Class C	192–223	$2^{21} = 2{,}097{,}151$	$2^8 - 2 = 254$

Since the address range of IPv4 is limited to 4.294.967.296 (2^{32}) possible unique addresses, the problem of IPv4 address exhaustion (Rekhter et al., 1996) became a major challenge in the last decade, especially after the limited use of IPv6 addresses. In order to address this problem, IANA reserved three IPv4 address ranges, one from each class, which are not used for Internet routing. These addresses are used in independent private networks. In order to connect the computers of such a network to the Internet, the private network is connected to the Internet through a router with Network address translation (NAT) functionality. NAT permits an Internet IP address to characterize the entire private network. When data is transmitted to one of the computers in the private network, NAT has mechanisms to route the data to the appropriate computer, although the last one has a private IP address. Table 7.3 depicts the IPv4 address range reserved for private networks.

IP addresses can be assigned to a computer in both automatic and manual ways based on the network's administration policy. Automatic IP address assignment appears in private LAN networks or computers directly connected to the Internet. The automatic assignment procedure is specified by Dynamic Host Configuration Protocol (DHCP). The idea behind DCHP is that when a new computer is connected to the network, it sends a request to a DHCP server which manages the available network's IP addresses and returns the assigned IP address.

TABLE 7.3 Reserved IPv4 Address Ranges for Private Networks

Referred Network Class	Address Range	Number of Addresses
A	10.0.0.0–10.255.255.255	16.777.216
B	172.16.0.0–172.31.255.255	1.048.576
C	192.168.0.0–192.168.255.255	65.536

7.6 Network Applications

Computer communication networks were created for assisting the computer users to execute various applications which demand interoperability and data exchange with other computers. These applications include services such as file transfer, electronic message exchange, and listen to audio or watch videos located on remote machines. Following the worldwide acceptance of the Internet, the number of network applications have increased and user requirements regarding performance and quality have become more demanding. Networks which are built based on OSI or TCP/IP model introduce various protocols that define how these applications must operate in the application layer. The following sections present the most common applications used and how they are implemented in the IP-based networks.

7.6.1 Domain Name System

In IP-based networks, each node's identifier is the IP address. IP address is a long number, thus it is difficult to be remembered by the applications' users. Moreover, in very large computer networks as Internet with a lot of private networks, some computers share the same IP address. The need for a memorable identifier led to the Domain Name System (DNS) development (Mockapetris, 1987). Domain name is a name that is associated with an IP address in order to identify a node in the network. Examples of domain names are usually available on the Web pages (www.mycompany.com, www.abc.org, service.university.edu, etc.) or in electronic mail addresses (joedoe@company.com, mary@goverment.gov, etc.).

DNS is based on a hierarchical, domain-based naming scheme and a distributed database system storing names and IP addresses association. The basic idea is that a network can be classified into groups of computers which may include another group recursively. This classification is made by social criteria such as country, type of organization, enterprise networks, and so on. These groups are called domains. Domain names consist of words separated with dots. Except the first word usually denoting the name of the node, the other words define the domain where it belongs and the domain's parent domains. The last word is usually either defining the country where the domain is located or the type of domain. An example to interpret the meaning of a domain name is *www.chu.cam.ac.uk,* which denotes that it refers to a Web server (www) of Churchill College (chu) of Cambridge (cam), which is an academic institute (ac) in the United Kingdom (UK). Identically, an electronic e-mail address refers to a person associated with a specific domain. For example, the mail address joedoe@company.com is possible to belong to a person named Joe Doe who works in an organization named Company, which is an enterprise (com).

The infrastructure of DNS is based on a distributed database system. Each database, which is called Name Server, is associated with one or more domains and keeps the matching of domain names to IP addresses. Moreover, it is able to publish its information to the other name servers and especially its parent domain name server. The resolve process of an IP address of a computer includes queries to the name servers serving the domains defined in the domain name until the node id is found.

7.6.2 File Transfer

File exchange between two machines is a mandatory service provided by a computer network. Most software applications are transfer files over a network. The challenge in file transfer service is fast transmission and reliability, meaning that a file will be correctly delivered despite its size and networks' possible failures. Although some applications use custom protocols on the application layer to transfer files, two are the dominant standards, the File Transfer Protocol (FTP) (Postel and Reynolds, 1985) and the Server Message Block/Common Internet File System (SMB/CIFS) (Hertel, 2003).

FTP is a network protocol used to transfer files between computers. FTP is built on a client–server architecture, where the client is the requester computer and the server is the one that sends the files. A characteristic feature of FTP is that it keeps separate connections for transmitting commands from client to server and for file transfer. FTP can transfer files in both ASCII mode as text or in Binary mode as bytes. Moreover, it supports resumption of transfer services when connection fails, and user authentication for private files. Furthermore, it is supported by the well-known Web browsers (Internet Explorer, Mozilla Firefox, etc.).

SMB/CIFS defines processes and services allowing a computer to have shared access to various resources on a computer network as files, printer, or serial ports. In most cases, SMB/CIFS is installed in an enterprise LAN and mainly used for file sharing, especially when the enterprise keeps all its files in a central file server. SMB/CIFS also operates based on a client–server architecture. SMB/CIFS is implemented by Microsoft Windows operating systems under the feature of "Microsoft Windows Network" and on Linux-based operating systems with the Samba service. In 2006, Microsoft introduced a new version of SMB/CIFS, the SMB2, with the commercial availability of Windows Vista, with significant upgrades on faster transfer rate and better management of large file transmission.

7.6.3 Electronic Mail

The need for text message exchanging among users in a computer network resulted in the development of the popular Electronic mail or e-mail. E-mail supports the transmission of a text message in a structure format defining the sender, recipient, subject, and so on. Furthermore, e-mails are able to include files as attachments to the main message.

Today's e-mail systems running on a computer network are based on an architecture with two major components: (a) User Agents, which provide the tools to the users in order to read and write e-mails and (b) Message Transfer Agents, which are responsible for transferring the messages from the sender to the recipients. E-mail systems support the following functions (Tanenbaum, 2003):

- *Composition* of the message by providing a text editor with additional features such as recipient selection, sender's e-mail address automatic completion, and so on.
- *Transfer* of the message to the recipient with the user's interference. The message is sent automatically to the recipient's computer or to a central e-mail server from where the recipient can retrieve it.
- *Reporting* of message delivery and informing of potential problems during the transmission (i.e., recipient's e-mail address is not valid or does not exist).
- *Displaying* incoming messages with a special viewer for reading.
- Messages *Disposition* refers to the ability the recipient retrieves and rereads old saved incoming messages.

There are two e-mail message formats used. The first is the RFC822, named by IETF's document standardizing the Internet e-mail (Crocker, 1982). RFC822 is used for sending messages in plain text. The second is the Multipurpose Internet Mail Extensions (MIME) format which allows messages with non-Latin characters, attachment files, and HyperText Markup Language (HTML) message formatting. Simple Mail Transfer Protocol (SMTP) is the Internet standard operating in the application layer which defines the e-mail's transmissions.

7.6.4 Hypertext Transfer Protocol

Hypertext Transfer Protocol (HTTP) is the dominant standard in the Internet for information exchange. Information can be represented in simple text or digital file such as images, videos, and sound. HTTP is the protocol mainly used for transferring HTML documents, which are interpreted and published as Web pages on Web browser software. The hypertext term is used to define text documents which, apart from the main text, include links to other text documents which can be immediately retrieved. Hypertexts can be enriched with images, sound, or video and then are called *hypermedia.*

HTTP is following a request–response model, where a computer sends a request for information and the other responses with the requested information. Information in HTTP can be considered as resources which are HTML files, image files, sound files, and so on. A key role in HTTP requests plays the Uniform Resource Locator (URL), which is the identifier used for denoting these resources. Examples of URL are http://www.company.com and http://service.uni.edu/sites/index.html. URLs are commonly used with the term *Web addresses.*

HTTP is used in various software applications in both Internet and enterprise LANs, because of its simplicity on development and maintenance. The main reason is that because Web browsers support data exchange using HTTP, these applications are installed in a single central computer (server) in the network and all the other workstations can use it without having to install additional software.

In the last few years, the need for application collaboration over networks resulted in the development of applications using Simple Object Access Protocol (SOAP). SOAP defines mechanisms of message and data transfer over HTTP, allowing the implementation of software applications which manage the interaction between two or more different systems. These applications, also referred in the literature as Web Services, are able to expose some internal functionalities of a software application installed on a machine to be invoked and used over a computer network by another application. Web Services is the base of the Service-Oriented Architecture (SOA), which aims to provide interoperability between different systems providing new automation services without the user's interference.

7.6.5 Multimedia

The term multimedia usually refers to an object consisting of multiple forms of media integrated together, like text, graphics, audio, video, and so on. Multimedia network software applications aim to deliver and display the multimedia content, as well as to provide the ability of user interaction with it. For example, a user can watch a video stored in another computer without downloading the entire file, and simultaneously be able to move forward the movie to a specific time stamp. The most used forms of multimedia are audio and video.

Because of the large file size of audio or video files, the data transmission was resulting in the user waiting for a long time until the end of file transfer. This situation motivated research and industry to introduce the concept of *media streaming,* which allows a user to start interacting with the media right after requesting it. The basic idea is the fragmentation of video/audio file in small-size pieces which are transferred very fast through the network. These pieces are transmitted consecutively, following the time line of the media, to the receiver where a software application combines and displays them to the user. The key concept in streaming is the *buffering* procedure, where the receiver waits for some time before starting to display the media content to the user, collecting and combining the first incoming pieces of video or audio. In that case, the media start displaying while the remaining pieces are delivered simultaneously. This is creating a virtual essence of real-time watching or listening a sound or video.

Moreover, another feature of streaming is the compression of the transmitted media. Audio or video files are usually demanding large storage space, for example a 2-h movie is stored in a DVD as files up to 4 GB total capacity. Compression is the process of media's digital content in order to decrease the file size, decreasing its quality in levels not annoying to the human. There are a lot of compression techniques for image, video, and audio compression. For media streaming, the most dominant techniques are MPEG-1 Audio Layer 3 (MP3) (from Moving Picture Experts Group) for audio compression, JPEG (from Joint Photographic Experts Group), Graphics Interchange Format (GIF) and Portable Network Graphics (PNG) for image compression, and MPEG-4 for video compression. There is a variety of protocols operating in the application layer of TCP/IP model such as Real-Time Streaming Protocol (RTSP), Real-time Transport Protocol (RTP), and the Real-Time Transport

Control Protocol (RTCP) which are responsible for managing the media streaming over networks. Moreover, most multimedia software applications such as Windows Media Player (Microsoft Corporation, Redmond, WA), Quick Time (Apple Corporation, Cupertino, CA), and Adobe Flash Player (Adobe Systems Corporation, San Jose, CA) are able to display streaming media using these protocols. Streaming technologies have also implementations in the area of Voice over IP (VoIP), which is the technology for executing phone calls over computer networks where the voice is traveling as data. Video streaming and VoIP are combined in order to provide video conference sessions where people from different locations can discuss and simultaneously "see" each other.

7.6.6 Network Applications Security

In the first few years of computer networks' appearance, data transmission security was not a popular issue. The last few decades of the growth of Internet raised the need for secured communications especially for sensitive personal or corporate information transfer. There are a lot of types of security attacks such as data theft during transmission, installation of harmful software to computers, which allow remote control and transmission of personal files to the network or even computer failure, and access to private information stored in software applications and databases of enterprises or public services. Network security problems can be classified in four areas:

- *Secrecy (or confidentiality)* which means that information stored in the network's computers or transmission over the network is protected from access by nonauthorized users.
- *Authentication* deals with the certification that a user or a system connected to another computer or application is the one who claims that it is.
- *Nonrepudiation* is a matter of business transactions and deals with the protection of a deal if a participant claims something different.
- *Data integrity* ensures that the information sent from one node to another is the one received without changes by some intruder.

Security is continuously evolving and new techniques appear since new types of attacks are discovered. However, some technologies offer preventive measures and general solutions to basic and common attacks.

Cryptography is a powerful tool for computer security. Data is encrypted using complex algorithms and they can be decrypted only by users owning the appropriate information to do that, in most cases this information is a cryptographic key. Cryptography is the basis of most of the defense techniques. It is used in order to transmit encrypted data over network preventing a third system to identify the contained information. Cryptographic keys are used in order to create certificates and digital signatures that, apart from permitting data encryption, uniquely identify a user or system. Today, most computer networks support the Public Key Infrastructure (PKI) approach which defines a set of hardware, software, people, policies, and procedures that manage digital signatures. Digital signatures are delivered to the users by an independent Certification Authority (CA) and contain Public Keys needed for data decryption. Digital signatures are also used for user authentication in software applications, replacing the traditional user-password authentication. Moreover, PKI is the basis of data exchange through secure channels following the specifications of Secure Shell (SSH) protocol. Also, cryptography-based protocols operate in the transport layer of TCP/IP model in order to secure data transmission such as Transport Layer Security (TLS) and its predecessor, Secure Socket Layer (SSL).

Apart from cryptography-based defense solutions, there are other techniques providing computer security to networks. Firewalls are able to detect and turn down suspicious connections to a computer or a private network. Antivirus software applications are mandatory to computers connected to a network, because they prevent the execution of harmful software such as computer viruses, malwares, or trojans that can cause file damages or system failures and they are easily spread over one network to other networks.

7.7 Conclusion

Computer networks consist of various technologies, techniques, methodologies, and industry standards defining specific requirements and solutions for the implementation of each component in computer communication procedure. Nowadays, computer networks research focuses on the Internet, the largest and most popular computer network. Academic and industry research efforts aim to optimize the existing technologies regarding the physical transmission medium, networks infrastructure, and the provided application services.

Phone line network is the basis of Internet, since it allows personal users and enterprises to connect to Internet on high data transfer rates with relatively low cost. DSL technology, which was the starting point of fast Internet, is continuously evolving such as the proposed Gigabit Digital Subscriber Line (GDSL) (Lee et al., 2007). Furthermore, the evolution of mobile Internet using cellular networks creates an opportunity for researchers to study the opportunity of worldwide wireless networks capable of reaching the performance and cost of today's cable-based networks. The main concept of research is the utilization of satellites as the transmission medium. Although there is already an approach called satellite Internet, its performance is not satisfying and thus of limited use. The majority of efforts are focused on the technology called Space Internet, especially from NASA which has proposed some architectures and technologies (Bhasin and Hayden, 2002). Today, Space Internet's architectures support IP-based addressing mechanisms and are able to transmit data from Earth to Deep Space (Khan and Tahboub, 2008).

The quality of services provided by computer networks and especially over the Internet is the major issue in current research. Security will always be an issue, since new attacks are discovered every day. Interenterprise and intraenterprise system interoperability is a hot issue in the Web services arena, since it is expected to automate and optimize the products' manufacturing, marketing, and delivery to the customer (Vernadat, 2010). Furthermore, next generation network applications are expected to support bandwidth-demanding services delivered in small devices through mobile and wireless networks (Sagan and Leighton, 2010). Audio and video streamings demand high transfer rates and their operation over low-speed networks is extremely challenging (Khan et al., 2009).

References

Bhasin, K. and Hayden, J. L. 2002. Space Internet architectures and technologies for NASA enterprises. *Int. J. Satellite Commun.*, 20(5), 311–332.

Braden, R. (Ed.) 1989. *Requirements for Internet Hosts—Communication Layers*. Internet Engineering Task Force, RFC-1122.

Day, J. 1995. The (un)revised OSI reference model. *ACM SIGCOMM Computer Commun. Rev.*, 25(5), 39–55.

Cerf, C. G. and Kahn, R. E. 1974. A protocol for packet network intercommunication. *IEEE Trans. Commun.*, 22(5), 637–648.

CISCO. 2003. Introduction to WAN technologies. *Internetworking Technologies Handbook*, 4th edition. IN, USA: Cisco Press.

Crocker, D. H. 1982. *Standard for the Format of ARPA Internet Text Messages*. Internet Engineering Task Force, RFC-822.

Haartsen, J., Naghshineh, M., Inouye, J., Joeressen, O. J., and Allen, W. 1998. Bluetooth: Vision, goals, and architecture. *Mobile Comput. Commun. Rev.*, 2(4), 38–45.

Hämäläinen, J. 1999. General packet radio service. In Zvonar, Z., Jung, P., and Kammerlander, K. (Eds.), *GSM Evolution towards 3rd Generation Systems*, pp. 65–80. Norwell, MA: Kluwer Academic Publishers.

Hertel, C. R. 2003. *Implementing CIFS: The Common Internet File System*. Upper Saddle River: Prentice-Hall.

IEEE. 1998. *ISO/IEC 8802-5: 1998E: Part 5: Token Ring Access Method and Physical Layer Specifications*. IEEE Standards Association, Institute of Electrical and Electronics Engineers Computer Society, NY, USA.

IEEE. 2002. *802-2001, IEEE Standard for Local and Metropolitan Area Networks: Overview and Architecture.* IEEE Standards Association, Institute of Electrical and Electronics Engineers Computer Society, NY, USA.

IEEE. 2007. *802.11-2007: IEEE Standard for Information Technology-Telecommunications and Information Exchange Between Systems-Local and Metropolitan Area Networks-Specific Requirements-Part 11: Wireless LAN Medium Access Control (MAC) and Physical Layer (PHY) Specifications.* IEEE Standards Association, Institute of Electrical and Electronics Engineers Computer Society, NY, USA.

IEEE. 2009. *IEEE Std 802.3-2008: Carrier Sense Multiple Access with Collision Detection (CSMA/CD) Access Method and Physical Layer Specifications Amendment 3: Data Terminal Equipment (DTE) Power via the Media Dependent Interface (MDI) Enhancements.* IEEE Standards Association, Institute of Electrical and Electronics Engineers Computer Society, NY, USA.

ISO/IEC. 1994. ISO/IEC 7498–1: *Information Technology—Open Systems Interconnection—Basic Reference Model: The Basic Model.* International Organization for Standardization & International Electrotechnical Commission, International Standard.

Khan, A., Sun, L., and Ifeachor, E. 2009. Content-based video quality prediction for MPEG4 video streaming over wireless networks. *J. Multimed.*, 4(4), 228–239.

Khan, J. and Tahboub, O. 2008. A reference framework for emergent space communication architectures oriented on galactic geography. *SpaceOps 2008 Conference*, May 12–16, Heidelberg, Germany.

Lee, B., Cioffi, J. M., Jagannathan, S., and Mohseni, M. 2007. Gigabit DSL. *IEEE Trans. Commun.*, 55(9), 1689–1692.

Liu Sheng, O. R. and Lee, H. 1992. Data allocation design in computer networks: LAN versus MAN versus WAN. *Ann. Oper. Res.*, 36(1), 124–149.

Mockapetris, P. 1987. *Domain Names—Concepts and Facilities.* Internet Engineering Task Force, RFC-1034.

Postel, J. 1981. *Internet Protocol: DARPA Internet Program—Protocol Specification.* Internet Engineering Task Force, RFC-791.

Postel, J. and Reynolds, J. 1985. *File Transfer Protocol (FTP).* Internet Engineering Task Force, RFC-959.

Rekhter, Y., Moskowitz, B., Karrenberg, D., de Groot, G. J., and Lear, E. 1996. *Address Allocation for Private Internets.* Internet Engineering Task Force Best Current Practice, RFC-959.

Sagan, P. and Leighton, T. 2010. The Internet & the future of news. *Daedalus*, 139(2), 119–125.

Tanenbaum, A. S. 2003. *Computer Networks*, 4th edition. Upper Saddle River: Prentice-Hall.

Vernadat, F. B. 2010. Technical, semantic and organizational issues of enterprise interoperability and networking. *Annu. Rev. Control*, 34(1), 139–144.

Yeh, J. W. and Siegmund, W. 1980. Local network architectures. In *Proceedings of the 3rd ACM SIGSMALL Symposium and the First SIGPC Symposium on Small Systems. SIGSMALL '80.* ACM, pp. 10–14.

Zimmermann, H. 1980. OSI reference model—The ISO model for architecture for open systems interconnection. *IEEE Trans. Commun.*, 28(4), 425–432.

8

Storage and Image Compression

Craig Morioka
UCLA Medical Imaging Informatics
VA Greater Los Angeles Healthcare System

Frank Meng
VA Greater Los Angeles Healthcare System

Ioannis Sechopoulos
Emory University

8.1 Introduction

A discussion on Picture Archiving and Communication Systems (PACS) is closely intertwined with the hospital information system (HIS), the overarching infrastructure for storing, viewing, and communicating the clinical data (see Figure 8.1). Starting in the 1960s, HIS was initially used for billing and accounting services, and HIS's capabilities have since grown considerably into the following areas: clinical care (e.g., medical chart review, computerized physician order entry, clinical protocol and guideline implementation, and alerts and reminders); administrative management (e.g., scheduling, billing, admission/discharge/transfer tracking); and as a data repository for clinical notes, discharge summaries, pharmacy orders, radiology reports, laboratory and pathology results, as well other structured information (Branstetter, 2007; Bui and Taira, 2010). Refer to Chapter 17 for a complete discussion on HIS.

Along with HIS is the Radiology Information System (RIS), a subcomponent of the HIS within the Veteran's Health Information Systems and Technology Architecture (VistA) electronic medical record system. The RIS system also interacts with the PACS through an HL7 interface. In most hospitals, the HIS is separate from RIS and also utilizes an HL7 interface to pass information between the two systems. Specific functions attributed to RIS include radiology patient scheduling, registering imaging procedures as the patient arrives at the modality, confirmation of the image exam after completion, and final radiology reporting. There is a tight coupling of the information in RIS with that of PACS. By way of illustration, RIS patient scheduling information or an imaging order is used to drive imaging study prefetching algorithms (from online archives) in anticipation of needed comparisons between studies for an individual; conversely, once the new study is acquired from the modality and within PACS, the radiologist's PACS work list is updated and an interpretative report must be generated for the corresponding imaging series. As imaging has become an integral part of the healthcare process, integration between HIS and RIS has been a concern for two main reasons: (1) to provide radiologists with a comprehensive context of the patient's history and presentation to reach a proper interpretation and (2) to ensure that radiology results are quickly disseminated to referring physicians within a medical enterprise.

An overview of the PACS components is presented in this chapter. Additionally, a discussion on the radiology workflow and how the image data and patient interact with the HIS/RIS/PACS infrastructure is provided. At the end of this section, a discussion on specific storage devices, magnetic, optical, and magneto-optical disk, as well as tape, is presented. Finally, image compression methods and algorithms, along with tests to characterize their impact on image quality, are discussed.

8.2 PACS Component Overview

A Digital Imaging and Communications in Medicine (DICOM) compliant PACS system consists of an image acquisition gateway (IAG), database (DB) server, workflow manager, HL-7/DICOM Broker, archive server, and display workstations integrated to allow the user to view images acquired at individual radiographic modalities (see Figure 8.2). The viewing workstations can have stand-alone applications running on the local computer or Web-based clients that deliver image studies through a Web browser.

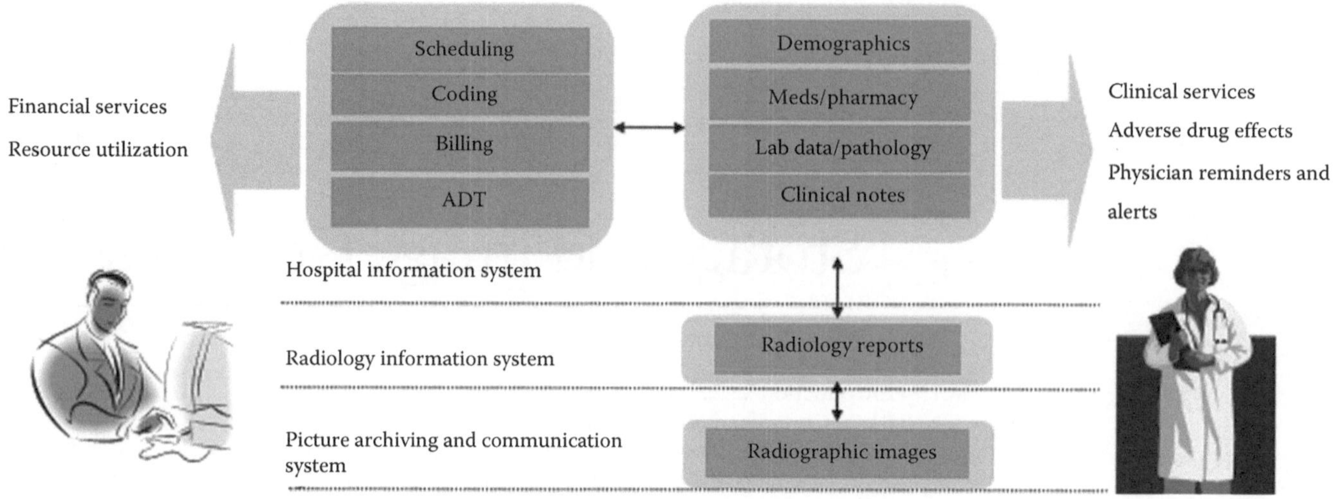

FIGURE 8.1 A high-level view of the data sources and functions of the HIS, RIS, and PACS.

8.2.1 Image Acquisition Gateways

The primary function of the IAG is to act as an interface between the modality and the PACS system. The IAG computer has three major functions: receives DICOM compliant image data from the modality, forwards the data to a local cache for temporary storage, and then sends a message to the DB controller that the image study information can be stored in the DB.

There may be multiple IAGs to avoid the bottleneck of multiple modalities sending images at the same time. Within a peer-to-peer network, the underlying interface uses Transmission Control Protocol/Internet Protocol (TCP/IP). The DICOM communication layer allows either a push of the image study from a client (Service Class User—SCU) to a server (Service Class Provider—SCP) or a pull of the image study from the client initiated by the server (DICOM, 2000). The preferred method of sending image studies to the PACS from the modality is to push the images to the IAG. In addition to receiving the image study, the IAG can also perform DICOM storage commit (Channin, 2001). DICOM Storage Commit sends a DICOM message back to the modality that the image study has been successfully transferred from the modality. Upon receipt of the DICOM storage commit, the image study can be deleted from the local storage of the modality. The IAG server also reads the DICOM image header and extracts patient demographics: accession number, study description, date and time of study, study, series and image unique identifiers, modality station name, and so on. An structured query language (SQL) update statement is then executed to update the DB server's image study information.

8.2.2 DB Server

The DB server contains the image study information extracted from the DICOM image header, as well as additional image study information such as number of images within the study, date/time image created within the DB, and date/time image

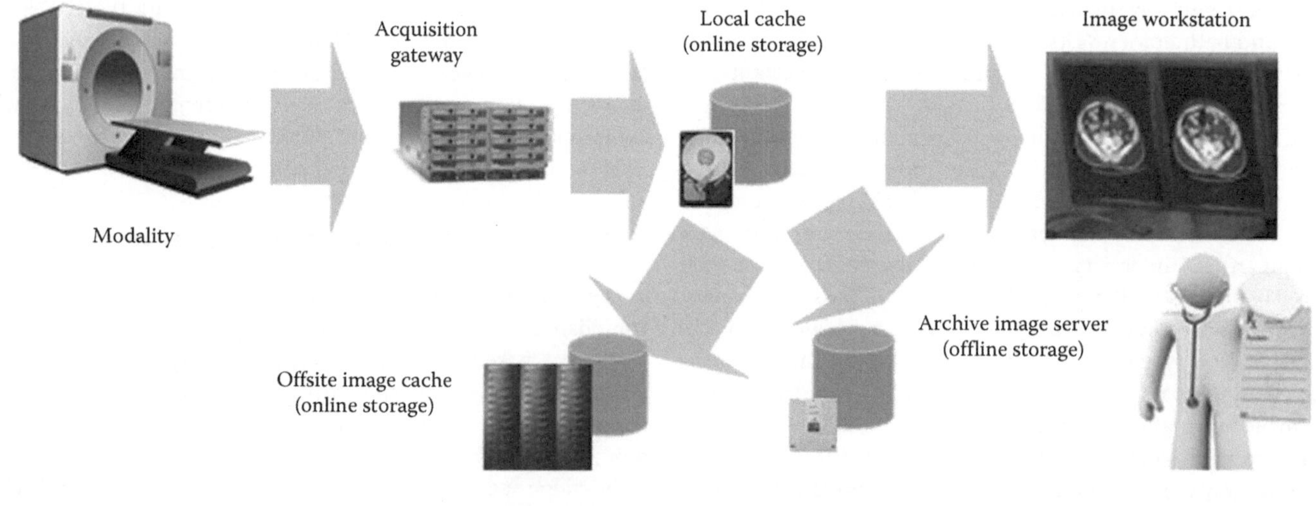

FIGURE 8.2 Components of a PACS. Arrows between the different components illustrate data flow.

study signed off within the RIS. In addition, the DB contains information about the location status of the image study. The image study can be situated in local cache, long-term archive, or as in offline study.

8.2.3 Workflow Manager

The workflow manager allows the PACS system to route images from location to location. The various states of storage include offline, local cache, and deep archive. The workflow manager retains the status of the image data. Local cache storage will depend on the amount of data that the institution would like to keep on hand. Typically this may be 2–3 years of image storage. As an example, within the Veterans Affairs Greater Los Angeles Healthcare System, approximately 2 years of local cache storage consisting of approximately 14 terabytes (TB) of formatted data storage is maintained. The workflow managers between the local PACS system, other Veterans Affairs (VA) hospitals within the Veterans Integrated Service Network (VISN), and long-term storage allow users to pull older image studies or studies from other institutions for comparison.

8.2.4 HL-7/DICOM Broker

The HL-7/DICOM broker manages information between the RIS and the PACS system. RIS messages are in HL-7 format and are decoded and routed to the PACS system as DICOM messages. One of the primary functions of the HL-7/DICOM broker is to receive the registered patient orders from the RIS system. After the patient arrives for the scheduled procedure, the x-ray technologist performing the procedure examines key patient information as they relate to the procedure. The x-ray technologist performs a quality control check to ensure that the proper procedure has been ordered, if the referring physician and/or radiologist has added important clinical notes or imaging procedure addendums. The technologist from the acquisition modality queries for the DICOM modality work list message from the HL-7/DICOM Broker and updates the patient work list on the local modality. The HL-7/DICOM Broker also transfers the patient's orders to the speech-recognition reporting system. The registered patient procedures are orders for new radiology reports that need to be dictated by the radiologist once the image studies are viewable on the radiologist's review workstation. The HL-7/DICOM Broker routes the registered patient procedures from HIS/RIS to the reporting system. These new orders are then routed to the radiologist's interpretation work list to select and then start dictating after reviewing the images on the PACS-viewing workstation. Once the report has been completed, it is sent back to the Radiology Reporting System. The Radiology Reporting System then forwards the final report as an HL-7 message that is sent to the RIS via the HL-7/DICOM Broker. If a radiologist wants to review a previous Radiology Report stored in the PACS, the HL-7/DICOM Broker will send an HL-7 message to the RIS to retrieve the previous report. The report is sent to the HL-7/DICOM Broker, which is then forwarded to the PACS system for review.

8.2.5 Archive Manager

The archive manager handles the offline storage of image studies. Stand-alone PACS systems will eventually run out of local cache and the offline storage of image studies is handled by the archive manager. Individual offline storage media are identified by a unique ID. The archive manager, through the DB manager, records which image study is stored on what offline storage media. The PACS system at the Veteran Affairs Greater Los Angeles Healthcare System (VAGLA) contains over 7000 magnetic optical disks (MODs). Each disk contains 2.3 gigabytes (GB) of image data. These MODs represent 10 years of offline storage. One of the goals at VAGLA hospital is to allow accessibility to all of our offline data, and the process of migrating the offline data into accessible long-term storage at a central archive, and transferring all image studies on MODs to the local cache. The archive server handles the first part of the data migration. The second part of the data migration requires transferring older studies in local cache through our local workflow manager to the remote workflow manager at the central archive. Currently, VAGLA has migrated approximately 95% of our offline MOD storage to the central archive Redundant Array of Independent Disks (RAID). This process is quite tedious, but well worth the effort as all of VAGLA offline storage will be accessible at the central archive.

8.2.6 Web Server

The Web server allows remote access to our image DB. The user interface is usually much simpler than a full radiology review workstation and can be quite cumbersome to use when reviewing large number of cross-sectional images. Web server utilization by both radiologists and referring clinicians has significantly increased over the past decade. The increase in Web accessible image study demand is driven by the ubiquitous availability of the Internet. Radiology in particular has been a support service to other areas of medicine, such as surgery, radiation therapy, orthopedics, and emergency medicine. Hospitals that provide 24 hours 7 days/week coverage have used teleradiology as a method to maintain clinical service without hiring additional staff. Teleradiology allows the transmission of medical images to any location on the Internet so that a radiologist and/or resident on-call can provide a real-time wet read or dictate a radiology report. Frequently, emergency room physicians require after-hour radiology consultations to assist in the diagnosis of the patient's condition based on the imaging evidence. The first documented use of teleradiology occurred in 1950 (Gershon-Cohen and Cooley, 1950), Dr. Gershon-Cohen and Dr. Cooley developed a system using telephone lines and a fax machine to transmit images between two hospitals over 27 m apart (45 km). In a 2003 American College of Radiology survey of 1924 professionally active postgraduate radiologists, results showed that 67% of active radiology practices use teleradiology (Ebbert et al., 2007). For those practices that reported using teleradiology, 82% reported transmitting images to a radiologist's home as the most common destination, while 15% of all practices use teleradiology to send images outside of their own practice to other facilities for external off hour reads. In a 2007 American College of Radiology survey, there was an increase in the number of radiology practices in the United States that used external off-hours teleradiology services (EOTs) to 44% (Lewis et al., 2009). The 44% of United States practices that utilize EOTs employ 45% of all United States radiologists. The latter half of this chapter has an extensive overview of Joint Photographic Experts Group (JPEG), and JPEG 2000 image compression algorithms utilized in the progressive transmission of radiographic images over a network.

8.2.7 Radiology Workflow

Without a complete understanding of the radiology workflow, one cannot fully appreciate the full functionality of the PACS system (Siegel and Reiner, 2002; Huang, 2010). This section describes the process by which a patient enters a hospital to be imaged to the final report generated by the radiologist (see Figure 8.3). The following section is specific to the workflow at VA Hospital utilizing VistA as the HIS/RIS and a commercial PACS system. It can be reasoned that everything that happens to a patient within the healthcare system can be summarized by an order. A patient enters the healthcare information system by first providing demographic information (e.g., patient's name, medical record number, birth date, gender, etc.). This information is contained within HIS upon registration. Ideally, the patient demographic information moves through the HIS from system to system without any human intervention. Patient demographic integrity is of utmost importance in any distributed information system. For radiology, the RIS receives the demographic information from the HIS. The patient visits their Primary Care Physician (PCP) and the clinician schedules an imaging exam through HIS/RIS. The patient then proceeds to radiology to obtain the examination. Before imaging is performed on the patient, the radiology clerk registers the patient into RIS for the scheduled procedure ordered by the PCP. Confirming that the correct patient has arrived for the impending imaging examination, the patient is then imaged by the x-ray technologist. Afterward, the x-ray technologist confirms that the imaging exam has finished by casing the registered order, which is to note any special procedures or modifications made while performing the imaging examination. The x-ray technologist then checks the image quality on the modality before the images are pushed to the IAG. The radiologist's work list is updated with this new study to be read after the study has been processed by the PACS DB server. The radiologist selects the new study and begins dictation. After a complete review of the current image study and any previous studies along with previous reports, the radiologist dictates the differential diagnosis on the patient's current condition. The final two steps for the radiologist involve assigning the diagnostic code for the severity of the findings, and signing off the report which is then sent to HIS/RIS for final storage. During the imaging process, the HIS/RIS system is continuously updated as the patient proceeds through the various stages of the radiology workflow: from *patient scheduled* to *patient examined* indicates the imaging procedure was *cased* and sent to PACS, from *imaging complete* to radiologist is *reading the exam* indicates the radiologist is generating the final report. A final status of *verified* indicates that the signed-off radiology report has been received by HIS/RIS. The PCP or referring physician can now review the final report (see Figure 8.4).

8.2.8 Image Storage and Archive

In the late 1990s, PACS image stores costs were once an impediment toward the purchase of digital imaging in radiology (Samei et al., 2004). Current trends in archival technology have

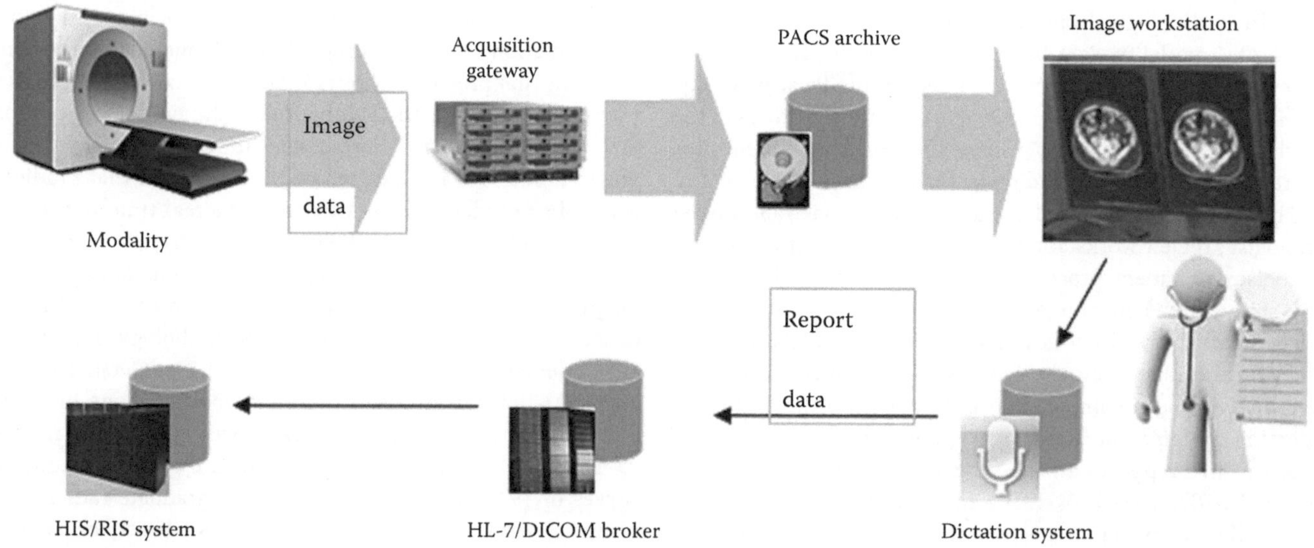

FIGURE 8.3 Image data flow from modality to PACS, then to radiologist workstation, and finally the finished case report is sent to HIS/RIS system.

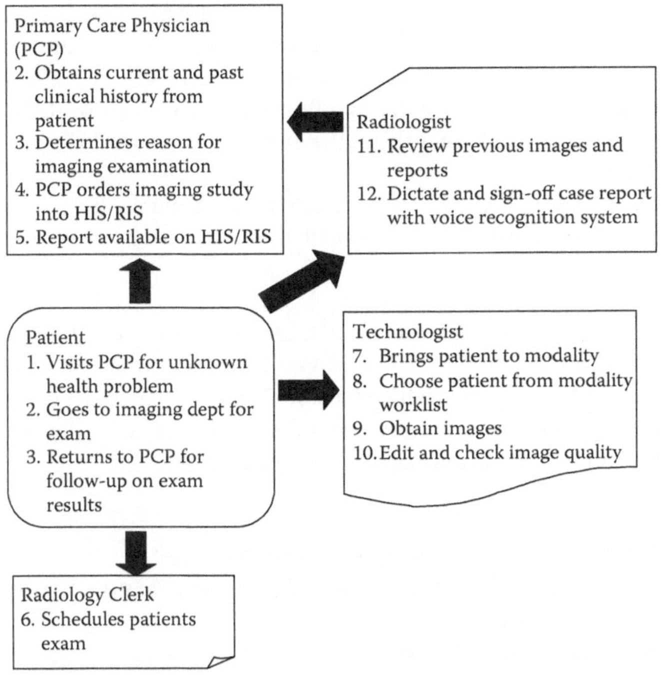

FIGURE 8.4 Radiology workflow for ordering, scheduling, image acquisition, case reporting, and review of patient's image exam on PACS.

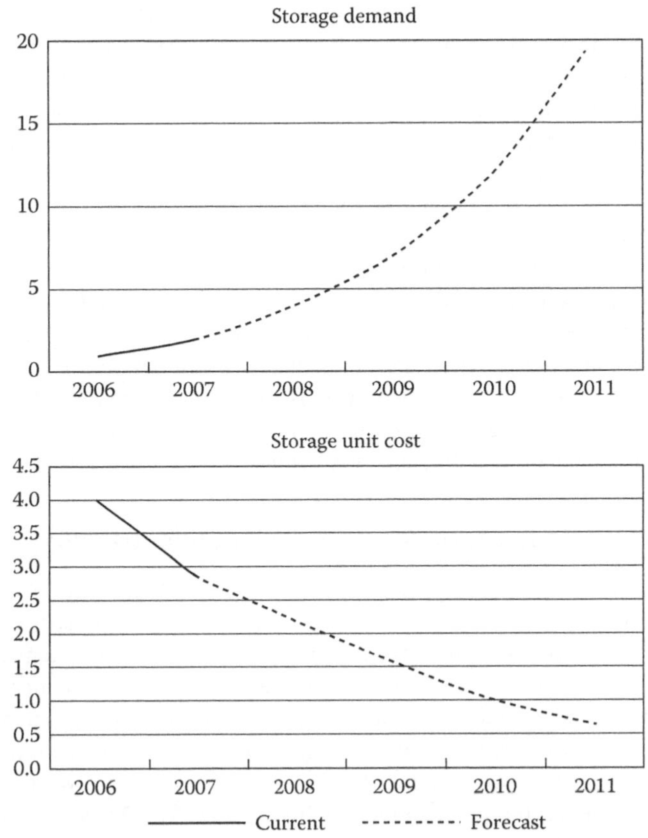

FIGURE 8.5 Storage demand, cumulative sum in thousands of petabytes (top). Unit cost of storage hardware per gigabyte in U.S. dollars (bottom).

shown that the demand for storage has grown by over 50% annually in recent years while the cost of storage has decreased over the same period (see Figure 8.5) (Samei et al., 2004; Kaplan et al., 2008). The exponential growth of digital information in healthcare provides an enormous challenge of trying to optimize storage demands, minimize cost, maximize resiliency of data, and maintain high-performance throughput. Hardware and software technology improvements for storage devices have removed digital archiving as a stumbling block for implementing PACS. Current PACS systems support magnetic, optical, and magneto-optical disks. However, optical and magneto-optical disks are becoming less popular because of lower storage capacity, slow access, and cost. They are still used as removable storage media in some legacy systems, and are typically stored in jukebox configurations that contain mechanical media movement devices.

Tape is also a viable removable storage medium that may also be kept in a jukebox or tape library. The individual tapes have very high capacity. The latest linear tape open format (LTO-5) has a capacity of 1.5 TB, and an uncompressed data transfer rate of 140 MB/s. In addition, LTO tapes have 15–30 years of archival life. The performance of tape limits its use for primarily disaster backup as well as long-term permanent storage.

The advent of multidetector computed tomography (CT) scanners that can produce scan thickness of less than 1 mm has resulted in image studies of over 2000 slices or approximately 1 GB of image storage. For instance, cardiology is an area that has seen increased growth in imaging in recent years. Over the past 5 years, cardiology PACS systems have also become quite prevalent. In comparison with radiology, where most images are usually static, cardiology has an overwhelming demand for dynamic image acquisition. Cardiac catheterization laboratories acquire x-ray cine sequences of the cardiac cycle. In addition, ultrasound echocardiograms can also acquire a large number of cine sequences. One area of concern at the VA is maintaining multiple copies of the same data. Within the VA's infrastructure, there are two image repository systems, VistA imaging and the PACS system, and both systems are backed up by two different departments. As with most hospitals, system administrators have difficulty aggressively purging outdated imaging information. According to Federal Regulations, U.S. hospitals are required to keep adult (>18 years old) images for at least 5 years from the last date of service. For mammograms, facilities are required to keep images for 10 years from the last date of service. Although removing noncritical data from the system would reduce the amount of data that needs to be stored, it is an expensive and time-consuming process.

There are four types of storage: local online cache, near-online cache, offline cache, and remote online cache (Dreyer et al., 2006; Kaplan et al., 2008). The local cache consists of a RAID storage system that can range from 100s of GBs to 100s of TBs in size. These types of storage systems typically have a high data transfer rate and can retrieve and store data very quickly. Current online storage costs (e.g., $0.13/GB) have dropped considerably over the past 10 years, making large capacity local cache systems more common. Near-online caches typically have a slower data transfer rate than the online systems, but are usually larger in storage capacity because they are comparatively lower in cost. Optical disk jukeboxes and automated tape cartridge systems are examples of near-online caches. The final storage tier is the offline cache. Once a near-online's (e.g., jukebox library) capacity is exceeded, any additional data must be written on offline media that is removed from the archive server and must be manually retrieved and loaded into the reader if needed. The main disadvantages to offline storage include the time delay in retrieving the data, human intervention is required, the media can be damaged through repeated use, and data could be lost due to misplacing or mislabeling of the media.

8.2.9 Redundant Array of Independent Disks

RAID is a commonly used technology for implementing local cache storage. The physical storage of the image data consists of two or three tiers of storage. In most common PACS systems, there are two tiers of storage. The short-term tier is the local cache. The second tier is usually a larger repository for long-term storage. The trend in PACS storage is to utilize RAID as both local tier 1 storage and tier 2 storage for long-term archive. The third tier represents redundant offline storage. RAID has several configurations that provided different levels of storage capacity, redundancy, and performance.

RAID 0 utilizes striping, which divides the data onto two hard drives simultaneously. Because the data are split between two drives, the controller can access the data in parallel thus reducing the access time. The disadvantage of RAID 0 is that there is no data redundancy. If you lose a drive, you have lost the data.

RAID 1 utilizes mirroring, which copies the data from one drive onto another. This method assures redundancy but offers no increase in performance.

RAID 10 is a combination of RAID 0 and RAID 1. RAID 10 stripes and mirrors the data on each drive. There is some performance gain since both disks can be accessed simultaneously. In addition, there is redundancy for protection of the data against drive failure.

RAID 3 and 4 utilize striping the data across multiple disks to optimizing parallel access to the data. In addition, RAID 3 and 4 utilize a parity bit to recover lost data. The parity bit acts as a checksum of the total number of bits in a single data word. The parity bit is stored on a separate drive.

RAID 5 is the most common configuration for PACS systems as it provides adequate redundancy and fault tolerance. The configuration for RAID 5 consists of striping and distributed parity across multiple drives. With distributed parity, a loss of a single drive will not cause any loss of data. There is also a spare drive available if a drive does go out, thus insuring that the system continues to work without any downtime. The time it takes to rebuild the spare hard drive within the RAID set is called mean time to recovery. RAID 5 allows read/write requests during the rebuild time, but the performance will be degraded because of the rebuild process going on in the background. As a result, the disk performance will suffer until the hard drive is replaced. Any subsequent loss of hard drives will result in data loss. In RAID 5, typically one of the disks in the grouping is assigned the parity disk.

RAID 6 is the highest level currently available. This level allows two drives to fail by calculating two parity bits instead of just one.

The VA Greater Los Angeles's PACS system has 14 TB of formatted RAID 5 local cache. In addition to our local cache, our central archive has 180 TB of RAID 5 remote storage. Within VISN 22 (Southern California and Nevada), ours' is one of five hospitals that utilize the central archive data center for our permanent and backup copy of our image studies.

8.2.10 Direct Access Storage, Storage Area Network, and Network Area Storage

First-generation PACS systems utilized direct access storage (DAS) image data onto local hard drives that were network accessible to all servers within the PACS cluster (see Figure 8.6). Storage area networks (SANs) allow multiple servers to a centralized pool of disk storage. Current PACS systems that utilize SAN allow the archive manager, workflow manager, DB manager, and IAGs access to the pool of disk storage through a dedicated backend Fibre Channel network (see Figure 8.7). This allows fast direct access to the RAID storage devices. The SAN network allows data transfer rates that are comparable to DAS that utilize high-speed peripheral channels. Network Area Storage (NAS) systems are file servers that direct connect via TCP/IP to the network (see Figure 8.8). The cost of NAS for the same amount of storage is usually less than SAN. A NAS system usually contains a minimal operating and file system. The main function of the NAS is to process input/output requests of either Common Internet File System (CIFS, part of Microsoft Windows, Redmond, WA) or NFS (Network File System, Sun Microsystems San Mateo, CA) file-sharing protocols.

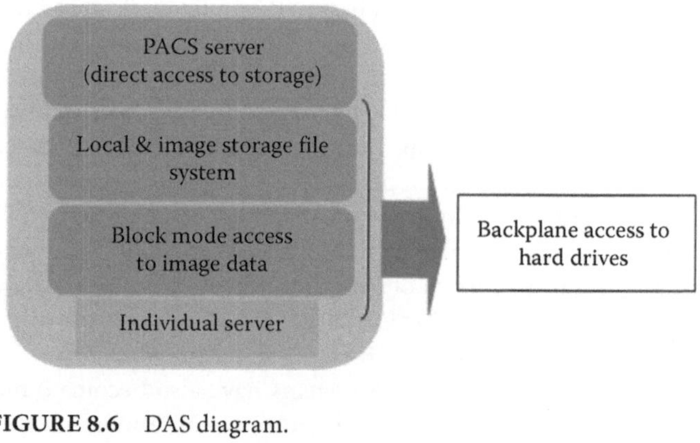

FIGURE 8.6 DAS diagram.

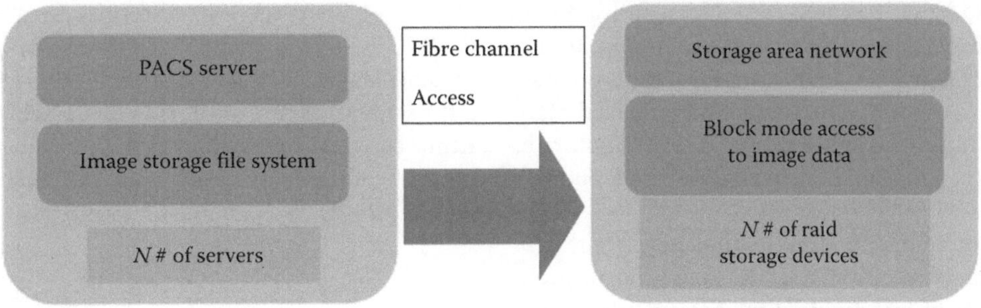

FIGURE 8.7 SAN diagram.

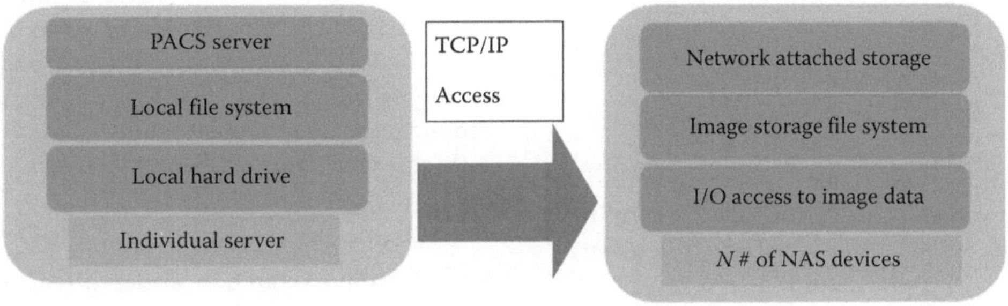

FIGURE 8.8 NAS diagram.

8.3 Medical Image Compression

8.3.1 Fundamentals

A digital grayscale image in its native format is stored in a computer as a succession of numbers, normally whole numbers (integers), with each number representing the brightness of the corresponding image pixel. In the case of color images, each pixel is normally represented by a set of numbers. For example, if an image is stored using the Red–Green–Blue (RGB) model, it takes three numbers to represent a single pixel. Since the values of the pixels are represented in order, one image row at a time, there is no need to store each pixel's location information in addition to its value. Although this ordering of information results in a reduction in the storage requirements for each image, many additional redundancies are normally present in a medical image that are not taken advantage of if storing or transmitting an image in its native format. In general, three types of redundancies can be identified that are relevant to compressed medical images:

- Coding
- Spatial, temporal, and bit depth
- Psychovisual

Image compression attempts to reduce or eliminate the presence of these redundancies to minimize the storage size or transmission time requirements for a given image. Depending on the methods used, coding and spatiotemporal redundancy reduction can be either reversible or irreversible, while psychovisual redundancy reduction is always irreversible. If the reduction is reversible, then the compression is said to be *lossless* since no information is lost by the process, and the original image can be reconstructed exactly. Compression algorithms which result in irreversible redundancy reduction are said to be *lossy*, and the reconstructed image after lossy compression is only an approximation of the original image. In general, lossy compression algorithms achieve higher compression levels than lossless algorithms.

8.3.1.1 Coding Redundancy

As mentioned above, an image is represented by an array of numbers, each one representing the intensity of its corresponding pixel. How these numbers are represented in the computer introduces *coding redundancy*. For example, a typical medical grayscale image may be composed of pixels with integer values between 0 and 4095 ($2^{12}-1$), so each pixel is represented by a 12-bit integer.* The use of the same number of bits to represent all pixels is called *fixed-length coding*. In this way, when a program reads an image file it knows that the first 12 bits represent the value of the first pixel, the next 12 bits represent the second pixel, and so on, with no need for any special symbol to represent the end of each pixel's data. However, fixed-length coding is inefficient, since, for example, 12 bits are used to represent pixels with values of 0 or 1, which could be represented by a single bit, pixels with values

* Typically, computers represent individual data with one or more sets of 8 bits (1 byte), so a pixel representing a 12-bit integer is stored using 16 bits (2 bytes). In this discussion, however, we consider that the pixel is represented with 12 bits, which is its true size.

of 2 or 3 which could be represented with 2 bits, and so forth, resulting in *variable-length coding*. The use of more bits that are needed to convey a given amount of information is called *coding redundancy*. To reduce this redundancy, special algorithms such as Huffman coding (Huffman, 1952) and arithmetic coding (Abramson, 1963) have been developed, and will be discussed later on.

To quantify the amount of coding redundancy in the representation of an image, we first need to know the theoretical minimum number of bits required to represent the image. This theoretical minimum is the actual amount of information included in the image, called its *entropy* (Shannon, 1948), which can be computed by

$$H = -\sum_{i=1}^{L} p_i \log_2(p_i) \quad (8.1)$$

where p_i is the probability of the value i appearing in the image which has L different values. By using the base 2 logarithm, the entropy is given in bits per pixel. The probability p_i is given by

$$p_i = \frac{n_i}{MN} \quad (8.2)$$

where n_i is the number of times that the value i appears in the image of size $M \times N$. With the computation of entropy, we can quantify the amount of coding redundancy present in a certain representation of an image using

$$R = b - H \quad (8.3)$$

where b is the average number of bits used to represent the image, and H is the image's entropy computed using Equation 8.1.

As an example, using Equation 8.1 on the 12-bit digital mammogram shown in Figure 8.9, we get an entropy of $H = 5.1144$ bits/pixel, which results in a coding redundancy of $R = 12 - 5.1144 = 6.8856$ bits/pixel. In theory, an optimal variable-length coding algorithm for this image should result in a file with an average of 5.1144 bits/pixel, and no coding algorithm could do better than this. If, in practice, this theoretical optimal coding could be achieved, the image in Figure 8.9 would be compressed by a compression ratio of

$$C = \frac{b_{\text{current}}}{b_{\text{compressed}}} = \frac{12}{5.1144} = 2.3463 \quad (8.4)$$

where b_{current} is the current number of bits/pixel used to store the image and $b_{\text{compressed}}$ is the average number of bits/pixel after compression. It is important to note that this theoretical maximum achievable compression is only applicable to reduction of coding redundancy; if other redundancies, such as spatial redundancy, are reduced, then the achieved compression could be higher than this theoretical maximum.

8.3.1.2 Spatial, Temporal, and Bit-Depth Redundancy

Spatial, temporal, and/or bit-depth redundancies are the consequences of the correlations present in most cases between pixels located in the same neighborhood in the image and, in time-sequence images, in consecutive frames. In general, adjacent pixels have similar, if not equal, values, and therefore the second pixel does not provide much information in addition to the first. The same is true for time-sequence data: the same pixel in two consecutive frames normally does not change values drastically. To reduce these types of redundancies, special *transforms* are applied to the image or sequence of images so as to reduce the spatial or temporal correlation between pixels. These transforms can be based on completely different concepts, such as *run-length encoding*, in which instead of pixel values representing the pixels' brightness, what is actually stored is the length of constant value pixel runs in images, or

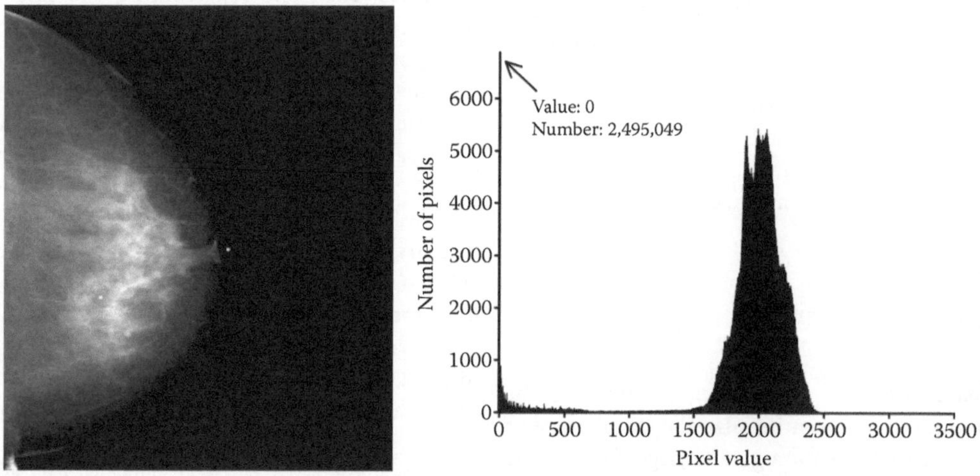

FIGURE 8.9 Typical digital mammogram with pixel values ranging from 0 to 4095, and therefore represented by 12 bits/pixel when stored with fixed-length coding. The size of this image is 1914 × 2294 pixels.

discrete cosine transforms (DCTs), which results in the coefficients of special basis functions that can fully represent the information in an image.

8.3.1.3 Psychovisual Redundancy

Not all information present in an image is seen by the viewer, or, specifically in medical imaging, used by the physician for diagnostic purposes. Therefore, the removal of this information, although irreversible and therefore resulting in a lossy compression algorithm, does not necessarily result in a reduction in diagnostic quality. Of course, which portions of image information can be removed, which in general can result in a loss of sharpness, change in texture, and/or introduction of artifacts, without affecting the diagnostic quality of an image is dependent on the clinical application and can only reliably be tested with human observer evaluation. This removal of image information is performed by *quantization*, which, although the details depend on the algorithm used, in general involves approximating certain image descriptors by limiting them to only a specific discrete set of values, and/or eliminating them altogether.

8.3.1.4 General Image Compression Algorithm

The mechanism of compression algorithms in general can be summarized as including two or three steps, as described in Figure 8.10. Lossless algorithms include only the steps of *transformation* and *encoding*, while lossy algorithms also perform *quantization* of the transformed data. The application of a *transform* during the transformation step aims to reduce the spatiotemporal redundancy present in the original representation of the image. The quantization of the transformed data, if performed, removes the psychovisual redundancies by an irreversible process. Finally, the data are encoded so as to minimize the coding redundancy in the final product. When an image is reconstructed from the compressed information, first the data is decoded by the inverse process of the encoding mechanism, and then the inverse of the transform used in the compression algorithm is applied. Note that the inverse of the quantization process is not performed, since there is no way to reverse this process. In the following sections, we discuss specific examples of common algorithms used to perform these three compression steps.

8.3.2 Basic Compression Methods

8.3.2.1 Huffman Coding

Huffman coding (Huffman, 1952) is one of the most common methods of *variable-length encoding* used to reduce the coding redundancy. As we have seen, representing every pixel of an image with a constant number of bits results in a suboptimal bit rate. Huffman coding is a simple algorithm that optimizes the bit rate by representing the pixel values that appear most times in an image with the shortest *codes* (symbols that require fewer bits), and using progressively longer codes for the pixel values that appear fewer times in the image. One challenge of variable-length encoding is that, during decoding, the symbols have to be unique, that is, without a constant number of bits representing each pixel, and without the use of a special "end-of-pixel" symbol (which would decrease the compression ratio), the decoding algorithm needs to know where the bits for one pixel end and the bits for the next pixel begin. Huffman coding guarantees the uniqueness of the decoding process so that a set of codes can only represent one set of image values.

For the digital mammogram in Figure 8.9, in its native format, 12 bits are used to represent each pixel. However, if we look at the histogram of the image, we see that the value 0 is the most common pixel value in the image, appearing with a probability of 0.568. Other values appear very few times, for example, the value 2583 appears in the image only five times. Therefore, if a shorter code (fewer number of bits) were used to represent a 0, and a longer code were used to represent values such as 2583, we would expect that the overall bit rate for the image would be smaller than 12 bits/pixel. Using Huffman coding, the value of 0 is represented with one bit, while the value of 2583 is represented with 20 bits. Of course, all values that appear in the image

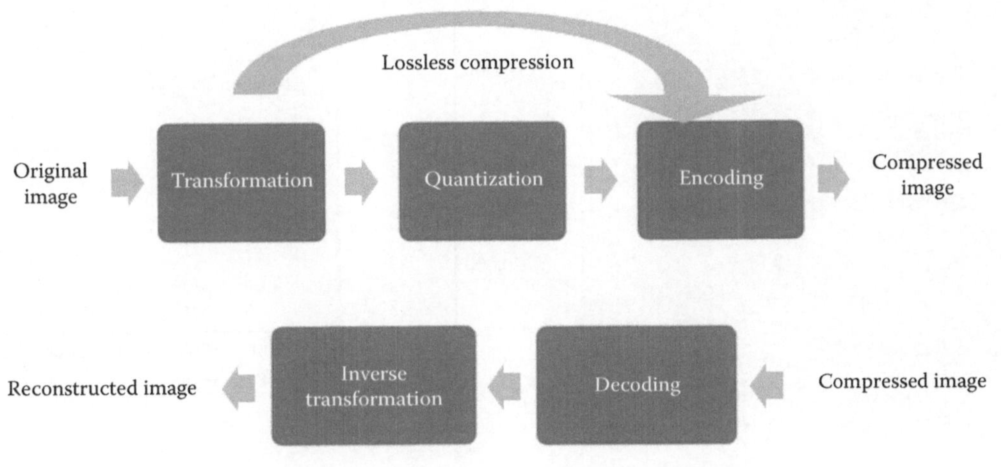

FIGURE 8.10 General algorithm for both lossless and lossy image compression and image recovery.

are assigned a code by the algorithm. The end result of Huffman coding this image is a bit rate of 5.19 bits/pixel, resulting in a compression ratio of 2.31:1. Remember that previously we calculated that the theoretical minimum bit rate achievable for this 12 bit image was 5.1144 bits/pixel, for a theoretical maximum compression ratio of 2.3463:1. As can be seen, Huffman coding can in many cases approximate the optimal theoretical maximum lossless compression ratio.

8.3.2.2 Pixel-Difference Encoding

In general, the values of adjacent pixels in an image (or the same pixel in adjacent frames in time-sequence images) do not vary considerably; consequently, each pixel does not provide much more information than its neighbor. Therefore, a simple method of reducing spatial redundancy is to store only the difference between pixels which are either adjacent by location (for two-dimensional [2D] and 3D images) or by time (time-sequence images).

Figure 8.11 shows the image obtained if the digital mammogram in Figure 8.9 is displayed as the difference between the original value at that position and the original value of the pixel to its immediate left. The first pixel in the image, at the top-left corner, has the same value as in the original image. The histogram of the difference image, shown in Figure 8.11, can be seen to be much more concentrated around a set of values (near zero) than the histogram of the original image shown in Figure 8.9. The impact of applying this transform to the mammogram is appreciated when Huffman encoding this difference image, which, due to its more compact histogram, results in a bit rate of 3.32 bits/pixel, which translates to a compression ratio of 3.61:1. This compression ratio is higher than that predicted by the entropy of the image because, as discussed above, entropy only takes into account the possible reduction in coding redundancy, and the encoding of pixel differences performed here reduces the spatial redundancy present in the image.

8.3.2.3 Arithmetic Coding

As opposed to Huffman coding which replaces each pixel value in an image with a special code on a one-to-one basis, another method of variable-length encoding, *arithmetic coding*, replaces a set of pixel values with one code (Abramson, 1963). As a simplified example to illustrate how arithmetic coding works, let us assume that an image includes only four different pixel values: *a*, *b*, *c*, and *d*. In the image that will be coded, the probabilities of each of these values appearing are those shown in Table 8.1. Then each of the four values is assigned a range of values between 0 and 1, as depicted in Table 8.1. To encode a certain four-pixel-long set of values (e.g., *adcb*), the process in Figure 8.12 is performed, in which each pixel value narrows the possible range of the code, until all four pixels are processed. The resulting code is any number included in the final range 0.134–0.136, and therefore could be 0.135. During reconstruction of the compressed image, it can be computed that only the equivalent set of values *adcb* could have resulted in any number between 0.134 and 0.136, making decoding of this coding algorithm also unique.

Again, using Figure 8.9 as an example, the arithmetic coding of this digital mammogram results in a bit rate of 5.1147, equivalent to a compression ratio of 2.3462:1. As can be seen, for this image, arithmetic coding improved slightly on the results of Huffman coding, and achieved almost the theoretical maximum compression ratio predicted by the image's entropy.

To avoid floating-point operations which would introduce inaccuracies in the coding process, actual implementations of arithmetic coding use binary values and processes such as *renormalization* to ensure accuracy and sufficient dynamic range.

8.3.2.4 Run-Length Encoding

It is common for certain areas of images to have the same value for a large number of contiguous pixels. For example, in the mammogram in Figure 8.9, the pixels outside the breast, in the open field area, all have the same value (in this case 0). Therefore, to

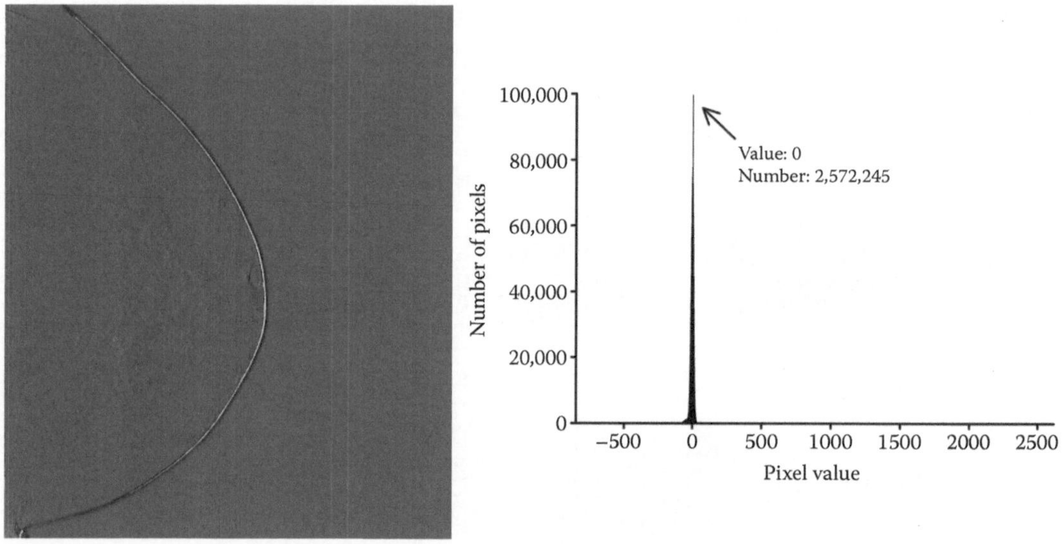

FIGURE 8.11 Difference image of the digital mammogram shown in Figure 8.9 and its histogram.

TABLE 8.1 List of Possible Values in Image to Be Compressed with Arithmetic Coding

Value	Probability	Range
A	0.20	0, 0.2
B	0.10	0.2, 0.3
C	0.20	0.3, 0.5
D	0.50	0.5, 1.0

represent the 1632 contiguous zeros in the first row of the image to the right of the breast tissue, instead of using 1632 zeros each occupying 12 bits ($1632 \times 12 = 19584$ bits), the same data could be represented by a *run-length pair*, denoting the number of contiguous pixels of the same value and their value. In this case, the run-length pair for the row of zeros at the top of the image would be (1632, 0), which occupies only $2 \times 12 = 24$ bits. If the entire image is run-length encoded in this manner so that all the pixels with zero values located at the end of each row are replaced by a 24-bit run-length pair, then for the image in Figure 8.9 the compression ratio achieved would be 2.3:1. This type of redundancy reduction in which only the last run of constant-value contiguous pixels is run-length encoded is used in JPEG compression, which will be discussed next.

8.3.3 JPEG Compression

The JPEG compression algorithm, an international standard, is one of the most commonly used image compression methods today, and is therefore an algorithm that is worth studying in detail. Here, we discuss its main compression stages that relate to the generic compression algorithm depicted in Figure 8.10, although some very specific implementation details are not addressed. The JPEG compression standard actually includes three different compression algorithms: (1) the standard algorithm, (2) the *progressive* JPEG algorithm for higher quality or compression ratios, and (3) a lossless algorithm (JPEG-LS). The first algorithm is the most commonly used one, so this is the one that is discussed here.

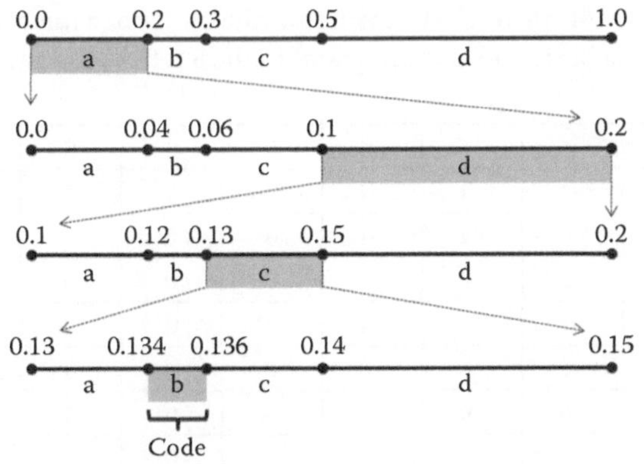

FIGURE 8.12 Simplified example of how arithmetic coding is performed.

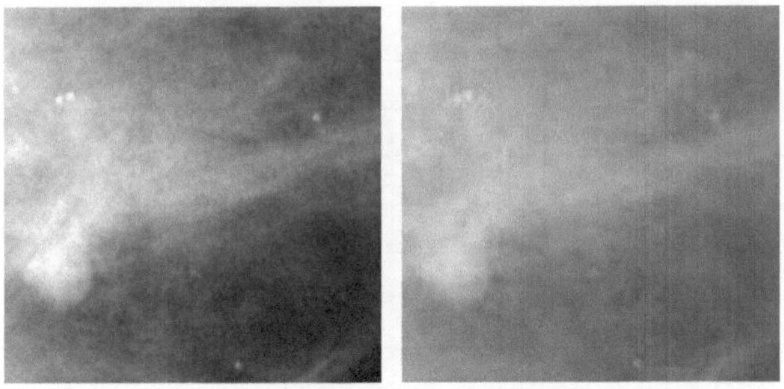

FIGURE 8.13 ROI of the digital mammogram in Figure 8.9 before and after JPEG compression, in which the blocking artifacts common to high compression ratios with JPEG can be seen.

8.3.3.1 8 × 8 Pixel Block Decomposition

The first step in JPEG compression of a grayscale image is the breakup of the image into 8 × 8 pixel blocks, which will be processed and stored separately and will not be recombined until the image is reconstructed. This block-based processing, although it improves computing efficiency, commonly introduces *blocking artifacts*, characterized by perceptible discontinuities at the block edges, especially at high compression ratios. Figure 8.13 shows an area of the digital mammogram shown in Figure 8.9, and the same area after aggressive JPEG compression (compression ratio of 24:1) in which the presence of the blocking artifacts can be clearly seen.

8.3.3.2 Discrete Cosine Transform

After dividing the image into 8 × 8 pixel blocks, each block is transformed (see Figure 8.10) using the DCT. The DCT* (Ahmed et al., 1974) is one of the most commonly used transforms in image compression. It is similar to the discrete Fourier transform, but its basis functions consist of cosine functions only. The DCT of an image $I(x, y)$ of size $M \times N$, denoted $I(u, v)$, is computed by

$$I(u,v) = a(u)a(v)I(x,y)\cos\left[\frac{(2x+1)\pi u}{2M}\right]\cos\left[\frac{(2y+1)\pi v}{2N}\right] \quad (8.5)$$

where

$$a(k) = \begin{cases} \sqrt{\dfrac{1}{N\ (\text{or } M)}} & \text{for } k = 0 \\ \sqrt{\dfrac{2}{N\ (\text{or } M)}} & \text{for } k = 1,2,3,\ldots,N\ (\text{or } M) - 1 \end{cases} \quad (8.6)$$

Similar equations can be formed for 1D data, 3D images, and 2D + t or 3D + t image sequences. The application of this transform to the image to be compressed results in the representation of the image information in a more compact form, allowing for

* There are actually eight types of DCT. Type II DCT is the one used in image compression algorithms, so we refer only to this one here.

Original image pixel values:

2123	2115	2113	2124	2116	2116	2129	2136
2120	2117	2114	2126	2153	2180	2138	2130
2136	2122	2125	2140	2203	2236	2169	2132
2143	2121	2137	2133	2165	2175	2149	2135
2112	2113	2105	2130	2146	2117	2114	2117
2105	2117	2091	2126	2150	2126	2120	2103
2085	2098	2106	2116	2120	2121	2121	2119
2108	2110	2103	2110	2108	2112	2132	2125

DCT coefficients:

17,021.2	−69.7	−45.8	63.6	2.5	−12.2	1.1	−4.0
80.5	−12.7	−8.7	35.7	−7.9	−3.8	26.0	−6.6
−51.8	−2.7	34.7	−32.3	−8.0	12.1	−2.4	−10.3
−71.7	26.5	8.3	−36.0	21.7	−0.9	−23.5	−4.0
−20.0	23.9	33.9	−33.6	11.7	13.1	−3.0	6.0
15.5	−6.8	0.9	−3.8	−7.3	7.1	11.2	4.3
23.9	−1.9	−6.4	23.9	−0.5	−1.9	−8.7	−0.5
−3.8	−9.4	−9.2	−1.4	2.3	−2.0	0.6	−1.4

FIGURE 8.14 Values of an 8 × 8 pixel block of the digital mammogram shown in Figure. 8.9 and of its DCT.

higher compression ratios when encoding the data and with lower correlation among the pixel values, reducing the spatiotemporal redundancy in the data (Rao and Yip, 1990).

As an example of the application of the DCT, Figure 8.14 shows the result of applying the DCT to an 8 × 8 pixel block of the mammogram in Figure 8.9. As can be seen, the same information in the DCT is largely concentrated in a small portion of the image at the top left, with the rest of the values being low values around zero, resulting in a block with lower spatial redundancy and making its posterior encoding more efficient.

8.3.3.3 Quantization

At this point, if the block shown in Figure 8.14 displaying the coefficients of the DCT is used to perform an inverse DCT, the original block could be recovered perfectly; that is, up to this point the compression method is lossless. It is in this *quantization* step that information loss is introduced, making JPEG a lossy compression algorithm. Specifically, the information loss is a consequence of two processes: psychovisual normalization and rounding to the nearest integer.

Normalization is performed with a set of values that reduce the presence of psychovisually redundant data. This is achieved by dividing the DCT coefficients by the corresponding values in a normalization matrix whose values vary according to how important each corresponding DCT frequency is to the perceived image quality. Figure 8.15 shows a typical normalization matrix used by JPEG compression, which has been established using the empirical methods.

16	11	10	16	24	40	51	61
12	12	14	19	26	58	60	55
14	13	16	24	40	57	69	56
14	17	22	29	51	87	80	62
18	22	37	56	68	109	103	77
24	35	55	64	81	104	113	92
49	64	78	87	103	121	120	101
72	92	95	98	112	100	103	99

FIGURE 8.15 JPEG psychovisual normalization matrix used in the quantization step.

As can be seen in Figure 8.15, the values to the right and bottom of the array get progressively larger, resulting in the progressively larger reduction of the DCT coefficient values representing the higher frequencies present in the 8 × 8 pixel block. Combined with the rounding off step, in which the result of the normalization division is rounded to the nearest integer, this results in the progressively stronger "filtering out" of the higher frequency components of the image. The scaling of the normalization array by multiplication with a constant allows for the adjustment of the degree of compression achieved by the algorithm by resulting in a higher reduction in the value of the DCT coefficients and therefore a higher number of these values being rounded to zero. Of course, this increase in the number of coefficients set to zero also results in stronger degradation of the signal, inherent in the inverse relationship between compression rate and image quality.

Figure 8.16 shows the result of normalizing and rounding the DCT coefficient array in Figure 8.14 using the JPEG normalization matrix in Figure 8.15. As can be expected, the presence of all the zero-valued DCT coefficients will result in a more compact representation of the data during the encoding step.

During the decoding process, the normalized coefficient arrays like the one shown in Figure 8.16 are multiplied by the normalization matrix and then used to perform an inverse DCT. For the 8 × 8 pixel block used as an example here, the pixel values for the decoded image would be those shown in Figure 8.17. To better visualize the effect of the compression, a graph of the horizontal profile through the third row (the center of a microcalcification visible in the mammogram) in the 8 × 8 pixel array before

1064	−6	−5	4	0	0	0	0
7	−1	−1	2	0	0	0	0
−4	0	2	−1	0	0	0	0
−5	2	0	−1	0	0	0	0
−1	1	1	−1	0	0	0	0
1	0	0	0	0	0	0	0
0	0	0	0	0	0	0	0
0	0	0	0	0	0	0	0

FIGURE 8.16 DCT coefficient array of Figure. 8.13 after the quantization step.

2120	2121	2121	2121	2122	2124	2127	2129
2127	2109	2104	2132	2171	2182	2153	2119
2141	2119	2114	2149	2199	2212	2176	2132
2136	2131	2132	2148	2167	2172	2158	2141
2112	2116	2123	2129	2131	2127	2120	2115
2101	2100	2105	2121	2135	2133	2113	2094
2104	2101	2104	2117	2132	2135	2124	2110
2100	2105	2107	2104	2100	2106	2122	2136

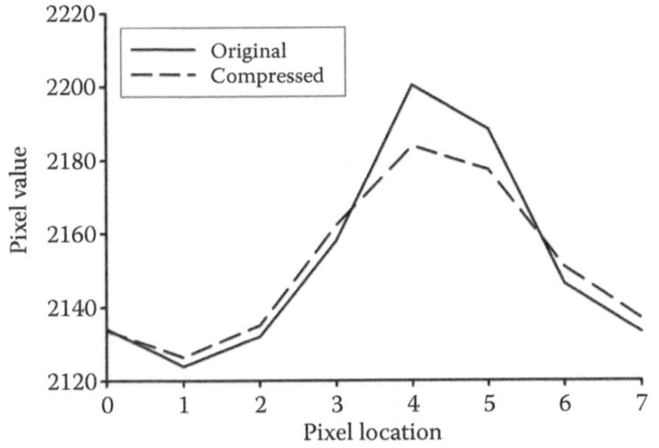

FIGURE 8.17 Pixel values in the 8 × 8 pixel array after compression and horizontal profile through the third row in the array, showing the loss of contrast and resolution introduced by lossy JPEG compression.

and after compression is also shown in Figure 8.17. As can be seen in the profile, the compression reduces both the contrast and the resolution of the image, by both lowering and widening the peak in grayscale values due to the presence of the microcalcification.

8.3.3.4 Encoding

During the final JPEG compression stage, the 8 × 8 pixel blocks of quantized DCT coefficients are encoded using a combination of Huffman-like coding and run-length encoding. Before undergoing these steps, however, the values in the array are organized as a vector maximizing the number of zero coefficients that are placed at the end of the vector to optimize the efficacy of the run-length encoding. For this, the array is ordered using a zigzag scheme starting at the top left corner (the lowest DCT frequency coefficients) and ending in the bottom right corner (the highest DCT frequencies). The resulting vector is encoded using an algorithm similar to Huffman coding but, as opposed to the Huffman coding system described in Section 8.3.2.1, JPEG compression uses a predetermined coding table that is built into the compression standard. Since it is known that each block involves 64 values, after the last nonzero coefficient is coded a simple end-of-block symbol is coded which indicates to the decoding algorithm that the remaining values of the 8 × 8 block of normalized DCT coefficients are zero.

In Figure 8.18, a comparison of a region of interest (ROI) of the original digital mammogram shown in Figure 8.9 is displayed for different JPEG compression ratios. As can be seen, for this particular image, at a compression ratio of 4.6:1 there is hardly any perceptible loss of image quality. At higher compression ratios, there is substantial loss of resolution (note two microcalcification specks pointed by the arrow that appear as one in the two highest compression ratios) and introduction of moderate-to-severe artifacts.

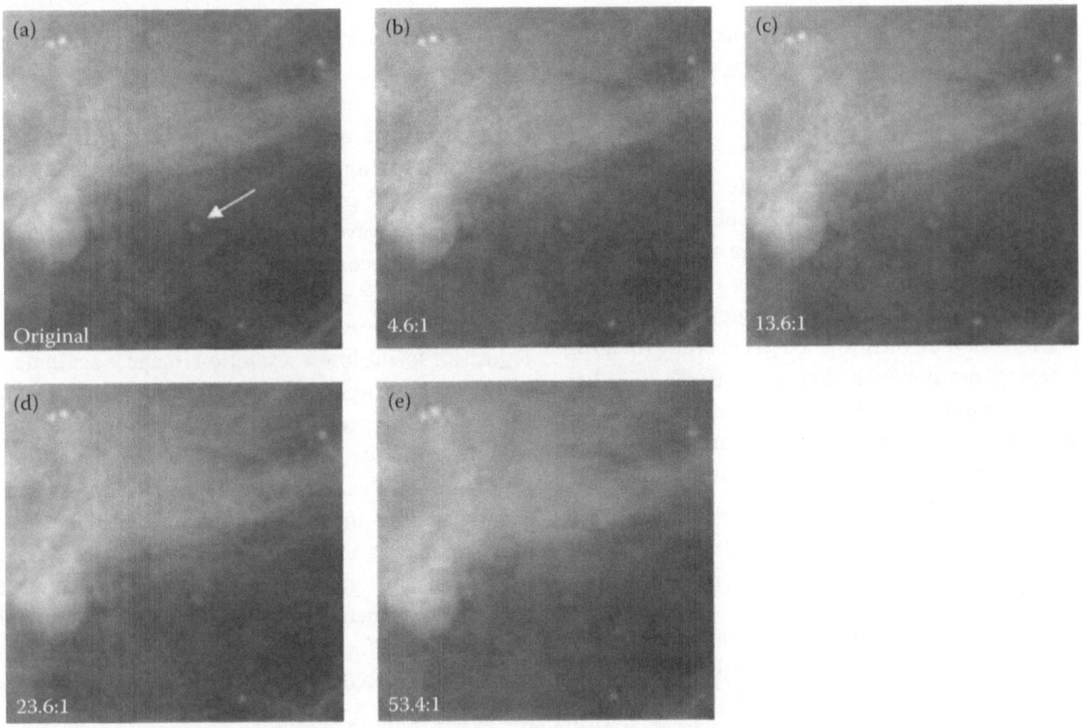

FIGURE 8.18 ROI of the digital mammogram in Figure 8.9 that includes regional punctuate microcalcifications, before and after JPEG compression with various compression rates.

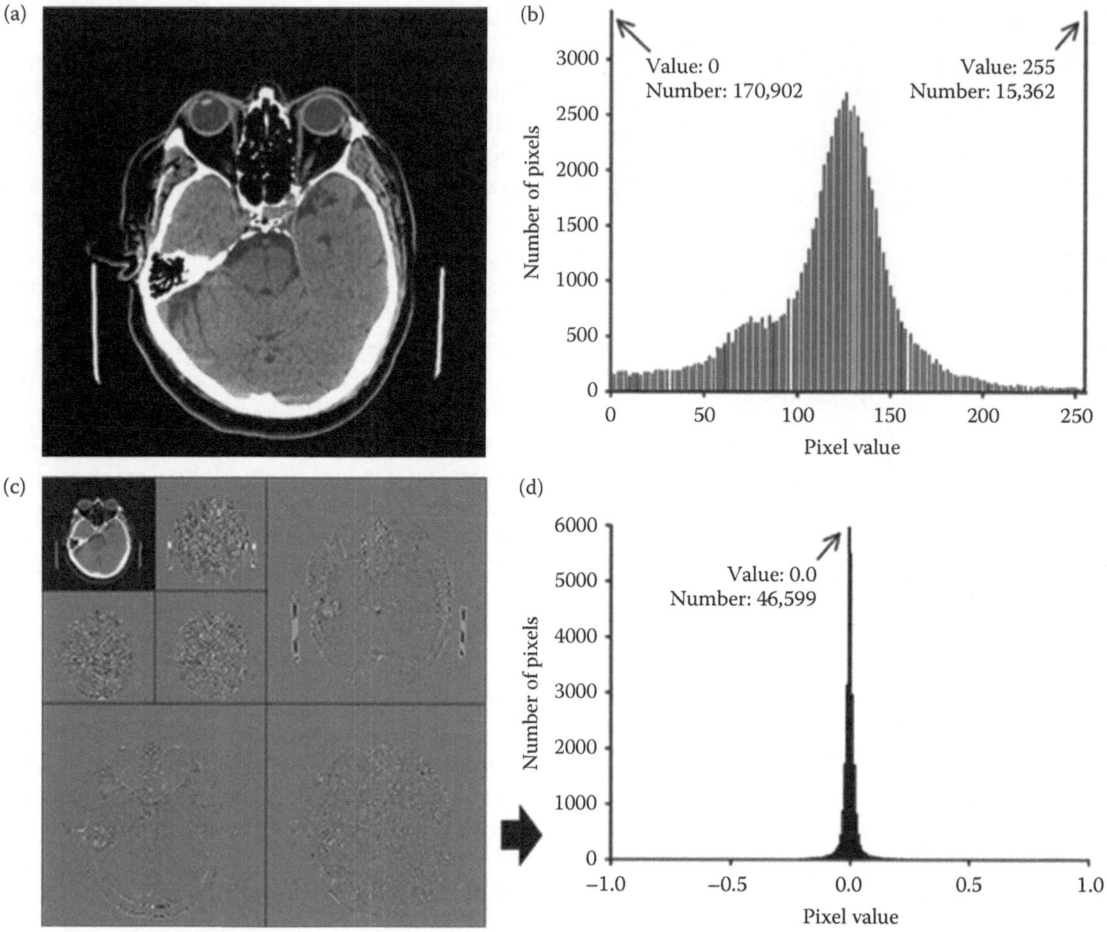

FIGURE 8.19 (a) 512 × 512 head CT image and (b) its histogram. (c) Two-level wavelet transform decomposition and (d) the histogram of the first level diagonal wavelet coefficient subband (bottom right image in [c]).

8.3.4 JPEG 2000 Compression

The JPEG 2000 compression algorithm was developed to take advantage of new image processing methods, specifically wavelet decomposition, which allow for higher image quality at equal or higher compression rates than those achievable with standard JPEG and other compression methods. Although JPEG is still the dominant compression algorithm used in everyday imaging (e.g., digital cameras, Internet Web sites, etc.), JPEG 2000 has many additional capabilities not present in standard JPEG that will surely result in its widespread use in the future. In addition, for specific high-tech applications such as medical imaging, currently the study of JPEG 2000 algorithms over older algorithms has become commonplace.

8.3.4.1 Wavelet Transform

JPEG 2000 compression follows the same generic algorithm used by other compression methods shown in Figure 8.10. In the case of JPEG 2000, the transform step is performed using wavelet-based decomposition rather than the DCT used in standard JPEG. In addition, due to their local nature and efficient computation, the wavelet decomposition is generally applied to the entire image, avoiding the blocking artifacts that are usual in JPEG due to its application of the transform and quantization to individual 8 × 8 pixel blocks.*

During the wavelet transform step, the image to be compressed is iteratively decomposed into a lower resolution version of the image and three *subbands* that contain the horizontal, vertical, and diagonal wavelet coefficients of the processed image (see Figure 8.19). The lower resolution image is again decomposed into its respective lower resolution version and its three subbands, and the process is repeated until the desired number of decomposition levels is reached. The JPEG 2000 standard does not specify the number of levels that should be performed. As can be seen from the wavelet decomposition in Figure 8.19, the horizontal, vertical, and diagonal wavelet coefficient subbands are similar to difference images, in which most of the pixel values are concentrated around zero, making them prime candidates for quantization and compression. Therefore, as in JPEG compression, the wavelet transform in JPEG 2000 is performed to obtain decorrelated images

* Separate processing of subimage blocks, called tiles in the JPEG 2000 standard, is an option in JPEG 2000, although in general the default and more common option is to compress the image in its entirety.

with a high number of small-valued pixels that are easy to quantize and therefore result in good compression ratios. Depending on the type of JPEG 2000 compression that is being performed, the wavelet decomposition is performed using a biorthogonal 5–3 wavelet transform (Le Gall and Tabatabai, 1988) (lossless) or a Daubechies 9–7 wavelet transform (Antonini et al., 1992) (lossy).

8.3.4.2 Quantization

After wavelet decomposition, the quantization of the wavelet coefficients is performed in a manner similar to that in JPEG compression, with the transform coefficients being divided by a scalar, whose value can be changed depending on the desired compression rate. In JPEG 2000, quantization is performed using the equation

$$q(u,v) = \text{sign}(c(u,v))\,\text{floor}\left[\frac{|c(u,v)|}{\Delta}\right] \tag{8.7}$$

where $c(u,v)$ is the wavelet coefficient to be quantized, $q(u,v)$ is the quantized value, and Δ is the quantization step size. The same quantization step size is used for all coefficients in a certain subband, and the magnitudes of the step sizes used for quantization of the whole wavelet decomposition can be varied depending on the desired compression rate. Of course, for lossless compression, the step size Δ is set to unity. Marcellin et al. (2002) provide a more thorough description of JPEG 2000 quantization and of a second quantization method incorporated into Part 2 of the JPEG 2000 standard.

8.3.4.3 Encoding

Encoding in JPEG 2000 has been implemented not only to maximize compression performance, but also to allow for several of JPEG 2000's special capabilities discussed later. After the quantization step, the image being compressed is still represented with a number of subbands, each being an image of wavelet coefficients. To encode these subbands, all the subbands are divided into the same regions called *precincts*. The precincts are selected in all the subbands such that all the data in one precinct corresponds to the same approximate area in the original image. Each precinct within each subband is then subdivided into smaller regions called *code blocks*. The encoding in JPEG 2000 is performed for each code block separately, in a bit-plane basis, so that, for example, the most significant bits of all the pixel values in a code block are encoded together, then the next lower significant bits throughout the code block are encoded, and so forth. The bits are encoded using a binary arithmetic coding similar to the method described above. Extensive details of the JPEG 2000 encoding method are given by Taubman et al. (2002).

Figure 8.20 shows a comparison of the results of compressing the digital mammogram in Figure 8.9 with the JPEG and JPEG 2000 algorithms at the same compression ratios. It can be seen that the JPEG 2000 compressed images are free of blocking artifacts, and that they retain higher resolution at the same compression ratios. At the highest compression ratio, while the JPEG image is practically unusable, the JPEG 2000 image still retains some image quality, although the pair of microcalcifications pointed out in Figure 8.18 at this compression ratio is seen as one larger speck.

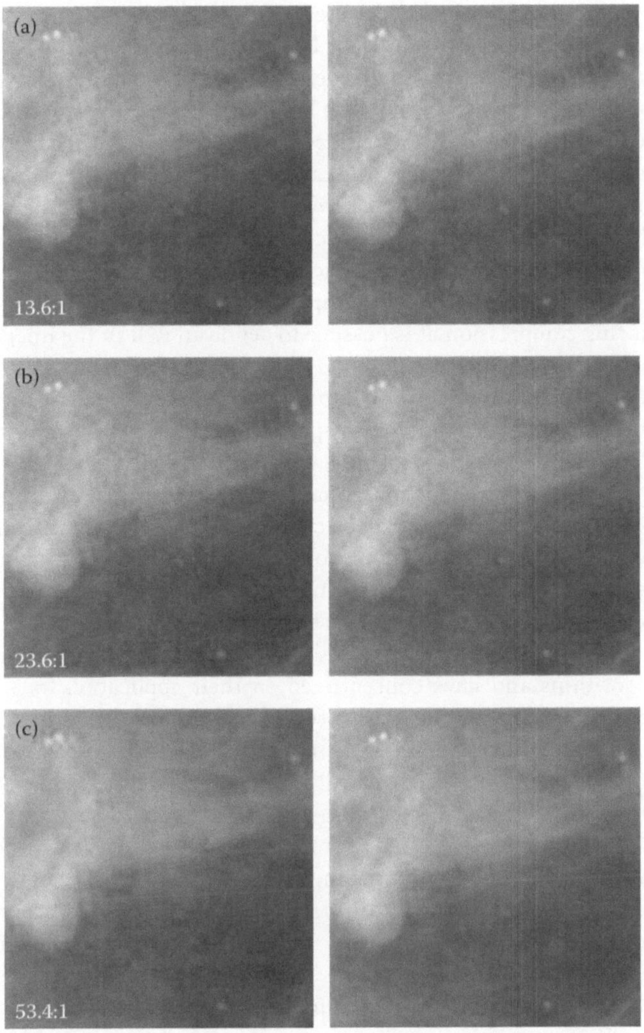

FIGURE 8.20 Comparison of image quality at equal compression ratios for the digital mammogram with JPEG compression (left column) and JPEG 2000 compression (right column).

8.3.4.4 Special Capabilities of JPEG 2000

In addition to overall better compression results with JPEG 2000 over JPEG, the former also incorporates some capabilities that are highly desired in image compression both in general and specifically for medical applications.*

One such capability useful for image transmission is *progressive decoding*, which allows for a lower quality version of the image to be decoded and displayed while additional image data

* Some of these features are actually present in standard JPEG but are either not used or not implemented often.

are communicated and decoded. Due to the bit-plane-based encoding, the most significant bits in each code block can be transmitted and decoded first resulting in a low-quality version of the image, while the bits with lower significance are transmitted and decoded later to add the remaining detail to the displayed image. Another capability that can be useful for image transmission in JPEG 2000 is *random access*, which allows for the selective decoding and displaying of specific ROIs in an image at higher quality than the rest of the image.

Finally, and possibly most important in medical applications, during compression, it is possible to define an ROI in the image being processed that should be compressed to a higher quality, and therefore a lower compression rate. It is apparent that in medical applications, this capability could be very useful in allowing the section of the image that includes the pathology of interest to retain the highest quality, while the normal portions of the depicted anatomy can be compressed to higher rates. This feature, however, may present some limitations in cases when it is necessary to retrospectively inspect supposedly normal areas of the image in search for a missed pathology.

Although here we have discussed only two compression algorithms and have concentrated on their application to 2D images, it should be noted that specific compression algorithms and modifications of these compression algorithms exist for other types of images, specifically 3D images and time-sequence images (2D + t and 3D + t).

For example, JPEG 2000 compression of 3D images can be performed by either compressing each individual 2D slice of the 3D image with the standard JPEG 2000 method, or by processing the entire 3D image at once, which is indirectly supported in the Part 2 extension of the JPEG 2000 standard, or using the specific extension for JPEG 2000 compression of volumetric images, the JP3D specification, described in Part 10 of the JPEG 2000 standard. In either case, processing a 3D image directly takes advantage of interslice redundancies that slice-by-slice 2D compression does not, potentially resulting in higher compression ratios at the same artifact rates or lower artifacts at the same compression ratios (Kimpe et al., 2007).

Time-sequence images, for example, echocardiography images, can be compressed using specific video compression algorithms, such as MPEG-1, -2, and -4. In their most basic form, video compression algorithms store only one full frame (an I-frame) every fixed number of frames (e.g., 15) in a compressed form similar to 2D compression, while for the next few frames (e.g., 14) only *motion vectors* that record how the features in the frames move through the images are stored. In this manner, video compression algorithms can achieve very high compression ratios since sections of images that do not change throughout the frames (features that do not move) occupy very little storage space. As examples of the capabilities of video compression algorithms for use in medical imaging, reports of MPEG-2 compression and MPEG-4 compression of echocardiography images have shown that compression ratios of ~50:1 up to ~1000:1 with little or no loss in diagnostic quality are achievable (Harris et al., 2003; Barbier et al., 2007).

8.4 Testing Image Quality

The ability to assess the image quality of an image that has been compressed with a lossy compression algorithm is very important. There are many ways to assess the image quality and the impact that compression has on it. Which method is used depends on what the assessment is used for and what resources (human, time, and economic) are available to perform the assessment. For example, during the development of a new compression algorithm or the optimization of the parameters of an existing compression algorithm for a certain type of image, it is probably impractical to perform a human observer study to determine clinical performance impact of the many options and parameters than can be changed, and therefore the use of objective metrics that can be determined with computer analysis is more feasible. However, since objective metrics have limited correspondence with human perception, studies that involve human observers should be performed in certain cases, such as in parameter optimization studies after the possible set of values has been narrowed to a manageable set, or, more commonly, during determination of the maximum compression ratio that results in clinically acceptable image quality.

8.4.1 Pixel Value Difference Metrics

The mean-squared error (MSE) (Fuhrmann et al., 1995) and peak signal-to-noise ratio (PSNR) (Said and Pearlman, 1996) are simple objective metrics to compare the similarity between two images. For the comparison between an original image $O(x, y)$ and its compressed version $C(x, y)$, both of size $M \times N$ and with grayscale values from $k = 0, 1, 2, \ldots, K - 1$, the MSE is given by

$$\mathrm{MSE} = \frac{1}{MN}\sum_{x=0}^{M-1}\sum_{y=0}^{N-1}\left[O(x,y) - C(x,y)\right]^2 \tag{8.8}$$

while the PSNR is given by

$$\mathrm{PSNR} = 10\log_{10}\left(\frac{K^2}{\mathrm{MSE}}\right) \tag{8.9}$$

The values of these two metrics behave in opposite ways: a low MSE signifies that the compressed image is similar to the original one, while this is true for an image that results in a high PSNR. The lower limit for MSE is zero, while PSNR has no maximum limit. These two metrics are attractive due to the simplicity to calculate them, but they are not directly related to the perception of image quality (Fuhrmann et al., 1995), which results in two limitations. In the first place, what is the maximum MSE or minimum PSNR that yields acceptable image quality? In the second place, different types of distortions could have a different impact in the values of these metrics resulting in a lower MSE (or higher PSNR) for an image that is perceptually of lower quality.

8.4.2 Structural Similarity Indices

The structural similarity index (SSIM) (Wang et al., 2004) has been proposed as an objective metric that more closely correlates with human perception of image quality than the objective metrics discussed above. The SSIM achieves this by comparing the two images' similarity in terms of the three characteristics that affect how the human visual system perceives an image: luminance, contrast, and structure. The multiscale extension of the SSIM (MS-SSIM) (Wang et al., 2003) was proposed to take into account in the image comparison the resolution of the display used to view the images and the viewing distance.*

In addition to correlating better with perceived image quality than the objective metrics discussed in the previous section, both SSIM and MS-SSIM are bounded metrics, with possible values between −1 and 1, with more similar images resulting in a higher value, but only the comparison of two identical images resulting in a value of unity with either of the two metrics.

8.4.3 Numerical Observers

Numerical observers, such as the nonprewhitening observer with an eye filter (NPWE) (Burgess et al., 2001), are mathematical models that attempt to replicate the human visual system, and therefore can be used to determine the detectability of specific features (e.g., nodules) in images. In general, numerical observer models analyze a large number of small simulated or hybrid real/simulated images, in which clinical backgrounds are used and realistic simulated lesions are superimposed for the positive cases to yield a *detectability index*. This index can be used to compare the visibility of lesions in sets of images compressed to different degrees. Since the detectability indices resulting from some of the numerical observers have been shown to correlate well with human detection performance, this computer-based method of comparing compressed image quality can yield useful results in controlled studies. For example, Suryanarayanan et al. (2005) used both the NPWE and the Laguerre–Gauss channelized Hotelling observer (LG-CHO) model to analyze the detection of masses in mammograms compressed with the JPEG 2000 algorithm and found no significant difference in detectability up to the maximum compression ratio studied (30:1). It should be noted that Suryanarayanan et al. also found good correlation between the results of the numerical observer studies and human observer studies.

8.4.4 Just Noticeable Difference

The aim of *just noticeable difference* (JND) experiments is to determine the compression rate at which the threshold between perceptually lossless and lossy compression is crossed. In other words, JND measurements attempt to identify what is the highest compression rate achievable which still results in no perceived loss of image quality by human observers.

To perform a JND experiment, a set of observers is shown a series of images both uncompressed and compressed to different compression ratios. Typically, many different images (e.g., 5–10 different radiographs) are included in the experiment, and each image is not only displayed at various levels of compression, but also it is displayed various times at each level. There are at least two basic methods to perform a JND experiment with this set of images.

In the first method, one image is displayed for a certain amount of time (e.g., 1 s) and after the time is up, the observer is asked if the image was compressed or not (Watson et al., 1997). The second method of performing a JND experiment is to use the two-alternative forced choice algorithm, in which the set of images is displayed in pairs, with always one of the images being the uncompressed version of the other image. Typically, the pair of images is displayed one at a time, in a random order, either for a predefined amount of time and not allowing the user to switch back and forth between images (Fuhrmann et al., 1995) or with no time limit and allowing the user to switch between images as many times as necessary (Eckert, 1997). In either case, the observer is asked which of the two images in the pair was compressed. For either experimental method, the JND threshold point is defined as the compression rate at which the observers correctly identified the compressed images a certain percentage of the time (e.g., 50%, although this percentage can vary).

JND experiments are more useful than the objective metrics discussed previously since they provide information on the actual perceived quality of the compressed images, and on at what compression rate quality loss is actually observable. However, JND experiments do have limitations. The chief among them is that the JND threshold point has been shown to vary depending on the expertise of or training given to the observers (Good et al., 1994; Fuhrmann et al., 1995). A second limitation is that for many medical applications although a certain compression rate is above the JND threshold point and therefore does result in a perceived change in the image, it might not necessarily affect the diagnostic quality of an image. Therefore, to better identify what compression rates are acceptable in each situation, a measurement that more closely reflects the diagnostic task is more useful.

8.4.5 Evaluation of Diagnostic Task

The most relevant measurement of the impact of compression on a medical image is one that relates to the diagnostic quality of the image. For example, if an aggressively compressed chest computerized tomography image does not decrease the number of true lung nodules that are detected and does not result in increased false positives, then it is not relevant if this compression rate is above the JND threshold point and/or the objective metrics translate to an unacceptable difference between the original and the compressed image. In other words, for measuring the impact of a compression algorithm on a medical image,

* A second extension of SSIM, complex wavelet SSIM (CW-SSIM) (Wang and Simoncelli, 2005), which allows for translations, rotations, and scaling of the image content, has also been proposed, but it is not relevant to image quality analysis in image compression.

the evaluation of the algorithm's impact on diagnostic quality is the gold standard. Of course, diagnostic quality evaluation is the more resource-intensive type of evaluation due to the number of images and the number and type of observers required. For example, although it could be argued that a JND experiment can be performed with nonclinical experts, diagnostic quality evaluations need to be performed with physicians trained in the interpretation of the specific image type being evaluated. Therefore, ideally the compression impact on mammograms should be evaluated by breast-imaging specialized radiologists while chest CTs for lung nodule detection should be evaluated by thoracic radiologists.

To determine the impact of image compression on clinical performance of a diagnostic task, a group of observers interprets a set of clinical images both before and after compression (at several ratios) in various sessions so that the same image at different compression ratios is not interpreted during the same session. During the interpretation of an image, the diagnostic task, for example, the presence of lesions suspicious for breast cancer in mammograms, is performed and recorded. The observers' interpretations are compared to the independently known truth, and one or a set of metrics is determined. Ideally, the questions asked to the observers should be designed so that a receiver operating characteristics (ROC) curve for the diagnostic task can be built and its area under the curve (AUC) determined, but other metrics such as sensitivity and specificity (which actually define one point on the ROC curve) could be used. Statistical comparison of the computed metric(s) between the uncompressed and the compressed images is then performed, and the maximum compression ratio that results in no significant difference is the highest ratio that can be used for the studied type of images for the studied diagnostic task. Note that this maximum acceptable ratio varies for different types of images (e.g., chest CT versus mammograms), and in many cases varies for the same types of images but different diagnostic task (e.g., lung nodule detection versus diffuse lung disease follow-up). Examples of this type of studies are provided in Ko et al. (2003) in which the use of JPEG 2000 compression was studied on chest CT images for lung nodule detection, and Kocsis et al. (2003) in which wavelet-based compression was studied on mammograms for microcalcification detection.

A related type of study in which the diagnostic task is considered when evaluating image quality but the clinical performance is not directly measured is the study which involves the subjective evaluation by expert observers of the visibility and overall quality of the display of specific, clinically relevant features. For example, Lucier et al. (1994) studied the impact of wavelet-based compression on mammograms. For this, a breast-imaging radiologist was asked to rate the visibility and number of microcalcifications, the degree of distortion of their morphology, and the possibility of false positives arising from compression artifacts, among other questions. The clinical performance of the diagnostic task, detection and diagnosis of suspicious microcalcifications, was not evaluated directly by measuring the AUC of the ROC curve, the sensitivity, and/or specificity. However, the study determined the impact of the compression algorithm on diagnostically relevant features, and therefore a compression algorithm that introduces artifacts that do not decrease the detection rate or increase the false positive rate is not penalized. Studies of this type do not provide as reliable information as the direct studies of clinical performance, but in general require fewer resources.

The use of compression algorithms in medical imaging, although possibly controversial when information loss is involved, provides definite advantages. The use of lossless compression, although the most conservative approach, results in limited benefit due to the inherently low compression ratios achievable in real medical images. Therefore, due to the exponential increase in the amount of image data generated every year, the use of lossy compression algorithms, with their substantially higher achievable compression ratios, sometimes with no perceptual and/or diagnostic quality loss, is probably unavoidable in the future. However, if the use of lossy compression becomes commonplace for storage, long-term archiving, and/or transmission of medical images, it will be important to perform comprehensive task-specific studies to optimize the image compression algorithms' parameter selection. These studies, such as the one performed by Zhang et al. (2004) for coronary angiograms, result in substantial increases in the degree of compression achievable with no significant loss in clinical performance, therefore both optimizing the resources available and ensuring the diagnostic quality of the compressed images.

References

Abramson, N. 1963. *Information Theory and Coding.* New York, NY: McGraw-Hill.

Ahmed, N., Natarajan, T., and Rao, K. 1974. Discrete cosine transfom. *IEEE Trans. Comput.*, 100, 90–3.

Antonini, M., Barlaud, M., Mathieu, P., and Daubechies, I. 1992. Image coding using wavelet transform. *IEEE Trans. Image Process.*, 1, 205–20.

Barbier, P., Alimento, M., Berna, G., Celeste, F., Gentile, F., Mantero, A., Montericcio, V., and Muratori, M. 2007. High-grade video compression of echocardiographic studies: A multicenter validation study of selected Motion Pictures Expert Groups (MPEG)-4 algorithms. *J. Am. Soc. Echocardiogr.*, 20, 527–36.

Branstetter, B. F. T. 2007. Basics of imaging informatics: Part 2. *Radiology*, 244, 78–84.

Bui, A. and Taira, R. 2010. *Medical Imaging Informatics.* New York, NY: Springer.

Burgess, A. E., Jacobson, F. L., and Judy, P. F. 2001. Human observer detection experiments with mammograms and power-law noise. *Med. Phy.*, 28, 419–37.

Channin, D. S. 2001. Integrating the healthcare enterprise: A primer. Part 2. Seven brides for seven brothers: The IHE integration profiles. *Radiographics*, 21, 1343–50.

DICOM. 2000. *Digital Imaging Communications in Medicine DICOM Part 3.* Rosslyn, VA: NEMA Standards Publications, National Electrical Manufacturers Association.

Dreyer, K., Hirschorn, D., Thrall, J., and Mehta, A. 2006. *PACS: A Guide to the Digital Revolution.* New York, NY: Springer.

Ebbert, T. L., Meghea, C., Iturbe, S., Forman, H. P., Bhargavan, M., and Sunshine, J. H. 2007. The state of teleradiology in 2003 and changes since 1999. *Am. J. Roentgenol.*, 188, W103–12.

Eckert, M. P. 1997. Lossy compression using wavelets, block DCT, and lapped orthogonal transforms optimized with a perceptual model. *Proc. SPIE*, 3031, 339–50.

Fuhrmann, D. R., Baro, J. A., and Cox, J. J. R. 1995. Experimental evaluation of psychophysical distortion metrics for JPEG-encoded images. *J. Electron. Imaging*, 4, 397–406.

Gershon-Cohen, J. and Cooley, A. G. 1950. Telognosis. *Radiology*, 55, 582–7.

Good, W., Maitz, G., and Gur, D. 1994. Joint Photographic Experts Group (JPEG) compatible data compression of mammograms. *J. Digit. Imaging*, 7, 123–32.

Harris, K. M., Schum, K. R., Knickelbine, T., Hurrell, D. G., Koehler, J. L., and Longe, T. F. 2003. Comparison of diagnostic quality of motion picture experts group-2 digital video with super VHS videotape for echocardiographic imaging. *J. Am. Soc. Echocardiogr.*, 16, 880–3.

Huang, H. 2010. *PACS and Informatics: Basic Principles and Applications.* Hoboken, NJ: Wiley-Blackwell.

Huffman, D. 1952. A method for the construction of minimum redundancy codes. *Proc. Inst. Elect. Radio Engineers*, 40, 1098–101.

Kaplan, J., Roy, R., and Srinivasaraghavan, R. 2008. Meeting the demand for data storage. *McKinsey on Business Technology*, Fall, 1–10.

Kimpe, T., Bruylants, T., Sneyders, Y., Deklerck, R., and Schelkens, P. 2007. Compression of medical volumetric datasets: Physical and psychovisual performance comparison of the emerging JP3D standard and JPEG2000. *Proc. SPIE*, 6512, 65124L-8.

Ko, J. P., Rusinek, H., Naidich, D. P., Mcguinness, G., Rubinowitz, A. N., Leitman, B. S., and Martino, J. M. 2003. Wavelet compression of low-dose chest CT data: Effect on lung nodule detection1. *Radiology*, 228, 70–5.

Kocsis, O., Costaridou, L., Varaki, L., Likaki, E., Kalogeropoulou, C., Skiadopoulos, S., and Panayiotakis, G. 2003. Visually lossless threshold determination for microcalcification detection in wavelet compressed mammograms. *Eur. Radiol.*, 13, 2390–6.

Le Gall, D. and Tabatabai, A. 1988. Sub-band coding of digital images using symmetric short kernel filters and arithmetic coding techniques. *IEEE Int. Conf. Acoust. Speech Signal Process.*, 2, 761–5.

Lewis, R. S., Sunshine, J. H., and Bhargavan, M. 2009. Radiology practices' use of external off-hours teleradiology services in 2007 and changes since 2003. *Am. J. Roentgenol.*, 193, 1333–9.

Lucier, B., Kallergi, M., Qian, W., Devore, R., Clark, R., Saff, E., and Clarke, L. 1994. Wavelet compression and segmentation of digital mammograms. *J. Digit. Imaging*, 7, 27–38.

Marcellin, M. W., Lepley, M. A., Bilgin, A., Flohr, T. J., Chinen, T. T., and Kasner, J. H. 2002. An overview of quantization in JPEG 2000. *Signal Process., Image Commun.*, 17, 73–84.

Rao, K. and Yip, P. 1990. *Discrete Cosine Transform: Algorithms, Advantages, Applications.* London: Academic Press.

Said, A. and Pearlman, W. A. 1996. A new, fast, and efficient image codec based on set partitioning in hierarchical trees. *IEEE Trans. Circuits Syst. Video Technol.*, 6, 243–50.

Samei, E., Seibert, J. A., Andriole, K., Badano, A., Crawford, J., Reiner, B., Flynn, M. J., and Chang, P. 2004. AAPM/RSNA tutorial on equipment selection: PACS equipment overview: General guidelines for purchasing and acceptance testing of PACS equipment. *Radiographics*, 24, 313–4.

Shannon, C. E. 1948. A mathematical theory of communication. *Bell Syst. Tech. J.*, 27, 379–423, 623–56.

Siegel, E. and Reiner, B. 2002. Work flow redesign: The key to success when using PACS. *Am. J. Roentgenol.*, 178, 563–6.

Suryanarayanan, S., Karellas, A., Vedantham, S., Waldrop, S. M., and D'orsi, C. J. 2005. Detection of simulated lesions on data-compressed digital mammograms. *Radiology*, 236, 31–6.

Taubman, D., Ordentlich, E., Weinberger, M., and Seroussi, G. 2002. Embedded block coding in JPEG 2000. *Signal Process., Image Commun.*, 17, 49–72.

Wang, Z., Bovik, A. C., Sheikh, H. R., and Simoncelli, E. P. 2004. Image quality assessment: From error visibility to structural similarity. *IEEE Trans. Image Process.*, 13, 600–12.

Wang, Z. and Simoncelli, E. 2005. Translation insensitive image similarity in complex wavelet domain. *Proc. IEEE Int. Conf. Acoust. Speech Signal Process.*, 2, 573–76.

Wang, Z., Simoncelli, E., and Bovik, A. 2003. Multiscale structural similarity for image quality assessment. *Proc. IEEE Asilomar Conf. Signals Syst. Comput.*, 2, 1398–402.

Watson, A. B., Taylor, M., and Borthwick, R. 1997. DCTune perceptual optimization of compressed dental x-rays. *Proc. SPIE*, 3031, 358–71.

Zhang, Y., Pham, B., and Eckstein, M. 2004. Automated optimization of JPEG 2000 encoder options based on model observer performance for detecting variable signals in x-ray coronary angiograms. *IEEE Trans. Med. Imaging*, 23, 459–74.

9

Displays

Elizabeth A. Krupinski
University of Arizona

9.1 Interpreting Medical Images: How Do Clinicians Process Image Information?

Informatics can be and has been defined in numerous ways, but a very useful definition is that it "is the discipline focused on the acquisition, storage, and *use* of information in a specific setting or domain" (Hersh, 2009, p. 10). Medical imaging informatics is a branch of informatics that deals in particular with the specific domain of medicine and the information contained in the wide variety of images encountered in the assessment and treatment of patients. Medical imaging informatics therefore intersects with every link in the imaging chain from image acquisition, to distribution and management, storage and retrieval, image processing and analysis, visualization and data navigation, and finally through to image interpretation, reporting, and dissemination of the report. This chapter deals with the *use* of image information by clinicians as they *visualize* and *interpret* medical images. Although the majority of imaging informatics work has been done in radiology, other image-based specialties such as pathology, ophthalmology, dermatology, and a wide range of telemedicine applications are beginning to consider the role of imaging informatics.

Central to any medical specialty that utilizes images is the image interpretation process. This process can be examined from two perspectives. First is the technology used to display the image information and how factors such as luminance and display noise can affect the quality of the image and thus the visibility of the information (i.e., diagnostic features needed to render a diagnosis). Second is the human observer relying on their perceptual and cognitive systems to accurately and efficiently process the information presented to them on the display. These components cannot be considered in isolation. Thus, it is important to understand some of the key issues involved in the image interpretation process and how to optimize the digital reading environment for effective and efficient information processing and image interpretation.

The technologies used in medical imaging are quite varied, so the study of image interpretation can be challenging. In general, medical images can be acquired with everything from sophisticated dedicated imaging devices (e.g., computed tomography (CT) or magnetic resonance imaging (MRI) scanners in radiology or nonmydriatic cameras in ophthalmology) to off-the-shelf digital cameras (e.g., for store-and-forward teledermatology). Even within a single field such as radiology, images can be grayscale or color, high-resolution or low-resolution, hardcopy or softcopy, uncompressed or compressed, and so on. Once acquired, the displays are just as variable. Radiology for the most part uses high-performance medical-grade monochrome displays calibrated to the Digital Imaging and Communications in Medicine Grayscale Standard Display Function (DICOM GSDF) (DICOM, 2010), for primary reading (Langer et al., 2004). However, ophthalmology, pathology, dermatology, and other specialties where color is important typically use off-the-shelf color displays (Krupinski et al., 2008; Krupinski, 2009a,b). All specialties to some degree or another are even investigating the potential of hand-held devices such as personal digital assistants for interpreting images (Toomey et al., 2010; Wurm et al., 2008).

Regardless of how they are acquired or displayed, medical images need to be interpreted because the information contained in them is not self-explanatory. In radiology, in particular, images vary considerably even within a particular exam type and/or modality. Anatomical structures can camouflage features of clinical interest, lesions often have very low-prevalence especially in screening situations, and there are notable

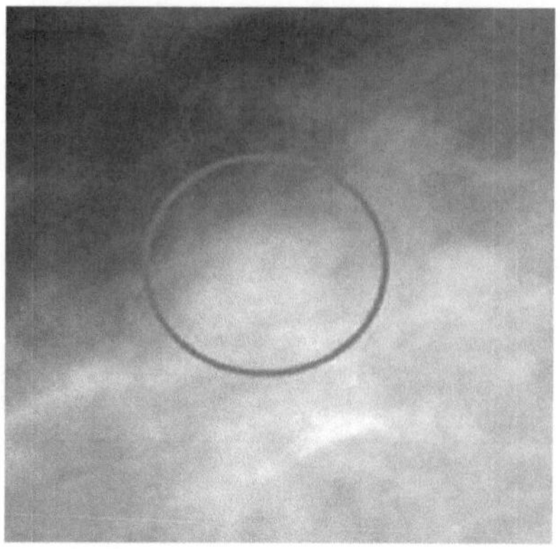

FIGURE 9.1 Portion of a mammogram showing a mass (in the circle) embedded in and camouflaged by surrounding parenchymal tissue. The edges of the mass are blurred or covered by breast tissue making it difficult to detect the lesion features.

variations from case to case with a multiplicity of abnormalities and normal features that the interpreter needs to be aware of (see Figure 9.1). The problem is that these complexities often can lead to interpretation errors (Berlin, 2005, 2007, 2009). In radiology, some estimates indicate that in some areas there may be up to a 30% miss rate with an equally high false-positive rate. These errors can have a significant impact on patient care causing delays in treatment or misdiagnoses that lead to the wrong treatment or no treatment at all.

Before methods can be developed to improve the delivery of image information and avoid or at least ameliorate errors, we need to understand the nature and causes of error. Numerous studies have been carried out over the years in radiology that have resulted in a categorization of errors. These studies have used eye-position recording to track the ways that radiologists search images for lesions (see Figure 9.2). It has been found that approximately one-third of errors are visual in nature. The radiologist does an incomplete search of the image data, missing a

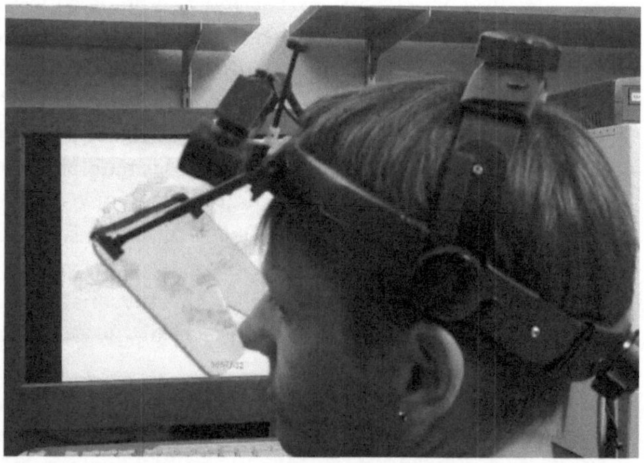

FIGURE 9.2 Typical eye-position recording set up. The head-mounted recording system uses special optics to reflect an infrared light source off the subject's pupil and cornea, sampling the eye every 1/60 s to record location and dwell information.

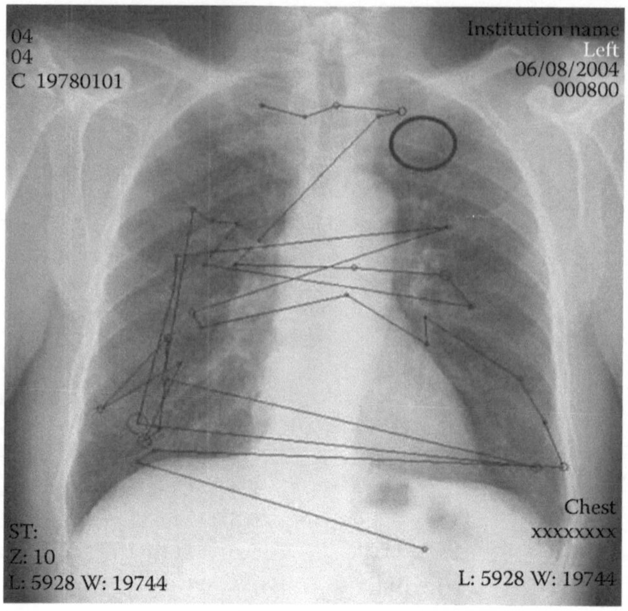

FIGURE 9.3 Example of typical search pattern generated by a radiologist searching a chest image for nodules as eye position is recorded. Each small circle indicates locations where the eyes land long enough to process information. The size of the circle indicates how long (dwell time) was spent at the location, with larger circles indicating longer dwells. The lines between the circles indicate the order in which they were generated. The larger circle in the upper left lung (right side of figure) is a nodule that the radiologist did not look at with high-resolution foveal vision and thus did not report as being present—a search error.

potential lesion (see Figure 9.3). About two-thirds of errors are of a cognitive nature and there are two types. About half occur when something suspicious is detected and scrutinized, but the radiologist fails to recognize it as a true abnormality and thus fails to report it. In the second half, an abnormality is scrutinized and recognized as a potential lesion but the radiologist makes an active decision-making error in calling the case negative or by calling the finding something other than what it is (Kundel, 1975, 1989; Kundel et al., 1989).

These interpretation errors are caused by a variety of psychophysical processes. For example, one reason is that the abnormalities can be camouflaged by normal structures (i.e., anatomical noise), which have been estimated to affect lesion detection thresholds by an order of magnitude (Samei et al., 1997). As noted above, visual search itself can also contribute to errors. Radiologists need to search images because high-resolution vision, which is needed to detect and extract subtle lesion features, is limited by the angular extent of the high-fidelity foveal vision of the human eye. Although it is generally agreed upon that interpretation is preceded by a global impression or gist that can take in quite a bit of relevant diagnostic information, it is generally insufficient to detect and characterize the majority of abnormalities. Visual search needs to occur in order to move the eyes around the image to closely examine image details (Kundel, 1975; Nodine and Kundel, 1987).

A number of studies (e.g., Hu et al., 1994; Krupinski, 1996; Krupinski and Lund, 1997; Manning et al., 2004) have revealed that there are characteristic dwell times associated with correct and incorrect decisions, and that these times are influenced by the nature of the diagnostic task, the idiosyncratic observer search patterns, the *nature of the display*, and the *way the information is presented* on that display (Krupinski and Lund, 1997; Kundel, 1989). True positives and false positives tend to be associated with longer dwell times than false negatives which in turn tend to have longer dwell times than true negatives. The fact that about 2/3 of missed lesions attract visual scrutiny has led to investigations that have successfully used dwell data to feed back these areas of visual interest to radiologists, resulting in significant improvements in detection performance—without associated increases in false positives (Krupinski et al., 1998; Nodine et al., 1999).

The information available has also been shown to affect the search and diagnostic performance. Moving to digital reading of radiographic images not only has created a number of significant reading benefits but it has also created some viewing challenges. In particular, the layout of information on the display (i.e., the Graphical User Interface [GUI]) is very important. One study showed that radiologists spend about 20% of their search time fixating the GUI (e.g., tool bar, image thumbnails) rather than fixating the radiograph (Krupinski and Lund, 1997).

It is not just information external to the image that can affect diagnoses. Information within the image can as well, and it is important to understand the nature of that information if we are going to utilize it to aid image interpretation. For example, Mello-Thoms (2006) found statistically significant differences in the spatial frequency representation of background areas in mammograms that were sampled before and after an observer first looked at the location of a verified mass. There is a shift in the way observers search images before and after they fixate a mass for the first time and report it. With true positives it appears that radiologists often detect the lesion very early (Kundel et al., 2008) then verify it by searching for background matches to the perceived finding. After fixating the lesion and deciding to report it, the search strategy changes and the radiologist analyzes more general background characteristics. With false negatives, a different pattern was seen. Before the eyes hit the mass, there seems to be no early perception of the lesion or its features, so there is no attempt to search the background for matching samples. If the miss is not due to a search error, the eyes do hit the mass and the background sampling strategy changes to sample similar background locations. It seems that the radiologist is trying to determine if the detection of a few suspicious mass features might indeed belong to a true mass. The spatial frequency information in the average background areas, however, were not sufficient to distinguish them from the potential lesion location (or vice versa) so the mass remains unreported.

In the phenomenon known as Satisfaction of Search (SOS), image information again seems to affect the interpretation process. In SOS, once an abnormality is detected and recognized, it takes additional diligence to search for other possible abnormalities (Berbaum et al., 2010; Smith, 1967; Tuddenham, 1962, 1963). Sometimes extra effort is not used and subsequent lesions in the same image or case are missed. SOS error estimates vary, but range from 1/5 to 1/3 of misses in radiology, and possibly as high as 91% in emergency medicine (Berbaum et al., 2010). SOS has been studied in depth by Berbaum and colleagues and they have found that ending search prematurely (i.e., not scanning the entire image but rather stopping once a lesion is detected) is generally not the root cause of SOS. Instead faulty pattern recognition and/or faulty decision making seems to be the more likely culprit. In some cases such as when contrast is used in a study, the presence of this distracting visual information is enough to draw search to the contrast containing region at the expense of the rest of the image information being searched (Berbaum et al., 1996).

9.2 The Radiology Workstation

How can imaging informatics help in the interpretation process? One way is in the design of the workstation environment in terms of how the information needed to render a diagnosis is displayed to the clinician. From the diagnostic point of view, the workstation needs to be able to provide adequate information to the clinician to maintain acceptable levels of sensitivity and specificity. From an ergonomic point of view, the workstation needs to make the viewing and interpretation process efficient, reducing stress and strain for the clinician and improving workflow. There is no single way to set up the "perfect" workstation and there is no single way to design a radiology reading room. However, over the years, some essential design considerations have emerged.

From a practical point of view, one question is how much information should be provided at once? Although circumstances may differ from setting to setting, most radiology workstations have two main display monitors for viewing clinical images. Typically they are medical-grade high-resolution (at least 2 MegaPixel) monochrome displays, although they are increasingly changing to high-resolution color displays (Geijer et al., 2007; Krupinski 2009a,b). For ergonomic considerations, they are generally positioned side-by-side and angled slightly inward toward so the radiologist can sit centrally and view each display. Positioning the displays slightly inward reduces head movements needed to direct the eyes to the screen, reducing neck, back, and shoulder strain that can result from prolonged sitting at the display (see Figure 9.4). In the early days of digital reading, display number varied considerably from one to six, but very quickly two displays became the norm as three or more tended to require too much movement to view all the displays at the right angle.

From an information perspective, one display is generally insufficient as most radiology exams have more than one image (multiple views) and there is often to need to compare images (different views or current vs. priors) simultaneously. As patients also tend to have images from more than one modality (e.g., digital mammography and Breast MRI) even if one exam (e.g., breast MRI) can be viewed on a single display, the presence of the second modality requires a second display for easier viewing.

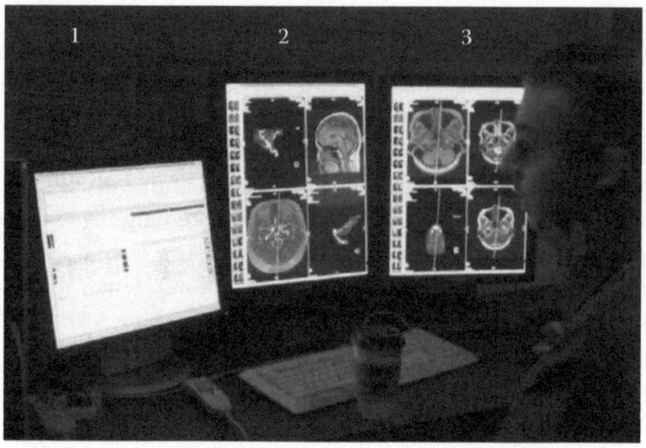

FIGURE 9.4 Typical radiology workstation set up with two high-performance monochrome monitors (2 and 3) tilted slightly inwards and a single off-the-shelf display (1) for reviewing the patient record and dictating the report.

Comparing images on a single display is feasible, but it often necessitates reducing the size of the images, making it difficult to discern fine details. With two displays, images can be viewed at full or near to full resolution and compared easily.

As images are not the only information required for interpretation, there is also typically a general purpose monitor used to display the work list (see Figure 9.4). It provides information such as which images are available on the Picture Archiving and Communications Systems, as well as patient information from the Radiology Information System (RIS) or Hospital Information System (HIS). This display can also be used for dictation, although sometimes a separate monitor is used and positioned on the other side of the diagnostic displays. Since most digital reading rooms have also moved from tape recording and transcriptionists to automatic speech recognition technologies, the radiologist needs a display for viewing and correcting their reports as they are generated. Thus, the radiologist is literally surrounded by displays containing information relevant to the interpretation process.

From a human factors perspective, it is generally recommended that the displays are set on a table that can be moved up and down in order to adjust the displays to the user's height (eyes about level with the display center). A separate desk level for the keyboard and mouse is also recommended (typically slightly lower than the display table) to avoid carpal tunnel syndrome and other repetitive stress computer injuries (Goyal et al., 2009; Ruess et al., 2003). Comfortable chairs that can be adjusted are also recommended for fine-tuning the user's height with respect to the displays. Chair wheels are useful so the user can move away from or closer to the display without bending over or leaning back to avoid strain injuries.

9.3 Displays: Presenting High-Quality Image Information for Interpretation

Displays may not be the first thing that comes to mind when thinking about imaging informatics, but when you consider that the display is how the image data is presented to the clinician it becomes clear that the display is a very critical component in the information chain. Quality assessment (QA) and quality control (QC) are playing increasingly important roles in medical imaging, since perception and diagnostic accuracy can be affected significantly by the quality of the image and thus the quality of information. The American College of Radiology has offered guidelines on image quality (ACR, 2010) as have other image-based specialties (American Telemedicine Association, 2004; Krupinski et al., 2008).

Guidelines for display performance and image quality testing have also been developed (Deutsches Institut fuer Normung, 2001; IEC, 2008; SMPTE, 1991; VESA, 2008). The most familiar in medicine are the DICOM, 2000 guidelines. It has been used in radiology since its creation and is being adopted by other clinical specialties as well (Kayser et al., 2008; Krupinski et al., 2008). The DICOM 14 GSDF determines the display function. It is based on the Barten Model and offers the advantage of perceptual linearization (Blume, 1996). Perceptual linearization optimizes a display by taking into account the capabilities of the human visual system. It produces a tone scale that equalizes changes in driving levels to yield changes in luminance that are perceptually equivalent across the entire luminance range. In other words, equal steps in perceived brightness represent equal steps in the acquired image data. Further, it has been demonstrated that perceptually linearized displays yield significantly better diagnostic accuracy and more efficient visual search than nonlinearized displays (Krupinski and Roehrig, 2000; Leong et al., 2010).

The American Association of Physicists in Medicine Task Group 18 has created a medical display QC program called "Assessment of Display Performance for Medical Imaging Systems" (Samei et al., 2005). The recommendations include two classes of tests. Visual or qualitative tests show an observer test patterns on a given display and requires them to decide if test objects (of a given size, contrast, etc.) are present or absent. Quantitative tests use an instrument such a photometer to make physical measurements on such display properties as luminance, resolution, noise, angular response, reflection, glare, distortion, color tint, artifacts, and contrast. The guidelines suggest minimum expected performance values and recommend a strategy to assess the maximum allowable illumination in the reading room by using the reflection and luminance characteristics of the display.

QA and QC are not simply about checking the physical performance of displays. There have been a number of studies verifying that many of these physical properties affect the quality of the images being displayed (i.e., the information contained in those images) and thus the quality of diagnostic decisions rendered and the efficiency with which they are generated. In addition to calibrating to the DICOM GSDF, display luminance (Krupinski et al., 1999), bit depth (Heo et al., 2008; Krupinski et al., 2007), ambient illumination (Heo et al., 2008; Mc Entee et al., 2006; McEntee and Martin, 2010), viewing angle (Krupinski et al., 2003), display size (Krupinski et al., 2006b; Toomey et al., 2010), veiling glare (Krupinski et al., 2006a), and a variety of other parameters have all been shown to influence diagnostic accuracy. Clearly, the quality of the display used to convey the image information to the clinician is critical to maintain high diagnostic performance.

9.4 Color Displays: Faithfully Maintaining Color Information

Although radiology probably still accounts for the widest use of medical images and displays, other medical specialties that utilize images in the diagnostic process are starting to acquire and view images digitally. In pathology (see Figure 9.5) and telemedicine (e.g., dermatology and ophthalmology), however, color information is often critical for accurate diagnosis. The ways in which color images are reproduced on color displays, the accuracy of color reproduction by the displays and the consistency of the color reproduction among color displays can all affect the interpretation. Thus, it is necessary to set up and calibrate color displays properly to prevent luminance and chrominance differences between displays from affecting the diagnosis.

A common but basic approach to consistent color display is use of the Gretag-Macbeth ColorChecker (McNeill et al., 2002), a pattern of 24 commonly used colors and gray steps. Color display calibration is done by adjusting on-screen color controls until there is *visually* little or no difference between the colors on the display and the actual physical chart held next to the display and illuminated by the suggested light source. The problem is that it is a visual match and therefore highly subjective, user dependent, and variable.

More sophisticated color display calibration techniques are based on the fact that the operation of color displays is based on a mixture of the three primary colors (red, blue, green) that in suitable quantities can produce many color sensations. Color coordinates and temperatures can be measured with colorimeters or spectroradiometers (Fetterly et al., 2008; Roehrig et al., 2010; Saha et al., 2010). Although there is clear consensus that color display calibration is a high priority, there has been very little progress in performing the basic research needed to develop and validate a color calibration standard for medical imaging. The fact that color displays can affect the quality of the image information and thus diagnostic performance has however been demonstrated, verifying the need for standard color calibration methods in medical imaging (Krupinski 2009a,b; Langer et al., 2006).

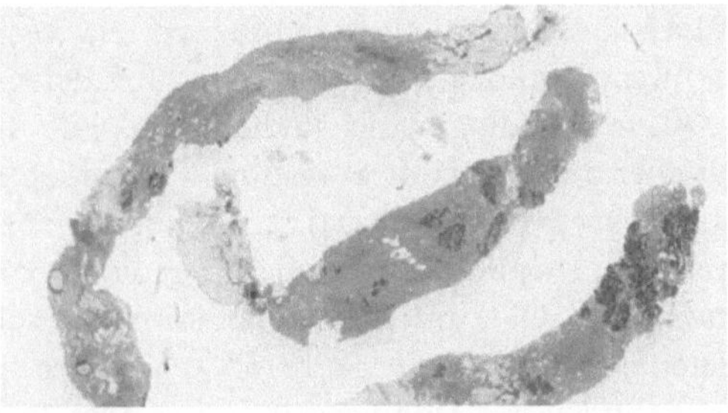

FIGURE 9.5 Example of a digital pathology specimen slide (frozen section breast biopsy). The purple-stained areas (black in this figure) tend to contain diagnostic information about malignancy while the pink-stained areas (gray in this figure) typically contain normal breast tissue.

9.5 The User Interface: Accessing Information

The design of the user interface for a clinical image viewing workstation is the core of the workstation and represents the portal through which the radiologist accesses the image information. The user interface should be fast, intuitive, user friendly, able to integrate and expand, and reliable. In radiology, one of the main issues is the "hanging protocol" or how to arrange the images on a computer display. The success of the image arrangement protocol relies on the quality of the default display (Moise and Atkins, 2005). Moise and Atkins demonstrated that the layout of the images in the default display affects the users' ability to extract the information needed to make accurate decisions and affected the speed with which they navigated through the images. For full-field digital mammography in particular (Zuley, 2010), having the "proper" hanging protocol is important because mammographers have eight critical images that need to be viewed and compared (CC and MLO right and left breasts for the current and prior exams). Being able to view images at full resolution (especially to detect subtle microcalcifications) is also important, making it necessary to toggle back and forth between viewing single images at full resolution and multiple images at the same time but at lower resolution (Zuley, 2010).

Interpretation speed is important because radiology services, especially high-technology modalities (Bhargavan and Sunshine, 2005), second opinion (DiPiro et al., 2002), and teleradiology (Ebbert et al., 2007) have increased significantly in recent years. Radiologists now read more studies, each containing more images. As a result, shortages of radiologists and increased workloads are common both in the United States and around the world (Lu et al., 2008; Nakajima et al., 2008; Sunshine and Maynard, 2008; Thind and Barter, 2008). The time needed to read the increased volume of imaging examinations has led to more studies being read after hours or by on-call radiologists, especially for CT and MRI.

Just having the images available on the workstation is not the whole story. Image processing tools (e.g., window/level), manipulation tools (e.g., rotate), and measurement tools are often used while viewing the images and this requires access to a menu or a tool bar to activate the tools. There is additional information as well however that needs to be integrated into the overall reading process and that can be accessed either through the diagnostic displays or through a separate display containing the RIS and/or HIS data. For both situations, the user should be able to use the basic navigation tools of the interface without any training and without any prior exposure. The systems need to be user friendly and easy to customize. Simple menus and file managers, single click navigation, visually comfortable colors or gray scales and an uncluttered workspace are all recommended. Images and other relevant information should be easily adjustable to meet personal visual preferences and interpretation patterns plus easy restoration of default values and set up.

One approach to organizing and accessing workstation information that has been advocated is the digital dashboard. The digital dashboard is both a portal to the information but it is also an active miner and integrator of information. Dashboards can be designed to integrate separate computerized information systems (e.g., RIS, HIS, and other clinical information systems) and summarize key work flow metrics in real time to facilitate informed decision making. Digital dashboards can alert radiologists to their unsigned report queue status, facilitate the transcription process by providing report templates, provide a link to the report signing application, and generally assess workflow throughout the chain from image acquisition to reporting (Khorasani, 2008; Minnigh and Gallet, 2009; Morgan et al., 2006, 2008; Nagy et al., 2009; Zhang et al., 2009). Digital dashboards have in some cases been shown to significantly improve workflow (Morgan et al., 2008) and potentially reduce image retakes (i.e., reduce excess dose to patients) by tracking technologist use patterns (Minnigh and Gallet, 2009).

The digital reading environment includes a variety of peripheral devices as well as the display devices. These devices are not only part of the information-processing environment but serve as a means of transmitting information to other users. One of the most important peripheral components is the digital voice recording system to generate a report or voice input systems into digital reporting forms. Advances in continuous voice recognition technologies have been important and many, although not all of the problems in terms of accuracy have been eliminated. Given the right system and enough training, voice recognition reporting systems can improve productivity by decreasing significantly report turn-around times (Bhan et al., 2008; Boland et al., 2008; DeFlorio et al., 2008; McGurk et al., 2008; Pezzullo et al., 2008).

9.6 Human Factors and Information Processing

There are a number of human factors issues related to the environment in which the workstation will be placed that are important in terms of the clinician being able to effectively process information. Surprisingly ambient noise levels do not seem to have much effect on the ability of radiologists to process image information and render correct diagnoses (McEntee et al., 2010). In the McEntee et al. study, noise levels were recorded 10 times in 14 environments in four hospitals. Thirty chest images were then presented to 26 radiologists who were asked to detect nodules in the absence and presence of noise at amplitudes recorded in the clinical environment. The noise amplitudes recorded rarely exceeded levels of normal conversation with the maximum being 56.1 dB. This noise level had no impact on the ability of radiologists to identify. In fact, performance was significantly better with noise than in the absence of noise.

Having some level of ambient of noise thus may help the radiologist focus on the task at hand and thus improve their ability to effectively process visual information. Not surprisingly, ambient auditory "noise" in the form of music may be useful as well. Mohiuddin et al. had eight radiologists listen to 1 h of classical chamber music from the baroque period while interpreting radiological studies during a typical workday and then they rated mood, concentration, perceived diagnostic accuracy, productivity, and work satisfaction. All the radiologists had a neutral or positive effect on mood, productivity, perceived diagnostic accuracy, and work satisfaction. Only one reported a negative effect of music on concentration. Women reported a greater effect on mood than men, and there was a greater effect on mood for those with experience playing instruments than those without. Those who listened to music for more than 5 h per week also reported greater scores for mood (Mohiuddin et al., 2009).

Other workstation environment issues that can affect information processing include how much heat does the workstation produce, how much noise does it produce, and what kind of ambient lighting is appropriate? It is recommended that 20 lux of ambient light be used since this is generally sufficient to avoid most reflections and still provide sufficient light for the human visual system to adapt to the surrounding environment and the displays (Krupinski et al., 2007). The ambient lighting should be indirect and backlight incandescent lights with dimmer switches rather than fluorescent are recommended. Light-colored clothing and lab coats can increase reflections and glare even with today's liquid crystal displays so they should be avoided if possible. The intrinsic minimum luminance of the device should not be smaller than the ambient luminance.

One concern that has not been considered very much is the visual fatigue that may result from the long hours that clinicians are spending in front of a computer every day. Close work of any kind for hours on end can overwork the eyes, resulting in eyestrain (known clinically as asthenopia) (Ebenholtz, 2001; MacKenzie, 1843). Being fatigued is likely to impact the clinician's ability to effectively and efficiently process image information and render correct diagnoses. In fact, radiologists have been found to report significant eye strain or fatigue as a function of hours spent reading exams (Krupinski and Kallergi, 2007; Vertinsky and Forster, 2005).

There have been only a few studies objectively examining the impact of fatigue on clinical performance (Christensen et al., 1997; Gale et al., 1984). In one recent study, the impact of fatigue was objectively measured along with other correlative measures of fatigue. Twenty radiology residents and 20 radiologists were shown 60 skeletal radiographic studies, half with fractures, before and after a day of clinical reading. Diagnostic accuracy was measured as was error in visual accommodation before and after each session. They also completed the Swedish Occupational Fatigue Inventory (SOFI) and the oculomotor strain subscale of the Simulator Sickness Questionnaire (SSQ) before each session. Diagnostic accuracy was significantly better prior to a day of work compared to after when measured using Receiver Operating Characteristic techniques. There was significantly greater error in accommodation at the end of the clinical workday, suggesting that there was a reduction in the ability to focus on the image and extract the information needed to render the correct diagnosis. The SOFI measures of lack of energy, physical discomfort, and sleepiness were higher after a day of clinical reading and the SSQ measure of oculomotor symptoms (i.e., difficulty focusing, blurred vision) was significantly higher after a day of clinical reading. It would appear that radiologists are visually fatigued by their clinical reading workday, reducing their ability to focus on diagnostic images, to extract information properly, and to accurately interpret them (Krupinski and Berbaum, 2010).

9.7 Conclusions

There are clearly a number of ways that the manner in which medical images are displayed and the reading environment can impact the flow and processing of information within the clinical environment. It is also clear that with an understanding of the perceptual and cognitive capabilities of the human being viewing and interpreting medical images we can better tailor the display and environment to facilitate and foster optimal decision-making strategies. As future changes occur in the ways that images are acquired and displayed, the information that they contain will undoubtedly impact the clinical decision-making process. For example, a wide variety of radiographic images are acquired as multiple slices through the patient (e.g., CT, MRI, digital breast tomosynthesis) and these slices can be viewed either in stack mode going through them sequentially or they can be reconstructed to create a 3D representation of the object (Getty, 2007).

From an informatics point of view, the question is whether the additional spatial information provided in the 3D representation actually improves the diagnosis. From the display perspective, the question is whether viewing a 3D representation on 2D display is effective or do we need to consider using new cutting-edge 3D display technologies that truly show the images in 3D in a sort of holographic representation?

These sorts of questions regarding the interplay between the way medical images are acquired, the platforms we use to display them, and the human observer who needs to interpret them will continue to arise as new technologies are developed. Answering them will continue to involve characterizing and tracking the flow of information from acquisition to interpretation.

References

American College of Radiology. 2010. *Guidelines and Standards*. Available at: http://www.acr.org/SecondaryMainMenu Categories/quality_safety/guidelines.aspx. Accessed March 9, 2010.

American Telemedicine Association. 2004. Telehealth practice recommendations for diabetic retinopathy. *Telemed. J. e-Health*, 10, 469–82.

Berbaum, K. S., Franken, E., Caldwell, R. et al. 2010. Satisfaction of search in traditional radiographic imaging. In Samei, E. and Krupinski, E. (Eds.), *The Handbook of Medical Image Perception and Techniques*, pp. 107–139. Cambridge: Cambridge University Press.

Berbaum, K. S., Franken, E. A., Dorfman, D. D. et al. 1996. Cause of satisfaction of search effects in contrast studies of the abdomen. *Acad. Radiol.*, 3, 815–26.

Berlin, L. 2005. Errors of omission. *Am. J. Roentgenol.*, 185, 1416–21.

Berlin, L. 2007. Accuracy of diagnostic procedures: Has it improved over the past five decades? *Am. J. Roentgenol.*, 188, 1173–8.

Berlin, L. 2009. Malpractice issues in radiology: res ipsa loquitur. *Am. J. Roentgenol.*, 193, 1475.

Bhan, S. N., Coblentz, C. L., Norman, G. R. et al. 2008. Effect of voice recognition on radiologist reporting time. *Can. Assoc. Radiol. J.*, 59, 203–9.

Bhargavan, M. and Sunshine, J. H. 2005. Utilization of radiology services in the United States: Levels and trends in modalities, regions, and populations. *Radiology*, 234, 824–32.

Blume, H. 1996. The ACR/NEMA proposal for a grey-scale display function standard. *Proc. SPIE Med. Imaging*, 2707, 344–60.

Boland, G. W. L., Guimaraes, A. S., and Mueller, P. R. 2008. Radiology report turnaround: Expectations and solutions. *Eur. Radiol.*, 18, 1326–8.

Christensen, E. E., Dietz, G. W., Murry, R. C. et al. 1997. The effect of fatigue on resident performance. *Radiology*, 125, 103–5.

DeFlorio, R., Coughlin, B., Coughlin, R. et al. 2008. Process modification and emergency department radiology service. *Emerg. Radiol.*, 15, 405–12.

Digital Imaging and Communications in Medicine Grayscale Standard Display Function. 2010. Available at: http://medical.nema.org/. Accessed August 3, 2010.

Digital Imaging and Communications in Medicine (DICOM) Part 14: Grayscale Standard Display Function. 2000. *NEMA PS 3.14*. Rosslyn, VA: National Electrical Manufacturers Association. Available at: http://medical.nema.org/. Accessed March 9, 2010.

DIN-6868-57. 2001. *Image Quality Assurance in X-ray Diagnostics, Acceptance Testing for Image Display Devices*. Berlin, Germany: Deutsches Institut fuer Normung.

DiPiro, P. J., vanSonnenberg, E., Tumeh, S. S. et al. 2002. Volume and impact of second-opinion consultations by radiologists at a tertiary care cancer center: Data. *Acad. Radiol.*, 9, 1430–3.

Ebbert, T. L., Meghea, C., Iturbe, S. et al. 2007. The state of teleradiology in 2003 and changes since 1999. *Am. J. Roentgenol.*, 188, W103–112.

Ebenholtz, S. M. 2001. *Oculomotor Systems and Perception*. New York, NY: Cambridge University Press.

Fetterly, K. A., Blume, H. R., Flynn, M. J. et al. 2008. Introduction to grayscale calibration and related aspects of medical imaging grade liquid crystal displays. *J. Digit. Imaging*, 21, 193–207.

Gale, A. G., Murray, D., Millar, K. et al. 1984. Circadian variation in radiology. In Gale, A. G. and Johnson, F. (Eds.), *Theoretical and Applied Aspects of Eye Movement Research*, pp. 313–22. London, England: Elsevier Science Publishers.

Geijer, H., Geijer, M., Forsberg, L. et al. 2007. Comparison of color LCD and medical-grade monochrome LCD displays in diagnostic radiology. *J. Digit. Imaging*, 20, 114–21.

Getty, D. J. 2007. Improved accuracy of lesion detection in breast cancer screening with stereoscopic digital mammography. Paper presented at the *93rd Annual Meeting of the Radiological Society of North America*, November 25–30, Chicago, IL.

Goyal, N., Jain, N., and Rachapalli, V. 2009. Ergonomics in radiology. *Clin. Radiol.*, 64, 119–26.

Heo, M. S., Han, D. H., An, B. M. et al. 2008. Effect of ambient light and bit depth of digital radiograph on observer performance in determination of endodontic file positioning. *Oral Surgery, Oral Med. Oral Pathol., Oral Radiol. Endodont.*, 105, 239–44.

Hersh, W. 2009. A stimulus to define informatics and health information technology. *BMC Med. Inform. Dec. Making*, 9, 24.

Hu, C. H., Kundel, H. L., Nodine, C. F. et al. 1994. Searching for bone fractures: A comparison with pulmonary nodule search. *Acad. Radiol.*, 1, 25–32.

IEC—International Electrotechnical Commission. 2008. Available at: http://www.iec.org. Accessed March 9, 2010.

Kayser, K., Gortler, J., Goldmann, T. et al. 2008. Image standards in tissue-based diagnosis (diagnostic surgical pathology). *Diag. Pathol.*, 3, 17.

Khorasani, R. 2008. Can metrics obtained from your IT databases help start your practice dashboard? *J. Am. Coll. Radiol.*, 5, 772–4.

Krupinski, E. A. 1996. Visual scanning patterns of radiologists searching mammograms. *Acad. Radiol.*, 3, 137–44.

Krupinski, E. A. 2009a. Virtual slide telepathology workstation of the future: Lessons learned from teleradiology. *Hum. Pathol.*, 40, 1100–11.

Krupinski, E. A. 2009b. Medical grade vs off-the-shelf color displays: Influence on observer performance and visual search. *J. Digit. Imaging*, 22, 363–8.

Krupinski, E. A. and Berbaum, K. S. 2010. Does reader visual fatigue impact interpretation accuracy? *Proc. SPIE Med. Imaging*, 7627, 76205.

Krupinski, E. A. and Kallergi, M. 2007. Choosing a radiology workstation: Technical and clinical considerations. *Radiology*, 242, 671–82.

Krupinski, E. A. and Lund, P. J. 1997. Comparison of conventional and computed radiography: Assessment of image quality and reader performance in skeletal extremity trauma. *Acad. Radiol.*, 4, 570–76.

Krupinski, E., Burdick, A., Pak, H. et al. 2008. American Telemedicine Association's practice guidelines for teledermatology. *Telemed. J. e-Health*, 14, 289–301.

Krupinski, E., Johnson, J., Roehrig, H. et al. 2003. On-axis and off-axis viewing of images on CRT displays and LCDs. Observer performance and vision model predictions. *Acad. Radiol.*, 12, 957–64.

Krupinski, E. A., Lubin, J., Roehrig, H. et al. 2006a. Using a human visual system model to optimize soft-copy mammography display: Influence of veiling glare. *Acad. Radiol.*, 13, 289–95.

Krupinski, E. A., Nodine, C. F., and Kundel, H. L. 1998. Enhancing recognition of lesions in radiographic images using perceptual feedback. *Opt. Eng.*, 37, 813–8.

Krupinski, E. A. and Roehrig, H. 2000. The influence of a perceptually linearized display on observer performance and visual search. *Acad. Radiol.* 7, 8–13.

Krupinski, E. A., Roehrig, H., Berger, W. et al. 2006b. Potential use of a large-screen display for interpreting radiographic images. *Proc. SPIE Med. Imaging*, 6146, 1605–7422.

Krupinski, E. A., Roehrig, H., and Furukawa, T. 1999. Influence of film and monitor display luminance on observer performance and visual search. *Acad. Radiol.*, 6, 411–8.

Krupinski, E. A., Siddiqui, K., Siegel, E., Shrestha, R., Grant, E., Roehrig, H., and Fan, J. 2007. Influence of 8-bit vs. 11-bit digital displays on observer performance and visual search: A multi-center evaluation. *J. Soc. Inf. Disp.*, 15, 385–90.

Krupinski, E. A., Williams, M. B., Andriole, K. et al. 2007. Digital radiography image quality: Image processing and display. *J. Am. Coll. Radiol.*, 4, 389–400.

Kundel, H. L. 1975. Peripheral vision, structured noise and film reader error. *Radiology*, 114, 269–73.

Kundel, H. L. 1989. Perception errors in chest radiography. *Semin. Resp. Med.*, 10, 203–10.

Kundel, H. L., Nodine, C. F., and Krupinski, E. A. 1989. Searching for lung nodules: Visual dwell indicates locations of false-positive and false-negative decisions. *Invest. Radiol.*, 24, 472–8.

Kundel, H. L., Nodine, C. F., Krupinski, E. A. et al. 2008. Using gaze-tracking data and mixture distribution analysis to support a holistic model for the detection of cancers on mammograms. *Acad. Radiol.*, 15, 881–6.

Langer, S., Bartholmai, B., Fetterly, K. et al. 2004. SCAR R&D symposium 2003: Comparing the efficacy of 5-MP CRT versus 3-MP LCD in the evaluation of interstitial lung disease. *J. Digit. Imaging*, 17(3), 149–57.

Langer, S., Fetterly, K., Mandrekar, J. et al. 2006. ROC study of four LCD displays under typical medical center lighting conditions. *J. Digit. Imaging*, 19, 30–40.

Leong, D. L., Haygood, T. M., Whitman, G. J. et al. 2010. DICOM GSPS affects contrast detection thresholds. *Proc. SPIE Med. Imaging*, 7627-07, 762708.

Lu, Y., Zhao, S., Chu, P. W. et al. 2008. An update survey of academic radiologists' clinical productivity. *J. Am. Coll. Radiol.*, 5, 817–26.

MacKenzie, W. 1843. On asthenopia or weak-sightedness. *Edinburgh J. Med. Surg.*, 60, 73–103.

Manning, D., Ethell, S., and Donovan, T. 2004. Detection or decision errors? Missed lung cancer from the PA chest radiograph. *Br. J. Radiol.*, 78, 683–5.

McEntee, M., Brennan, P., Evanoff, M. et al. 2006. Optimum ambient lighting conditions for the viewing of softcopy radiological images. *Proc. SPIE Med. Imaging*, 6146, 1–7.

McEntee, M. F. and Martin, B. 2010. The varying effects of ambient lighting on low contrast detection tasks. *Proc. SPIE Med. Imaging*, 7627, 76270N.

McGurk, S., Brauer, K., Macfarlan, T. V. et al. 2008. The effect of voice recognition software on comparative error rates in radiology reports. *Br. J. Radiol.*, 81, 767–70.

McNeill, K. M., Major, J., Roehrig, H. et al. 2002. Practical methods of color quality assurance for telemedicine systems. *Med. Imaging Technol.*, 20, 111–6.

Mello-Thoms, C. 2006. The problem of image interpretation in mammography: Effects of lesion conspicuity on the visual search strategy of radiologists. *Br. J. Radiol.*, 79, S111–116.

Minnigh, T. R. and Gallet, J. 2009. Maintaining quality control using a radiological digital X-ray dashboard. *J. Digit. Imaging*, 22, 84–8.

Moise, A. and Atkins, S. 2005. Designing better radiology workstations: Impact of two user interfaces on interpretation errors and user satisfaction. *J. Digit. Imaging*, 18, 109–15.

Morgan, M. B., Branstetter, B. F., Lionetti, D. M. et al. 2008. The radiology digital dashboard: Effects on report turnaround time. *J. Digit. Imaging*, 21, 50–8.

Morgan, M. B., Branstetter, B. F., Mates, J. et al. 2006. Flying blind: Using a digital dashboard to navigate a complex PACS environment. *J. Digit. Imaging*, 19, 69–75.

Mohiuddin, S., Lakhani, P., Chen, J. et al. 2009. Effect of Baroque classical music on mood, concentration, perceived diagnostic accuracy, productivity, and work satisfaction of diagnostic radiologists. *Am. J. Roentgenol.*, 195(S), 72.

Nagy, P. G., Warnock, M. J., Daly, M. et al. 2009. Informatics in radiology: Automated Web-based graphical dashboard for radiology operational business intelligence. *Radiographics*, 29, 1897–906.

Nakajima, Y., Yamada, K., Imamura, K. et al. 2008. Radiologist supply and workload: International comparison—Working Group of Japanese College of Radiology. *Radiat. Med.*, 26, 455–65.

Nodine, C. F. and Kundel, H. L. 1987. Using eye movements to study visual search and to improve tumor detection. *RadioGraphics*, 7, 1241–50.

Nodine, C. F., Kundel, H. L., Mello-Thoms, C. et al. 1999. How experience and training influence mammography expertise. *Acad. Radiol.*, 6, 575–85.

Pezzullo, J. A., Tung, G. A., Rogg, J. M. et al. 2008. Voice recognition dictation: Radiologist as transcriptionist. *J. Digit. Imaging*, 21, 384–9.

Roehrig, H., Rehm, K., Silverstein, L. D. et al. 2010. Color calibration and color-managed medical displays: Does the calibration method matter? *Proc. SPIE Med. Imaging*, 7627, 76270K.

Ruess, L., O'Connor, S. C., Cho, K. H. et al. 2003. Carpal tunnel syndrome and cubital tunnel syndrome: Work-related musculoskeletal disorders in four symptomatic radiologists. *Am. J. Roentgenol.*, 181, 37–42.

Saha, A., Kelley, E. F., and Badano, A. 2010. Accurate color measurement methods for medical displays. *Med. Phys.*, 37, 74–81.

Samei, E., Badano, A., Chakraborty, D. et al. 2005. Assessment of display performance for medical imaging systems. Report of the American Association of Physicists in Medicine (AAPM) Task Group 18. Madison, WI: Medical Physics Publishing AAPM on-line Report No. 03. Available at: http://deckard.duhs.duke.edu/~samei/tg18.htm. Accessed March 9, 2010.

Samei, E., Flynn, M. J., and Kearfott, K. J. 1997. Patient dose and detectability of subtle lung nodules in digital chest radiographs. *Health Phys.*, 72, 6S.

Smith, M. J. 1967. *Error and Variation in Diagnostic Radiology*. Springfield, IL: Charles C. Thomas.

SMPTE Specifications for medical diagnostic imaging test pattern for television monitors and hard-copy recording cameras. 1991. SMPTE RP 133. White Plains, NY: Society of Motion Picture and Television Engineers.

Sunshine, J. H. and Maynard, C. D. 2008. Update on the diagnostic radiology employment market: Findings through 2007–2008. *J. Am. Coll. Radiol.*, 5, 827–33.

Thind, R., Barter, S., Service Review Committee. 2008. The service review committee: Royal College of Radiologists. Philosophy, role, and lessons to be learned. *Clin. Radiol.*, 63, 118–24.

Toomey, R. J., Ryan, J. T., McEntee, M. F. et al. 2010. Diagnostic efficacy of handheld devices for emergency radiologic consultation. *Am. J. Roentgenol.*, 194, 469–74.

Tuddenham, W. J. 1962. Visual search, image organization, and reader error in Roentgen diagnosis: Studies of psychophysiology of roentgen image perception. *Radiology*, 78, 694–704.

Tuddenham, W. J. 1963. Problems of perception in chest roentgenology: Facts and fallacies. *Radiol. Clin. North America*, 1, 227–89.

Vertinsky, T. and Forster, B. 2005. Prevalence of eye strain among radiologists: Influence of viewing variables on symptoms. *Am. J. Roentgenol.*, 184, 681–6.

VESA—Video Electronics Standards Association. 2008. Available at: http://www.VESA.org. Accessed March 9, 2010.

Wurm, E. M. T., Hofmann-Wellenhof, R., Wurm, R. et al. 2008. Telemedicine and teledermatology: Past, present and future. *Journal der Deutschen Dermatologischen Gesellschaft*, 6, 106–12.

Zhang, J., Lu, X., Nie, H. et al. 2009. Radiology information system: A workflow-based approach. *Int. J. Comput. Assist. Radiol. Surg.*, 4, 509–16.

Zuley, M. 2010. Perceptual issues in reading mammograms. In Samei, E. and Krupinski, E. (Eds.), *The Handbook of Medical Image Perception and Techniques*, pp. 365–379. Cambridge: Cambridge University Press.

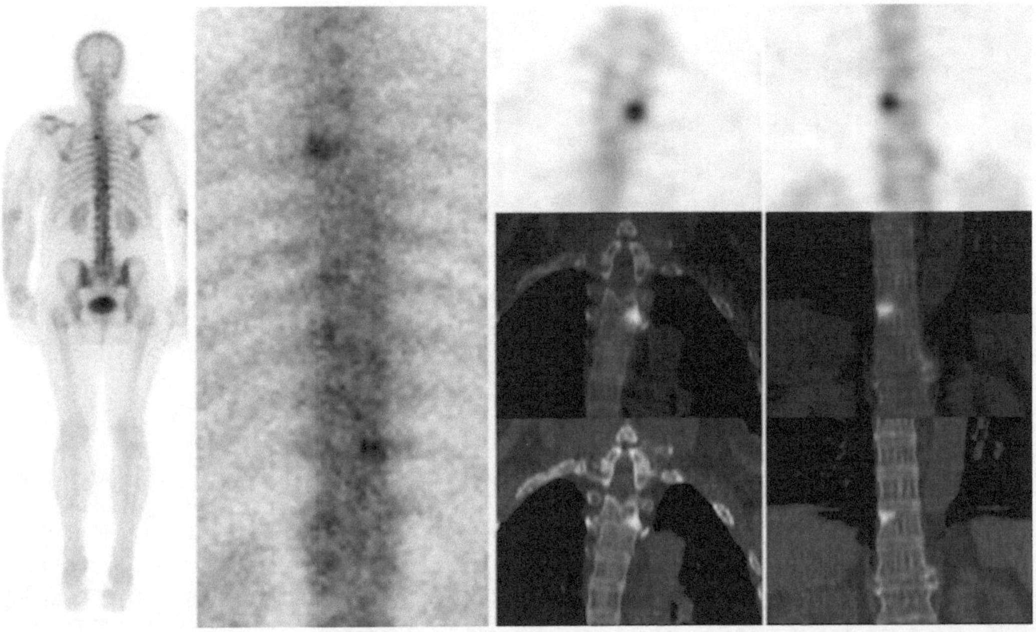

FIGURE 14.1 Example of MMI in oncology. Bone SPECT/CT in the research for bone metastases in breast cancer; the whole body (left) and centered (second image) planar images show two abnormal foci on the spine without immediate differentiation between osteoarthrosis and metastases; the combined SPECT/CT (right columns: top: SPECT, bottom: CT; middle: fused SPECT/CT) shows without need for further imaging that the two foci correspond to malignant osteocondensation. (From Papathanassiou, D. and Liehn, J. C. 2008. *Crit. Rev. Oncol./Hematol.*, 68: 60. With permission.)

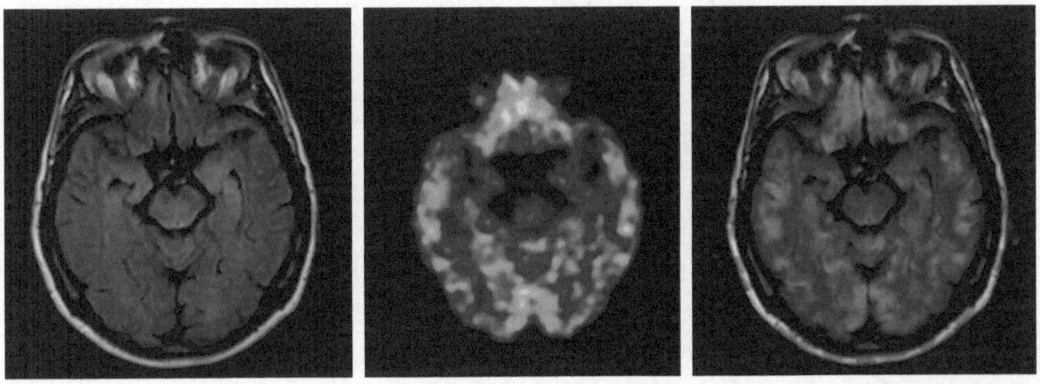

FIGURE 14.2 MR (FLAIR sequence), (18F) FDG-PET and fused images of the brain acquired simultaneously with an integrated MR/PET system. (From Schlemmer, H. P. et al. 2009. *Abdom. Imaging*, 34, 668–74. With permission.)

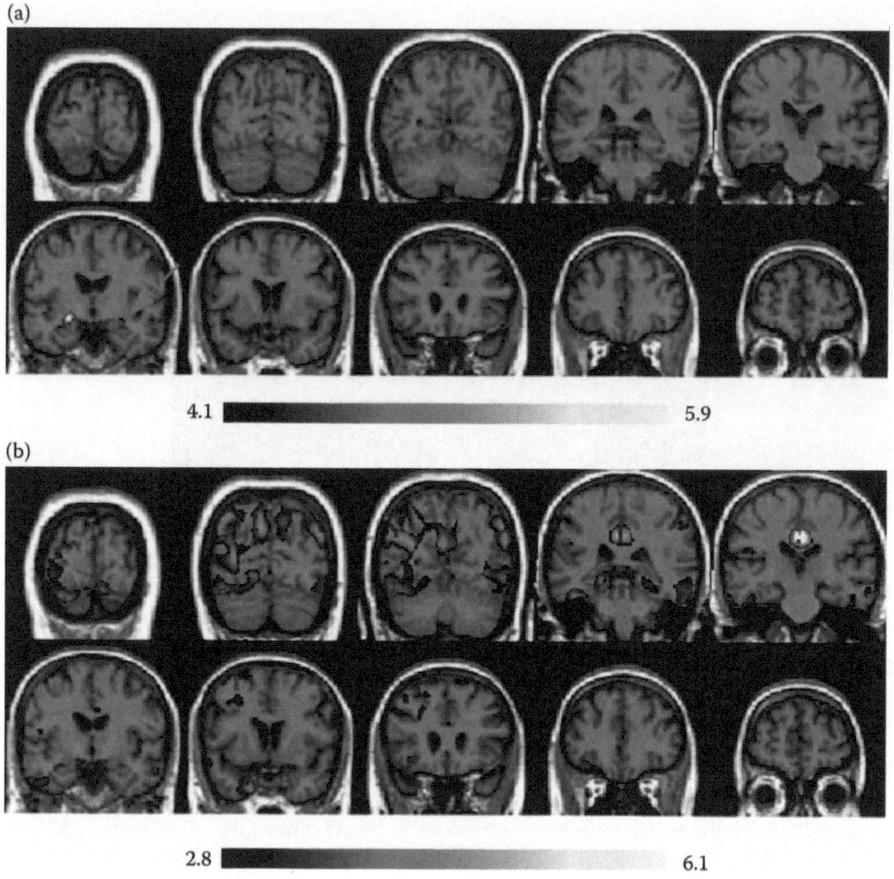

FIGURE 14.3 (a) Metabolic compensation map, showing regions of significant relatively preserved metabolism; (b) Functional depression map, showing regions of hypometabolism exceeding atrophy in a group of 25 AD patients as compared to a group of 21 normal controls. Both maps were computed through BPM. (From Caroli, A. et al. 2010b. *Dementia Geriatr. Cogn. Disord.*, 29, 37–45. With permission.)

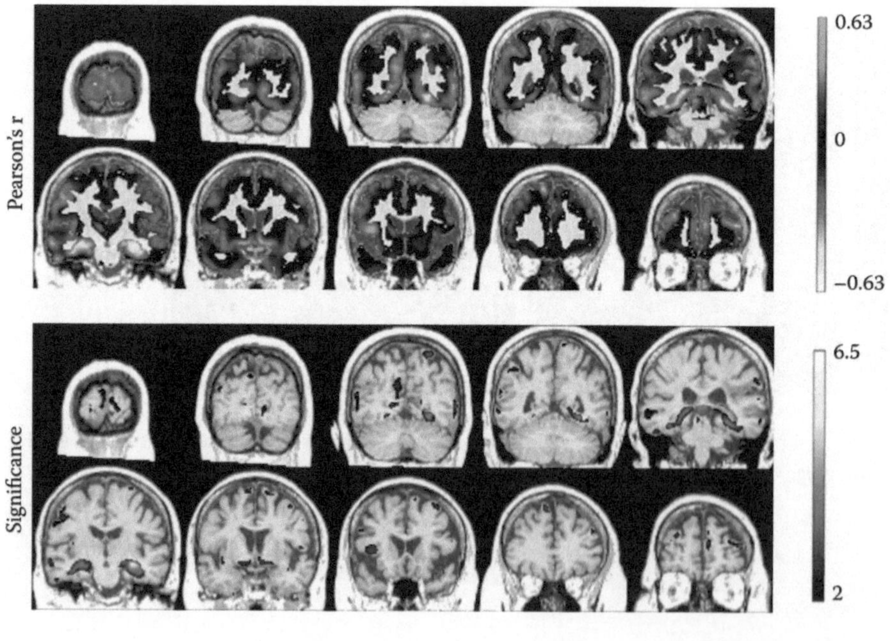

FIGURE 14.4 Pearson's correlation between gray matter and (11C)-PiB uptake in 23 AD patients and 17 normal controls and associated significance map. (Modified from Frisoni, G. B. et al. 2009. *Neurology*, 72, 1504–11. With permission.)

10

Digital X-Ray Acquisition Technologies

John Yorkston
Carestream Health

Randy Luhta
Philips Medical Systems

10.1 Introduction

The acquisition, processing, and distribution of a medical image can be viewed as a chain of sequential events. As with any complex system, the strength of this imaging chain is determined by the system's weakest link. It is therefore arguable that the front-end image acquisition is the most fundamentally important stage in the chain since it determines the upper limit on information content of the image. It is not uncommon for the imaging "signal" at this point to be composed of extremely small signal differences measured in thousands rather than millions of electrons. Great care must be exercised to tease out this very weak "signal" without the introduction of additional noise originating within the imaging system itself. Information that is lost at this stage can never be recaptured, and any later step in the chain can, at best, only maintain this initial level of information content. The design and operation of this initial stage is therefore crucial in ensuring the most efficient transfer of information to the rest of the image analysis and distribution system. In this chapter, we describe the different stages involved in the acquisition of a medical x-ray image and highlight the aspects of system design that affect the quality of the information transfer between stages.

X-ray imaging is, by far, the most commonly performed medical imaging procedure, so we limit our discussion to x-ray acquisition modalities, specifically two-dimensional (2D) radiography and 3D computed tomography (CT), to illustrate some of the considerations for the optimal design and operation of the image acquisition stage. Many of the concepts that are discussed are common to other imaging applications that utilize ionizing radiation, such as radiation oncology (Antonuk, 2002) and nuclear/molecular imaging (Lewellen, 2008; Nikiforidis et al., 2008; Kagadis et al., 2010), but there are fundamental differences between the acquisition stages of other modalities such as ultrasound imaging, magnetic resonance imaging (MRI), and optical imaging. Interested readers are directed to the review papers by Carson and Fenster (Carson and Fenster, 2009), Pickens (2000), and others for more information on the specific details of these other modalities.

Projection radiography and CT are similar in that they use ionizing radiation (i.e., x-rays) to create the information that is the basis for the diagnostic interpretation of the image. This information is encoded into the x-ray field in the form of intensity variations in space caused by the differences in x-ray attenuation at different locations in the patient's body. Projection radiography creates a 2D map of this intensity variation while CT analyzes multiple projection images, taken at numerous different orientations through the patient, to reconstruct a full 3D map of the x-ray attenuation coefficients of the material components within the patient.

For both projection radiography and CT, the fundamental limit on the information content is determined by the statistical nature of the fluctuations in the x-ray field being imaged. The inherent noise associated with the quantized nature of the x-ray photons is Poisson in nature and increases as the square root of the total number of x-rays. This means that the signal-to-noise ratio (or alternatively the inherent information content) improves as the number of x-rays increases. In other words, the more radiation that is used, the higher the "quality" of the resulting image. However, there is a cost for this increased image quality in that ionizing radiation is generally detrimental to living tissue. The higher the radiation exposure used, the more likely is the induction of future deleterious effects in the patient (Brenner et al., 2003). The risks associated with exposure to x-rays must be balanced against the benefit that is expected from the medical imaging procedure being undertaken. The desire to use as low as reasonably achievable exposure levels (the ALARA concept) provides strong motivation to extract the maximum amount of information from the x-ray field. In the context of the ALARA principle, it is also important to recognize that there is a lower limit of exposure below which a given subject contrast from two neighboring anatomical features will be obscured by the statistical uncertainty in the signal at the different locations. Even with a "perfect" detector, for a given clinical task this inherent uncertainty sets a fundamental lower limit on patient exposure that must be used to accomplish the task.

In addition to the statistical uncertainty due to the inherent stochastic nature of the x-rays, the detection system itself introduces noise into the image. The optimization of the acquisition stage is therefore key in the determination of how efficiently the image information can be extracted from the incident x-ray fluence.

In the following sections of this chapter, we divide x-ray image acquisition into a number of stages. We discuss some of the important considerations for the optimization of each stage and describe how they affect the "quality" of the image information as it passes through a generic x-ray detector. The unique features that lead to the increase in image quality achievable with the recently introduced flat-panel detectors are discussed. The details of a generic CT system, and the specialized image processing required to create the 3D image information, are then reviewed. The recent move toward quantitative rather than qualitative image interpretation using multimodality and multispectral image information will demand more sophisticated data-handling capabilities in the coming years. In the final section, we review some of the developments in these advanced imaging applications and discuss some of the implications for the future demands on information-handling capabilities that will be required to fully utilize the enhanced information these developments can offer.

10.2 Image Acquisition

The x-ray image acquisition process can be divided into a number of fundamental stages. One such division includes the production of the x-rays; the transfer of the x-rays through the object; the absorption of the resulting x-ray field and the transfer of their energy into a secondary quantum field (either light photons or electron–hole pairs); the transfer of this secondary quantum field to the input plane of the secondary quantum field detector; and finally the detection of this secondary quantum field. The next few sections discuss each of these stages in turn but focus mainly on the x-ray absorption and secondary quanta detection stages, since these are the components that are currently undergoing rapid development and are also the stages that ultimately differentiate between the capabilities of different commercially available systems.

10.2.1 The X-Ray Beam

A number of new approaches to the design of x-ray production equipment are being reported in the literature. These include distributed sources made from carbon nanotubes (Qian et al., 2009) and the creation of clinically practical coherent x-ray sources suitable for phase-contrast imaging that can generate totally new types of x-ray imaging information (Donath et al., 2010). These approaches hold great promise for improving the capabilities of modern x-ray equipment but are still at a relatively early stage of development, and are therefore not discussed further here.

X-rays for medical imaging are most commonly produced by accelerating electrons using a high-voltage electric field and focusing them onto a heavy metallic target such as tungsten, molybdenum, or rhodium. The x-ray imaging process starts with the choice of accelerating voltage and hence the energy of the initial x-rays, known as the kVp of the beam. In some applications, such as external beam radiation therapy, it is the treatment beam that is being imaged and the choice of beam energy is determined by considerations other than imaging. In radiology, however, the choice of beam energy is determined by the body part being imaged and the imaging task under consideration. Historically, in mammography, low-energy x-rays (25–30 kVp with molybdenum or rhodium targets and filters) have been used due to the low attenuation differences between the different tissues being imaged (Johns and Yaffe, 1987). At these energies, the attenuation differences between adipose and glandular tissue are sufficient for imaging, but patient-absorbed dose can still be kept within reasonable limits. The recent introduction of high-efficiency digital detectors (Pisano and Yaffe, 2005) is changing this beam selection to higher average energies (30–35 kVp with tungsten and silver targets and filters), where comparable image quality can be maintained but with a lower patient-absorbed dose (Williams et al., 2008).

For general radiography, higher energies are more common. However, for a given body part, the specific imaging task can also influence the kVp choice. A normal PA chest exam is performed at ~120 kVp. This choice is partly due to the desire to reduce the contrast of the ribs which can obstruct the detection of important pulmonary lesions. However, chest exams, where the ribs are the primary anatomy of interest, are performed at ~70 kVp, which enhances the visibility of the bone trabeculae.

The choice of beam energy is therefore a careful balance between ensuring enough x-rays pass through the patient to create an image of sufficient contrast, while simultaneously ensuring as low a patient dose as possible. Once the beam energy has been decided upon, the x-ray detector must be designed to optimally absorb the x-rays penetrating the patient while maintaining spatial information on the x-ray intensity variations. The different beam energies have an impact on the suitability of different absorption materials used for this task. This will be discussed in more detail in Section 10.2.2.

One complicating factor is the presence of scattered radiation in the x-ray fluence that is incident on the detector. This scattered radiation originates from the patient and can be a significant source of additional signal in the detector. In general, scattered radiation carries no traditional imaging information, acts to reduce image contrast, and is a source of noise that reduces image quality. The magnitude of this scattered radiation can be significant, with scatter-to-primary ratios of 2:1 or higher possible in common imaging configurations (Smans et al., 2010). With screen–film systems, antiscatter grids were used that preferentially transmitted the primary beam while blocking much of the scattered radiation. They necessitated an increase in patient exposure to maintain the optimal density on the resulting film; however, their use has become less common in many applications where digital imaging has been introduced. This is partly due to the belief that digital image processing can restore the contrast in the image that is lost due to the presence of scatter. While this is true for image contrast, this postprocessing cannot account for the additional noise introduced by the scattered x-ray photons. Consequently, the contrast-to-noise ratio of an image containing scatter will always be degraded, even after digital image processing. In other words, the presence of scattered photons will always reduce the information content of the image. The question of whether or not the use of a grid is advantageous is determined by the scatter-to-primary ratio and the scatter rejection capabilities of the grid. Other issues such as workflow and the physically cumbersome nature of a grid can also affect this decision. The issue of fundamental information content available in the x-ray distribution, after passing through the patient, has been investigated by various researchers but the early paper by Motz and Danos is an excellent introduction into this topic (Motz and Danos, 1973).

The intensity and energy of the scattered radiation in the beam can also have an effect on the performance of the next stage of the imaging chain: the x-ray absorption layer.

10.2.2 The X-Ray Absorption Layer

The function of the x-ray absorption layer is to transform the energy and spatial intensity distribution of the incident x-rays into a distribution of lower energy "secondary" quanta that can be more readily measured than the x-rays themselves. This is perhaps the most important stage of the image formation chain and the material that is used to convert the incident x-ray energy into these secondary quanta must be chosen with care. There are two main types of x-ray absorber currently used in x-ray imagers; those that convert the x-ray energy into visible light photons, known as phosphors; and those that convert the x-ray energy into electron–hole pairs, known as photoconductors.

Historically, there have been a large number of phosphors used for x-ray detection. Two of the more common prompt-emitting phosphors are gadolinium oxysulfide (Gd_2O_2S, also known as GOS) and cesium iodide (CsI). They have been used in projection radiography, fluoroscopy, and CT for many years and are still widely used in many commercial systems. These phosphors emit their light photons immediately on absorption of an x-ray. The color of the emitted photons is determined by the doping materials such as terbium, thallium, or sodium, which are introduced in trace amounts into the crystalline structure of the phosphor. A related type of phosphor "stores" a fraction of the absorbed x-ray energy in latent excitation sites. This latent image information is then "read out" by scanning the phosphor with a laser (Rowlands, 2002). This scanning is typically done seconds or minutes after the x-ray exposure. These materials are known as storage phosphors and form the basis of computed radiography (CR) systems. The most common examples of these types of materials are BaFBr(I) and CsBr(Eu).

GOS was extensively used in screen–film systems, where its high light output (typically thousands of light photons per absorbed x-ray) was advantageous in creating the systems that required reduced amounts of radiation to optimally expose the x-ray film. GOS is also used in CT systems (see Section 10.6), where it is formed into individual ceramic blocks approximately 1 mm or smaller in dimension. The dopants used for CT applications (most commonly Praseodymium) are somewhat different than with planar radiography, due to the need for fast response and recovery times that are required for high-speed CT applications.

In modern flat-panel detectors, a GOS phosphor layer is typically fabricated in a particle-in-binder configuration where the phosphor grains are held in a plastic binder layer that is ~100–300 μm thick. This configuration allows the light that is emitted isotropically from the phosphor grains, to spread laterally from the x-ray interaction site. This causes a decrease in spatial resolution of the final image. The amount of lateral spread is generally proportional to the thickness of the particle-in-binder layer with thicker screens displaying more spreading. High-resolution applications therefore used thinner layers (~100 μm thick), which do not absorb as many of the incident x-rays, resulting in a less-efficient imaging system, albeit one with higher spatial resolution properties. These higher resolution systems normally used higher levels of x-ray exposure than those designed for applications with lower spatial resolution requirements.

The lateral light spreading phenomenon introduces an unavoidable trade-off between increased x-ray absorption (i.e., thicker layers of phosphor) and improved spatial resolution (i.e., thinner layers of phosphor). This has been a significant issue in screen–film systems since their inception. Various approaches were implemented to help this situation, including the use of asymmetric screen configurations where a double-sided film was sandwiched between two phosphor layers of different thicknesses, one layer providing a high-resolution image of patient anatomy, while the other thicker layer providing enhanced visualization of low-contrast anatomy (Van Metter and Dickerson 1994). The high cost of current flat-panel detectors has prevented this approach being implemented, although developments to reduce both the cost and thickness of the flat-panel detector substrate may make this a feasible configuration in the future.

In modern digital imaging systems, the use of structured phosphor layers has significantly improved this trade-off. Under suitable conditions, CsI(Tl) naturally grows in columnar structures that inhibit the lateral spread of the light photons generated within them. This property allows the use of significantly thicker phosphor layers (~500 μm or more) while maintaining the spatial resolution of the image information. CsBr(Eu) is an example of a storage phosphor that exhibits the same needle-structured morphology and has also been shown to allow the use of significantly thicker layers that absorb large percentages of the incoming x-ray photons while maintaining their spatial resolution when read out by the scanning laser (Cowen et al., 2007). These structured phosphors form the basis of a number of commercially available flat-panel detector and storage phosphor systems and due to this high x-ray absorption coupled with their high spatial resolution, their image quality is generally among the best in their class.

An alternative to a phosphor for an x-ray absorption layer is a photoconductor. With this type of material, the absorption of an x-ray generates a large number of electron–hole pairs. A high voltage (typically ~1–10 V/μm) is applied across the thickness of the photoconductor (~100 to >500 μm thick depending on the application) to quickly sweep these electrical charges to the opposing surfaces of the material where they are read out as image signal. This happens fast enough that there is little or no appreciable lateral travel of the generated charge cloud and the spatial resolution of the resulting image is almost entirely dominated by the dimensions of the readout pixel. This confinement of the secondary electrical charges by the applied electric field is similar to the optical confinement of the light photons within the structured phosphor. Both allow thick layers of material to be used without the usual reduction in spatial resolution.

The most common photoconductor material used in modern x-ray detectors is amorphous selenium (a-Se) but newer materials including HgI, PbI, CdTe, CdZnTe, and PbO, among others, are being investigated for various applications. In general, the x-ray attenuation properties of a material are proportional to the cube of its effective *Z* number. The relatively low *Z* value of a-Se ($Z = 34$) means it has lower x-ray stopping power at higher energies than the competing phosphor materials. Figure 10.1 shows the attenuation coefficients of a-Se and CsI(Tl) as a function of energy plotted along with a typical x-ray beam spectrum used in mammography (28 kVp, Mo/Mo with 4 cm of added polymethyl-methacrylate (PMMA) filtration) and one similar to that used in chest radiography (120 kVp with 40 mm added Al filtration, RQA-9 beam (IEC, 2005)). Figure 10.2 shows the absorption as a function of thickness for a-Se and CsI(Tl) for these two different x-ray spectra. It can be seen that the inherent absorption

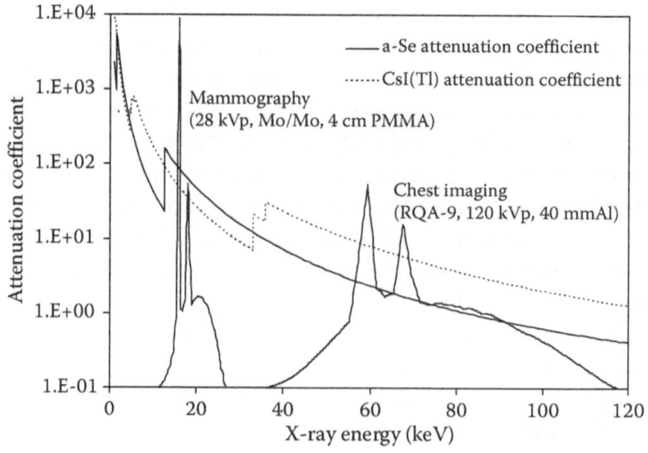

FIGURE 10.1 Attenuation curves for a-Se and CsI(Tl) as a function of x-ray energy plotted with typical x-ray beam energy distributions for a representative mammographic beam (28 kVp, Mo/Mo with 4 cm PMMA filtration) and a chest-imaging beam (RQA-9, 120 kVp with 40 m Al filtration).

characteristics of a-Se make it more suited to the lower energies used in mammography and as a consequence, it currently forms the basis of a number of commercially successful mammographic systems where ~200 μm of a-Se absorbs the majority of incident x-rays. The CsI(Tl) absorption properties make it more suited to higher-energy applications such as chest imaging where ~500–600 μm of material is typically used. The newer, higher *Z* photoconductors have higher x-ray stopping power than a-Se and can produce significantly more electron–hole pairs per absorbed x-ray. This increased absorption and signal generation gives them the potential to produce higher-quality images than a-Se at the higher energies used in general radiography. However, for most of them there still remain a number of important materials property issues yet to be resolved before they will be suitable for cost-effective, commercial implementation in large-area detectors.

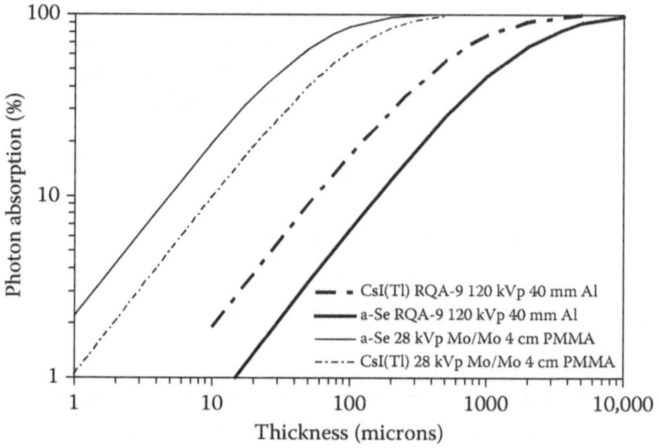

FIGURE 10.2 X-ray photon absorption as a function of material thickness for a-Se and CsI(Tl) for the representative x-ray beams shown in Figure 10.1.

In addition to the spatial resolution of an x-ray absorber, its noise transfer properties are also key in determining the materials ability to convey image information. Ideally, all x-rays of the same energy would create the identical number of light photons or electron–hole pairs. As with x-ray production however, the process of absorbing an x-ray and converting its energy into either light or electrical charge is statistical in nature and introduces additional uncertainty in the determination of the x-ray intensity at a given spatial location. Identical energy x-rays produce varying amounts of light or electrical charge and these quanta are collected with varying efficiency by the secondary quantum detector. This added uncertainty in signal intensity, arising from the inherent properties of the absorbing material, is generally known as Swank noise (Swank, 1973). The physical processes leading to Swank noise at this stage of the image chain are many and varied, but include fundamental properties of the material such as k-fluorescence generation as well as properties of the absorbing layer associated with its manufacture, such as the refractive index matching between the binder and the phosphor grains in particle-in-binder materials. Signals generated at different depths in the absorption layer can also exhibit different amounts of spatial spreading. This is a particular concern for phosphors and results in noise from different depths of the phosphor having different spatial frequency components, which ultimately affects the "texture" of the noise appearance. This is known as the Lubberts effect (Lubberts, 1968). Both Swank noise and the Lubberts effect increase the relative noise level in the image and so degrade the information content.

Once the secondary quanta have been produced and transported to the output surface of the x-ray absorber, they need to be transferred to the input surface of the secondary quantum detector where their intensity and spatial location are measured. The factors affecting the design and performance of this component will be the subject of Section 10.2.3.

10.2.3 The Secondary Quantum Detector

There have been a number of important developments in the design of the component that measures the intensity and location of the secondary quanta produced by the x-ray absorption layer. For many years, the only approach routinely used in the clinic was light-sensitive film. The development of electronic detectors that produced analog output voltages proportional to the intensity and location of the incident x-rays led to the creation of fluoroscopic systems that allowed real-time visualization of anatomical structures and surgical procedures. The digitization of these electrical signals ultimately led to the advent of practical CT systems, where the digital information could be processed by a computer to create 3D slices through an object (see Section 10.6.10). In fluoroscopy, the digitization of the image also allowed new procedures such as digital subtraction angiography to become clinically feasible. More recently, the introduction of charge-coupled devices/complementary metal-oxide semiconductors (CCDs/CMOS) and flat-panel detector technology has further accelerated the placement of digital systems in the general radiology environment. Many of these new systems are characterized by imaging performance that is significantly better than the systems they have replaced. Much of this improvement is due to the characteristics of the secondary quantum detectors used in these systems. We discuss some of the more important aspects of their design and operation in this section.

The design of the secondary quantum detector depends on whether it must register light photons or electron–hole pairs. A photodiode is typically used to turn incident light photons from a phosphor into an electrical signal, while a capacitive element is used to collect the electrical charge that is generated by a photoconductor. Until recently one of the main challenges with designing the secondary quantum detection stage was how to take the information from the large, 2D area required to image the human anatomy (up to the size of a 14 × 17 in. film or larger) and reduce it to the smaller dimensions of the secondary quantum detectors that were available for electronic readout of this information (typically centimeters in size).

Since 1D detectors (e.g., linear CCDs) were readily available and could relatively easily be configured to cover long dimensions, an early approach was to acquire the image in a linear fashion with the x-ray field collimated into a 1D profile, and scan this profile across the patient. Different versions of this linear scanning approach have been implemented for mammography and general radiography (Chotas et al., 1990; Piccaro and Toker, 1993; Villies and Jager, 2003; Samei et al., 2004; Despres et al., 2005). One significant benefit of this approach is the inherent scatter rejection achieved by the tight x-ray collimation. This resulted in images that have been deemed to be of exemplary quality (Samei et al., 2004) but the perceived mechanical complexity and long exposure times have generally prevented their widespread clinical implementation. In contrast to this general experience however, one modern mammographic system that utilizes photon counting in a scanned configuration is gaining attention for its high image quality and low patient exposure (Aslund et al., 2007).

Modern CR systems can also be regarded as 1D scanning systems, with the readout being performed by a rastered laser spot or a scanned laser line. The stimulated light emitted from the storage phosphor is collected by a plastic light guide or optical collection cavity and guided onto one or more photomultipliers (Rowlands, 2002). However, for this approach the x-ray exposure stage is still 2D so these CR systems do not benefit from the scatter rejection advantages present in other scanned x-ray systems.

An alternative to imaging a 1D section of the patient at a time is to minify the full 2D image to the size of the available 2D detectors. The efficiency of this demagnification tends to be one of the limiting stages in the information transfer through these systems. For CCD-based systems, where the readily available active area dimensions are typically multiple centimeters in size, demagnifications of ~×5 to ×10 are necessary. This can be achieved with lens or fiber optic-based systems (Hejazi and Trauernicht, 1997), but both approaches have a fundamental issue with transfer efficiency of the light photons to the photosensitive surface of the CCD. Transfer efficiencies of much less than 1% are common. This normally results in what is known as a secondary quantum sink in the transfer of the image information. At this point in the imaging chain, the number of secondary quanta associated with an absorbed x-ray drops to a level where the statistical uncertainty in the number of secondary quanta is larger than the relative uncertainty in the incident x-ray flux. This link in the image chain then becomes the dominant stage in determining the noise content of the final image. A general rule of thumb is that one needs at least 10 or more secondary quanta per absorbed x-ray photon to avoid this secondary quantum sink. Even for CCDs with extremely low electronic noise levels, this secondary quantum sink can irreversibly reduce the quality of the image. Recent developments in large-area CMOS detector fabrication, using 12 in. diameter crystalline silicon wafers, offer the possibility of creating tiled, "large" area secondary quantum detectors of a size suitable for clinical applications that require smaller detectors, such as mammography where 8 × 10 in. or 10 × 12 in. detectors are acceptable. However, this technology is only recently being introduced into the marketplace (Naday et al., 2010).

In fluoroscopy, a different approach to demagnification of the image has been implemented. Traditional image intensifier-based fluoroscopy systems convert the light from the input CsI phosphor layer, into electrons using a photocathode. These electrons are then accelerated, demagnified, and focused, using electric fields, onto another phosphor that generates higher-intensity light signals (i.e., the electron acceleration serves as a gain stage in the number of light photons in the signal). This bright, small area image is then focused (with only minimal or no demagnification) using optical lenses onto the input plane of a light-sensitive camera (such as a vidicon, or more recently a CCD). This camera finally converts the light into electrical signals that are amplified and digitized. The signal-to-noise imaging capabilities of these systems are exceptionally high but they tend to be bulky and susceptible to external influences that can affect the quality of the final image.

The practical issues associated with reading out a large area photoconductor have, until recently, been similar to those with phosphors described above, namely, there was not a viable, robust method for reading out signals from large areas of the photoconductor. 1D scanning approaches were implemented in mammography and general radiography (Boag, 1973; Neitzel et al., 1994), but neither is still commercially available.

This situation for phosphor and photoconductor-based systems changed dramatically in the early 1990s with the advent of flat-panel detector technology. Due to the significant changes brought about by their introduction, Section 10.3 is dedicated to a more detailed description of their configuration and use.

10.3 Flat-Panel Detectors

Flat-panel detectors combine traditional x-ray absorption materials (GOS, CsI(Tl), a-Se) with readout arrays fabricated from a material known as hydrogenated amorphous silicon (a-Si:H). This a-Si:H material was originally developed for large-area photovoltaic and liquid crystal display applications. It is fabricated in large vacuum chambers using a technique called plasma enhance chemical vapor deposition. It is this fabrication approach, using deposition from a gas/plasma, which enables the creation of extremely large-area readout arrays. This large-area fabrication capability is the unique advantage of this technology over other approaches for creation of the secondary quantum detector. It is possible to make a secondary quantum detector for either light or electric charge that has pixel dimensions of around a hundred microns with a physical dimension of greater than 40 × 40 cm on a single monolithic substrate. Since the dimensions of the a-Si:H readout circuitry can be made comparable to those of human anatomy, there is no need for the demagnification stage that affects the fundamental image quality capabilities of other approaches. Signal transfer efficiencies of 50–90% or higher between the x-ray absorption layer and the secondary quantum a-Si:H detector are possible. This high transfer efficiency removes the image quality limitations associated with a possible secondary quantum sink within the device. In addition, since the x-ray absorber is coupled directly to the surface of the a-Si:H readout component, these systems can be extremely compact. Modern systems have been introduced that have the same form factor as a traditional screen–film cassette (see Figure 10.3, which shows a selection of different flat-panel detectors). One important difference is that unlike a screen–film system, where the GOS phosphor has to withstand the physical abrasion associated with the continual insertion and extraction of the film, the x-ray absorption layer in a flat-panel detector is protected by the detector housing. This means that it is feasible to use materials that are less physically robust but can provide improved imaging capabilities compared to GOS, in particular CsI(Tl) and a-Se. The use of these high efficiency x-ray converters coupled to the large-area flat-panel detector readout is largely responsible for the dramatic improvement in image quality reported from these flat-panel-based systems.

10.3.1 Flat-Panel Detector Configuration

A flat-panel detector is comprised of three main components: the x-ray absorption layer; the a-Si:H readout panel; and the peripheral electronics required to control the readout and digitization of the signal information from the a-Si:H panel. As previously discussed, these devices use x-ray absorption materials that have been known for many decades. The novel feature of their design is the a-Si:H readout array. The a-Si:H panel itself is a rectilinear array of pixels on a ~75–500 micron pitch, depending on the application. Arrays have ~2000–3000 pixels along each dimension and an image typically has 2 bytes of data per pixel. This results in images that are ~10–15 Mbytes in size. The choice of pixel dimensions is an important aspect of array design. While many systems quote the size of their pixels as the defining aspect of their spatial resolution capabilities (by listing the Nyquist frequency of their pixel sampling), it is usually the x-ray absorber that defines the clinically relevant spatial resolution capabilities of the detector. Using a pixel size smaller than is warranted by

FIGURE 10.3 Examples of the variety of flat-panel detectors. From the right the detectors are a projection radiography detector (Trixel 4600), a small format fluoroscopic detector (Varian Paxscan 2520), and one of the new film-cassette-sized, wireless, battery-powered detectors recently introduced into the market (Carestream DRX-1 shown with battery charger).

the capabilities of the x-ray absorption layer can result in images that increase dramatically in size with little or no increase in clinically relevant information.

On an a-Si:H flat-panel detector, each pixel is comprised of a detection/storage element and a switching element. The detection element is either a photodiode, when the array is used with a phosphor, or a storage capacitor, when it is coupled to a photoconductor. The array of pixels is read out, one row at a time, by controlling the voltage applied to the switching element. This read out can be performed at frame rates compatible with static projection radiography, or at higher rates for fluoroscopic or volumetric imaging, ~30 fps or higher (Colbeth et al., 2005).

With faster frame rates, the typical imaging signal is lower in magnitude than with static imaging and the electronic noise levels associated with the peripheral readout electronics becomes increasingly important. With the current levels of electronic noise, flat-panel detectors still lag behind the image quality capabilities of state-of-the-art image intensifiers for the lowest exposure applications. However, their lack of image distortion, compact form factor, and relative insensitivity to external electromagnetic fields make them an attractive alternative for all but the most demanding dose-sensitive applications in fluoroscopy.

Increasing the signal associated with low x-ray exposures is a subject of active research. Two main approaches are being pursued: increasing the signal generated per absorbed x-ray by using new x-ray absorption materials with lower energy requirements per electron–hole pair or light photon generated; and providing pixel-level amplification that will increase the pixel signal to levels that make the additional electronic noise from the readout circuits insignificant. Future developments in these areas promise to enhance the capabilities of these detectors and hold the intriguing possibility that individual x-ray photon imaging may eventually be possible (Antonuk et al., 2009).

To date flat-panel detectors have found application in most, if not all, projection radiography applications from mammography, through general radiology and fluoroscopy to radiation oncology imaging. They are also being used in volumetric imaging applications in dentistry, ENT, orthopedics, and breast imaging. Their large area and high image quality can capture sufficient information to reconstruct a large imaging volume in a single rotation of the detector and source. This simplifies the design of the mechanics of the acquisition system compared to a traditional CT system (see Section 10.6) and commercial cone beam CT imaging systems are already available in many of these fields. Their large area make these systems more susceptible to image degradation from scattered x-rays than diagnostic CT scanners and while their in-plane soft tissue imaging quality is currently inferior to that of a diagnostic CT scanner, their small footprint, low cost, and isotropic imaging resolution make them particularly useful for many specialist applications.

10.4 Image Processing

Once the raw image data have been read out from the secondary quantum detector, the pixel data typically has to undergo a number of corrections. These can be separated into two types: those necessary to account for the nonideal behavior of the detector and those that optimize the image data for display to the viewer.

Detailed discussion on the latter type of *for presentation* image processing is outwith the scope of this chapter, but interested readers should consult one of the many review articles for more information on this important topic (e.g., Prokop et al., 2003). However, there is one aspect of this *for presentation* image processing that is worth commenting upon. Modern digital detectors free the user from the constraints imposed by screen–film systems, where the detection device (i.e., the film) was also the display modality. The density of the final digital image is no longer associated with the exposure level used to acquire the data, but is determined by the image processing software. This separation allows for the individual optimization of both the acquisition and the display stages, resulting in a more versatile system. This separation will undoubtedly lead to more customization of the exposure levels used for different clinical exams. The signal-to-noise ratio required for the specific clinical task will be the determining factor in how much radiation to use, rather than the requirement to achieve a certain density or contrast on a piece of film. This task-specific optimization has already begun and it is possible that the move from contrast-limited imaging to noise-limited imaging, inherent in the move from analog-to-digital radiology, will have a positive effect on patient exposure levels in the future.

In terms of the image processing associated with detector performance limitations, these are necessary to account for: spatial variability in the sensitivity and transfer efficiency of the x-ray absorption layer's output signal; differences in sensitivity of the individual pixels of the secondary quantum detector; differences in the signal charge produced by the inherent dark currents present with many types of the detection element present in each pixel (offset corrections); and the distracting visual impact of defective pixels and lines. The offset corrections can also include dark noise contributions from the photoconductor layer if this is used for x-ray absorption. The data used for both offset and gain sensitivity corrections are themselves subject to contamination from the various noise components present in the system, including stochastic x-ray noise and dark current shot noise. Consequently, the normal process for determining the appropriate level of correction is to average together a number of dark-field images (i.e., images taken with no x-rays applied to the detector) to determine a low-noise offset correction, and to average a number of fixed exposure x-ray images (known as flat fields) to determine a low-noise gain correction image. The reduction in the noise contained in the correction data is usually proportional to the square root of the number of images averaged together.

Corrections to remove the distracting influence of defective pixels and lines can be performed in a number of different ways. Many of these are vendor specific and are not discussed here. The relative importance of these corrections depends on the clinical task at hand. Applications where subtle variations in feature structure, such as calcification characterization in mammography, or the characterization of bone metastases will be less tolerant of defective pixels than applications where more gross features of the anatomy are more important, such as the determination of vertebrae alignment in scoliosis assessment. The acceptable number of defective pixels present on an array has a direct effect on the yield of the array manufacture and consequently the final cost of the detector. The issue of determining acceptable levels of pixel defects for different clinical applications is a topic that may well become more significant as the pressure for lower cost detectors becomes more important in the future.

10.5 Imaging Performance

The fundamental imaging capabilities of an x-ray detector have traditionally been investigated by evaluation of the detector's signal-to-noise transfer efficiency. The figure of merit is known as the detector's detective quantum efficiency (DQE) and it measures the efficiency of transfer of the signal-to-noise through the different stages of the detector as a function of both spatial frequency and exposure level. A detailed discussion on this topic is outwith the scope of this chapter, but a few observations about the interpretation of DQE are warranted. DQE is a single figure of merit that takes into account both the detector's spatial resolution capabilities and its noise transfer capabilities. Both these properties must be considered when evaluating a detectors performance.

Figure 10.4 shows the spatial resolution (measured by the modulation transfer function [MTF]) for two flat-panel detectors, one utilizing 500 μm thick a-Se, and the other 500 μm of CsI(Tl) as their x-ray absorber. Figure 10.5 shows a section of an x-ray image of a high contrast lead bar pattern test phantom taken with these two different detectors under exactly the same exposure conditions. It clearly illustrates the difference in spatial resolution indicated in Figure 10.5. Figure 10.6 shows the difference in "clinical" image quality produced by the same two detectors. These images show a lateral chest of the same patient imaged under identical acquisition conditions. From this clinical example, it can be seen that although the a-Se detector has better spatial resolution capabilities (as shown in Figures 10.4 and 10.5), the lower levels of noise present with the CsI(Tl) system result in superior image quality under these clinical acquisition conditions (120 kVp, high scatter environment). The spinous processes and other vertebral structures are much more easily visualized in the CsI(Tl) image. With digital imaging, it is no longer sufficient to rely only on a detector's spatial resolution

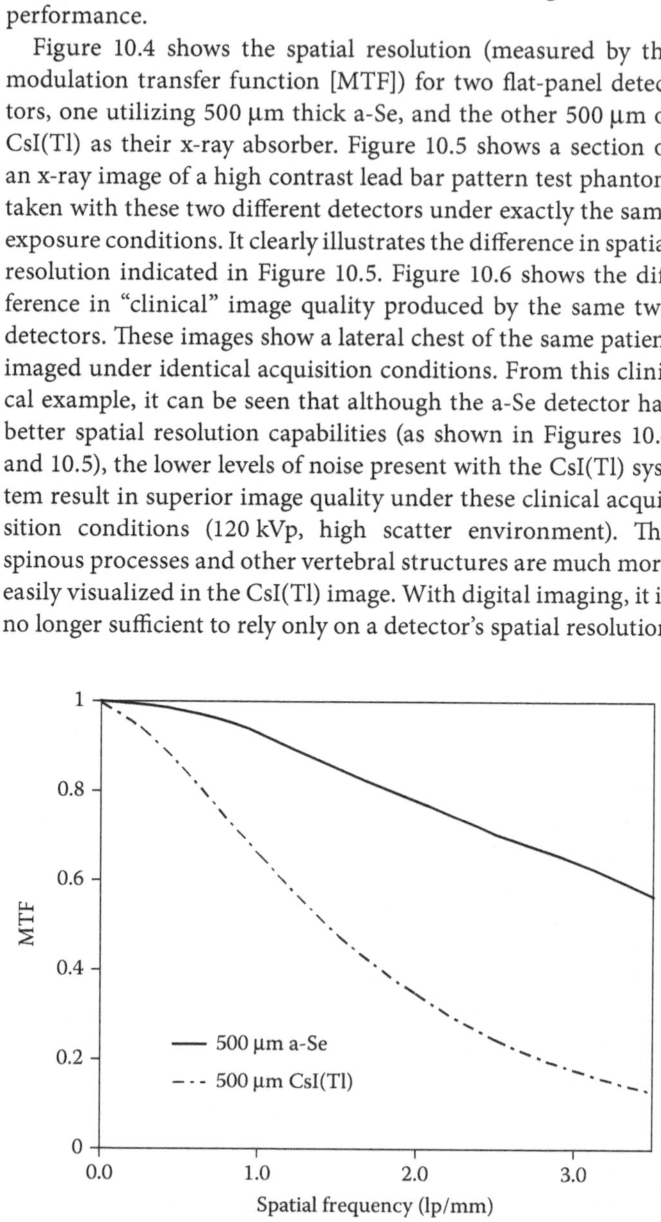

FIGURE 10.4 Graph of the MTF for a 500 μm thick layer of a-Se and CsI(Tl).

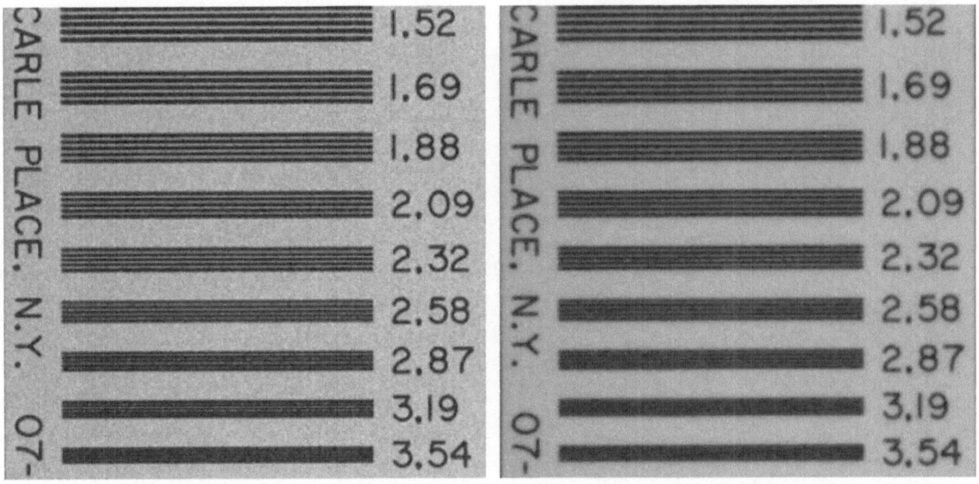

FIGURE 10.5 X-ray image of a bar-pattern spatial resolution phantom showing the visibly higher-resolution capabilities of an a-Se detector (left-hand side image) compared to the "smoother" but less noisy image from a CsI(Tl) detector (right-hand side image). These images show native capabilities of the two x-ray absorption layers with no additional image processing to reduce noise or enhance spatial resolution.

as a metric for image quality. Figure 10.7 shows the experimentally measured DQE for the two detectors mentioned above. It is clear that the CsI(Tl) has a higher performance (i.e., higher DQE) than the a-Se detector for the x-ray beam quality tested (RQA-9, 120 kVp with 40 mm added Al filtration). However, in a clinical setting, the detector's DQE is only one factor in the performance of the complete system. Practical issues such as choice of antiscatter grid, exposure level used for acquisition, and the quality of the *for presentation* image processing can significantly affect the final image quality produced.

While the sections above have discussed the various image acquisition stages in general terms, they have been somewhat focused on 2D projection radiography. Much of the discussion is also pertinent to 3D volumetric imaging but there are sufficient differences in the equipment and acquisition technique that a separate discussion on CT, its underlying concepts, technology, and image processing requirements is presented in Section 10.6.

10.6 Computed Tomography

10.6.1 Introduction to CT Scanners

A CT scanner is an x-ray imaging device that produces a cross-sectional image of a patient, where the image pixel values are related to the x-ray attenuation properties of tissue and sometimes injected contrast agents inside the patient. The major

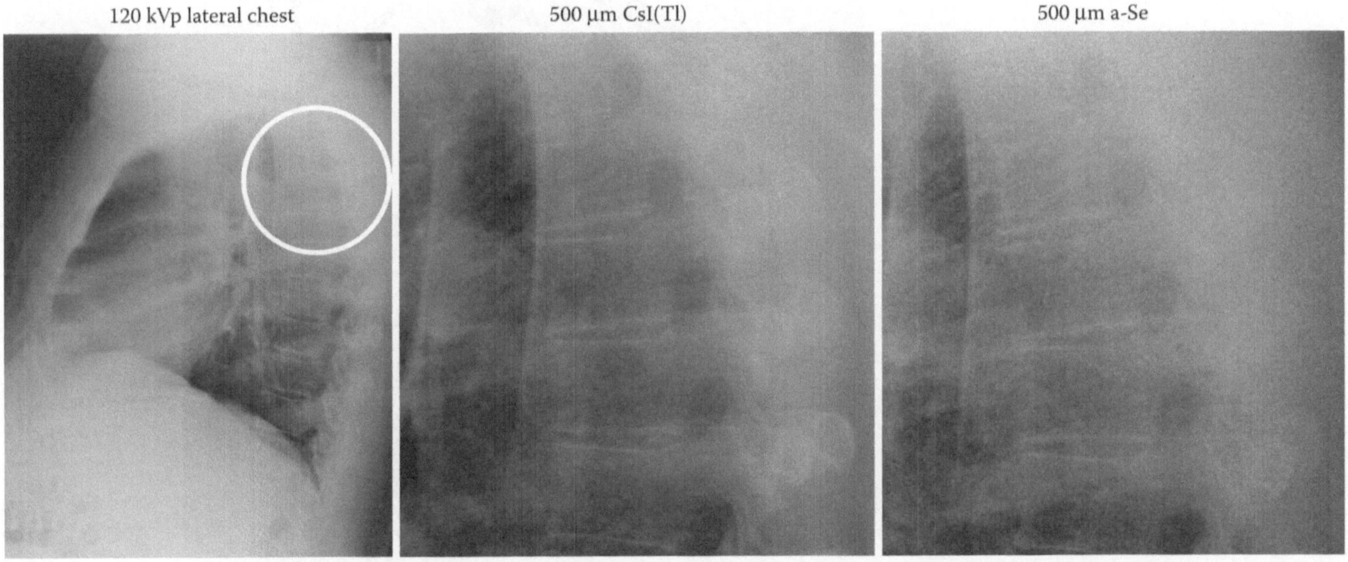

FIGURE 10.6 Lateral chest image of the same patient acquired with a CsI(Tl) an a-Se detector. The images were acquired with identical acquisition techniques and patient setup (although slight changes in patient orientation are visible). The circle in the left-hand side image shows the general region enlarged for enhanced visibility of the image quality differences between the two acquisitions in the center and right-hand side images.

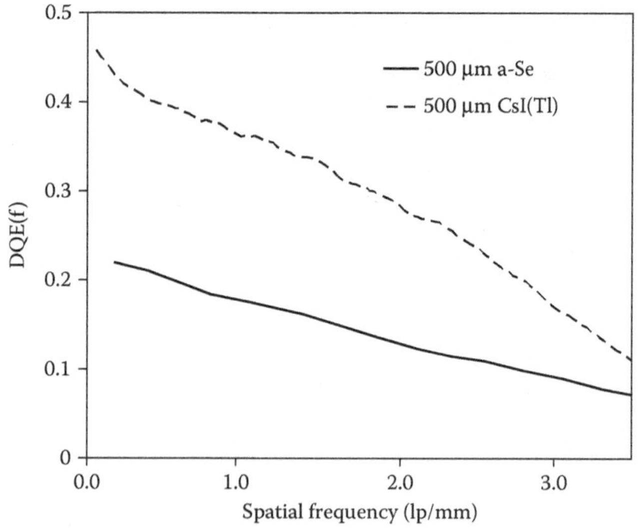

FIGURE 10.7 Graph of the DQE for the detectors used to acquire the images in Figure 10.6. Measurements were performed with an RQA-9 beam (120 kVp, 40 mm Al filtration) and an exposure level of ~0.5 mR input to the detector surface.

components of a modern CT scanner are shown in Figure 10.8. The patient is positioned on a couch in the bore of the scanner. Inside the scanner, an x-ray tube emits x-rays which pass through the patient and are detected on the opposite side of the patient with an x-ray detector system. The detector system consists of many individual x-ray detectors with each detector on the order of 1 mm^2. The signal measured on an individual x-ray detector is determined by the intensity of the x-ray focal spot inside the x-ray tube and the attenuation of the x-ray beam as it passes through the patient along a line from the focal spot to the detector. The x-ray detector system simultaneously measures the x-ray attenuation along many individual straight line paths from the x-ray focal spot through the patient to the detector elements. During a CT scan, the x-ray tube and the detector system rotate around the patient allowing the detectors to measure the attenuation through the patient from multiple angles known as views. The data from the detectors are sent from the rotating frame of the scanner to the stationary frame through a device called a data slip ring. The data are then sent to a computer system, which does specialized calculations to form the cross-sectional image. The computer system often contains custom digital hardware to accelerate the massive number of calculations required.

10.6.2 CT Scanner Reference Frame

The conventional coordinate system used when describing a CT scanner is shown in Figure 10.8 (Hsieh, 2003). The *X*–*Y* plane of the scanner is the plane in which the cross-sectional image is made. It is also the plane in which the detectors and x-ray tube focal spot rotate. The detectors are positioned along an arc approximately along the *X*-direction. The *Y*-direction is the line from the center of the detector system to the focal spot. The *X*–*Y* coordinates thus rotate as the x-ray tube and detectors rotate. The *Z*-axis of the scanner is the axis of rotation. It is also the direction that the couch holding the patient moves.

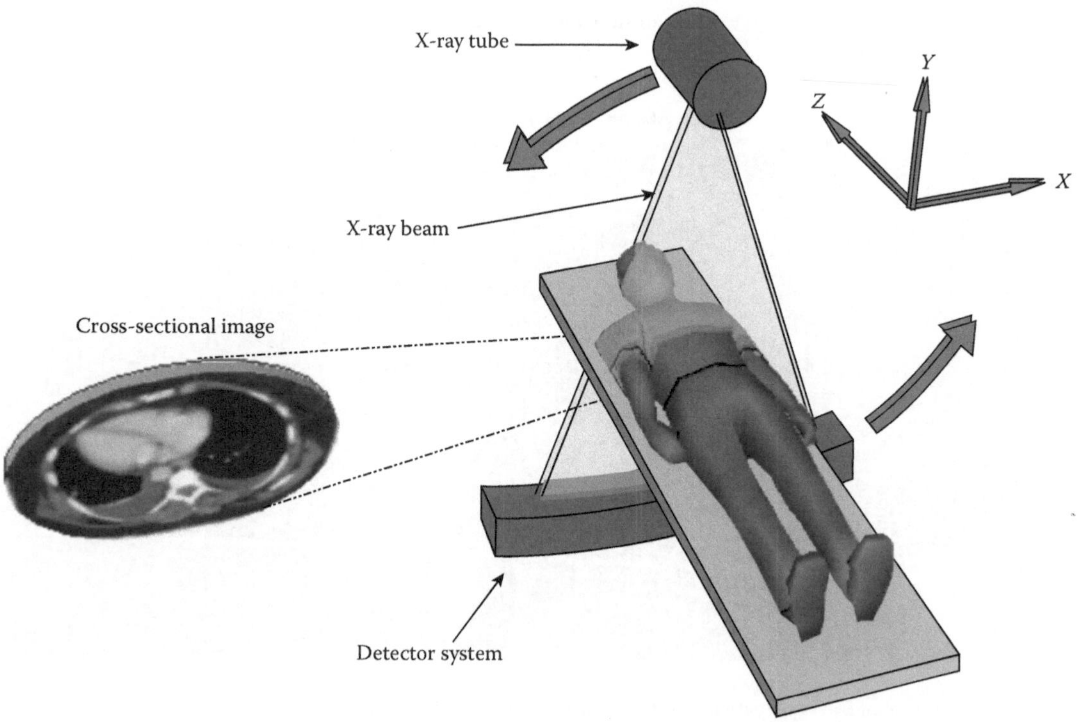

FIGURE 10.8 A basic CT scanner.

10.6.3 A Basic Single-Slice, Axial Mode CT Scanner

A basic or minimal CT scanner is described first before moving on to describe more advanced forms of CT scanner. A CT scanner known as a single-slice scanner has a single row of detectors inside the detector system that lie along the X-axis and rotate in the X–Y plane perpendicular to the rotation axis. Typically, there are 600–1000 detectors along X with a pitch of 1–1.4 mm. A modern CT scanner typically has a rotation speed of 0.25–3 revolutions per second and 600–4800 views or measurement angles per revolution.

A basic mode of scanning known as axial mode occurs when the patient/couch is stationary while the x-ray tube and detectors rotate to create one cross-sectional image. To create multiple cross-sectional images in axial mode, the couch can be incremented between each scan. Axial mode is shown in Figure 10.9.

10.6.4 Spiral Mode

To increase the speed in collecting images for multiple cross sections, the couch can be made to move continuously while the images are being acquired. This is known as spiral mode since the x-ray tube and detectors follow a spiral (helical) trajectory relative to the patient (Kalender, 2005). In spiral mode, an interpolation step is used during image reconstruction to take the data collected along a spiral and reorient it into planes perpendicular to the rotation axis. All modern CT scanners allow both spiral mode and axial mode with spiral mode being the most commonly used.

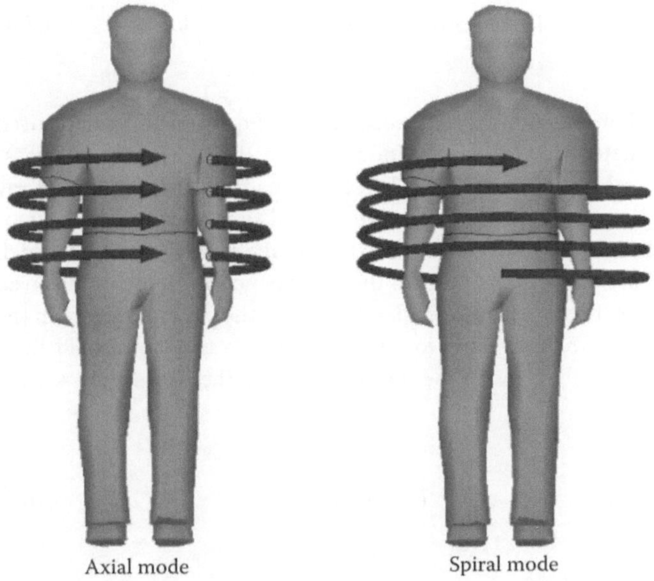

FIGURE 10.9 Axial and spiral modes of scanning.

10.6.5 Multislice CT Scanners

In the late 1990s, technology had advanced such that many CT companies introduced CT scanners known as multislice scanners which have more than one row of detectors and could create multiple cross-sectional images per rotation. While a single-slice scanner has one row of detectors, a multislice scanner has a 2D array of detectors in the X–Z plane, as shown in Figure 10.10. The number of detectors along the X-direction is the same as in the single-slice scanner (~600–1000), but the number of detectors along the Z-direction is greater than one. The number of detectors along the Z-direction has steadily increased from 2 in 1998 to as high as 320 at the present time (2010). Scanners with 128 or more slices offer the potential to image whole organs in one single revolution.

Currently, the most common multislice scanner has 64 detectors in Z-direction. Multislice scanners greatly increase the speed that one can scan a given volume of the patient. This speed in turn greatly improves the scanners diagnostic ability since images contain less image artifacts due to patient motion and the greater speed can more effectively image x-ray contrast agents that are injected into the blood flow. With this speed comes a great increase in the data rate and total quantity of data created by the CT scanner. Another approach that has been introduced to improve the speed of acquisition of data over an extended imaging volume is the use of dual x-ray sources and detectors. This new development has allowed the acquisition of high-quality cardiac images with no discernible motion artifacts from the heart motion.

10.6.6 Data Rates and Quantity of Data Produced for a CT Scan

To give the reader an appreciation for the data rates and quantity of data produced by a modern CT scanner, we choose typical values for the scanner parameters and show how these multiply

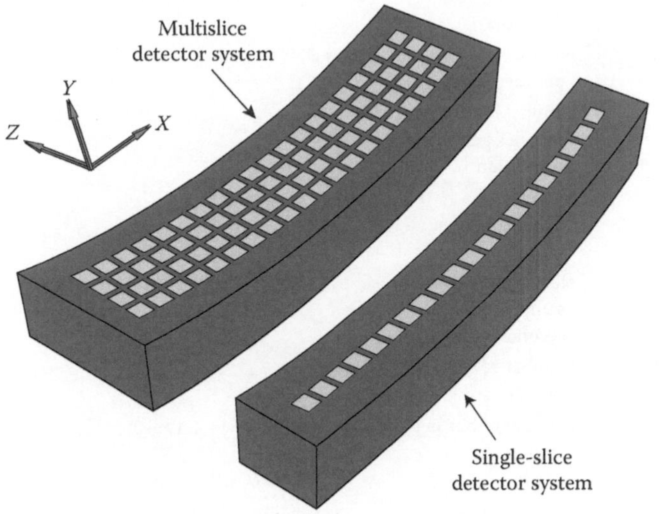

FIGURE 10.10 Single-slice and multislice detectors.

to give the total rate and quantity of data. This example calculation is shown in Table 10.1. For each scanner parameter, a typical range is given (ImPACT, 2009) and from this range is chosen a common value for the example calculation. The example thus represents a fictitious scanner but one that is representative. From the table, one can see that for the 64-slice scanner we have chosen for our example the data rate out of the detector system is 4.4 Gbits/s. Given that the scan time for the study is 10 s, the quantity of data is thus 44 Gbits. Given the values in the typical range column, the reader can also get an appreciation for how the data rate and quantity of data might change as parameters are changed. A word of caution, one cannot calculate the maximum data rate and quantity of data in use today by multiplying the maximum value in each of the typical ranges as these parameters are interrelated. For example, a scanner with a very high rotation speed has a shorter scan time for the same study.

10.6.7 The Measurement of X-Ray Attenuation

As stated previously, the CT scanner produces cross-sectional images of the patient, where the image pixels are related to the x-ray attenuation properties of the tissues within the patient. The x-ray detectors cannot measure the x-ray attenuation value of a particular point inside the patient directly. Instead they measure the total x-ray attenuation of the patient along lines passing through the patient. Figure 10.11 illustrates an x-ray beam

TABLE 10.1 Calculation Using Typical Parameter Values to Illustrate Data Rates and Total Data for a CT Scan

	CT Scanner Parameter	Typical Range	Typical Value for Calculation	Note
a	Number of detectors in X	600–1000	800	
b	Number of detectors in Z	1–320	64	
c	Total number of detectors	—	51,200	$c = a \cdot b$
d	Number of views per revolution	600–4800	2400	
e	Number of detector samples per revolution	—	1.2×10^8	$e = c \cdot d$
f	Number of revolutions per second	0.25–3	2	
g	Number of detector samples per second	—	2.4×10^8	$g = e \cdot f$
h	Number of bits per sample	16–22	18	
i	Number of bits per second	—	4.4×10^9	$i = g \cdot h$
j	Scan time (s)	1–100	10	
k	Total number of bits in scan		44×10^9	$k = i \cdot j$

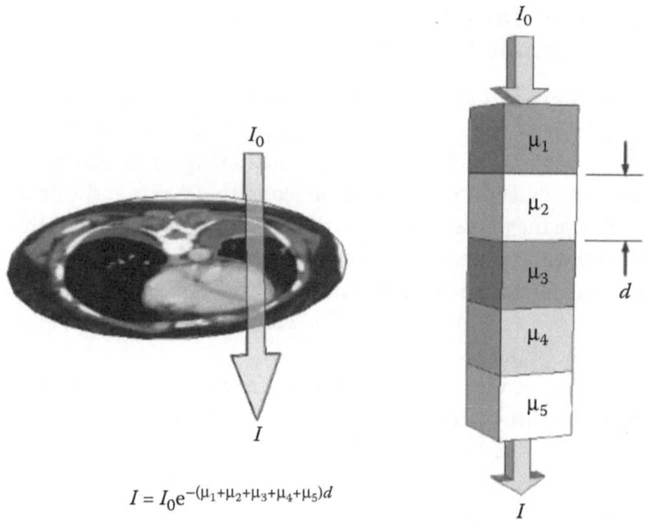

FIGURE 10.11 The attenuation of x-rays.

passing through a length of material and then being detected at the exit end. This material represents the material present along one line through the patient. The material is divided into a number of subregions along its length with each subregion having a different value of x-ray attenuation. The width of each subregion is d and the attenuation coefficients of the subregions are $\mu_1, \mu_2, \mu_3, \ldots, \mu_N$. The value of d is effectively the width of a voxel inside the patient. The equation which gives the total attenuation of the x-ray beam as it passes through the length of material is given in Figure 10.11, where I_0 is the x-ray flux entering the material and I is the x-ray flux exiting the material. The detectors measure the value of I directly. The value of I_0 is obtained by taking a detector measurement with no material present. For a given value of d, the sum of attenuation coefficients can therefore be calculated using

$$\mu_1 + \mu_2 + \mu_3 + \cdots + \mu_N = \sum_{i=1}^{N} \mu_i = \frac{1}{d} \ln\left(\frac{I_0}{I}\right) \tag{10.1}$$

For each detector measurement (I), Equation 10.1 can be used to determine the sum of individual μ values along the line passing through the patient. These μ sums from many different ray paths through the patient are passed to a reconstruction algorithm, which determines what the individual values of μ for each voxel are. A description of the reconstruction algorithm is given later. The μ values for each voxel inside the patient or pixel in the image are converted to a scale known as the Hounsfield scale named after one of the inventors of CT. The Hounsfield scale is defined as

$$H = 1000\left(\frac{\mu - \mu_w}{\mu_w}\right) \tag{10.2}$$

where μ is the attenuation coefficient of the material and μ_w is the attenuation coefficient of water. Hounsfield numbers are also referred to as CT numbers. In this scale, water has a value of zero. Materials that attenuate x-rays less than water have an H value less than zero and materials that attenuate more than water have an H value higher than zero. Water is chosen as the basis of the Hounsfield scale since the human body comprises mostly water. In a CT image, each pixel has a Hounsfield value or CT number associated with it, but the actual gray level or brightness of the pixel displayed may be a linear or nonlinear value derived from the Hounsfield value using *for presentation* image processing.

10.6.8 The CT Detector

The x-rays that pass through the patient are absorbed in an x-ray detector, whose function is to accurately measure the x-ray intensity, which is in turn used to determine the attenuation of the x-ray beam through the patient. This process is essentially the same as described in Sections 10.2.2 and 10.2.3. A diagram of a modern CT x-ray detector is shown in Figure 10.12. The part of the detector which absorbs the x-rays is called the scintillator. When x-rays are absorbed, the scintillator gives off light in proportion to the x-rays absorbed. This light is then collected by a silicon photodiode, which converts the light into an electrical current. This type of detector is referred to as indirect since there is an x-ray to light conversion followed by light to electrical current conversion. The electrical current is then passed to an analog/digital (A/D) converter or in some cases an amplifier followed by an A/D converter. The A/D converter integrates the current from the photodiode over a period of time known as the sampling time or integration period and produces a digital number representing the signal in that time period. The number of samples produced per second is typically between 1000 and 10,000. While all modern CT scanners use scintillator plus photodiode-type indirect detectors, other types of detector have been used in the past such as scintillator plus photomultiplier tube indirect detectors and xenon gas direct conversion detectors.

The scintillator used in a CT detector should have a fast speed of response compared to the sampling rate and a very low residual signal known as afterglow. Only a few scintillators meet these requirements for CT. The four common scintillators used in modern CT scanners are shown in Table 10.2. Each scintillator

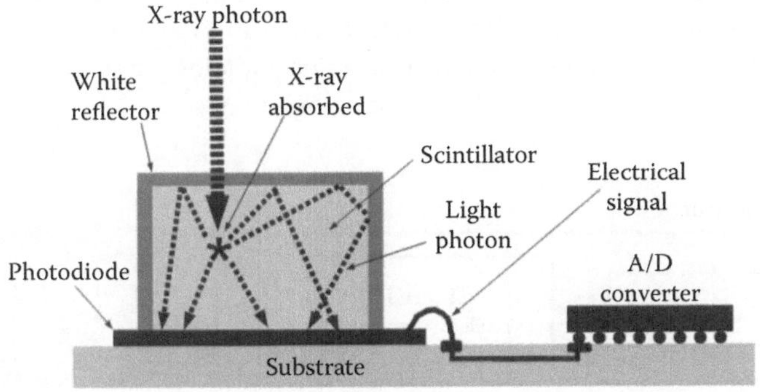

FIGURE 10.12 A basic CT detector.

TABLE 10.2 Modern CT Scintillators

Common Name	Chemical Formula	Reference
GOS, UFC	$Gd_2O_2S:Pr^{3+}$	Ronda (2008)
Highlight	$(Y,Gd)_2O_3:Eu^{3+}$	Ronda (2008)
Cadmium tungstate	$CdWO_4$	Ronda (2008)
Gemstone	Rare earth garnet	ACerS (2010)

element has a white reflector on five sides with the sixth side facing the photodiode in order to maximize the amount of light entering the photodiode. The detector elements are on the order of 1 × 1 mm in the X–Z plane and on the order of 1–2 mm thick in the Y-direction (x-ray absorbing direction). For each x-ray photon absorbed by the scintillator, the number of light photons that are produced and get absorbed by the photodiode is on the order of 1500 (Luhta et al., 2006). For each light photon absorbed in the photodiode, there is approximately a 90% conversion to electrons produced. This means that there is typically no secondary quantum sink in these detector systems.

The number of x-ray photons incident on each detector element depends on the intensity of the x-ray tube and the attenuation of the patient. Without a patient, in the x-ray beam, the number of x-ray photons at the detector surface is on the order of 10^9 photons/mm²/s. Assuming a typical sampling rate of 2500 samples/s and a typical detector area of 1 mm², the number of x-ray photons absorbed per detector element per sample with no patient is then 400,000. The attenuation caused by the patient can vary over a wide range. At the edges of the patient where the x-ray path length is small, the attenuation of the x-ray beam is also small. For x-rays passing through 20 cm of water which would be similar to 20 cm of tissue, the x-rays would be reduced by a factor of about 50.

For 30 cm of water, the factor would be about 300. For extreme high attenuation, such as through the long bones of the shoulders, the factor could be as much as 30,000. Therefore, the typical range of the number of x-ray photons detected per sampling period is on the order of hundreds of thousands down to less than 10.

10.6.9 Noise in CT

As in any electronic imaging system, noise limits the accuracy with which measurements can be made which in turn limits the image quality achieved. The two main types of noise limiting a CT scanner are quantum noise and electronic noise. Quantum noise arises from the fact that an x-ray beam is made up of a finite number of x-ray photons and the number of these photons emitted by the x-ray tube in a given time frame is random and has a statistical fluctuation associated with it. As previously described, these statistical fluctuations obey Poisson statistics, where the variability is given by the square root of the total number of photons. It is quantum noise which in most cases gives the mottled noise look in a CT image. As with projection radiography, the "quality" of a CT image generally improves as the amount of radiation used to create it increases.

The second main type of noise in a CT scanner is electronic noise. This is noise originating from the detector electronics commonly made from MOSFET transistors in an integrated circuit. This noise gets added to the signal during the A/D converter stage (or amplifier+ A/D converter stage). Unlike quantum noise which varies with the signal level, electronic noise has a constant level. It is the goal of the CT detector design to keep the electronic noise lower than the quantum noise at the lowest practical signal level. That is, the goal is to keep electronic noise lower than the quantum noise associated with only a few x-rays being detected. Similar to projection radiography, electronic noise becomes important at low signal levels when the attenuation due to the patient is high.

10.6.10 CT Image Reconstruction

A generic diagram showing the various steps involved in transforming detector data into a CT image is shown in Figure 10.13. Data acquired by the detectors must first be moved from the rotating frame to the stationary frame of the gantry through a data slip ring. Next corrections must be made to the data in order to correct for the nonideal properties of the scanner. Each CT manufacturer will have its own proprietary set of algorithms for doing these corrections, which to some extent are specific to a certain scanner design. Examples of some corrections are (1) gain correction to account for differences in the gains of individual detectors; (2) offset correction to account for detector values having an added constant error; (3) off-focal correction to account for the x-ray tube having radiation emanating from points outside the focal spot region; (4) scatter correction to account for x-rays that scatter inside the patient and into detectors where they are not wanted. There are many more and the reader may consult the references for more information about them. The corrections attempt to make the data like it would be if it had been collected on an ideal scanner.

As described earlier, calculating the logarithm of the detector data is a step that must be performed to convert from the measurement of x-ray intensity by the detectors to a number representing the sum of attenuation coefficients along a line through the patient. The data then pass to the filtered back-projection algorithm. The data from the detectors effectively have the image voxel data encoded into them in a known mixed way. The filtered back-projection algorithm unmixes the data and "reconstructs" the image.

Although other algorithms exist, the most common algorithm used for CT reconstruction is called filtered back-projection (known as FBP) since it is the one that can be computed with the least computation and thus with the greatest speed (Kak and Slaney, 2001; Natterer, 2001). Future increases in computer power and memory will allow the implementation of practical iterative reconstruction approaches that can help reduce many of the image reconstruction artifacts currently seen with FBP approaches. Filtered back-projection can be split into two operations, filtering and back-projection. We explain back-projection first as this makes the reason for filtering more apparent. A diagram illustrating back-projection is shown in Figure 10.14. On the left is shown how the detectors measure patient x-ray attenuation along lines through the patient. A set of detector measurements at one angle of the rotating frame is called a projection. On the right is shown a 2D matrix of squares, which represent image pixel memory locations. Outside the memory array is shown a projection after logarithm and corrections. For a given position on the projection, a line is drawn through the 2D memory array. For each of the pixels along this line, the value of the projection is added to the memory location. In effect, the projection is smeared along the image memory at the same angle it was measured in the CT scanner. This operation is repeated for all the angles until an image is built up. The formation of an image as projections are added is shown in Figure 10.15. It is the nature of the mathematics of the back-projection operation that if it is done without filtering the resulting image is blurred as shown in the lower-right side of Figure 10.15. More precisely, when an image is formed by projection followed by back-projection, the high spatial frequencies which correspond to the fine detail in the image are reduced in magnitude making the image appear blurry. The filter step in filtered back-projection is an operation which enhances the high spatial frequencies (fine detail) so that the resulting image contains the high-frequency components. The filtering operation is most commonly done on the projections before back-projection as a kind of prefiltering so that after back-projection the image appears correctly. It is possible however to do the filtering after back-projection. There are many texts which can describe the reconstruction process in more detail.

Although CT technology is significantly different to that used in projection radiography, the issues affecting the final image quality have many similar considerations, with the capabilities of the x-ray detector stage determining the fundamental imaging capabilities of the system. Section 10.7 describes a number of new developments in projection radiology systems that also have consequences for CT imaging.

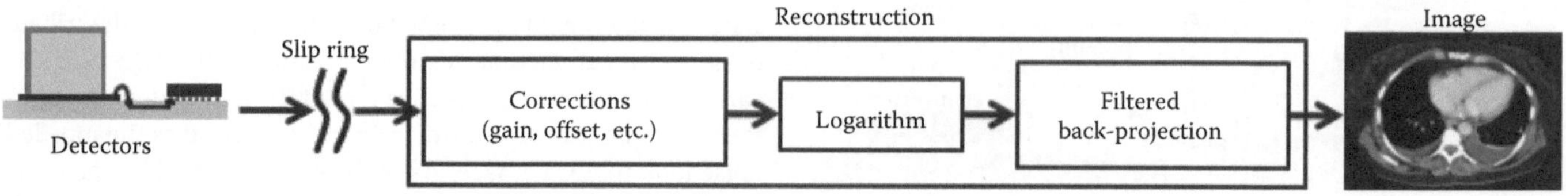

FIGURE 10.13 Generic image reconstruction steps in CT.

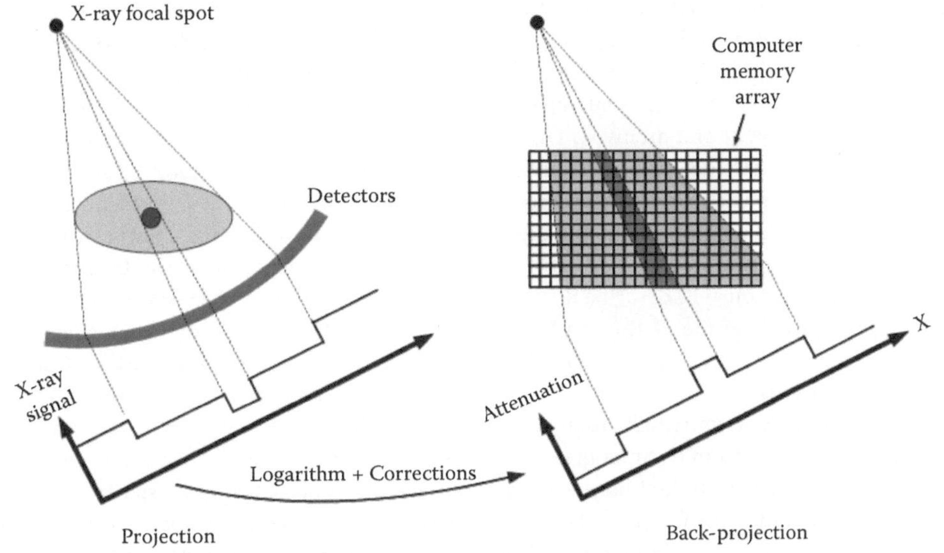

FIGURE 10.14 Projection and back-projection.

10.7 Advanced Applications and Future Directions

The introduction of new digital acquisition technologies into the clinical environment enables a large number of new imaging applications. Many of these applications are focused toward multidimensional imaging, where the additional dimensions can be spatial (e.g., tomosynthesis and cone beam computed tomography (CBCT)), spectral (e.g., dual energy and photon counting detectors), or temporal (e.g., temporal subtraction). There is also a move from qualitative image evaluation to the extraction of quantitative data from the different types of digital images available, with measurements from different modalities being integrated to provide enhanced information on the patient's anatomical and physiological condition. Many of these developments are taking place outside the radiology department with imaging playing an

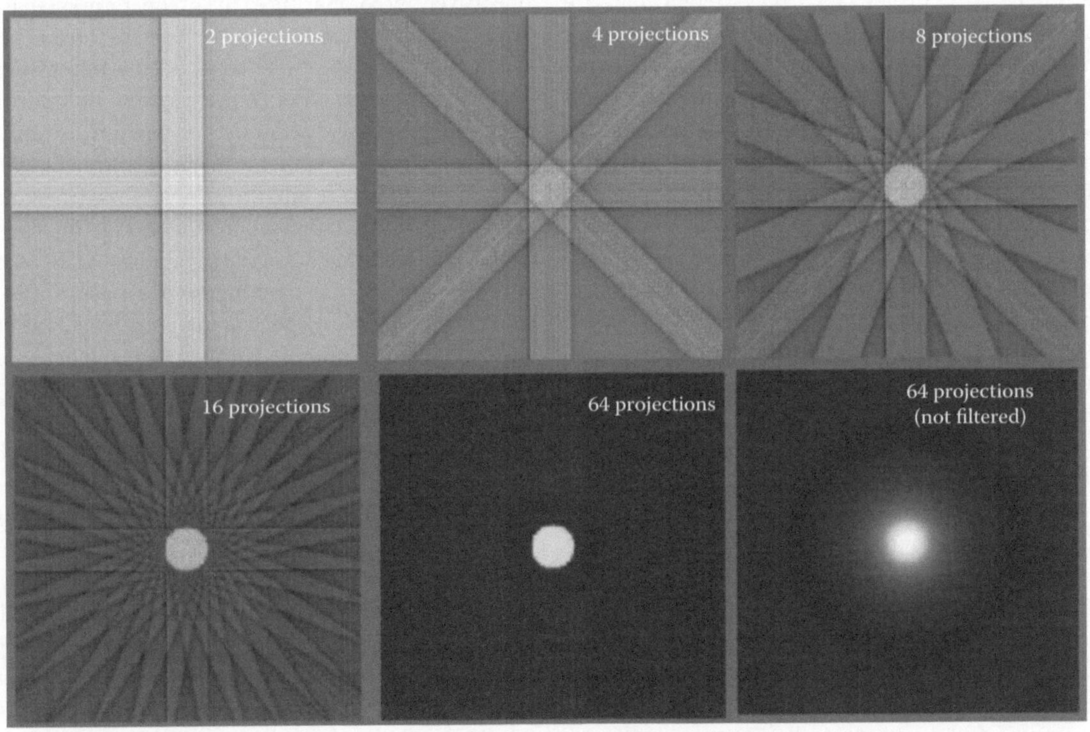

FIGURE 10.15 How the image of a circular object is formed with back-projection.

ever-increasing role in surgical procedures, orthopedic practices, dental offices, and many other medical specialties. In these environments, much of the emphasis is on procedure guidance rather than patient diagnosis and may require that different types of image information be extracted, tracked, and archived. Digital imaging is also reinvigorating the use of older acquisition methodologies by improving the efficiency with which these procedures can be carried out in the clinical environment. These new imaging capabilities offer many opportunities and challenges for the informatics community. This section will describe some of these new developments and highlight some of the challenges created by the varied types of information being generated by these new capabilities and applications.

One seemingly inconsequential, but extremely important, development with the new digital acquisition technologies is the integration of the x-ray delivery system with the image acquisition device. This has been the situation with CT since its inception, but it is a new development in projection radiography. The link between the control of the x-ray beam (kVp, filter, mAs, etc.), the preparation and readout of the detector, and the synchronized control of mechanical motion of the x-ray tube and detector opens up a multitude of possibilities for new applications and novel acquisition procedures.

One example of a simple application made possible by the integration of the different system components is the opportunity for previous technique recall. In the intensive care unit, where patients undergo multiple exams over extended periods of time, it is often difficult to assess the changes to the patient's condition due to the inconsistency in image quality and contrast from day to day. Much of this inconsistency is caused by changes in image acquisition technique (i.e., images taken on different days are acquired with different kVp, imaging distance, and mAs settings). If this information could be automatically loaded into the x-ray generator at the patient's bedside, based on prior exams, it could significantly improve the consistency of image quality and help with the accurate evaluation of the patient's condition from day to day. If the technique information were available from the patient's record, this capability would be straightforward to implement when the x-ray generator is controlled by the computer that handles the patient data and the image acquisition. As a side note, the issue of consistency of image presentation is of more widespread concern than this one application. For institutions with equipment from multiple vendors, the differences in vendor-specific *for presentation* image processing can result in significant variation of the "look" of images of the same patient acquired on different systems. These differences can be problematic for the correct display of the data on third-party workstations, as well as causing potentially distracting variations in image contrast and density even when correctly displayed. With the proprietary nature of much of this image processing, it is difficult to see how this problem can be solved in the current situation, where the result of the vendor-specific processing is typically irreversibly burned into the image sent from the acquisition modality.

Another example of the power of the integration of the x-ray delivery and image acquisition is in dual energy and tomosynthesis. These applications have been known for many years, but have lacked widespread clinical implementation due to a number of factors. They either took a significant effort to implement in a clinical environment, suffered from reduced image quality, or could not be performed at all due to the limitations of the digital technology available at the time. With the new digital systems, these procedures can be performed with a single button push, in a manner that can be virtually transparent to the workflow of the technologist. The image quality achievable with these new systems also means that they can be implemented with little or no increase in patient exposure. In many situations, the image clutter associated with overlapping anatomical structures is the dominant noise source hindering the clinical assessment of the underlying pathology. These two methods for separating overlying anatomy through tissue separation (dual energy) or spatial separation (tomosynthesis) may have a significant impact on the practice of general radiology in the not too distant future. Indeed, mammography has already seen the introduction of the first clinical tomosynthesis systems. The methodology for how to most efficiently process, store, transmit, and display the additional information inherent in these new approaches is still being developed.

In areas outwith general radiology, these new system capabilities are also being utilized in various ways. The use of 3D imaging (particularly CBCT) in the operating room is becoming more pervasive as the capabilities of the commercially available systems improve. The ability to create accurate 3D representations of the patient that can be used for guidance during surgery, evaluate this information compared to preoperative images that were used for planning, and then evaluate the progress of the patient postoperatively by comparing to the preoperative plan is an extremely attractive proposition. How the data sets from potentially different technologies can be consolidated into a single image set, such that the extraction of the appropriate information can be achieved quickly and accurately, will require careful development of robust data-handling protocols (e.g., for image registration, deformation, and reformatting) as well as efficient methodologies for display of this enhanced information to the viewer.

Recent developments in the capabilities of the secondary quantum detector are also opening up new possibilities for information extraction. Multispectral imaging, where the energy of each individual x-ray photon is measured and recorded, may well have a significant impact on the practice of radiology. In remote sensing, the ability to detect light photons within different frequency bands allows enhanced analysis of the data by inspecting the relative amplitudes of the signal from the different frequency bands. With multispectral x-ray imaging, it is possible that similar signal comparisons can yield additional information about the composition of the tissue being imaged by inspection of the relative intensities of the image data from the different x-ray energy ranges. This capability is already appearing in state-of-the-art CT systems, where such spectral information is being used to classify the constituents of kidney stones (Wang et al., 2010). The photon-counting capabilities necessary for this spectral imaging have already been demonstrated in mammography

(Aslund et al., 2007; Fredenberg et al., 2010), and it is likely only a matter of time before a similar capability, that is, x-ray photon counting with energy resolution, is introduced in general radiography (Antonuk et al., 2009). The handling of these images and the accompanying data analysis will likely prove challenging for today's methods of image handling and distribution.

When one considers these new capabilities, in concert with recent developments in functional imaging made possible with molecular imaging, it is clear that the traditional notion of an image that is subjectively evaluated by a viewer as being the endpoint for a medical imaging exam will come under increasing pressure. As quantitative imaging becomes more pervasive, it is conceivable that for certain exam types and patient conditions, a viewer may never look at an actual image but at a series of numerical biomarkers that have been extracted from a range of image data acquired from different modalities to diagnose the presence or evaluate the progress of a given disease. This will require a reassessment of the image data associated with a patient as being more than a collection of disconnected imaging procedures. Future information systems will need to handle these different pieces of information as different facets of a single entity, where cross-correlations and similarities (or disparities) between different data sets will be sought by sophisticated data-mining software. This may allow a more accurate assessment of patient condition than evaluation of the individual image data alone. The acquisition of this type of data is already underway with PET&SPECT/CT systems experiencing ever-increasing popularity. It is likely that this fusion of different acquisition modalities will continue in the future and it will be important for information systems to evolve to handle the ever-increasing amounts and variety of data in efficient ways.

10.8 Conclusions

In this chapter on data acquisition technologies, we have focused on the use of x-rays to create a patient image. We have separated the image acquisition chain into a series of sequential stages and tried to provide the reader with an appreciation of the flow of the image information through each stage. The different aspects of each stage that are most important in determining the image quality have been described and some of the limitations of different systems have been identified. The discussion has been illustrated with examples of projection radiography systems, but we have reviewed the unique aspects of CT systems that characterize their data acquisition and their image processing methodologies. We concluded the chapter with a brief description of some of the new developments that are changing the traditional approach to image acquisition and highlighted the importance of the trend toward quantitative imaging that will undoubtedly bring significant changes to the way "image" information is reviewed and assessed. It is likely that the coming years will also see a blurring of the historical divisions between the different methods for acquiring patient images. Projection x-ray systems capable of both static and dynamic imaging as well as volumetric capture are already feasible, ultrasound and x-ray tomography are being combined into a hybrid system for mammography (Carson, 2010), research is underway to incorporate x-ray imaging capabilities into MRI scanners (Fahrig et al., 2005), and the combination of nuclear medicine imaging and CT is already enjoying considerable commercial success. When one considers the variety of other imaging modalities not discussed in detail here, such as functional MRI, 3D ultrasound, molecular imaging, and optical tomography, it is clear that the opportunity for innovation in the development of hybrid imaging systems is enormous. New methodologies for handling the many disparate types of patient information will probably be necessary if these new multimodality approaches are to live up to their full potential. It will be interesting to see how the informatics community responds to the ever-increasing demands on data handling and data mining that will accompany these developments.

References

ACerS. 2010. Novel GE scintillator delivers CT imaging revolution. *Am. Ceram. Soc. Bull.*, 89(8), 43–44.

Antonuk, L. E. 2002. Electronic portal imaging devices: A review and historical perspective of contemporary technologies and research. *Phys. Med. Biol.*, 47(6), R31–65.

Antonuk, L. E., Koniczek, M., El-Mohri, Y., and Zhao, Q. 2009. Active pixel and photon counting imagers based on ploy-Si TFTs—Rewritting the rule book on large area, flat-panel x-ray devices. *SPIE Phys. Med. Imaging*, 7258, 7525814-1-10.

Aslund, M., Cederstrom, B., and Danielsson, M. 2007. Physical characterization of a scanning photon counting digital mammography system based on Si-strip detectors. *Med. Phys.*, 34(6), 1918–27.

Boag, J. W. 1973. Xeroradiography. *Phys. Med. Biol.*, 18, 3–37.

Brenner, D. J., Doll, R., Goodhead, D. T. et al. 2003. Cancer risks attributable to low doses of ionizing radiation: Assessing what we really know. *Natl. Acad. Sci. USA*, 100, 13761–6.

Carson, P. L. and Fenster, A. 2009. Anniversary paper: Evolution of ultrasound physics and the role of medical physicists and the AAPM journal in that evolution. *Med. Phys.*, 32(2), 411–28.

Carson, P. L. 2010. Multi-modality breast imaging systems: Tomo/ultrasound/optics, ultrasound. *Med. Phys.*, 37(6), 3371–2.

Chotas, H. G., Floyd, C. E., Dobbins, J. T., Lo, J. Y., and Ravin C. E. 1990. Scatter fractions in AMBER imaging. *Radiology*, 177(3), 879–80.

Colbeth, R. E., Mollov, I. P., Roos, P. G. et al. 2005. Flat panel CT detectors for sub-second volumetric scanning. *SPIE Phys. Med. Imaging*, 5745, 387–98.

Cowen, A. R., Davies, A. G., and Kengyelics, S. M. 2007. Advances in computed radiography systems and their physical imaging characteristics. *Clin. Radiol.*, 62(12), 1132–41.

Despres, P., Beaudoin, G., and Gravel, P. 2005. Evaluation of a full-scale gas microstrip detector for low-dose X-ray imaging. *Nucl. Inst. Methods Phys. Res. Sect. A*, 536(1–2), 52–60.

Donath, T., Pfeiffer, F., Bunk, O. et al. 2010. Toward clinical X-ray phase-contrast CT: Demonstration of enhanced soft-tissue contrast in human specimen. *Invest. Radiol.*, 45(7), 445–52.

Fahrig, R., Ganguly, A., Pelc, N. et al. 2005. Performance of a Static-anode/flat-panel X-ray fluoroscopy system in a diagnostic strength magnetic field: A truly hybrid X-ray/MR imaging system. *Med. Phys.*, 32(6), 1775–84.

Fredenberg, E., Cederstrom, B., Danielsson, M. et al. 2010. Contrast-enhanced spectral mammography with a photon-counting detector. *Med. Phys.*, 37(5), 2017–30.

Hejazi, S. and Trauernicht, D. P. 1997. System considerations in CCD-based x-ray imaging for digital chest radiography and digital mammography. *Med. Phys.*, 24(2), 287–97.

Hsieh, J. 2003. *Computed Tomography: Principles, Design, Artifacts and Recent Advances*. Bellingham, WA: SPIE Press.

IEC Standard 61267. 2005. Medical diagnostic x-ray equipment—Radiation conditions for use in the determination of characteristics. Geneva, Switzerland.

ImPACT Report CEP08007. 2009. *Buyers' Guide: Multislice CT Scanners*. The ImPACT Group, St. Georges Healthcare Trust, Medical Physics Department, Bence Jones Offices, Perimeter Road, Tooting, London.

Johns, P. C. and Yaffe, M. J. 1987. X-ray characterization of normal and neoplastic breast tissue. *Phys. Med. Biol.*, 32(6), 675.

Kagadis, G. C., Loudos, G., Katsanos, K., Langer, S., and Nikiforidis, G. C. 2010. *In-vivo* small animal imaging: Current status and future prospects. *Med. Phys.*, 37(12), 6421–42.

Kak, A. C. and Slaney, M. 2001. *Principles of Computerized Tomographic Imaging*. Philadelphia, PA: SIAM.

Kalender, W. A. 2005. *Computed Tomography: Fundamentals, Systems Technology, Image Quality, Applications*. Erlangen: Publicis Corporate Publishing.

Lewellen, T. K. 2008. Recent developments in PET detector technology. *Phys. Med. Biol.*, 53(17), R287.

Lubberts, G. 1968. Random noise produced by x-ray fluorescent screens. *J. Opt. Soc. Am.*, 58(11), 1475–83.

Luhta, R., Chappo, M., Harwood, B., Mattson, R., Salk, D., and Vrettos, C. 2006. A new 2D-tiled detector for multislice CT. In Flynn, M. J. and J. Hsieh (Eds.), *Medical Imaging 2006: Physics of Medical Imaging*, Proceedings of SPIE Vol. 6142, Bellingham, WA: SPIE, 6142OU.

Motz, J. W. and Danos, M. 1973. Image information content and patient exposure. *Med. Phys.*, 5(1), 8–22.

Naday, S., Bulard, E. F., Gunn, S. et al. 2010. Optimised breast tomosynthesis with a novel CMOS flat panel detector. In Marti, J., Oliver, A., Freixenet, J., and Marti, R., (Eds.), *Digital Mammography, 10th International Workshop IWDM 2010*, pp. 428–435. New York: Springer. ISBN 978-3-642-13665-8.

Natterer, F. 2001. *The Mathematics of Computerized Tomography*. Philadelphia, PA: SIAM.

Neitzel, U., Maack, I., and Gunther-Kohfahl, S. 1994. Image quality of a digital chest radiography system based on a selenium detector. *Med. Phys.*, 21(4), 509–16.

Nikiforidis, G. C., Sakellaropoulos, G. C., and Kagadis, G. C. 2008. Molecular imaging and the unification of multilevel mechanisms and data in medical physics. *Med. Phys.*, 35(8), 3444–52.

Piccaro, M. F. and Toker, E. 1993. Development and evaluation of a CCD based digital imaging system for mammography. *SPIE Cameras, Scanners, Image Acquis. Syst.*, 1901, 109–19.

Pickens, D. 2000. Magnetic resonance imaging. In Beutel, J., Kundel, H., and Van Metter, R. L. (Eds.), *Handbook of Medical Imaging:Vol.1 Physics and Psychophysics*, pp. 373–458. Bellingham, WA: SPIE Press.

Pisano, E. D. and Yaffe, M. J. 2005. Digital mammography. *Radiology*, 234(2), 353–62.

Prokop, M., Neitzel, U., and Schafer-Prokop, C. 2003. Principles of image processing in digital chest radiography. *J. Thoracic Imaging*, 18, 148–64.

Qian, X., Rajaram, R., Calderon-Colon, X. et al. 2009. Design and characterization of a spatially distributed multi-beam field emission x-ray source from stationary digital breast tomosynthesis. *Med. Phys.*, 36(10), 4389–99.

Ronda, C. 2008. *Luminescence*. Weinheim: Wiley-VCH.

Rowlands, J. A. 2002. The physics of computed radiography. *Phys. Med. Biol.*, 47(23), R123–166.

Samei, E., Saunders, R. S., Lo, J. Y. et al. 2004. Fundamental imaging characteristics of a slot-scanned digital chest radiographic system. *Med. Phys.*, 31, 2687–98.

Smans, K., Zoetelief, J., Verbrugge, B. et al. 2010. Simulation of image detectors in radiology for determination of scatter-to-primary ratios using Monte Carlo radiation transport code MCNP/MCNPX. *Med. Phys.*, 37, 2082–91.

Swank, R. W. 1973. Absorption and noise in x-ray phosphors. *J. Appl. Phys.*, 44, 4199–203.

Van Metter, R. and Dickerson, R. 1994. Objective performance characteristics of a new asymmetric screen-film system. *Med. Phys.*, 21(9), 1483–90.

Villiers, M. and Jager, G. 2003. Detective quatntum efficiency of the Lodox system. *SPIE Phys. Med. Imaging*, 5030, 955–60.

Wang, J., Qu, M., Leng, S., and McCullough, C. H. 2010. Differentiation of uric acid versus non-uric acid kidney stones in the presence of iodine using dual-energy CT. *SPIE Phys. Med. Imaging*, 7622, 76223O-1–9.

Williams, M. B., Raghunathan, P., More, M. J. et al. 2008. Optimization of exposure parameters in full field digital mammography. *Med. Phys.*, 35(6), 2414–2423.

Zysk, A. M., Nguyen, F. T., Oldenberg, A. L. et al. 2007. Optical coherence tomography: A review of clinical development from bench to bedside. *J. Biomed Opt.*, 12(5), 051403–424.

11

Efficient Database Designing

John Drakos
University of Patras

11.1 Introduction

By the term Database in Informatics, we refer to a structured collection of records which are stored in a computer. Every database is based on specific software for managing and storing the data. The architecture under which a Database is built characterizes its structure or "data model." The most commonly used data model is the Relational Model.

The software for managing a Database is known as Data Base Management Systems (DBMSs) and is usually categorized according to the data model it supports. A Relational Data Base Management System (RDBMS), for example, is the software for managing a Relational Database. The model tends to determine the query languages that are available to access a database.

While the terms "Database" and DBMS refer to different concepts, they are often misused. One example of DBMS is MySQL, whereas an example of a database is the medical record.

11.2 History

The term Database (Wikipedia) was used for the first time in November 1963 when System Development Corporation (SDC; the first software company, in Santa Monica, CA) organized a symposium titled *Development and Management of a Computer-centered Data Base.*

The first DBMS (Swanson, 1963) was developed in the 1960s. Charles Bachman, a pioneer in this field, started research for the most effective usage of the new, random access, storage mediums that arose in the beginning of the 1960s. Until then, data management was accomplished using punch cards or magnetic tapes, so sequential access was the only way for reading the data.

Two major data models were created in 1960: The Network Model, based on Bachman's ideas, was developed by Conference on Data Systems Languages (CODASYL, an Information technology industry consortium formed in 1959) and the hierarchical model was developed by North American Rockwell and was quickly adopted by IBM (company world headquarters).

The two largest databases, or more properly DBMS, conquering the 1960s, were the Information Management System (IMS, developed by IBM and based on the hierarchical model) and the Integrated Database Management System (IDMS, developed by CODASYL) and based on the network model. Several databases were also born in the same decade and are still being used even today. Pick (a demand-paged, multiuser, virtual memory, time-sharing operating system based around a unique database) and Massachusetts General Hospital Utility Multi-Programming System (MUMPS created in the late 1960s, originally for use in the healthcare industry, was designed for the production of multiuser database-driven applications) are worth mentioning, as they were initially developed as operating systems with embedded DBMSs and were later transformed into platforms for developing health-oriented databases.

The relational model was proposed by E. F. Codd in 1970. The main criticism of Codd on the existing models was about causing confusion between the substantial representation of information and the physical form of data storage. At first, only academics were interested in the relational model, due to its hardware and software requirements, which were not available in early 1970s. However, the comparative advantages of the relational model kept its theoretical development alive, hoping that future technologies will allow its use in production environments.

Among the first implementations of the relational model is Ingres that was developed by Michael Stonebraker in the University of Berkeley and System R that was developed by IBM. Both efforts were genuine research projects, presented in 1976. The first commercial products, based on the relational model, named Oracle (Software Development Laboratories) and DB2 (IBM), appeared on the market in 1980 and were built for supercomputers. dBASE was the first DBMS for personal computers (PC), published by Ashton-Tate (a US-based software company) for CP/M (Control Program for Microcomputers, an operating system originally created for Intel 8080/85-based microcomputers by Gary Kildall of Digital Research, Inc.) and MS-DOS (MicroSoft Disk Operating System, an operating system for x86-based PC) operating systems.

In the 1980s, research activity was focused on distributed Databases. An important theoretical achievement of that era was the Functional Data Model, but its implementation was limited to the specific areas of genetics, molecular biology, and investigating financial crimes. As a result, this model never became widely known.

Research interest was moved to object-oriented Databases in the 1990s. Such Databases gained partial success in areas of interest that required dealing with more complex data than the ones easily managed by relational Databases. Examples of areas that benefited from object-oriented Databases are Geographical Information Systems (GIS, spatial databases), engineering applications (i.e., software or document repositories), and multimedia applications. Some of the ideas introduced by object-oriented Databases, for data management, were adopted by the manufacturers of relational Databases and appeared on their products. During the 1990s, open source DBMSs like PostgreSQL (http://www.postgresql.org/) and MySQL (http://www.mysql.com/) appeared for the first time.

During the 2000s, the innovation "trend" in data management was XML databases. As with basic features of object-oriented Databases, companies developing relational model Databases adopted the key ideas of the XML model.

11.3 The Data Base Management System Concept

A DBMS (Kroenke and Auer, 2007) is a set of software programs controlling organization, storage, management, and retrieval of data in a database. DBMSs are classified in accordance to their data structures or types. They receive data requests from an application and instruct the operating system to transfer the appropriate data. All queries and responses must be submitted and received in a format conforming to one or more specific protocols.

A DBMS consists of many subsystems, each one responsible for addressing different tasks.

11.3.1 Engine

A DBMS Engine accepts logical requests from various DBMS subsystems, converts them into physical equivalents, and directly accesses the database and data dictionary as they exist on a storage device.

11.3.2 Data

11.3.2.1 Definition

The Data Definition Subsystem helps the user to create and maintain the data dictionary as well as to define the file structure in a database.

11.3.2.2 Management

The Data Management Subsystem helps the user to add, change, and/or delete information in a database and also to query it for stored information. The primary interface between the user and the information contained in a database is one of the software tools available within the data management subsystem. It allows users to specify their logical information requirements.

11.3.3 Administration

The Data Administration Subsystem helps users to manage the overall database environment by providing the necessary tools for backup and recovery, security management, query optimization, concurrency control, and general tasks.

11.3.4 Developing Tools

The Application Generation Subsystem (AGS) contains tools necessary for developing transaction-intensive applications. The AGS facilitates easy-to-use data entry screens, programming languages, and interfaces.

11.3.5 Common Features

11.3.5.1 Answering Questions

Querying is the process of requesting information from various sources and combinations of data. For example a typical query could be: "How many patients are adults and have low white cell count?" A database query language and report writer allows users to interactively interrogate the database, analyze its data and update it in accordance with the user privileges.

11.3.5.2 Making Calculations

Many common computations regarding data stored in a database, such as counting, summing, averaging, sorting, grouping, and cross-referencing can be provided directly from the DBMS, relieving higher level applications from the task of implementing such functions.

11.3.5.3 Enforcing the Law

Applying rules to data is necessary so that data are clean and reliable. For example, one may set a rule that says each patient can have only one Social Security Number (SSN). If somebody tries to associate a second SSN with a specific patient, the DBMS will deny such request and respond with an error message.

11.3.5.4 Authentication, Authorization, and Auditing

It is always necessary to limit users who can read and/or change specific data or groups of data. This may be entirely managed by an individual (a DB administrator), or by the assignment of individuals to groups with specific privileges, or through the assignment of individuals and groups to roles which are then granted entitlements (in the most elaborate models).

It is also often needed to record who accessed what data, what was changed, and when it was changed. Logging services allow this by keeping a record of access occurrences and changes.

11.3.5.5 Making Life Easier and More Secure

In case there are frequently occurring usage patterns or requests, some DBMS can adjust themselves to improve the speed of such interactions. In some cases, the DBMS will merely provide tools to monitor the performance, allowing a human expert to make the necessary adjustments after reviewing the collected statistics.

11.4 Common Database Models

There are several methodologies used for building the data models.

Most DBMSs support a particular model, although increasingly more systems support more than one model.

11.4.1 The Hierarchical Model

In a hierarchical model (Teorey et al., 2005), records are organized in a tree structure (Figures 11.1 and 11.3), which includes a unique upward link for each data node. That is, each node is allowed to have many child nodes, but a sole parent node. Additional information is stored in the nodes of every level in order to classify the records.

11.4.2 The Network Model

The network model (Teorey et al., 2005) stores records in nodes (Figures 11.2 and 11.3), which retain links with nodes that are directly related. That is, every node can have multiple "children" and "parents."

11.5 The Relational Model

The basic structure of the relational data model (Teorey et al., 2005) consists of a table that stores data, in columns and rows, under which a particular entity is referred (i.e., patient demographics). Columns in each table depict different characteristics of each entity (i.e., File No., Name, Address, Phone, Sex, Age, Habits, etc.), while each row depicts/reflects a snapshot of a particular entity (i.e., 1524, John Williams, Sunset Blvd 86, 2610123456, Male, 32, nonsmoker). In other words, every row of "patient demographics" represents various characteristics of a particular patient.

All tables in a relational Database must follow certain basic rules, axioms of the mathematical theory of the relational model:

- The order that columns and rows will be arranged does not affect the described entity.
- Every column contains a single data type.
- There should not be two identical rows in a table.
- Every row must contain unique values for each attribute (column). Therefore, each cell should store only one value.

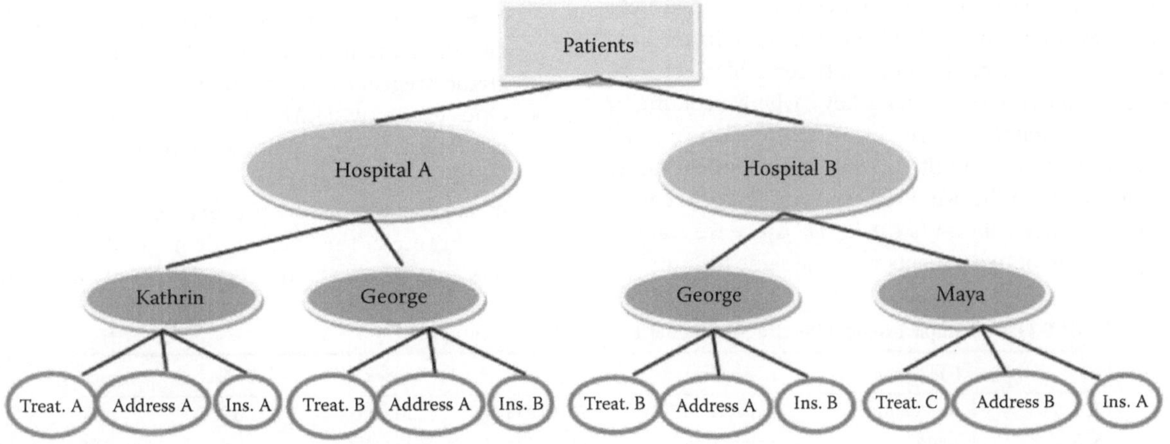

FIGURE 11.1 Sample chart of Hierarchical model. Notice the data recurrence. ("George," a common patient between the two hospitals. "Address A," a common address between Kathrin and George and some insurance companies.)

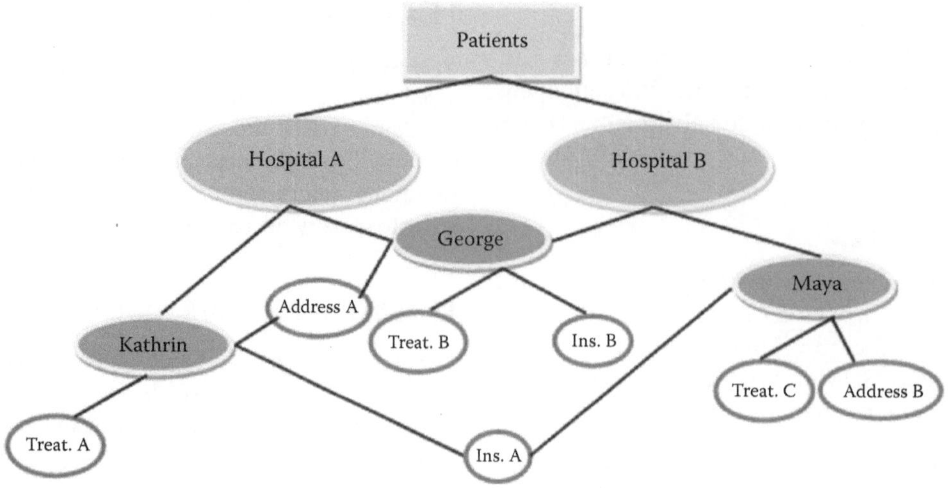

FIGURE 11.2 Sample chart of Network model. Notice that data recurrence is not present in this model, but chart's complexity and readability is highly boosted.

Although these rules are not obligatory, regarding the limitations of each RDBMS, they dramatically increase efficiency and constitute the definition of a relational Database.

Most relational Databases contain more than one table, each of which follows a flat hierarchy (the order of columns and rows does not affect the quality of information that its data represent). The relational model's key point, provided by the four basic rules, is that values contained in two different records (rows), of the same or different tables, automatically define the relation between the two records. That is, the tables of "patient demographics" (Table 11.1) and "biochemical blood tests" (Table 11.2).

Anyone can easily notice that a common link between the two tables is the column "PID" (Patient ID). By reading the "PID" of a specific patient from the demographics table, we can retrieve the patient's tests from the biochemical table and vice versa.

Despite its simplicity, a relational data model may raise data integrity problems like having multiple patients with the same "PID" or assigning some tests to nonexistent "PIDs."

It is notable that none of the above integrity problems violate the basic rules of the relational data model. In order to achieve more accurate data integrity, RDBMS allow the establishment of rules, which impose stringent relations between tables.

The first set of rules are the primary keys, where a column or a group of columns define a unique identifier for each row and do not allow the creation of multiple rows with identical values in the column/s defining the primary key. At the "patient demographics" table the key column is the "PID," since we want to avoid the registration of two patients with the same file number. At the "blood tests" table, we could use for primary key the combination of the "PID" and "date" columns (in case we do not need multiple records for the same patient in one specific date).

The second set of rules defines a parent–child relation between columns of two tables. The columns involved in a parent–child relation must store the same data type. If the rule includes only the primary key of the parent table and the corresponding column or columns of the child table, then it is called primary-foreign key relation. The column or columns participating in the rule, on the child table side, are considered as the foreign-key and in order to obtain a value, it must already exist in the corresponding primary-key column or columns of the parent table.

The parent–child relations in detail:

- One to one (1:1). Every row of the parent table is allowed to correspond with one and only one row of the child table and vice versa. For example, the "patient demographics" table and the table that stores the historical average of the blood tests (Table 11.3). Both tables contain at most one row for each patient and are connected through the "PID" column. The "blood test average" table is the child table, and therefore it is not allowed to contain PID values not already registered in the "patient demographics" table.
- One to many (1:*M*). Each row of the parent table is allowed to correspond with more than one rows of the child table, but every row of the child table is allowed to correspond with only one row of the parent table. For example, the "patient demographics" table and the one that stores the detailed values of the "blood tests."

TABLE 11.1 Sample Medical Record Storing Patients' Basic Demographic Data

PID	Full Name	Address	Phone	Sex	DoB	Smoker
1524	John Phantom	Street A, 1172	2610123456	Male	1/1/1970	No
1525	Helen Doe	Street B, 882	2101234567	Female	2/2/1954	Yes
1526	Stacy Rio	Street C, 2371	2310123456	Female	3/3/1981	Yes

Note: DoB, date of birth; PID, patient ID.

TABLE 11.2 Sample Blood Test Results

PID	Date	HCT	WBCC	LDH	PLT	TKE
1524	10/1/2008	40	12.000	248	198	33
1525	15/1/2008	37	1.500	77	135	15
1526	16/1/2008	35	2.300	114	126	20
1524	10/2/2008	41	9.000	257	205	35
1524	10/3/2008	40	10.000	212	199	32

The first table includes one unique row for every patient, whereas the second one includes one row for every examination of each patient. The link is established again through the "PID" column. "Patient demographics" is the parent table, thus, registering a blood test to a nonexistent file number is impossible.

- Many to many (*M:M*). Each row of the parent table is allowed to correspond with many rows of the child table and vice versa. The detailed tables of blood tests and urine tests are an example.

Table relations are either originating from the data structure or imposed by the rules. They describe the way of linking of different entities, which, when combined, create logical tables that describe more complex entities. That is, the demographics table combined with the blood test and the urinal test tables describe partially the patient's case. Moreover, they provide information which none of the tables can provide alone.

11.5.1 Relational Transaction

Searching data among a relational Database is accomplished by querying. Such queries are written in a special language, originally named SeQuel Language (Gray and Reuter, 1992) and then Structured Query Language (SQL). Although SQL was originally targeted to end-users, now is increasingly often replaced by high-level applications that undertake compiling SQL queries on behalf of the user. Most Web applications run SQL queries to create the content of their Web pages, depending on the visitors' requests. When a visitor of Wikipedia searches for an article using keywords, the Information System will be responsible to compile a particular query, then send it to the Database and then format and display the results on the visitor's screen.

The basic SQL command upon which all data-retrieval queries are built is "SELECT." The syntax of the command "SELECT" is as follows:

`SELECT` Column1, Column5, Column3, ... (names of columns separated with a comma or with symbol* to display all the columns) `FROM` Table1 (the name of the table that includes the records we search for) `WHERE` Column1 = 1528 and Column3 > "1/1/1980" and ... (the conditions that the requested records should meet) `ORDER BY` Column1 `ASC` (the column based on which the results will be sorted, ascending "ASC" or descending "DESC")

TABLE 11.3 Table Containing the Average Blood Test Measurements

PID	HCT	WBCC	LDH	PLT	TKE
1524	40,5	10.333,3	239	200,7	33,3
1525	37	1.500	77	135	15
1526	35	2.300	114	126	20

Examples of queries using the SELECT command:

- Selecting all the fields (columns) and all the records (rows) of the "patients" table.
```
SELECT * FROM Patients;
```
- Choosing all the fields of the patients who were born after 1979.
```
SELECT * FROM Patients WHERE d.o.b.] >
   ['31/12/1979'
```
- Choosing "PID," "name," and "age" of the patients who were born after 1979 and sorting the records by "name."
```
SELECT [File No.], Name, [d.o.b.] FROM
   Patients WHERE [d.o.b.] > '31/12/1979'
   ORDER BY name ASC
```
- Choosing "PID," "name," and "age" of patients whose name includes the term "John," born after 1979 and sorting the records by "name," descending
```
SELECT [File No.], Name, [d.o.b.] FROM
   Patients WHERE [d.o.b.] > '31/12/1979'
   AND name LIKE '*John*' ORDER BY Name
   DESC
```

Note that character *, in the syntax of command LIKE, is used for searching text patterns. '*John *' = includes the term "John," 'John *' = starts with the term "John," '*John' = ends with the term "John."

11.5.2 The Object-Oriented Model

Recently the object-oriented model appeared among the field of Databases. Object-oriented Databases tend to reduce the distance that separates the real world from the Databases world; they accomplish this by storing directly in a Database the objects that are presented by the real-world applications. This way, the additional cost of converting the stored information (table rows) back to the real-world data is reduced. In other words, a medical image is stored in an object-oriented Database in the same format (i.e., TIFF or DICOM) that the real-world display application requires to project it. Finally, object-oriented Databases introduce the basic ideas of object-oriented programming, like encapsulation and polymorphism.

11.6 Database Engineering

11.6.1 Physical Structure and Storage

Database tables and data are usually stored in hard disks and occupy a large portion of the main memory during operation. The usual physical structures that Database storage files follow are: sorted or not, files of flat hierarchy, Indexed Sequential Access Method (ISAM), Heaps, Hash Buckets, and B+ Trees. The selection of each structure is undertaken by the DBMS and the most common are B+ (define) trees and ISAM (Speel out and define) method.

Other important design choices related to the physical structure of a Database are grouping of related data (clustering), creation of preprocessed views, and segmentation of data depending on a range of values (partitioning).

All the techniques that are described in the previous two paragraphs concern "internal" features of Databases and their optimization. They do not interact in any case with the end user.

11.6.2 Indexing

Every Database, regardless of the model it is structured on, uses indexing techniques (Lightstone et al., 2007) to increase its performance. The need for indexing rises because of the way the records of a Database are stored. In order for a database to be able to receive large amounts of new records and modifications on short notice (the record for serving requests is 4,092,799 requests per minute), most DBMSs do not consume resources during creating new records and store data in a serial way (Table 11.4).

Then, to accelerate the process of searching and retrieving data they create "indexes," for the columns that are most commonly used among searches.

In the example above (Tables 11.5 and 11.6), searching for "file" 1522 will be made using the sorted file numbers and not the real data. The next step of the search will be to find the position of the file (row 3) and then the data will be retrieved from the physical storage medium. Using the same technique, searching for the name XXXXX will be held on the index of column "name" and given its alphabetical sorting, the search will end (without any records) after the first step (because XXX < YYYY).

The indexing method of this sorted column, that was just described, is the most known, but has been now overtaken technologically by indexing methods that are able to achieve faster searches. The most commonly used "indexing" methodologies are binary trees (B-Trees) and Hash Tables.

To sum up, indexes are flexible structures created for the columns on which we are searching usually for records and point the physical position (usually the row number) where each record is stored.

The disadvantages of using "indexes" for improving the efficiency of a Database are three:

- Increased storage space for the same amount of data (table storage space + "index" storage space).
- Reduction of free system memory, since during operating a Database, large amount of "indexes" are transferred to the main memory.
- The delay, introduced by the recalculation of the "index" when a new record is created or when one is modified.

Excessive use of "indexes" leads to a decrease, instead of an increase of the efficiency of a Database and for this reason they should be used after analyzing certain needs.

"Primary Keys," except for the functions that were described during the previous section, serve also as an index for the column or columns that created them ("primary key" = pointer + constraint for unique records).

11.6.3 Transactions and Concurrency

By the term "transaction," we refer to a set of commands, which were selected arbitrarily by the user and must be handled as a single process. For example, the following sequence of commands:

```
Delete all the patients with date of birth
after "1/1/1980";
Show the patients' average age;
```

is a transaction. As a transaction, it could also be considered every command on its own or a larger sequence of commands. It depends exclusively on the grouping that the user will define, to determine the way the commands will be executed.

Most transactional Databases are trying to impose the rules of Atomicity, Consistency, Isolation, and Durability (ACID) model to the commands of the users and to the way they are executed. The rules of the ACID model that define the transactions in Databases follow in the consequent subsections.

11.6.3.1 Atomicity

Either all the commands of a transaction will be executed successfully or none of them should be. In case one of the commands fail, a rollback must be possible so that the database will return to its state before the transaction. Otherwise the data will be in a random state (i.e., partial completion of a "delete" command).

11.6.3.2 Consistency

Every transaction must comply with all the rules and restrictions set during the creation of tables and their in-between relations (primary—foreign keys, field lengths, data types, unique or not values, etc.).

11.6.3.3 Isolation

Two transactions running simultaneously must not interfere, in any way, with one another (i.e., modifying shared data). The intermediate results of a transaction are not visible to others.

TABLE 11.4 Table Rows Displayed in the Serial They Are Stored in a Database

PID	Name	Address	Phone	Sex	DoB	Smoker
1524	John Phantom	Street A, 1172	2610111111	Male	1/1/1970	No
1525	Helen Doe	Street B, 882	2103333333	Female	2/2/1954	Yes
1522	John Rock	Street D, 3821	2310555555	Male	3/3/1981	Yes
1411	Peter Mountain	Street E, 2216	2102222222	Male	4/4/1964	No
2480	Stacy Rio	Street C, 2371	2610444444	Female	5/5/1990	Yes

TABLE 11.5 Index Sample for Column "PID"

PID	Row Number
1411	4
1522	3
1524	1
1525	2
2480	5

11.6.3.4 Durability

The successfully completed transactions are not able to be cancelled. The results of the successful transactions must be preserved even in a case of a planned or not restart of the DBMS.

In practice, most of DBMSs allow selective looseness of the ACID rules to increase efficiency.

11.6.3.5 Parallelism

The control of parallel operations running in a Database targets to the safe execution of the transactions and compliance with the ACID rules.

11.7 The Database Schema

The schema (Kroenke, 1997) (pronounced skee-ma) of a database system is its structure described in a formal language supported by the DBMS.

11.7.1 Levels of Database Schema

11.7.1.1 Conceptual Schema

A conceptual schema (Halpin, 1995) or conceptual data model is a map of concepts and their relationships. It describes the semantics of an organization and represents a series of assertions about its nature. Specifically, it describes the things of significance to an organization (entity classes), about which it is inclined to collect information, and characteristics of (attributes) and associations between pairs of those things of significance (relationships).

Since a conceptual schema represents the semantics of an organization, and not a database design, it may exist on various levels of abstraction. The original ANSI four-schema architecture began with the set of external schemas that represent one person's view of the world around him or her. These are consolidated into a single conceptual schema that is the superset of all those external views. A data model can be as concrete as each person's perspective, but this tends to make it inflexible. If a person's world changes, the model must consequently change.

TABLE 11.6 Index Sample for Column "Name"

Name	Row Number
Stacy Rio	5
John Rock	3
John Phantom	1
Helen Doe	2
Peter Mountain	4

Conceptual data models take a more abstract perspective, identifying the fundamental things, things that an individual deals with. The model allows also inheritance in object-oriented terms. The set of instances of a specific entity class may be subdivided into entity classes in their own right. Thus, each instance of a subtype entity class is also an instance of the entity class's supertype. Each instance of the supertype entity class, then is also an instance of one of the subtype entity classes. Supertype/subtype relationships may be exclusive or not. A methodology may require that each instance of a supertype may only be an instance of one subtype. Similarly, a supertype/subtype relationship may be exhaustive or not. It is exhaustive if the methodology requires that each instance of a supertype must be an instance of a subtype.

11.7.1.2 Logical Schema

A Logical Schema is a data model of a specific domain problem expressed in terms of a particular data management technology. Without being specific to a particular database management product, it is object-oriented classes, or XML tags (in terms of either relational tables or columns). This is, as opposed to a conceptual data model, what describes the semantics of an organization without reference to technology, or a physical data model, which describe the particular physical mechanisms used to capture the data in a storage medium.

11.7.1.3 Physical Schema

The next step in creating a database, after the logical schema is produced, is to create the physical schema. Physical Schema is a term used in relation to data management. In the ANSI four-schema architecture, the internal schema was the view of data that involved data management technology. This was opposed to the external schema that reflected the view of each user in the organization, or the conceptual schema that was the integration of a set of external schemas. Subsequently, the internal schema was recognized to have two parts: The logical schema (the way data were represented to conform to the constraints of a particular approach to database management). At that time, the choices were hierarchical and network.

Describing the logical schema, however, still did not describe how physically data would be stored on disk drives. That is the domain of the physical schema. Now logical schemas describe data in terms of relational tables and columns, object-oriented classes, and XML tags. A single set of tables, for example, can be implemented in dozens of different ways, up to and including the architecture where some rows are on a computer in Cleveland and others are on a computer in Warsaw. This is the physical schema.

11.7.2 Entity Relationship Diagram

An Entity Relationship (ER) diagram (Bagui and Earp, 2003) is an abstract and conceptual data representation.

ER modeling is a database modeling method, used to produce a type of conceptual schemas or a semantic data model of a system, often a relational database, and its requirements in a

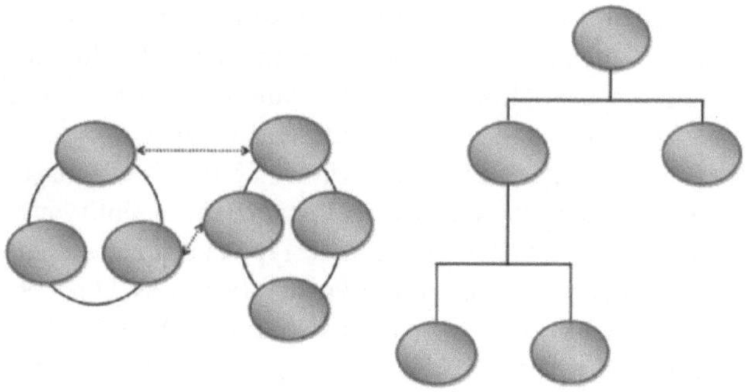

FIGURE 11.3 Comparison figures between the Network (left) and Hierarchical (right) model.

top-down approach. Diagrams created by this process are called entity–relationship diagrams, ER diagrams, or ERDs.

The building blocks of an ER diagram are the entities, the relationships, and the attributes.

11.7.2.1 Entities

An entity may be defined as something being capable of an independent existence and uniquely identified. An entity is an abstraction from the complexities of some domains. When we refer to an entity, we normally refer to some aspects of the real world which can be distinguished from others. An entity may be a physical object such as a patient or a medication, an event such as an examination or an operation, or a concept such as diagnosis or prognosis. Although the term entity is the most commonly used, we must differentiate when speaking for entities and entity-types. An entity-type is a category. An entity, strictly speaking, is an instance of a given entity-type. There are usually many instances of an entity-type. Because the term entity-type is somewhat cumbersome, most people tend to use the term entity as a synonym for this term. Entities can be considered as nouns.

11.7.2.2 Relationships

A relationship captures how two or more entities relate to each other. Relationships can be thought of as verbs, linking two or more nouns. Examples: an attendant relationship between a doctor and a patient, a treatment relationship between a medication and a disease. The model's linguistic aspect described above is utilized in the declarative database query language ERROL, which mimics natural language constructs.

11.7.2.3 Attributes

Entities and relationships can both have attributes. For example: an employee entity might have an SSN attribute; the proved relationship may have a date attribute. Every entity (unless it is a weak entity) must have a minimal set of uniquely identifying attributes, which are called the entity's primary keys.

11.7.2.4 Conventions

Entity sets are drawn as rectangles (Figure 11.4), relationship sets as diamonds. If an entity set participates in a relationship

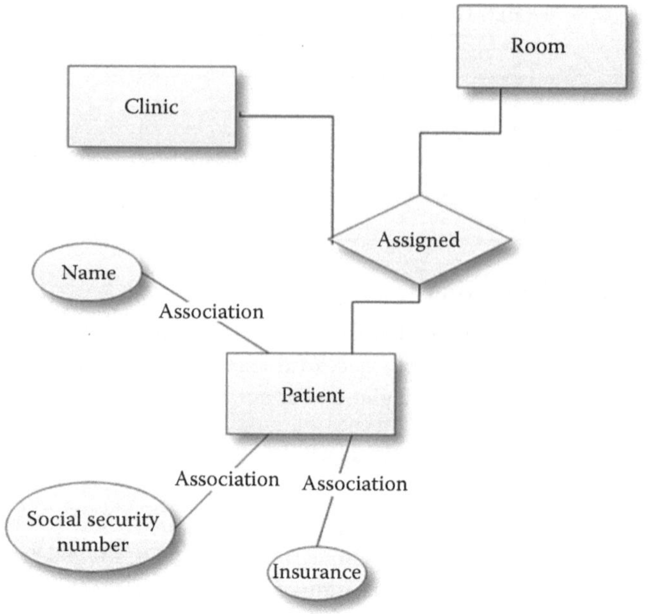

FIGURE 11.4 Sample ER diagram.

set, they are connected with a line. Attributes are drawn as ovals and are connected with a line to exactly one entity or relationship set. Cardinality constraints are expressed as follows: a double line indicates a participation constraint, totality, or subjectivity: all entities in the entity set must participate in at least one relationship in the relationship set; an arrow from an entity set to a relationship set indicates a key constraint, that is, injectivity: each entity of the entity set can participate in, at most, one relationship in the relationship set; a thick line indicates both, that is, bijectivity: each entity in the entity set is involved in exactly one relationship. An underlined name of an attribute indicates that it is a key: two different entities or relationships with this attribute always have different values for this attribute. Attributes are often omitted as they can mess up a diagram; other diagram techniques often list entity attributes within the rectangles drawn for entity sets.

11.8 Conclusions

These days the importance of Medical and Hospital Information Systems is proven beyond question. From medical practice to clinical research, concrete data storage and management system is required for success in everyday practice.

The success or failure of an information system is mainly based upon the design of the database it is built on. Filling a database with data is the easy part. However, designing a database in such way that it's fast, reliable, accurate, and productive is a challenge.

Have in mind that all Internet search engines have, more or less, the same amount of data in their databases. It is the one that has a superior design the one you use the most.

11.9 Appendix I: DBMS Examples

- 4D
- ADABAS
- Alpha Five
- Apache Derby (Java, also known as IBM Cloudscape and Sun Java DB)
- BerkeleyDB
- CSQL
- dBase
- FileMaker
- Firebird (database server)
- Hsqldb (Java)
- IBM DB2
- IBM IMS
- IBM UniVerse
- Informix
- Ingres
- Interbase
- MaxDB (formerly SapDB)
- Microsoft Access
- Microsoft SQL Server
- Model 204
- MySQL
- Nomad
- Objectivity/DB
- OpenLink Virtuoso
- OpenOffice.org Base
- Oracle Database
- Paradox (database)
- PostgreSQL
- Progress 4GL
- RDM Embedded
- ScimoreDB
- SQLite
- Superbase
- Sybase
- Teradata
- Visual FoxPro

References

Bagui, S. and Earp, R. 2003. *Database Design Using Entity-Relationship Diagrams*. Florida: Auerbach Publications.

Gray, J. and Reuter, A. 1992. *Transaction Processing: Concepts and Techniques*. California: Morgan Kaufmann Publishers.

Halpin, T. A. 1995. *Conceptual Schema and Relational Database Design*. Australia: Prentice Hall.

Kroenke, D. M. 1997. *Database Processing: Fundamentals, Design, and Implementation*. New Jersey: Prentice-Hall, Inc.

Kroenke, D. M. and Auer, D. J. 2007. *Database Concepts*. New York, NY: Prentice-Hall.

Lightstone, S., Teorey, T., and Nadeau, T. 2007. *Physical Database Design: The Database Professional's Guide to Exploiting Indexes, Views, Storage, and More*. San Francisco: Morgan Kaufmann Press.

Swanson, K. 1963. *Development and Management of a Computer-Centered Database* [Online]. Available at: http://www.dtic.mil. Accessed July 20, 2010.

Teorey, T., Lightstone, S., and Nadeau, T. 2005. *Database Modeling & Design: Logical Design*. San Francisco: Morgan Kaufmann Press.

Wikipedia. *Wikipedia* [Online]. Available at: http://www.wikipedia.org. Accessed September 1, 2010.

12

Web-Delivered Interactive Applications

John Drakos
University of Patras

12.1 History

12.1.1 World Wide Web

There are numerous ways to benefit from the Internet, either by finding information on any subject imaginable or by interacting with people and organizations from all over the world. In other words, the World Wide Web (WWW) is the way for people to share resources, at any location and time. The environment that makes this kind of interaction possible is the WWW, which is commonly known as the Web. The Web is actually a system of interlinked hypertext documents that can be accessed via the Internet. Simply by using a Web browser, one can be navigated through a wide range of text, images, videos, and other multimedia.

The foundation of the Web has been credited to the European Laboratory for Particle Physics (CERN) (WWW). However, it was the National Center for Supercomputing Applications, NCSA, that developed the tools that made it user-friendly. Suddenly, a tool created merely for research purposes became fun. Also, when Mosaic and the other graphical Web browsers were introduced, free Internet communication became a fact. The contents of a Web page vary from text information, pictures, sounds, and video to FTP links for downloading software, and much more. The documents are "alive" as never before, since they can be programmed into a weekly, daily, or even hourly refresh to meet the demands of the Web surfers. As the technology constantly evolves, it is safe to expect more and more amazing applications in the future.

12.1.2 Web 1.0

In the early 1990s, the British scientist Tim Berners-Lee (Wikipedia) built the first HTTP client for transmitting and sharing data among researchers. Since then, the WWW has gone through a fascinating path, becoming the ultimate information network that more than a billion humans use globally and making a dramatic impact on how IT and technology in general is used by society. It is universally agreed that the WWW has gone through two basic phases: Web 1.0 and Web 2.0, while we are anticipating the next wave of innovation (Web 3.0).

The period right after the creation of the WWW did not have much to show, since Web browsers were rather simple. They acted as information relay and presentation mediums while the end user ("Web surfer") could only consume the information provided and not contribute in any way. It was not long though, until rich media interactions were introduced and WWW became the first and foremost medium for information distribution and collaboration, making the way for new business models, offering unparallel networking capabilities and being the key enabler for the era of Web services and cloud computing that is emerging.

12.1.3 Web 2.0

The "New Internet" (Exforsys) or the second wave of the WWW, in other words Web 2.0 (Wikipedia), cannot be described by one specific application or technology. It explains two paradigm shifts within Information Technology: "user-generated content" and "thin client computing."

Facebook, MySpace, YouTube, blogs, vlogs, or any other Web application that enables users with no prior programming qualifications to easliy create Web contact, is what user-generated content is all about. It is practically reforming the way we use the Internet since the users are making the WWW a pool of knowledge and news that is created and reported on by "citizen journalists."

12.1.4 Web 1.0 vs. Web 2.0

- Web 1.0 sites are static.

  A personal Web page that gives information about the site's owner and never changes can be a good example of the Web 1.0 sites. There is no reason for a visitor to return to the site later (HowStuffWorks), whereas a Web 2.0 version (e.g., a blog or MySpace account) involves frequent updates by the owners.
- Web 1.0 sites are not interactive.

  Visitors can only explore the site. Most organizations prefer profile pages that the visitors can look at but not impact or alter in any way. A wiki, on the other hand, allows anyone to contribute while visiting.
- Web 1.0 applications are proprietary.

  In a Web 1.0 environment, the user is free to download an application without being able to change it or see how it works. The source code is freely accessible, however, in Web 2.0 applications (open source program). Not only can the users have a clear understanding of the application and make modifications, but also expand the existing applications with new ones. A good example of the Web 1.0 era is the proprietary Web browser Netscape Navigator, while Firefox following the Web 2.0 philosophy provides all the tools that developers need to create new Firefox applications.

Creating a Web page that visitors can impact plays an important role in the Web 2.0 philosophy (HowStuffWorks). Visitors of the Amazon Web site, for example, have the chance to post product reviews that will be later read by other visitors during their research on a product they want to buy. Although this kind of interaction can be helpful, there are cases where the Webmaster would not want the visitors to impact the Web page. It is not the place either for reviews or changes when, for example, someone is visiting the Web site of a restaurant.

12.2 Interface

The Web is being enriched with more and more applications every day. The software-as-a-service model is very appealing to companies since there are no platform constraints or installation requirements. Web application interface design is, basically, Web design that mainly focuses on function. In order to rise above desktop applications, Web apps must let their user get things done with less effort and time. Hence, they offer simple, intuitive, and responsive user interfaces (SmashingMagazine).

What a user can do with an application and how they go about doing it, in other words the interaction between a user and the application, are called Interface Design. On one hand, the Interface should translate the desires and intentions of humans into executable, logical machine instructions and, at the same time, turn computer-generated data into meaningful human-readable information.

As you might imagine, this translation is not easily achieved. While creating Web applications, Web developers need to focus on building a solid front-end as well as developing a dependable backend.

The use and manipulation of "UI Widgets" or "controls" should be the main concern, while developing a Web application. Widgets are interface tools that include buttons, scroll bars, grids, or even clickable images, and their purpose is to make the clients' intentions and desires known to the program.

The Web interface makes client functionality almost unlimited. Because of Java, JavaScript, DHTML, Flash, and other technologies, many application-specific methods are possible (e.g., playing audio, drawing on the screen, access to the keyboard, and mouse). A combination of services have been used, in order to make a more familiar interface system, while general purpose techniques such as drag and drop are also supported by these technologies.

12.3 Structure

Applications are usually divided into parts called "tiers" (Wikipedia), where every tier is assigned a role. Traditional applications once again differ from Web applications. They consist only of 1 tier residing on the client machine, while Web applications are created on an *n*-tiered approach. The most common structure is the three-tiered application, although many different versions are possible.

In the three-tiered form, the tiers are called presentation, application, and storage, in this order.

The first tier (presentation) is a Web browser.

The middle tier (application or business logic) consists of an engine that uses some dynamic Web content technology (such as ASP, ASP.NET, CGI, ColdFusion, JSP/Java, PHP, Perl, Python, Ruby on Rails or Struts2).

The third tier is a database (storage). Requests are being sent to the middle tier from the Web browser. The middle tier services the requests by making queries and updates against the database and generates a user interface.

As the complexity of the business logic grows so are the tiers of an application. In some cases, the 3-tier approach may fall short and a 4-tier or even an *n*-tiered model is the best solution. A 4-tier, in most cases, is an integration mechanism that resides between the application tier and the data tier. An example of such integration mechanism is having a group of higher level functions to access a patient's health record, rather than using a Structured Query Language (SQL) query to retrieve the patient's data. The integration tier allows the redesign or the complete replacement of the underlying database without affecting the other tiers.

Extra tiers may also derive from the scalar analysis of the business logic, which divides the application level into more than one, fine-grained, tiers.

12.4 Business Use

One of the newest trends in the software industry is to provide Web access to applications that were previously distributed as stand-alone software (Wikipedia). The transition of an application's user interface to a Web-based form may require the complete redesign of that application to be able to function as multitier service or just the replacement of the OS-based user interface with a Web-based one. Most billing scenarios of such applications are quite different from the way traditional software is charged and instead of having a one-time purchase fee they include small periodic fees. Furthermore, the end-user is able to access the Web application without having to install it on a local hard drive. A company that provides Web access to applications is known as an application service provider (ASP) and ASPs are currently receiving much attention in the software industry.

The main reason that ASPs are gaining each year a bigger piece of the software market is because they can save companies millions of dollars in software, hardware, and maintenance costs. The concept behind ASPs is the "centralized processing" or the "cloud computing." The idea of cloud computing is to have one central installation of the software in a distributed system that acts as a single computer, as far as the end-used is concerned. Cloud computing enables a company to have low-end PCs with a Web browser, instead of thousands of workstations with thousands of different copies of the software. This login behind cloud computing is not new and it dates back to mainframes of the 1960s, but only in the last few years did the ASPs become sophisticated enough to earn the trust of large companies. Today ASPs have grown to the point that they are able to provide software solutions, similar to the ones of stand-alone or client–server applications, that dramatically reduce the cost of installation, maintenance, upgrades, and support desks. Upgrades are seamless and quietly done, and problems like viral infections and conflicts over your system's registry do not affect the Web-based software.

Advantages of using ASPs:

- ASP software solutions have almost zero installation time and are simpler to maintain than conventional software.
- ASP software upgrades are transparent to the end-user.
- ASP maintenance and support requires significant fewer resources than a local IT department.
- End-users have fewer crashes, because there is no installed software that will conflict with other installed software.
- It is easier and costs less to migrate from one ASP to another than switching between conventional software solutions.

Disadvantages of using ASPs:

- Software performance and availability depends on the Internet connection.
- The feeling of the user interface of an ASP software may seem slower and clumsier than conventional software.

12.5 Architectures

In software engineering, the pull model and the push model designate two well-known approaches for exchanging the data between two distant entities (Martin-Flatin). A nice metaphor that describes both models is an ill person before and after entering the hospital. Before entering the hospital a person is communicating with a doctor to describe the symptoms of a disease and get information and instructions. After entering a hospital, the doctor is deciding whenever to give instructions and/or information to the patient. The "before" case is an example of the pull model (the patient is "pulling" data from the doctor), while the "after" case is an example of the push model (the doctor is pushing the data to the patient).

During the first years of Web-based applications, people developed the pull model by using HTML forms to standardize and automate problem reporting mechanisms for helpdesk departments. After that, the push model appeared when network administrators start using the corporate's internal Web servers to publish, electronically periodic reports that used to be printed.

Nowadays, many new technologies have appeared on the Web that made modern Web-based application reach the Web 2.0 era. Besides form and online reports, we can also use applets, Java, servlets, RMI, AJAX, scripting languages, and so on.

12.5.1 Pull-Based

The pull model is based on sending requests and getting responses and is also called data polling. When a client sends a request to the server, the server is processing the request and producing a response that corresponds to that request. After that, the server sends the response back to the client, either synchronously or asynchronously according to the design of the Web application. This is functionally equivalent to the client "pulling" the data off the server. In this approach, the data transfer is always initiated by the client.

12.5.2 Push-Based

The push model, on the other hand, is based on distribution, publication, and subscription scenarios. In this model, the server is "advertising" which part of the data and/or services it is hosting and the clients are able to "subscribe" to those services. Then, whenever a service produces data the server is pushing the data to the clients who are subscribed to the specific service. This is functionally equivalent to the server "pushing" the data to the clients. In this approach, the data transfer can be initiated either by the server or by the client.

12.6 Writing Web Applications

12.6.1 Security

People often think of Web (application) security as an area that deals with hackers messing-up Web pages, hitting sites with denial of service attacks, or finding security holes for stealing the credit card numbers (Microsoft). These common, popular, threats that all Web applications have to face are not the only security subjects that a Web developer must have in mind. An application server must be protected against computer viruses, worms and Trojan horses like any other computer and, furthermore, against venomous employees and rogue administrators. Sometimes, even a simple user that is misusing the application may be a significant threat to the system.

Web Application Security can be divided into the following subjects:

- *Authentication:* Authentication is the process of uniquely identifying the users of an application. To gain access into the system, each user must provide valid credentials, like a username/password combination, a smartcard, or some biometric information (e.g., fingerprints). An authentication paradigm from the medical world is the doctor's ID tag.
- *Authorization:* Authorization is the process of deciding what an authenticated user can or cannot do. Each time a user is trying to access a resource or perform an action, the authentication mechanism is checking if the resource or the action is available to that specific user and decides either to allow it or not. An authorization paradigm from the medical world is the security personnel that is screening which personnel may access an intensive care unit.
- *Auditing:* Detailed auditing and logging is the key to tracking and nonrepudiation. Nonrepudiation guarantees that a user cannot deny performing an operation or initiating a transaction. For example, in an HIS, nonrepudiation mechanisms are required to make sure that a doctor cannot deny prescribing a medication to a patient.
- *Privacy:* Privacy is the process of making sure that data remains private and confidential, and that it cannot be viewed by unauthorized users or get stolen by hackers. The most common and secure way to enforce privacy is an encryption mechanism. Privacy is a key concern in HIS', especially when access is permitted from remote locations.
- *Integrity:* Integrity is the process of protecting data against accidental or deliberate (malicious) modification. Like privacy, integrity is a significant factor in any HIS.
- *Availability:* Having in mind that a Web application has to be available to authenticated users at all time, the security system must guarantee such availability. For example, some malicious users may try to overload or crash an application by performing denial of service attacks, and the security arsenal of a Web application must be able to prevent system downtime.

The security of a Web application must be addressed across the full spectrum of the architecture that the application is based on (e.g., multiple tiers). A weak spot in any level will make the whole system vulnerable (Microsoft).

12.6.2 Database Access and Mapping

Most, if not all, of the modern Web services are database driven (DevShed). Web-banking, online shops and auctions, Web-based email, forums, blogs, corporate Web sites, news portals, and Web-based social networks are all build upon databases.

Information stored in a database can be presented using numerous ways through a Web server. A Web developer has many solutions to choose from, regarding the DBMS, the operating system, and the development platform.

If the application displays static information from a database that is updated periodically, then a solution is to manually create Web-presentable reports and post them on the Web.

If the application handles dynamic information that has to do with user interaction, the solution is to use a Web application server.

Web applications are most likely to be developed having a relational database as a backend. In order to access and manipulate the relational database, a standard computer language, SQL has to be used. SQL statements play an important role when developing the database application.

Taking a HIS as an example, if the doctor (end-user) wants to update a patient's prognosis record, the system has to retrieve the corresponding data from the Prognosis table and display it to the doctor. Then the doctor will make the desired changes to the record and the system has to update the record accordingly. It is noticeable that a Web application requires a lot of coding for communicating with the database and handling SQL statements so as to access and manipulate the data.

A clear picture of the importance of databases in Web applications is that developers spend almost 50% of development time for implementing SQL queries. Moreover, mapping between the persistent code and database table is maintained throughout the development life cycle. Once there is a change in the structure of a database table, SQL statements which related to the modified table have to be rewritten. Developers have to keep an eye on every change in the database schema.

12.6.3 URL Mapping

URL mapping helps you map a specified URL to another URL and automatically provide to the user the content of the second (mapped) URL (Dotnetspider). To give an example of URL mapping, let us say you have a page called "diagnosis.html" in your site for the doctors to access the diagnosis submission form. Due to some reason, you changed the name of the diagnosis page to "PatientDiagnosis.html." Using URL mapping, instead of informing the doctors that they have to update their bookmarks and start using a new URL to access the diagnosis form, you just have to map the old URL to the new one.

Advantages of URL mapping:

- End-users do not have to change their bookmarks each time there is a change to a URL on the application server.
- A big and complicated URL may be mapped to a user-friendly one.
- URL mapping can be used as an extra level of security, since the user is not able to see the real page name on the URL.

12.6.4 Web Template System

Dynamic Web pages usually consist of a static part (HTML) and a dynamic part, which is code that generates HTML (Wikipedia). The code that generates the HTML can do this based on variables in a template, or on code. The text to be generated can come from a database, thereby making it possible to dramatically reduce the number of pages in a site.

Consider the example of a hospital with 5000 patients' records. In a static Web site, the hospital would have to create 5000 pages in order to make the patients' information available to the doctors. In a dynamic Web site, the hospital would simply have to design a template container for patients' records and then connect the dynamic page to a database table of 5000 records.

In a template, variables from the programming language can be inserted without using code, thereby losing the requirement of programming knowledge to make updates to the pages in a Web site. A syntax is made available to distinguish between HTML and variables.

Many template engines do support limited logic tags, like IF and FOREACH. These are to be used only for decisions that need to be made for the presentation layer, in order to keep a clean separation from the business logic layer.

12.6.4.1 Caching

The user of a Web application is able to look for and retrieve all kinds of information, without having any knowledge of the topology of the network between the client and the server. From the user's point of view, it is not important if the desired information, for example, an HD video of an operation, is hosted on a server located inside the hospital or on the other side of the world.

An effective technique to improve the quality of service and the response times of a Web application is to decrease the network load by using a Web caching service (Forskingnett). Caching effectively migrates copies of popular documents from Web servers closer to the Web clients.

12.6.5 Asynchronous JavaScript and XML

Asynchronous JavaScript and Extensible Markup Language (XML) (Ajax) is the main standard of the software industry for developing highly responsive interactive Web applications (Sun). Ajax is a foundation technology that Web 2.0 is based upon. It is almost impossible for a developer of Web applications not be aware of Ajax. That is because Ajax is the technological key behind the success of the most popular Web applications, like Facebook, Twitter, Google maps, Hotmail, and many more. These applications are representative of the new generation (Web 2.0) of highly responsive, highly interactive Web applications that often involve users collaborating online and sharing content.

Ajax enables high responsiveness because it supports asynchronous and partial refreshes of a Web page. A partial refresh means that when an interaction event fires—for example, a user moves the cursor across a Google map—a Web server processes the information and returns a limited response specific to the data it receives. Significantly, the server does not send back an entire page to the client of the Web application—in this case a Web browser—as is the case for conventional "click, wait, and refresh" Web applications. The client then updates the page based on the response.

Asynchronous means that, after sending data to the server, the client can continue processing while the server does its processing in the background. This means that a user can continue interacting with the client without noticing a lag in response. For example, a user can continue to move the mouse over a Google map and see a smooth, uninterrupted change in the display because extended parts of the map have been loaded asynchronously. The client does not have to wait for a response from the server before continuing, as is the case for the traditional synchronous approach.

12.6.6 Web Services

Web services, like most modern Web technologies, are becoming more and more popular. Web services provide a standard means of interoperating between different software applications, running on a variety of platforms and/or frameworks (Patel). Web services are characterized by their great interoperability and extensibility, as well as their machine-processable descriptions thanks to the use of XML. They can be combined in a loosely coupled way, in order to achieve complex operations. Programs providing simple services can interact with each other in order to deliver sophisticated added-value services.

Today, WWW is full of services like search engines, social networks, email providers, online maps, traveling guides, booking sites, language translators, weather guides, dictionaries, directories, and many more.

When developing Web applications, the service developer must provide presentation logic coupled with the business logic. That is not always good. Today, not all people browse the Internet with some PC-based software. Browsers are now found in cell-phones which require specialized presentation. The user may also wish to integrate the result from a service inside his/her own software, without all the presentation stuff coming from the server. Therefore, to provide services over the Web, standards have to exist. But, whose and which standards?

To avoid the definition of proprietary interfaces and the development of compatible connectors, the Universal Description,

Discovery, and Integration (UDDI) was founded. UDDI is a platform-independent, XML-based registry for businesses worldwide to list themselves on the Internet. UDDI is an open industry initiative, sponsored by the Organization for the Advancement of Structured Information Standards, enabling businesses to publish service listings and discover each other and define how the services or software applications interact over the Internet.

12.7 Conclusions

12.7.1 Benefits

Web-based applications have evolved significantly over the past years and it is not exaggeration to say that they are now entering their mature era (Lazakidou, 2009). The integration level in conjunction with the stability and security improvements of the provided Web technologies is pushing toward the migration of many traditional software-based applications and systems to a Web-based platforms.

Below are some of the core benefits of Web-based applications.

- *Compatibility among more operating systems and hardware:* Web-based applications are by far more compatible across different operating systems than traditional software. Typically, the minimum requirement for a Web application to run is a Web browser, of which there are many (Internet Explorer, Firefox, Netscape, Chrome to name but a few). These Web browsers are available for most, if not all, of the operating systems and so whether you use Windows, Linux, Mac OS, or FreeBSD you can still run the Web application.
- *More manageable:* The installation steps of a Web-based system include only the server side and therefore the end-user has no or minimal requirements for the part of the workstation. Having all the application components on the server makes maintaining and updating the system much simpler and any client updates can be pushed to the workstation via the Web server with relative ease.
- *Easily deployable:* Owing to the manageability and cross platform support, deploying Web applications to the end user is far easier. They are also ideal where bandwidth is limited and the system and data are remote to the user. At their most deployable, you simply need to send the user a Website address to log in to and provide them with Internet access.

  This has huge implications allowing you to widen access to your systems, streamline processes and improve relationships by providing more of your customers, suppliers, and third parties with access to your systems.
- *Secure live data:* Typically in larger more complex systems, data is stored and moved around separate systems and data sources. In Web-based systems, these systems and processes can often be consolidated reducing the need to move data around.

  Web-based applications also provide an added layer of security by removing the need for the user to have access to the data and back end servers.
- *Reduced costs:* Web-based applications can dramatically lower costs due to reduced support and maintenance, lower requirements on the end user system and simplified architecture.

12.7.1.1 Advantages for Users

- No installation and updating.
- Access from anywhere with the Internet.
- Data is stored remotely.
- Cross-platform compatibility.
- More suitable for low-end computers and require little disk space.
- Client computer is better protected from viruses as the app is sandboxed inside a browser.

12.7.1.2 Advantages for Developers

- Easier to monitor every user actions, get full statistics and feedback.
- You can choose to completely control the server-side code making it impossible to pirate.
- Easier to add collaboration possibilities as data is stored on the server.
- Easier to make a mobile version if you use HTML and JS.
- Easier integration with Web services.

12.7.2 Drawbacks

The use of Web applications has a prerequisite that no user can bypass: a permanent Internet connection. Even though our days most users have access to a permanent Internet connection, this need makes a Web application vulnerable to more technical problems than a traditional one.

The development of Web applications is a complicate task that also exceeds the needs of common programming (basic Web applications). Not only does it require quite a serious amount of effort to design and develop the functions of the program, but also demands tremendous care for the development of the user-interface, using much simpler tools than those provided in the traditional graphical environments.

Analyzing a Digital Imaging and Communications in Medicine (DICOM) image online raises concerns on the evasive issue of file sharing and collaboration. What must be noticed is that Web 2.0 applications are used by accessing the data through remote Web servers. It is therefore threatening for the image, if the connection is suddenly lost or interrupted. Chances are that the analysis being done online will be lost and in extreme cases the image may become irretrievable. Another concern regarding remote data is the extra security measures needed to provoke the unauthorized access and/or data loss due to possible attacks to the Web servers.

This kind of disadvantage poses a threat to the existence of the Web applications, hence, companies such as Google and Microsoft have made preliminary solutions to this problem. However, for now, only prototypes have been developed to repress the threat raised by this problem.

12.7.2.1 Disadvantages for Users

- Traditional (desktop) applications have better user interface and usually provide more functionality to the user.
- Permanent Internet access is mandatory.
- An attack to the remote server could leak sensitive/private information.

12.7.2.2 Disadvantages for Developers

- Since the developing tools for Web application are much simpler than the ones provided for developing native application, the Web developer has a lot of restrictions and limitations.
- Web developing platforms contain less tools and frameworks like every other (relatively) newborn technology.

12.7.3 Epilogue

There is no doubt that the Web applications are growing fast and this trend will continue for the visible future. Applications that were impossible a few years ago such as browser-based DICOM analysis and editing software are now good enough for professional use. Email, collaboration, office, and project management Web apps are starting to replace desktop applications. There is a high chance, in the next years, to experience a major transition from traditional software to Web 2.0 applications. Some major players of the software industry, like Google, have already uncovered their plans of personal computers running only a Web browser instead of an operating system.

There is no scientific way for someone to predict the future, but Web applications will definitely be a part of it.

References

BASICWEBAPPLICATIONS. *Basic Web Applications* [Online]. Available at: http://basicwebapplications.com. Accessed September 1, 2010.

DEVSHED. *Database Applications and the Web, O'Reilly Media* [Online]. Available at: http://www.devshed.com. Accessed http://www.devshed.com.

DOTNETSPIDER. *Take Advantage of URL Mapping with the Help of ASP.NET 2.0* [Online]. Available at: http://www.dotnetspider.com. Accessed September 1, 2010.

EXFORSYS. *Exforsys* [Online]. Available at: http://www.exforsys.com. Accessed September 1, 2010.

FORSKINGNETT. *Web Caching Architecture, The Norwegian Research Network* [Online]. Available at: http://forskningsnett.uninett.no. Accessed September 1, 2010.

HOWSTUFFWORKS. *How Stuff Works* [Online]. Available at: http://www.howstuffworks.com. Accessed September 1, 2010.

Lazakidou, A. 2009. *Web-Based Applications in Healthcare and Biomedicine*, Sparti, Greece: Springer.

Martin-Flatin, J. P. *Push vs. Pull in Web-Based Network Management* [Online]. Available at: http://www.sscwww.epfl.ch. Accessed September 1, 2010.

MICROSOFT. *Microsoft Developer Network* [Online]. Available at: http://msdn.microsoft.com. Accessed September 1, 2010.

NCSA. *National Center for Supercomputing Applications (NCSA)* [Online]. Available at: http://www.ncsa.illinois.edu. Accessed September 1, 2010.

Patel, A. S. *Web Services Explained* [Online]. Available at: http://www.object-ideas.com. Accessed September 1, 2010.

SMASHINGMAGAZINE. *Smashing Magazine* [Online]. Available at: http://www.smashingmagazine.com.Accessed September 1, 2010.

SUN. *Sun Developer Network* [Online]. Available at: http://java.sun.com. Accessed September 1, 2010.

WIKIPEDIA. *Wikipedia* [Online]. Available: http://www.wikipedia.org.Accessed September 1, 2010.

WWW. *World Wide Web Consortium (W3C)* [Online]. Available at: http://www.w3.org. Accessed September 1, 2010.

13

Principles of Three-Dimensional Imaging from Cone-Beam Projection Data

Frédéric Noo
University of Utah

13.1 Introduction

This chapter discusses the problem of image reconstruction in modern x-ray computed tomography (CT) systems. X-ray CT aims at noninvasive visualization of structures inside a three-dimensional (3D) object using the x-ray linear attenuation coefficient (LAC) (Hubbell, 2006) as the physical parameter that distinguishes these structures from each other (Buzug, 2008; Hsieh, 2009). The LAC cannot be measured directly; access to this quantity can only be achieved indirectly by first measuring attenuation effects and then solving an inverse problem that links the desired quantity to these measurements. The solution of the inverse problem is the image reconstruction process (Defrise and Gullberg, 2006; Herman, 2009; Natterer and Wubbeling, 2007). Each measurement is a line integral of the spatial distribution of the LAC, that is, the sum of the values taken by the LAC on a line in space along which a beam of x-ray photons is transmitted through the object. The measurement is essentially obtained as the logarithm of the ratio between the number of photons that enters the object and the number of photons that exits (Buzug, 2008; Hsieh, 2009; Hubbell, 2006).

A large number of measurements are typically needed to allow accurate reconstruction of the spatial distribution of the LAC and these measurements must correspond to a wide range of line directions through the object. In medical imaging, the imaged object is rarely steady. When the object moves during data acquisition, the measurements taken before and after the motion occured are typically inconsistent; they correspond to two different LAC distributions. Such a data inconsistency prevents accurate imaging. Hence, a significant effort in x-ray CT is always being spent on speeding up the data acquisition process (Kalender, 2006). In the early age of CT, the desired line integrals were measured sequentially, one after the other, leading to long data acquisitions times, of several minutes. In modern CT scanners, 2D sets of line integrals are measured sequentially, with each set obtained in the time that was previously needed for a single line integral, thereby allowing the data acquisition for accurate full thorax or abdomen imaging in less than 10 s (Buzug, 2008; Hsieh, 2009; Kalender, 2006). Each 2D set is obtained by letting the x-ray source emit a cone-shaped beam of x-rays toward a pixelated flat-panel detector, with the patient being naturally placed between the source and the detector. The 2D x-ray image measured on the detector is called a cone-beam (CB) projection of the object. A set of CB projections is obtained by moving the source–detector assembly relative to the patient. This set represents the CB projection data from which the spatial distribution of the LAC has to be reconstructed.

Nowadays, x-ray CB tomography is extensively applied in healthcare, not only for the diagnosis of many diseases (Buzug, 2008; Hsieh, 2009; Kalender, 2006), but also to assist with minimally invasive surgical procedures (Lauritsch et al., 2006; Orth et al., 2008; Zellerhoff et al., 2005), or to monitor the treatment in radiation therapy (Cho et al., 2009; Jaffray et al., 2002). Technological advances are continuously allowing improvements in this scanning method, in terms, for example, of data quality and speed of data acquisition, so that the method is likely to see further increase in usage in the future. The main objective of this chapter is to discuss what conditions have to be met to allow accurate reconstruction, how the desired spatial distribution of the LAC can be obtained from complete CB projection data, and how the fundamental practical issue of the so-called data truncation can be solved. Many practical reconstruction theories have been developed over the years, some approximate and others theoretically exact and stable (TES). This chapter does not attempt at performing an exhaustive review of these theories. Instead, the discussion is restricted to TES methods, and more particularly to reviewing fundamental tools and to presenting from these tools a filtered-backprojection (FBP) method that is, in many data acquisition geometries, a solid starting point for the development of efficient reconstruction algorithms.

13.2 Mathematical Formulation

In this section, we first explain our mathematical notation for integrals and related tools. Then, we formulate the image reconstruction problem of CB tomography.

13.2.1 Integrals and Related Tools

Image reconstruction theory for CB tomography involves various types of integrals in the 3D Cartesian space, some over (half) lines, and others over surfaces and volumes. We always use the Lebesgue integral (Burk, 1998), and refer to any point in space by the vector that connects it to the origin. Also, all vectors are underlined to help in distinguishing them from scalars and other quantities.

Consider a scalar function f that changes its value according to position $\underline{x}$ in the 3D Cartesian space. The integral of f on the half-line L that starts at location $\underline{x}_0$ and stretches out in the direction of unit vector $\underline{\alpha}$ is

$$\int_L f(\underline{x})\,\mathrm{d}\underline{x} = \int_0^{\infty} f(\underline{x}_0 + t\underline{\alpha})\,\mathrm{d}t. \tag{13.1}$$

The integral of f on a given volume Ω is

$$\int_\Omega f(\underline{x})\,\mathrm{d}\underline{x} = \int_{(x,y,z)\in\Omega} f(\underline{x})\,\mathrm{d}x\,\mathrm{d}y\,\mathrm{d}z, \tag{13.2}$$

where (x, y, z) are the Cartesian coordinates of x. The integral of f on a given surface Σ is

$$\int_\Sigma f(\underline{x})\,\mathrm{d}\underline{x} = \int_{(u,v)\in\Gamma} f(\underline{x}(u,v))J(u,v)\,\mathrm{d}u\,\mathrm{d}v, \tag{13.3}$$

where u and v are parameters that together describe the position $\underline{x}(u, v)$ on Σ, with the understanding that a single position is associated with a fixed value of (u, v) and all of Σ is covered by letting (u, v) vary over a certain set, called Γ. Quantity $J(u, v)$ is the Jacobian of the transformation from (u, v) to $\underline{x}(u, v)$ and is given by

$$J(u,v) = \left\| \frac{\partial \underline{x}(u,v)}{\partial u} \times \frac{\partial \underline{x}(u,v)}{\partial v} \right\|. \tag{13.4}$$

Two surface integrals are of particular interest: integrals on planes and integrals on the unit sphere. Let $\Pi(\underline{n},s)$ be the plane orthogonal to $\underline{n}$ at signed distance s from the origin, with s measured positively in the direction of $\underline{n}$, as shown in Figure 13.1. Following Equation 13.3, we have

$$\int_{\Pi(\underline{n},s)} f(\underline{x})\,\mathrm{d}\underline{x} = \int_{-\infty}^{\infty} \mathrm{d}t_1 \int_{-\infty}^{\infty} \mathrm{d}t_2\, f(s\underline{n} + t_1\underline{m}_1 + t_2\underline{m}_2), \tag{13.5}$$

where $\underline{m}_1$ and $\underline{m}_2$ are two arbitrary unit orthogonal vectors that form a basis in $\Pi(\underline{n},s)$ and are thus orthogonal to $\underline{n}$. Next, let S^2 be the sphere of unit radius centered on the origin. According to Equation 13.3, integration on S^2 can be written as

$$\int_{S^2} f(\underline{\alpha})\,\mathrm{d}\underline{\alpha} = \int_{\|\underline{\alpha}\|=1} f(\underline{\alpha})\,\mathrm{d}\underline{\alpha} = \int_0^{2\pi} \mathrm{d}\phi \int_0^{\pi} \mathrm{d}\theta\,|\sin\theta|\, f(\underline{\alpha}(\theta,\phi)), \tag{13.6}$$

where θ and ϕ are the spherical coordinates of unit vector $\underline{\alpha}$ in any preferred right-handed system of Cartesian coordinates, that is

$$\underline{\alpha}(\theta,\phi) = \cos\phi\sin\theta\,\underline{w}_1 + \sin\phi\sin\theta\,\underline{w}_2 + \cos\theta\,\underline{w}_3, \tag{13.7}$$

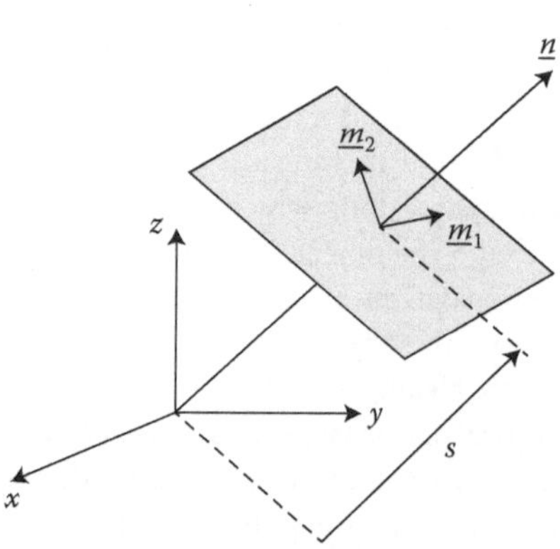

FIGURE 13.1 Depiction of $\Pi(\underline{n}, s)$, the plane orthogonal to unit vector $\underline{n}$ at signed distance s from the origin.

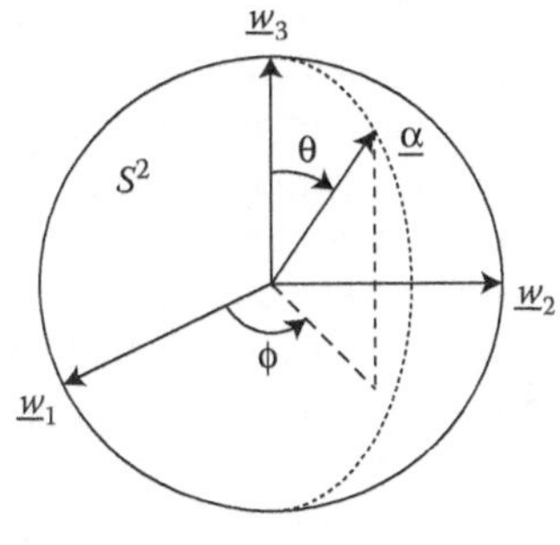

FIGURE 13.2 Spherical coordinates. In this figure, angles ϕ and θ are the spherical coordinates of unit vector $\underline{\alpha}$ relative to $\underline{w}_1$, $\underline{w}_2$, and $\underline{w}_3$, which are three mutually orthogonal unit vectors defined such that $\underline{w}_3 = \underline{w}_1 \times \underline{w}_2$.

where $\underline{w}_1$, $\underline{w}_2$, and $\underline{w}_3$ are three mutually orthogonal unit vectors defined with $\underline{w}_3 = \underline{w}_1 \times \underline{w}_2$, as shown in Figure 13.2. Note that there is considerable flexibility in the range of variation selected for θ and ϕ. For instance, instead of the ranges selected in Equation 13.6, we could have used $\theta \in [-\pi, \pi]$ with ϕ restricted to $[\pi/4, 3\pi/4]$. For such a case, $\sin\theta$ is not always positive, and the absolute value over $\sin\theta$ in Equation 13.6 is crucially needed.

An important change in variables is that from Cartesian coordinates to spherical coordinates. In our notation, the following expressions can be used

$$\begin{aligned}\int_{\mathbb{R}^3} f(\underline{x})\,\mathrm{d}\underline{x} &= \int_{S^2} \mathrm{d}\underline{\alpha} \int_0^{\infty} \mathrm{d}t\, t^2 f(t\underline{\alpha}) \\ &= \int_{S^2} \mathrm{d}\underline{\alpha} \int_{-\infty}^{0} \mathrm{d}t\, t^2 f(t\underline{\alpha}) \\ &= \frac{1}{2}\int_{S^2} \mathrm{d}\underline{\alpha} \int_{-\infty}^{\infty} \mathrm{d}t\, t^2 f(t\underline{\alpha}),\end{aligned} \tag{13.8}$$

where the integral on the unit sphere can be performed using any preferred parameterization for unit vector $\underline{\alpha}$.

Most mathematical proofs given in this chapter are formal and involve the Dirac impulse, $\delta(t)$, and its derivative, $\delta'(t)$, which satisfy the following important properties for our purposes:

- Let $k(t)$ be a smooth integrable function, then

$$\int_{-\infty}^{\infty} \delta(t-u)k(u)\,\mathrm{d}u = k(t) \tag{13.9}$$

and

$$\int_{-\infty}^{\infty} \delta'(t-u)k(u)\,\mathrm{d}u = k'(t), \tag{13.10}$$

where $k'(t)$ is the derivative of $k(t)$.

- Let $w(t)$ be a smooth function with N nontangential zeros at locations t_i, $i = 1, \ldots, N$, and let $k(t)$ be again a smooth integrable function, then

$$\int_{-\infty}^{\infty} \delta(w(t))\,k(t)\,\mathrm{d}t = \sum_{i=1}^{N} \frac{k(t_i)}{|w'(t_i)|} \tag{13.11}$$

and

$$\int_{-\infty}^{\infty} \delta'(w(t))\,k(t)\,\mathrm{d}t = -\sum_{i=1}^{N} \frac{\mathrm{sign}(w'(t_i))}{\left(w'(t_i)\right)^2} k'(t_i), \tag{13.12}$$

where $w'(t)$ is the derivative of $w(t)$, and $\mathrm{sign}(t) = 1$ if $t > 0$ and 0 otherwise.

- Homogeneity property:

$$\delta(at) = \frac{1}{|a|}\delta(t) \tag{13.13}$$

and

$$\delta'(at) = \frac{\mathrm{sign}(a)}{a^2}\delta'(t), \tag{13.14}$$

where $a \neq 0$ is independent of t, which can be used along with Taylor's series expansions to prove Equations 13.11 and 13.12 from Equations 13.9 and 13.10, respectively.

13.2.2 Data Model and Reconstruction Problem

Throughout this chapter, the spatial distribution of the LAC that is to be reconstructed is denoted as f or $f(\underline{x})$ depending on the context, and the Cartesian coordinates of the point indicated by $\underline{x}$ are (x, y, z). We assume that function f is smooth and compactly supported within a given convex set Ω. Typically, Ω is pictured as a cylinder of radius R centered on the z-axis, that is, as

$$\Omega = \left\{(x, y, z) \mid \quad x^2 + y^2 < R^2\right\}, \tag{13.15}$$

which physically corresponds to assuming that the patient is lying along the z-axis. Parameter R is commonly referred to as the field-of-view radius.

The measurements may be acquired in a step-and-shoot mode or using continuous x-ray emission. In either case, the data acquisition takes place while the x-ray source moves along a given trajectory relative to the patient. We call this trajectory the vertex path and describe any position on this path as $\underline{a}(\lambda)$ with $\lambda \in \Lambda$ being some parameter. The vertex path is required to lie outside Ω, and may consist of either one curve or a finite union of curves, each of which is assumed to be smooth and of finite

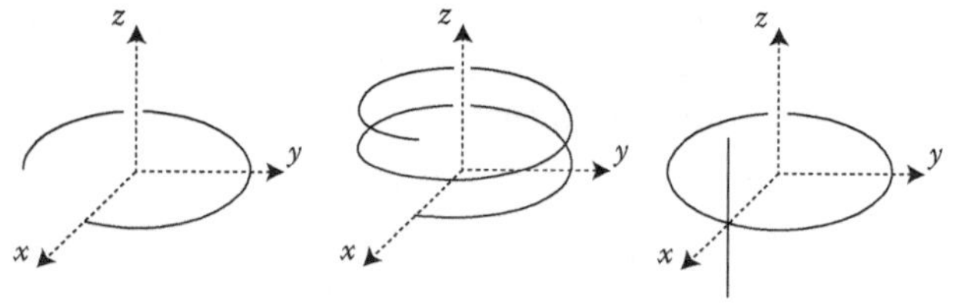

FIGURE 13.3 Examples of vertex paths. Left: the circular arc. Middle: the helical trajectory. Right: the circle-plus-line trajectory. See the text for a mathematical description of these paths.

length. A few examples of vertex paths are given below for the case where Ω is defined as in Equation 13.15; these examples are illustrated in Figure 13.3:

1. The circular vertex path of radius R_0 in the equatorial plane:

$$\underline{a}(\lambda) = [R_0 \cos\lambda, R_0 \sin\lambda, 0], \quad \lambda \in [0, \lambda_m). \tag{13.16}$$

Generally, either $\lambda_m = 2\pi$ or $\lambda_m = \pi + 2 \arcsin(R/R_0)$ is chosen. In the former case, a full scan is said to be performed. In the latter case, the expression "short scan" is used.

2. The helical vertex path of radius R_0 and pitch P:

$$\underline{a}(\lambda) = \left[R_0 \cos\lambda, R_0 \sin\lambda, P\frac{\lambda}{2\pi} \right], \quad \lambda \in [0, \lambda_m), \tag{13.17}$$

where λ_m specifies the amount of rotation being performed. For example, two full rotations are being considered if $\lambda_m = 4\pi$ is chosen.

3. The circle-plus-line vertex path (Zeng and Gullberg, 1992):

$$\underline{a}(\lambda) = \begin{cases} \left[R_0 \cos\lambda, R_0 \sin\lambda, 0\right] & \lambda \in [0, 2\pi) \\ \left[R_0, 0, (\lambda - 3\pi)H/(2\pi)\right] & \lambda \in [2\pi, 4\pi], \end{cases} \tag{13.18}$$

which consists in the union of two curves, namely a full circle and a segment of line of length H, drawn orthogonally to the circle through the point at $\lambda = 0$. If preferred, this vertex path could also be described using two distinct parameters, λ and λ', with $\lambda \in [0, 2\pi)$ parameterizing the circle as above, whereas $\lambda' \in [-H/2, H/2]$ would be used for the line segment, with $\underline{a}(\lambda') = [R_0, 0, \lambda']$.

Many other vertex paths can be found in the literature. Other popular examples include the circle-plus-many-lines (Noo et al., 1996), the circle-plus-arc (Hoppe et al., 2007; Katsevich, 2005; Wang and Ning, 1999), the circle-plus-helix (Noo et al., 1998; Yang et al., 2009), and the saddle (Lu et al., 2009; Pack et al., 2004; Yang et al., 2006; Yu et al., 2005) trajectories. The choice for a given vertex path generally depends on the application and the specifics of the imaging apparatus.

As discussed in the case of the circle-plus-line path, using more than one parameter to describe the vertex path may sometimes be preferred. The text of this chapter could have been written in a more general manner, allowing the use of a number of parameters for the vertex path, instead of a single one. However, such a choice would have unduly complicated the exposition; thus, for the sake of clarity, we opted for the use of a single parameter. The most important change to keep in mind when using more than one parameter is that all integrals in λ that will appear later in the text should be replaced by a sum of integrals, with each integral being associated with one of the parameters.

We always assume that λ is selected so that $\underline{a}'(\lambda) \neq 0$. This condition is easily understood in the context of continuous x-ray emission. If $\underline{a}'(\lambda)$ were equal to 0 at a given location on the vertex path, the x-ray source would basically be stalling at this location while continuously emitting x-rays, which is not realistic.

The CB measurements are described by the divergent-beam transform of f, which is a scalar-valued function of $\lambda \in \Lambda$ and of unit vector $\underline{\alpha} \in W_\lambda \subset S^2$. The expression for this transform is

$$g(\lambda, \underline{\alpha}) = \int_0^\infty f(\underline{a}(\lambda) + t\underline{\alpha})\, dt, \tag{13.19}$$

which means that $g(\lambda, \underline{\alpha})$ is the integral of f along the half-line that starts at $\underline{a}(\lambda)$ and stretches out in the direction of $\underline{\alpha}$. Since the vertex path lies outside Ω and Ω is convex, one of the following two relations always holds for any given $\underline{\alpha}$: $g(\lambda, \underline{\alpha}) = 0$ or $g(\lambda, -\underline{\alpha}) = 0$.

The set of values taken by function g at any fixed value of λ is the mathematical definition of a CB projection. There are two types of CB projections: complete (nontruncated) and incomplete (truncated). The projection at position λ is said to be complete when W_λ is identical to S^2 and to be incomplete when W_λ is only a subset of S^2 with $g(\lambda, \underline{\alpha}) \neq 0$ for some $\underline{\alpha} \notin W_\lambda$. In the first case, the (half) line integrals are known for all lines that diverge from the vertex point $\underline{a}(\lambda)$; in the second case, they are only known over a subset of these lines.

The CB reconstruction problem is formulated as the problem of computing f in a given region-of-interest (ROI), $\Omega_{ROI} \subset \Omega$, from knowledge of $g(\lambda, \underline{\alpha})$ for $\lambda \in \Lambda$ and $\underline{\alpha} \in W_\lambda$. Solution of this problem strongly depends on the definition of the vertex path and the expression of W_λ. To allow for an accurate reconstruction, the vertex path must satisfy Tuy's condition, which will be discussed in Section 13.3. Usually, image reconstruction from nontruncated CB projections is significantly easier than that from truncated projections. Unfortunately, projection incompleteness is a common feature in medical imaging, due, on the one hand, to the high cost of detectors and, on the other, to being most of the time only interested in a portion of the human body, which implies that the dose should be mostly focussed on this region, given the negative health effects of x-ray radiations.

Reconstruction algorithms are either iterative or analytical. In this chapter, we focus on analytical methods, which perform the reconstruction through discretization of a continuous-form formula that aims at relating the value of f over the desired ROI to the divergent-beam transform of f. Computer power has been growing fast over time, but so has the size of data sets in CT;

for this reason and others (Pan et al., 2009), analytical methods remain nowadays the preferred approach in CT and they are the focus of this chapter.

Analytical algorithms can either be TES or approximate. In a TES method, the underlying continuous-form formula exactly yields f and does it in a manner that is robust to discretization errors and to statistical uncertainties (noise) in the measurements. A TES method can only be devised when Tuy's condition is satisfied. Approximate algorithms may be used when the data set does not provide enough information for a TES reconstruction, when the computational effort for a TES reconstruction appears unattractive in comparison with what can be achieved by allowing some bias in the reconstruction, or when a TES reconstruction method that makes effective use of the data in terms of noise control remains elusive. The development of approximate algorithms often derives insight from TES methods. This chapter is restricted to the development of TES methods of the FBP type.

A CB projection is usually measured using a pixelated, flat-panel detector that is along the path of x-ray photons transmitted through the object, as depicted in Figure 13.4. Geometrically, the detector lies in a plane that is intersected by all half-lines that start at $\underline{a}(\lambda)$ and pass through Ω. During data acquisition, the detector is commonly moved together with the x-ray source so as to maintain this intersection condition. The data measured on the flat-panel detector is a function g_m of two Cartesian coordinates, u and v, measured along two orthogonal unit vectors in the detector plane, called $\underline{e}_u(\lambda)$ and $\underline{e}_v(\lambda)$. The location $(u, v) = (0,0)$ is selected at the orthogonal projection of $\underline{a}(\lambda)$ onto the detector plane, and the distance from $\underline{a}(\lambda)$ to this plane is called $D(\lambda)$. Letting $\underline{e}_w(\lambda) = \underline{e}_u(\lambda) \times \underline{e}_v(\lambda)$, and assuming that $\underline{e}_w(\lambda)$ points toward $\underline{a}(\lambda)$, we have

$$g_m(\lambda,u,v) = g\left(\lambda,\underline{\alpha}_m(\lambda,u,v)\right) \tag{13.20}$$

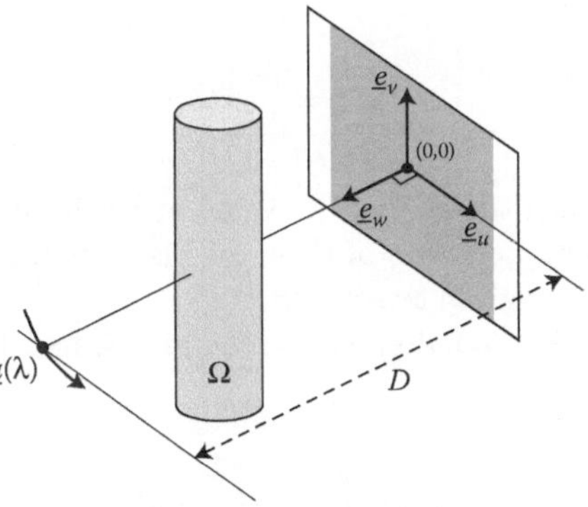

FIGURE 13.4 CB geometry with a flat-panel detector. The shaded area on the detector highlights the region of the detector where the object is projected. The detector is too short in v to capture all lines that pass through Ω, but not too short in u. The projection shown here is said to be axially truncated. Although not indicated in the figure, the vectors $\underline{e}_u$, $\underline{e}_v$, and $\underline{e}_w$ generally depend on λ, and so does D, as emphasized in the text.

with

$$\underline{\alpha}_m(\lambda,u,v) = \frac{u\,\underline{e}_u(\lambda) + v\,\underline{e}_v(\lambda) - D(\lambda)\,\underline{e}_w(\lambda)}{\sqrt{u^2+v^2+(D(\lambda))^2}}. \tag{13.21}$$

At fixed λ, function $g_m(\lambda, u, v)$ is measured on a given (u, v)-region that is defined by the (constant) size of the detector and its orientation. If $g_m(\lambda, u, v)$ is known to be zero outside that region, the projection is complete, otherwise it is truncated. The region is a simple interval $[u_1(\lambda), u_2(\lambda)] \times [v_1(\lambda), v_2(\lambda)]$ when the detector pixels are aligned along the u and v axes.

When $\underline{a}(\lambda)$ lies on a cylinder as in the examples above, it is common to choose $\underline{e}_v(\lambda)$ along the z-axis and $\underline{e}_w(\lambda)$ along the projection of $\underline{a}(\lambda)$ onto the (x, y)-plane. For the circle and the helical trajectories, this choice yields

$$\begin{aligned} \underline{e}_w(\lambda) &= \left[\cos\lambda, \sin\lambda, 0\right], \\ \underline{e}_u(\lambda) &= \left[-\sin\lambda, \cos\lambda, 0\right], \\ \underline{e}_v(\lambda) &= \left[0,0,1\right]. \end{aligned} \tag{13.22}$$

In such a case, truncation in v is usually referred to as axial truncation, whereas truncation in u is called transaxial or transverse truncation. Figure 13.4 illustrates a case of axial truncation. Usually, axial truncation is unavoidable, whereas transverse truncation is avoided. TES reconstruction with transverse truncation is only possible in very particular cases, whereas TES reconstruction with axial truncation is manageable in most situations. In this chapter, when discussing how to handle truncation, we focus on axial truncation.

In the case where the projections are truncated, it may be tempting to think that each CB projection should at least include all the lines that pass through the ROI. Such a condition is far too restrictive and should not be assumed. TES reconstruction of the ROI has been shown to be feasible in many circumstances without this condition holding.

13.3 Reconstruction from Nontruncated Projections

In this section, we review the theory established by the works of Tuy (1983) and Grangeat (1991) for image reconstruction from CB projections. This theory, developed in the 1980s, is designed for complete projections, but turned out to be valuable even for reconstruction from truncated projections. We start with Tuy's condition that identifies the vertex paths for which accurate (that is to say, TES) reconstruction from nontruncated projections is possible. Next, we discuss the 3D Radon transform and present Grangeat's formula, which allows linking the CB projection data to this transform in a local manner. This link helps understanding how Tuy's condition comes into play, although Tuy did not use Grangeat's formula. Finally, we explain how all these results can be combined together to obtain a practical reconstruction algorithm.

Note that the entire section is dedicated to complete projections. The assumption that the projections are nontruncated is implicitly made throughout the section. Thus, all statements being made are only valid for complete projections.

13.3.1 Tuy's Condition

Accurate reconstruction from complete projections is not always possible. The vertex path must be properly shaped and oriented relative to the desired ROI within the object. In 1983, Tuy established that accurate reconstruction is possible whenever the following condition is satisfied (Tuy, 1983, p. 547):

Every plane passing through the ROI must intersect the vertex path in a nontangential manner.

Later, Finch (1985) showed that this condition is not only sufficient but also necessary, except for the nontangentiality requirement, which does not affect stability, but represents nevertheless a thorny numerical problem to be careful with when dealing with discretized data.

For piecewise-smooth vertex paths, and thus, for most practical data acquisition geometries, Tuy's condition can be expressed in the following, more convenient, local form:

Accurate reconstruction at position $\underline{x} \in \Omega$ is possible if and only if every plane passing through $\underline{x}$ intersects the vertex path.

Moreover, the following result holds for any connected vertex path: Tuy's condition is satisfied for any point $\underline{x}$ that belongs to the convex hull of the vertex path (Finch, 1985).

One consequence of Tuy's condition is that accurate reconstruction using a circular vertex path is only possible within the plane of this path. For illustration, consider the vertex path of Equation 13.16 with $\lambda_m = 2\pi$ and a point $\underline{x}$ on the positive side of the z-axis, that is, $\underline{x} = (0, 0, \eta)$ with $\eta > 0$, as illustrated in Figure 13.5. Then, for any angle ϕ, all planes normal to

$$\underline{n} = \left[\cos\phi \sin\theta, \sin\phi \sin\theta, \cos\theta\right] \tag{13.23}$$

with $|\theta| < \arctan(\eta/R)$ will fail intersecting the vertex path. The larger the η, the larger the set of planes, where Tuy's condition is not met and the more difficult it is to achieve a satisfactory reconstruction. This observation is unfortunate because using a circular source trajectory is highly practical.

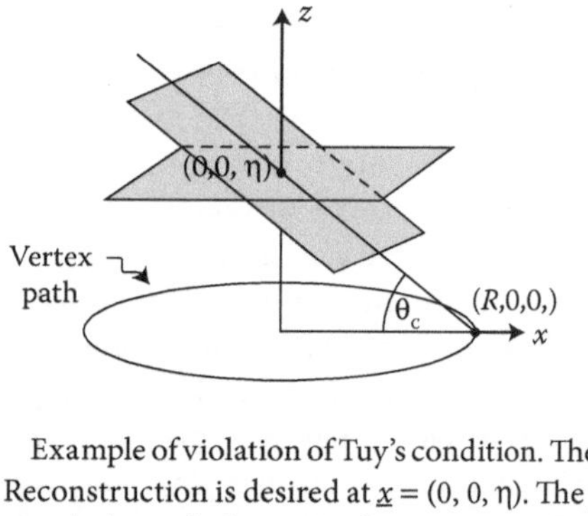

FIGURE 13.5 Example of violation of Tuy's condition. The vertex path is a full circle. Reconstruction is desired at $\underline{x} = (0, 0, \eta)$. The plane parallel to the vertex path through this point does not meet the vertex path, and so do the planes that make an angle smaller than $\theta_c = \arctan(\eta/R)$ with the z-axis.

To satisfy Tuy's condition while using a circular vertex path, it is needed to perform data acquisition on an additional segment of curve, such as a line orthogonal to the circle. This observation motivated the introduction of the circle-plus-line(s) trajectory and other similar paths. However, it is not required to have a circle as part of the vertex path for Tuy's condition to be satisfied. For example, the helical trajectory of Equation 13.17 satisfies this condition over a large region defined by the helix pitch, P, and the angular coverage, λ_m.

13.3.2 3D Radon Transform

The 3D Radon transform of f is a scalar-valued function that associates a number to each plane of $\mathbb{R}^3$, namely the integral of f on the plane (Deans, 2007). Let $\underline{n}$ be a unit vector, let $s \in \mathbb{R}$, and let $\Pi(\underline{n}, s)$ be the plane orthogonal to $\underline{n}$ at signed distance s from the origin, with s measured positively in the direction of $\underline{n}$, as shown earlier in Figure 13.1. Then, the 3D Radon transform of f can be described as a function of $\underline{n}$ and s given by

$$r(\underline{n}, s) = \int_{\Pi(\underline{n}, s)} f(\underline{x})\, d\underline{x}, \tag{13.24}$$

where the right-hand side of the equation can be rewritten as a simple 2D integral if desired, as explained by Equation 13.5.

For our purposes, the most important aspect of the 3D Radon transform is that it can be inverted, to provide $f(\underline{x})$ from $r(\underline{n}, s)$. Many inversion formulas can be formulated. Here, the discussion is restricted to a preferred expression:

$$f(\underline{x}) = -\frac{1}{8\pi^2} \int_{S^2} r''(\underline{n}, \underline{x} \cdot \underline{n})\, d\underline{n} \tag{13.25}$$

with

$$r''(\underline{n}, s) = \frac{\partial^2}{\partial s^2} r(\underline{n}, s). \tag{13.26}$$

This formula shows that the value of f at location $\underline{x}$ can be obtained from the values of r'' on the planes that contains $\underline{x}$, as these planes are given by $s = x \cdot n$.

The proof of Equation 13.25 requires using a Fourier slice theorem: let

$$\hat{f}(\underline{\nu}) = \int_{\mathbb{R}^3} e^{-i2\pi \underline{x} \cdot \underline{\nu}} f(\underline{x})\, d\underline{x} \tag{13.27}$$

be the Fourier transform of f, defined with $\underline{\nu} \in \mathbb{R}^3$, and let

$$\hat{r}(\underline{n}, \sigma) = \int_{-\infty}^{\infty} e^{-i2\pi s \sigma} r(\underline{n}, s)\, ds \tag{13.28}$$

be the Fourier transform of r in s, defined with $\sigma \in \mathbb{R}$. Then,

$$\hat{f}(\sigma \underline{n}) = \hat{r}(\underline{n},\sigma). \quad (13.29)$$

This theorem shows that the Fourier transform of f along a line of fixed direction $\underline{n}$ through the origin can be calculated by applying a Fourier transform to the values taken by r on the planes orthogonal to $\underline{n}$. Obtaining Equation 13.29 can be achieved by first using expression 13.5 for $r(\underline{n}, s)$ in Equation 13.28, which gives

$$\int_{-\infty}^{\infty} e^{-i2\pi s\sigma} r(\underline{n},s)\,ds = \int_{-\infty}^{\infty} ds \int_{-\infty}^{\infty} dt_1 \int_{-\infty}^{\infty} dt_2\, e^{-i2\pi s\sigma} f(s\underline{n} + t_1\underline{m}_1 + t_2\underline{m}_2). \quad (13.30)$$

Next, the (s, t_1, t_2)-triplet of variables is transformed into (x, y, z) according to the equation

$$\underline{x} = s\,\underline{n} + t_1\,\underline{m}_1 + t_2\,\underline{m}_2 \quad (13.31)$$

with $\underline{n}$, $\underline{m}_1$, and $\underline{m}_2$ being regarded as fixed quantities. This change of variables defines a rotation in $\mathbb{R}^3$, so the Jacobian of the transformation is equal to one, and we obtain

$$\int_{-\infty}^{\infty} e^{-i2\pi s\sigma} r(\underline{n},s)\,ds = \int_{\mathbb{R}^3} e^{-i2\pi(\underline{x}\cdot\underline{n})\sigma} f(\underline{x})\,d\underline{x}, \quad (13.32)$$

which is equivalent to Equation 13.29.

Now, we can prove Equation 13.25. First, use the inverse Fourier transform operation to write

$$f(\underline{x}) = \int_{\mathbb{R}^3} e^{i2\pi\underline{x}\cdot\underline{v}} \hat{f}(\underline{v})\,d\underline{v}. \quad (13.33)$$

Next, apply a change of variables from Cartesian coordinates to spherical coordinates, mimicking Equation 13.8 with $\underline{v} = \sigma\underline{n}$ (instead of $\underline{x} = t\underline{\alpha}$), to get

$$f(\underline{x}) = \frac{1}{2}\int_{S^2} d\underline{n} \int_{-\infty}^{\infty} d\sigma\, \sigma^2 e^{i2\pi\underline{x}\cdot(\sigma\underline{n})} \hat{f}(\sigma\underline{n}). \quad (13.34)$$

Then, the Fourier slice theorem can be invoked to obtain

$$f(\underline{x}) = \frac{1}{2}\int_{S^2} d\underline{n} \int_{-\infty}^{\infty} d\sigma \sigma^2 e^{i2\pi\sigma\underline{x}\cdot\underline{n}} \hat{r}(\underline{n},\sigma), \quad (13.35)$$

which is equivalent to the sought expression, thanks to the differentiation properties of the Fourier transform, namely

$$r''(\underline{n},s) = \int_{-\infty}^{\infty} (i2\pi\sigma)^2 e^{i2\pi s\sigma} \hat{r}(\underline{n},\sigma)\,d\sigma$$

$$= -4\pi^2 \int_{-\infty}^{\infty} \sigma^2 e^{i2\pi s\sigma} \hat{r}(\underline{n},\sigma)\,d\sigma \quad (13.36)$$

since

$$r(\underline{n},s) = \int_{-\infty}^{\infty} e^{i2\pi s\sigma} \hat{r}(\underline{n},\sigma)\,d\sigma. \quad (13.37)$$

13.3.3 Grangeat's Formula

The observation made in the previous section that f can be reconstructed at $\underline{x}$ from the values of r'' on the planes containing $\underline{x}$ yields a line of reasoning for reconstruction of f at $\underline{x}$ from CB projection data. Specifically, the observation leads to the following question: can the CB data be linked to the values of the 3D Radon transform of f, so as to allow subsequent application of Equation 13.25 to obtain f at any desired location? The answer to this question turns out to be positive. Moreover, more than one link can be found. Here, we restrict the discussion to the link made by Grangeat (1991).

It should not be too surprising that there exists a link between the 3D Radon transform of f and CB projections. Consider a fixed source position and a flat-panel detector opposing the source. Then, draw a line in the detector plane, as depicted in Figure 13.6. Together with the source position, this line defines a plane, so that summing together the detector values taken along the line is bound to yield a result close to the 3D Radon transform of f on the plane. Actually, the result would be exact if the source position was far away, so that the CB measurements would seem to have been made on lines parallel to each other. The finite distance from the source to the object is the main cause for the summation yielding only an approximation.* Fortunately, as observed by Grangeat, this limitation can be overcome by focusing on calculation of $r'(n, s)$ instead of $r(n, s)$, where $r'(n, s)$ is the first partial derivative of r with respect to s.

* The situation is comparable to integration using polar coordinates in 2D. Let p be a scalar-valued function of Cartesian coordinates, x and y, in $\mathbb{R}^2$, and consider the calculation of

$$I = \int_{-\infty}^{\infty} dy \int_{-\infty}^{\infty} dx\, p(x_0 + x, y_0 + y), \quad (13.38)$$

where x_0 and y_0 are constants, which is basically the sum of the values of p over the entire $\mathbb{R}^2$ plane. Going to polar coordinates, θ and u, with $x = u \cos\theta$ and $y = u \sin\theta$, we get

$$I = \int_0^{2\pi} d\theta \int_0^{\infty} du\, u\, p(x_0 + u\cos\theta, y_0 + u\sin\theta). \quad (13.39)$$

If the Jacobian, u, were not there, this last expression would be the sum over θ of the integrals of p over the half-lines that start at (x_0,y_0). The Jacobian is what makes the summation of CB measurements over a line in the detector plane different from an integration over a plane.

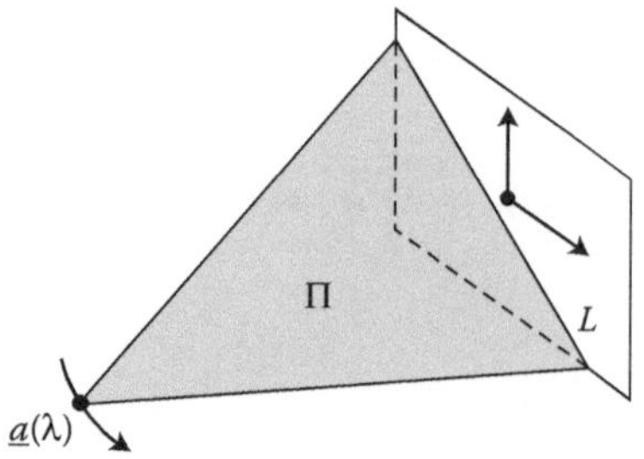

FIGURE 13.6 The link between a CB projection and the 3D Radon transform. A line L in the detector plane and the source position define together a plane Π, such that summing the values taken by the measurements on L yields a result that is approximately equal to the value of the 3D Radon transform of f on Π. The result is made exact by involving a differentiation step.

The link found by Grangeat can be expressed as follows. Let

$$G(\lambda,\underline{n}) = -\int_{S^2} \delta'(\underline{n}\cdot\underline{\alpha})\, g(\lambda,\underline{\alpha})\,d\underline{\alpha}. \tag{13.40}$$

Then

$$G(\lambda,\underline{n}) = r'(\underline{n},\underline{a}(\lambda)\cdot\underline{n}), \tag{13.41}$$

that is, $G(\lambda,n)$ is equal to the value taken by r' on the plane that is orthogonal to $\underline{n}$ through the vertex point $\underline{a}(\lambda)$, since $s = \underline{a}(\lambda)\cdot\underline{n}$ for this plane. To relate this result to the comments in the previous paragraph, note that the right-hand side of Equation 13.40 would just be a summation of CB projection data if the minus sign were omitted and δ were used instead of δ', and this summation would be over the measurements corresponding to the lines that are in the plane orthogonal to $\underline{n}$ through $\underline{a}(\lambda)$, since $\delta(\underline{n}\cdot\underline{\alpha})$ would restrict the summation to vectors $\underline{\alpha}$ that are orthogonal to $\underline{n}$. But this summation would not yield $r(\underline{n}, \underline{a}(\lambda)\cdot\underline{n})$ because the lines being involved are not parallel to each other. Inserting a differentiation through the use of δ' solves the problem.

To prove Equation 13.41, we first insert definition (13.19) for $g(\lambda, \underline{\alpha})$ into Equation 13.40, which yields

$$G(\lambda,\underline{n}) = -\int_{S^2} d\underline{\alpha} \int_0^\infty dt\, \delta'(\underline{n}\cdot\underline{\alpha})\, f(\underline{a}(\lambda)+t\underline{\alpha}). \tag{13.42}$$

Next, property (13.14) for δ' is invoked to rewrite the result as

$$G(\lambda,\underline{n}) = -\int_{S^2} d\underline{\alpha} \int_0^\infty dt\, t^2\, \delta'(\underline{n}\cdot t\underline{\alpha})\, f(\underline{a}(\lambda)+t\underline{\alpha}), \tag{13.43}$$

which is fine even for $t = 0$ because f is zero in the neighborhood of $\underline{a}(\lambda)$. This last expression then recalls integration using spherical coordinates, as in Equation 13.8, so that we may write

$$\begin{aligned} G(\lambda,\underline{n}) &= -\int_{\mathbb{R}^3} \delta'(\underline{n}\cdot\underline{x})\, f(\underline{a}(\lambda)+\underline{x})\,d\underline{x} \\ &= -\int_{\mathbb{R}^3} \delta'(\underline{n}\cdot(\underline{x}-\underline{a}(\lambda))\, f(\underline{x})\,d\underline{x}, \\ &= -\int_{\mathbb{R}^3} \delta'(\underline{x}\cdot\underline{n}-\underline{a}(\lambda)\cdot\underline{n})\, f(\underline{x})\,d\underline{x}, \end{aligned} \tag{13.44}$$

which is equal to $r'(\underline{n}, \underline{a}(\lambda)\cdot\underline{n})$. Indeed, applying change of variables Equation 13.31 yields

$$\begin{aligned} G(\lambda,\underline{n}) &= -\int_{-\infty}^{\infty} ds\delta'(s-\underline{a}(\lambda)\cdot\underline{n}) \int_{-\infty}^{\infty} dt_1 \int_{-\infty}^{\infty} dt_2\, f(s\underline{n}+t_1\underline{m}_1+t_2\underline{m}_2) \\ &= -\int_{-\infty}^{\infty} ds\delta'(s-\underline{a}(\lambda)\cdot\underline{n}) r(\underline{n},s) \\ &= r'(\underline{n},\underline{a}(\lambda)\cdot\underline{n}) \end{aligned} \tag{13.45}$$

due to the properties of δ'.

13.3.4 General Reconstruction Scheme

We are now ready to present a general scheme for reconstruction of f inside Ω_{ROI} from its CB projections. This scheme is summarized by the flowchart in Figure 13.7. Reading of the chart starts with the top left box, which contains the CB projections. The first step is to transform each CB projection from this box into values of r', the derivative of the 3D Radon transform of f. This step is performed using Grangeat's formula. Thus, the CB projection at position λ yields the values of r' on all planes that contain $\underline{a}(\lambda)$,

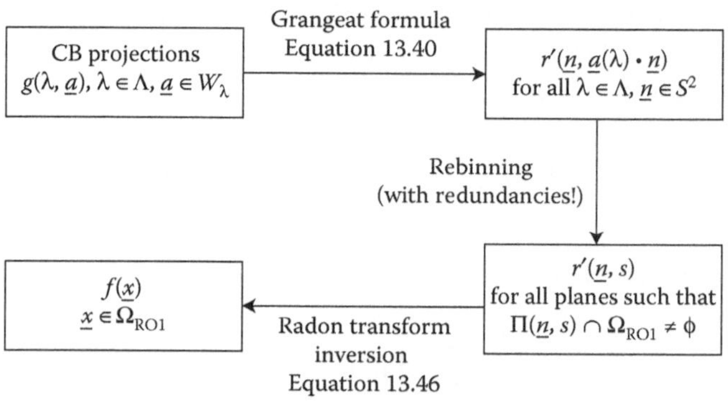

FIGURE 13.7 General scheme for reconstruction of f over Ω_{ROI}. On the top row, Grangeat's formula is used to convert the CB projections into samples of the derivative of the 3D Radon transform of f on the planes that intersect the vertex path. On the bottom row, the inversion formula for the 3D Radon transform is used to obtain f over Ω_{ROI} from samples of the derivative of this transform over all planes that intersect Ω_{ROI}. TES reconstruction is enabled when the samples needed for the operation on the bottom row are found within the samples created on the top row. With discretized data, the conversion from one set of samples to the other is a 3D interpolation (rebinning) step.

as highlighted by the top right box in the chart. The next step is a search procedure that aims at finding the values of r' on the planes $\Pi(\underline{n},s)$ that intersect Ω_{ROI} from the values of r' on the planes that intersect the vertex path (viz., the values in the top right box). When the searching step can be successfully completed, computation of $f(\underline{x})$ is enabled using the following variant of Equation 13.25 for inversion of the 3D Radon transform, which gives $f(\underline{x})$ from $r'(\underline{n}, s)$:

$$f(\underline{x}) = -\frac{1}{8\pi^2}\int_{S^2} \left.\frac{\partial r'(\underline{n},s)}{\partial s}\right|_{s=\underline{x}\cdot\underline{n}} d\underline{n}. \quad (13.46)$$

By design, the reconstruction scheme does not work for an arbitrary vertex path. The scheme requires the bottom right box to be a subset of the top right box, which is basically Tuy's condition: "every plane that intersects Ω_{ROI} must intersect the vertex path." When Tuy's condition is violated, Equation 13.46 cannot be exactly applied because values of $r'(\underline{n}, s)$ are missing. Then, only an approximate reconstruction may be obtained, using a guess for the missing values.

An important aspect of the search procedure is that it includes redundancies. Indeed, any given plane $\Pi(\underline{n},s)$ that intersects the vertex path usually intersects this path more than once. In such a case, $r'(\underline{n}, s)$ can be computed in more than one way: each CB projection that corresponds to a source position lying in $\Pi(\underline{n},s)$ yields access to $r'(\underline{n}, s)$. This redundancy may be used to combat noise in the CB data, by computing $r'(\underline{n}, s)$ as the average of all available estimates. Alternatively and interestingly, this redundancy may be used to circumvent the problem of truncation. However, this alternative use is not trivial and requires further mathematical machinery exposed in Section 13.4.

As mentioned at the beginning of this section, it is assumed here that the projections are nontruncated. When a projection is truncated, there always exist planes for which Grangeat's formula cannot be applied; see Figure 13.8. As a result, values of r' on the planes passing through the vertex path may not be all obtainable, so that the search procedure will either fail or require undesired extrapolation.

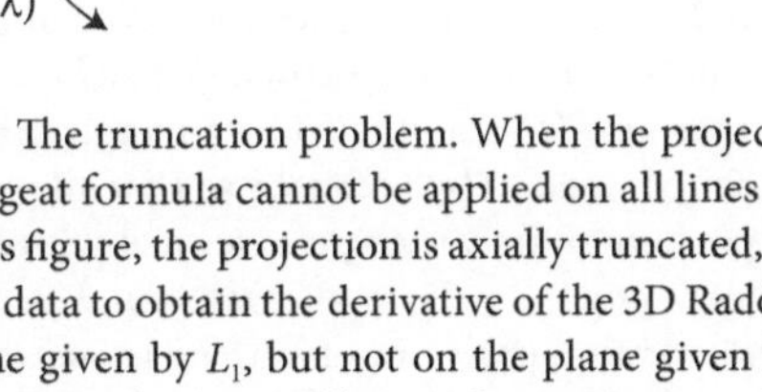

FIGURE 13.8 The truncation problem. When the projection is truncated, the Grangeat formula cannot be applied on all lines in the detector plane. In this figure, the projection is axially truncated, so that there remain enough data to obtain the derivative of the 3D Radon transform of f on the plane given by L_1, but not on the plane given by L_2. If this L_2-based plane cannot be covered by another projection, the rebinning process will require extrapolation preventing TES reconstruction.

When dealing with sampled data, the search procedure becomes a 3D interpolation (rebinning) step that aims at obtaining values of r' sampled on planes that contain the source positions into values of r' that are uniformly sampled in s and n, the latter being expressed using spherical coordinates. A number of algorithms have been suggested for implementation of this step; see Noo et al. (1997) and references therein.

13.4 Reconstruction from Truncated Projections

This section presents a reformulation of the general reconstruction scheme given in Section 13.3.4 into an FBP format, and explains thereby how and under which circumstances TES reconstruction can be achieved from the truncated projections. Throughout this section, it is understood that $\underline{x} \in \Omega_{ROI} \subset \Omega$.

13.4.1 Local Reconstruction Scheme

To develop the announced FBP method, the reconstruction scheme of Section 13.3.4 must be first rewritten into a local form that focuses on computation of f at a single point $\underline{x}$, and that avoids the use of r' in favor of using r''.

Avoiding the use of r' is made possible by introducing the following quantity:

$$g'(\lambda,\underline{\alpha}) = \frac{\partial}{\partial\lambda} g(\lambda,\underline{\alpha}), \quad (13.47)$$

which is the derivative of g with respect to λ at fixed line direction $\underline{\alpha}$. From g', we define

$$G'(\lambda,\underline{n}) = -\int_{S^2} \delta'(\underline{n}\cdot\underline{\alpha})\, g'(\lambda,\underline{\alpha})\, d\underline{\alpha}, \quad (13.48)$$

which can be seen to be the partial derivative of $G(\lambda, \underline{n})$ of Equation 13.40 in λ. Then, by differentiating each side of Grangeat's formula in Equation 13.41 with respect to λ, we obtain

$$G'(\lambda,\underline{n}) = (\underline{a}'(\lambda)\cdot\underline{n})\, r''(\underline{n},\underline{a}(\lambda)\cdot\underline{n}), \quad (13.49)$$

or more interestingly

$$r''(\underline{n},\underline{a}(\lambda)\cdot\underline{n}) = \frac{G'(\lambda,\underline{n})}{\underline{a}'(\lambda)\cdot\underline{n}}. \quad (13.50)$$

This last equation shows that, by using g' as input data instead of g, the second derivative of r in s can be directly obtained on any plane that intersects the vertex path. Thus, there is no need to involve r'.

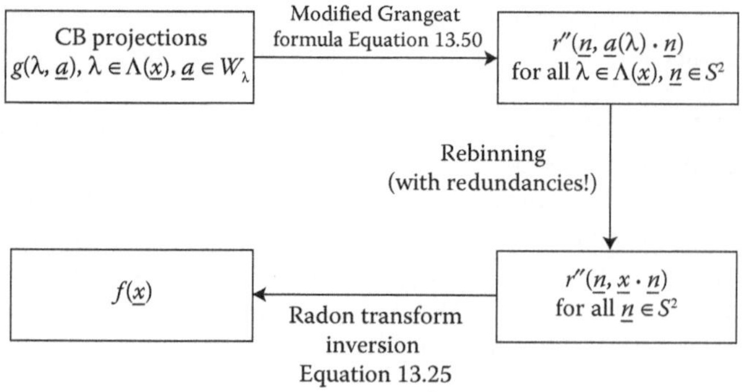

FIGURE 13.9 Local reconstruction scheme. Here, the goal is to obtain f at a single location, $\underline{x}$. On the top row, Grangeat's formula is used to obtain the second derivative of the 3D Radon transform of f on the planes that intersect a portion of the vertex path defined by $\Lambda(\underline{x})$. On the bottom row, the inversion formula for the 3D Radon transform is used to obtain f at $\underline{x}$ from the values of the second derivative of this transform over all planes that contain $\underline{x}$. TES reconstruction at $\underline{x}$ is enabled when these planes are among those that contain the source positions given by $\lambda \in \Lambda(\underline{x})$.

Note that a mathematical problem arises with Equation 13.50 when $\underline{n}$ is such that the plane orthogonal to $\underline{n}$ through $\underline{a}(\lambda)$ is tangent to the vertex path at $\underline{a}(\lambda)$. In this case, $\underline{a}'(\lambda) \cdot \underline{n} = 0$ and a division by zero occurs in the right-hand side of Equation 13.50. This difficulty is the reason why Tuy introduced a requirement of nontangentially in his condition. Tuy considered that r'' could not be exactly computed on a plane that is tangent to the vertex path at each of its intersections with it (because then all possible ways to compute r'' on that plane would involve a division by a zero quantity) and that this situation would preclude accurate reconstruction. Although a sensitive issue in numerical implementation, this difficulty is in general not a mathematical problem because the delicate planes identified by Tuy usually form a set of measure zero that has no impact on our calculations given that we use the Lebesgue integral. Interestingly, the celebrated helical vertex path does not satisfy the nontangentiality condition of Tuy. Hence, it is fortunate that this condition could be discarded.

Using Equation 13.50, the general reconstruction scheme of Figure 13.7 can be modified into the local scheme outlined in Figure 13.9. Note that in this local scheme, Λ was replaced by $\Lambda(\underline{x})$, a subset of Λ. As indicated, $\Lambda(\underline{x})$ depends on $\underline{x}$; this set may be arbitrarily chosen among the subsets of Λ that meet the following property: the portion of the vertex path defined by $\lambda \in \Lambda(\underline{x})$ must be piecewise smooth and must be intersected by all planes that contain $\underline{x}$. This freedom of restricting the reconstruction scheme to a subset of the vertex path is generally made possible by the presence of redundancies in coverage of values of r'', and represents a key ingredient in the design of algorithms that allow some level of truncation in the CB projections.

13.4.2 FBP Formula

By definition, an exact FBP formula yields f at $\underline{x}$ using a backprojection of filtered projections, where the filter is defined as a linear operation. In CB tomography, an FBP method may or may not be computationally efficient depending on the nature of the filtering step, which may be shift invariant or not. Here, the issue of computational complexity is seen as secondary to that of handling data truncation, and will thus only be discussed later, in Section 13.4.5.

Analyzing closely the reconstruction scheme in Figure 13.9, we observe that the value of f at $\underline{x}$ could be obtained in the following two steps if we did not need to consider redundancies: (i) for each λ, compute the value of r'' on each plane that contains $\underline{a}(\lambda)$ and $\underline{x}$ using Equation 13.50, then sum these values together, and (ii) add together the summation results obtained from (i) for each $\lambda \in \Lambda(\underline{x})$. Such a technique is definitely of FBP type, but, as it stands, it would not work, because some values of r'' would be counted more than once. However, the method can be rescued using proper weights during the summation process in (i).

To formalize the procedure just outlined, a parameterization of planes that contain $\underline{x}$ and $\underline{a}(\lambda)$ is first needed. This parameterization is obtained using three orthogonal unit vectors:

$$\underline{\omega}(\lambda) = \frac{\underline{a}(\lambda) - \underline{x}}{\|\underline{a}(\lambda) - \underline{x}\|}, \tag{13.51}$$

$$\underline{e}_1(\lambda) = \frac{\underline{a}'(\lambda) - (\underline{a}'(\lambda)\cdot\underline{\omega}(\lambda))\underline{\omega}(\lambda)}{\|\underline{a}'(\lambda) - (\underline{a}'(\lambda)\cdot\underline{\omega}(\lambda))\underline{\omega}(\lambda)\|}, \tag{13.52}$$

$$\underline{e}_2(\lambda) = \underline{\omega}(\lambda) \times \underline{e}_1(\lambda). \tag{13.53}$$

These vectors are depicted in Figure 13.10, where a unit sphere centered on $\underline{x}$, called Σ, is drawn. As λ varies over $\Lambda(\underline{x})$, $\underline{\omega}(\lambda)$ follows a curve on Σ that is called the vertex path trace. Vector $\underline{e}_1(\lambda)$ is tangent to this trace, and vector $\underline{e}_2(\lambda)$ is defined such that the

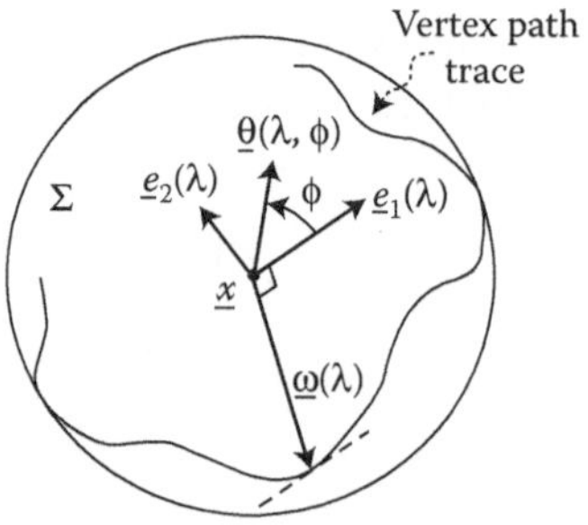

FIGURE 13.10 The unit sphere, Σ, centered on $\underline{x}$. Vector $\underline{\omega}(\lambda)$ is the unit vector from $\underline{x}$ toward $\underline{a}(\lambda)$. As λ varies over $\Lambda(\underline{x})$, $\underline{\omega}(\lambda)$ follows a curve on Σ called the vertex path trace. Vector $\underline{e}_1(\lambda)$ is the unit vector tangent to this trace, and $\underline{e}_2(\lambda)=\underline{\omega}(\lambda) \times \underline{e}_1(\lambda)$. Vector $\underline{\theta}(\lambda, \phi)$ is obtained by rotating $\underline{e}_1(\lambda)$ about $\underline{\omega}(\lambda)$ by an angle ϕ in the counterclockwise direction. Since $\Lambda(\underline{x})$ is chosen so that all planes through $\underline{x}$ intersect the portion of the vertex path defined with $\lambda \in \Lambda(\underline{x})$, there is, for any unit vector $\underline{n}$, at least one value of $\lambda \in \Lambda(\underline{x})$ that gives $\underline{\omega}(\lambda)\cdot\underline{n} = 0$, and, for any such value of λ, there is one angle $\phi \in [0, 2\pi)$ yielding $\underline{\theta}(\lambda, \phi) = \underline{n}$. Hence, $\underline{\theta}(\lambda, \phi)$ covers the whole unit sphere with $(\lambda, \phi) \in \Lambda(x) \times [0, 2\pi)$.

three vectors form together a right-handed system of mutually orthogonal unit vectors. Note that all three vectors depend on $\underline{x}$, but this dependence is not written explicitly to simplify the notation. Any plane that contains $\underline{x}$ can be identified by its normal, which is one of the vectors that connects the center of Σ to a point on the boundary of Σ. Conversely, any vector that connects the center of Σ to a point on the boundary of Σ can be seen as the normal to a plane that contains $\underline{x}$. Thus, all planes that contain $\underline{x}$ and $\underline{a}(\lambda)$ can be described using the unit vector

$$\underline{\theta}(\lambda,\phi) = \cos\phi\,\underline{e}_1(\lambda) + \sin\phi\,\underline{e}_2(\lambda), \quad \phi \in [0,2\pi). \tag{13.54}$$

Since $\Lambda(\underline{x})$ is selected so that every plane through $\underline{x}$ intersects the portion of the vertex path defined with $\lambda \in \Lambda(\underline{x})$, $\underline{\theta}(\lambda, \phi)$ will cover the entirety of Σ as (λ, ϕ) varies over $\Lambda(\underline{x}) \times [0, 2\pi)$. Given the definition of $\underline{e}_1(\lambda)$ and $\underline{e}_2(\lambda)$, $\underline{\theta}(\lambda, \phi)$ is orthogonal to $\underline{\omega}(\lambda)$, therefore, from Equation 13.51,

$$\underline{x}\cdot\underline{\theta}(\lambda,\phi) = \underline{a}(\lambda)\cdot\underline{\theta}(\lambda,\phi), \tag{13.55}$$

which is in agreement with our expectation that the plane orthogonal to $\underline{\theta}(\lambda, \phi)$ through $\underline{a}(\lambda)$ contains $\underline{x}$.

Note that $\underline{\theta}(\lambda, \phi)$ and $\underline{\theta}(\lambda, \phi + \pi)$ are opposite vectors. Therefore, they do not represent different planes through $\underline{x}$, and the range for ϕ could be restricted to $[0, \pi)$, but this restriction would complicate the mathematical developments hereafter due to the fact that any plane through $\underline{x}$ can be characterized by not one but two normal vectors, which are opposite to each other. This approach is consistent with our formula (Equation 13.25) for inversion of the Radon transform, which also counts each plane twice as the integration is performed over the whole of S^2.

Because the portion of the vertex path defined with $\lambda \in \Lambda(\underline{x})$ may be intersected more than once by any given plane, $\underline{\theta}(\lambda, \phi)$ may be expected to take the same value for several (λ, ϕ)-pairs. A weighting function is now introduced to allow normalization to one when summing the values of r'' over the planes through $\underline{x}$ described by $(\lambda, \phi) \in \Lambda(\underline{x}) \times [0, 2\pi)$. This function of λ and ϕ is specific to $\underline{x}$ in the sense that completely different expressions can be chosen for different points $\underline{x}$, with no concerns regarding smoothness in $\underline{x}$. We denote this as $M(\lambda, \phi; \underline{x})$. There is high flexibility in the selection of $M(\lambda, \phi; \underline{x})$, as it is only required to satisfy the following condition:

Let $\underline{n} \in S^2$, and let $N(\underline{n}, \underline{x})$ be the number of intersections between the plane orthogonal to $\underline{n}$ through $\underline{x}$ and the portion of the vertex path defined with $\lambda \in \Lambda(\underline{x})$. Let λ_k, $k = 1, \ldots, N(\underline{n}, \underline{x})$ be the source locations where the intersections occur, and $\phi_k \in [0, 2\pi)$ be such that $\underline{\theta}(\lambda_k, \phi_k) = \underline{n}$. Then, function M must be such that $M(\lambda, \phi + \pi; \underline{x}) = M(\lambda, \phi, \underline{x})$ for any $\phi \in [0, \pi)$ and

$$\sum_{k=1}^{N(\underline{n},\underline{x})} M(\lambda_k,\phi_k;\underline{x}) = 1 \tag{13.56}$$

for any $\underline{n} \in S^2$.

In this condition, the requirement that $M(\lambda, \phi + \pi; \underline{x}) = M(\lambda, \phi; \underline{x})$ reflects the earlier-noted fact that $\underline{\theta}(\lambda, \phi)$ and $\underline{\theta}(\lambda, \phi + \pi)$ characterize the same plane through $\underline{x}$. The simplest solution to Equation 13.56 is

$$M(\lambda,\phi;\underline{x}) = \frac{1}{N(\underline{\theta}(\lambda,\phi),\underline{x})}. \tag{13.57}$$

Selecting this solution amounts to giving equal weights to all estimates that are available for $r''(\underline{n}, \underline{x} \cdot \underline{n})$, independently of $\underline{n} \in S^2$, which is a noise-effective strategy when the projections are nontruncated. On the other hand, when data truncation is present, equal weighting is not necessarily optimal. Fortunately, many other solutions to Equation 13.56 can be formulated. Observe that solution (13.57) is a piecewise constant function because the number of intersections between a curve and a varying plane can only change under two conditions: (1) the plane becomes tangent to the curve and (2) the plane meets one of the endpoints of the curve.

Using property (Equation 13.11) for the Diract impulse, the condition on M can be rewritten in the following integral form:

$$\int_{\Lambda(\underline{x})} M(\lambda,\phi^*;\underline{x}) \frac{|\underline{a}'(\lambda)\cdot\underline{n}|}{\|\underline{x}-\underline{a}(\lambda)\|} \delta(\underline{\omega}(\lambda)\cdot\underline{n})\, d\lambda = 1 \tag{13.58}$$

where $\phi^* \in [0, 2\pi)$ is such that

$$\underline{\theta}(\lambda,\phi^*) = \frac{\underline{n}-(\underline{n}\cdot\underline{\omega})\underline{\omega}}{\|\underline{n}-(\underline{n}\cdot\underline{\omega})\underline{\omega}\|}. \tag{13.59}$$

Since the left-hand side of Equation 13.58 is required to be equal to 1 for any $\underline{n} \in S^2$, it may be introduced as part of the integrand inside Equation 13.25 for inversion of the 3D Radon transform. Changing the order of integration, this insertion yields

$$\begin{aligned} f(\underline{x}) &= -\frac{1}{8\pi^2}\int_{S^2} r''(\underline{n},\underline{x}\cdot\underline{n}) \\ &\quad\times\left[\int_{\Lambda(\underline{x})} M(\lambda,\phi;\underline{x})\frac{|\underline{a}'(\lambda)\cdot\underline{n}|}{\|\underline{x}-\underline{a}(\lambda)\|}\delta(\underline{\omega}(\lambda)\cdot\underline{n})\,d\lambda\right]d\underline{n} \\ &= -\frac{1}{8\pi^2}\int_{\Lambda(\underline{x})}\frac{g_F(\lambda,\underline{x})}{\|\underline{x}-\underline{a}(\lambda)\|}d\lambda \end{aligned} \tag{13.60}$$

with

$$g_F(\lambda,\underline{x}) = \int_{S^2} r''(\underline{n},\underline{x}\cdot\underline{n})\,|\underline{a}'(\lambda)\cdot\underline{n})|\,M(\lambda,\phi^*;\underline{x})\,\delta(\underline{\omega}(\lambda)\cdot\underline{n})\,d\underline{n}. \tag{13.61}$$

Now, observe that the condition $\underline{\omega}(\lambda) \cdot \underline{n} = 0$ is equivalent to $\underline{x} \cdot \underline{n} = \underline{a}(\lambda) \cdot \underline{n}$, therefore, by Equation 13.50, Equation 13.60 can be rewritten in the form

$$g_F(\lambda,\underline{x}) = \int_{S^2} \text{sign}(\underline{a}'(\lambda)\cdot\underline{n})\,M(\lambda,\phi^*;\underline{x})\,G'(\lambda,\underline{n})\,\delta(\underline{\omega}(\lambda)\cdot\underline{n})\,d\underline{n}, \tag{13.62}$$

$$= \int_0^{2\pi} d\breve{\phi} \int_0^{\pi} d\breve{\theta}\, \sin\breve{\theta}\, \text{sign}(\underline{a}'(\lambda)\cdot\underline{n})\, M(\lambda,\phi^*;\underline{x})\, G'(\lambda,\underline{n})\, \delta(\cos\breve{\theta}), \tag{13.63}$$

where $\breve{\theta}$ and $\breve{\phi}$ are spherical coordinates such that

$$\underline{n} = \cos\breve{\theta}\, \underline{\omega}(\lambda) + \sin\breve{\theta}\,(\cos\breve{\phi}\, \underline{e}_1(\lambda) + \sin\breve{\phi}\, \underline{e}_2(\lambda)). \tag{13.64}$$

Again invoking property (Equation 13.11), this time to calculate the integral in $\breve{\theta}$, and substituting afterward ϕ for $\breve{\phi}$, we get

$$g_F(\lambda,\underline{x}) = \int_0^{2\pi} q(\lambda,\phi;\underline{x})\, G'(\lambda,\underline{\theta}(\lambda,\phi))\, d\phi \tag{13.65}$$

with

$$q(\lambda,\phi;\underline{x}) = \text{sign}(\underline{a}'(\lambda)\cdot\underline{\theta}(\lambda,\phi))M(\lambda,\phi;\underline{x}) = \text{sign}(\cos\phi)M(\lambda,\phi;\underline{x}), \tag{13.66}$$

where the last equality is easily found from inspection of Equations 13.51 through 13.54.

Equations 13.48, 13.54, 13.60, 13.65, and 13.66 define together the desired FBP formula, as it was first presented in Chen (2003), Katsevich (2003), Pack et al. (2004), and Zhao et al. (2005). At first glance, this result does not look very impressive, as the only gain it seems to offer is the allowance for reconstruction of $f(\underline{x})$ through sequential processing of the CB projections. However, it is a marvel, as it comes with tremendous flexibility in the definition of $\Lambda(\underline{x})$ and $M(\lambda, \phi; \underline{x})$, but this strength is difficult to appreciate without knowledge that the filtering expression given by Equation 13.65 is in fact a weighted linear combination of 1D convolutions. We demonstrate this aspect in Section 13.4.3.

13.4.3 Nature of the Filtering Step

Since Equation 13.65 involves G' and the calculation of G' is affected by truncation in the same way as G, which was illustrated by Figure 13.8, Equation 13.65 could easily be thought to be not immune to truncation. However, such a line of reasoning does not account for the integration in ϕ. To appreciate the impact of this integration, we now replace G' by its definition in terms of g', which was given by Equation 13.48. This substitution yields

$$g_F(\lambda,\underline{x}) = \int_0^{2\pi} d\phi\, q(\lambda,\phi;\underline{x}) \int_{S^2} d\underline{\alpha}\, \delta'(\underline{\alpha}\cdot\underline{\theta}(\lambda,\phi))\, g'(\lambda,\underline{\alpha}). \tag{13.67}$$

Then, $\underline{\alpha}$ is parameterized using spherical coordinates, $\hat{\theta} \in [-\pi,\pi]$ and $\hat{\phi} \in [0,\pi]$, such that

$$\underline{\alpha} = -\cos\hat{\theta}\, \underline{\omega}(\lambda) + \sin\hat{\theta}\, \underline{\theta}^{\perp}(\lambda,\hat{\phi}) \tag{13.68}$$

with $\underline{\theta}^{\perp}(\lambda,\hat{\phi})$ being the result of rotating $\underline{\theta}(\lambda,\hat{\phi})$ about $\underline{\omega}(\lambda)$ by 90° in the counterclockwise direction, that is

$$\underline{\theta}^{\perp}(\lambda,\hat{\phi}) = -\sin\hat{\phi}\, \underline{e}_1(\lambda) + \cos\hat{\phi}\, \underline{e}_2(\lambda). \tag{13.69}$$

This parameterization leads to

$$\begin{aligned} &g_F(\lambda,\underline{x}) = \\ &\int_0^{2\pi} d\phi q(\lambda,\phi;\underline{x}) \int_0^{\pi} d\hat{\phi} \int_{-\pi}^{\pi} d\hat{\theta}\, |\sin\hat{\theta}|\, \delta'(\sin\hat{\theta}\sin(\phi-\hat{\phi}))\, g'(\lambda,\underline{\alpha}) \\ &= \int_0^{\pi} d\hat{\phi} \int_{-\pi}^{\pi} d\hat{\theta}\, |\sin\hat{\theta}|\, g'(\lambda,\underline{\alpha}) \int_0^{2\pi} d\phi\, q(\lambda,\phi;\underline{x})\, \delta'(\sin\hat{\theta}\sin(\phi-\hat{\phi})) \\ &= \int_0^{\pi} c(\lambda,\hat{\phi};\underline{x})\, g_H(\lambda,\hat{\phi},\underline{x})\, d\hat{\phi}, \end{aligned} \tag{13.70}$$

where

$$g_H(\lambda,\phi,\underline{x}) = \int_{-\pi}^{\pi} \frac{1}{\sin\theta}\, g'\left(\lambda, \cos\theta\, \underline{\alpha}^*(\lambda,\underline{x}) + \sin\theta\, \underline{\theta}^{\perp}(\lambda,\phi)\right) d\theta \tag{13.71}$$

with

$$\underline{\alpha}^*(\lambda,\underline{x}) = -\underline{\omega}(\lambda) = \frac{\underline{x}-\underline{a}(\lambda)}{||\underline{x}-\underline{a}(\lambda)||} \tag{13.72}$$

and

$$c(\lambda,\phi;\underline{x}) = \left(\frac{\partial q}{\partial \phi}\right)(\lambda,\phi+\pi;\underline{x}) - \left(\frac{\partial q}{\partial \phi}\right)(\lambda,\phi;\underline{x}) = -2\left(\frac{\partial q}{\partial \phi}\right)(\lambda,\phi;\underline{x}) \tag{13.73}$$

since $q(\lambda,\phi+\pi;\underline{x}) = q0(\lambda, \phi; \underline{x})$. Note that we intentionally dropped the hat symbol on θ and ϕ in Equations 13.71 and 13.73, because this symbol was only needed as an intermediate notation; angle θ in Equation 13.71 is a dummy variable, and so is angle $\hat{\phi}$ in Equation 13.70, which can be more simply written as

$$g_F(\lambda,\underline{x}) = \int_0^{\pi} c(\lambda,\phi;\underline{x})\, g_H(\lambda,\phi,\underline{x})\, d\phi. \tag{13.74}$$

Note also that the integrand in Equation 13.71 includes a singularity at $\theta = 0$; to handle this issue the integral needs to be calculated as a Cauchy principal value, namely as

$$\begin{aligned} g_H(\lambda,\phi,\underline{x}) &= \lim_{\varepsilon\to 0}\left[\int_{-\pi}^{-\varepsilon} + \int_{\varepsilon}^{\pi}\right] \frac{1}{\sin\theta} \\ &\times g'\left(\lambda, \cos\theta\, \underline{\alpha}^*(\lambda,\underline{x}) + \sin\theta\, \underline{\theta}^{\perp}(\lambda,\phi)\right) d\theta. \end{aligned} \tag{13.75}$$

(The fact that $\sin\theta = 0$ at $\theta = \pi$ and $\theta = -\pi$ is not a problem because g' is then equal to zero.)

Equation 13.74 is the end result we wanted to reach. This equation shows that $g_F(\lambda, \underline{x})$ is a weighted linear combination of quantities $g_H(\lambda, \phi; \underline{x})$ in $\phi \in [0, \pi)$ with the weight given by $c(\lambda, \phi; \underline{x})$. As illustrated in Figure 13.11, computing $g_H(\lambda, \phi; \underline{x})$ only requires the CB data along one line in the detector plane, namely the line $L(\phi)$ that lies in the plane orthogonal to $\underline{\theta}(\lambda,\phi) = \underline{\alpha}^*(\lambda,\underline{x}) \times \underline{\theta}^{\perp}(\lambda,\phi)$ through $\underline{x}$ and $\underline{a}(\lambda)$. If the weight $c(\lambda, \phi; \underline{x})$ can be chosen to be zero when $L(\phi)$ is affected by data truncation, that is, when $L(\phi)$ is for instance oriented like line L_2 in Figure 13.8, then the truncation problem is solved. Line $L(\phi)$ is called a filtering line.

Inspection of Equations 13.66 and 13.73 shows that $c(\lambda, \phi; \underline{x})$ can be zero over large intervals in ϕ by just ensuring that $M(\lambda, \phi; \underline{x})$ is constant over these intervals. For example, if we select $M(\lambda, \phi; \underline{x})$ as in Equation 13.57, then $c(\lambda, \phi; \underline{x})$ will reduce to a sum of Dirac impulses located at $\phi = 0$, where the signum of $\cos\phi$ changes, and at the angles ϕ, where $N(\underline{\theta}(\lambda, \phi), \underline{x})$ changes value, which only happens when the plane orthogonal to $\underline{\theta}(\lambda, \phi)$ through $\underline{x}$ is either tangent to the portion of the vertex path defined with $\lambda \in \Lambda(\underline{x})$ or contains an endpoint of this portion. Note that the number of endpoints is two times the number of disconnected curves that form the portion of the vertex path.

In general, if $q(\lambda, \phi; \underline{x})$ changes value only at J locations, denoted ϕ_j, $j = 1, \ldots, J$, then Equation 13.74 simplifies to

$$g_F(\lambda,\underline{x}) = \sum_{j=1}^{J} c_j(\lambda,\underline{x})\, g_H(\lambda,\phi_j,\underline{x}), \qquad (13.76)$$

where

$$c_j(\lambda,\underline{x}) = -2\lim_{\varepsilon\to 0}\left[q(\lambda,\phi_j+\varepsilon;\underline{x}) - q(\lambda,\phi_j-\varepsilon;\underline{x})\right]. \qquad (13.77)$$

The importance of having wide flexibility in selecting $\Lambda(\underline{x})$ and $M(\lambda, \phi; \underline{x})$ has now become apparent: these quantities define the number and direction of filtering lines involved in the definition of $g_F(\lambda, \underline{x})$, and offer thereby a means to control the amount of data needed to reconstruct $f(\underline{x})$. How much truncation is allowed will depend on the vertex path and the creativity associated with the selection of $\Lambda(\underline{x})$ and $M(\lambda, \phi; \underline{x})$. Of course, the detector must always be large enough to cover at minimum the projection of $\underline{x}$ for all $\lambda \in \Lambda(\underline{x})$.

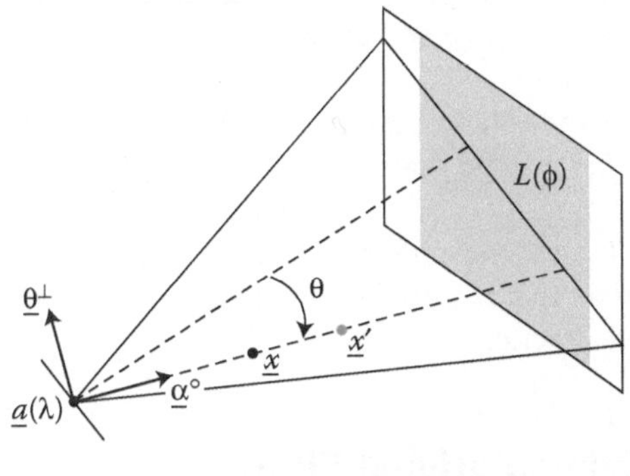

FIGURE 13.11 Filtering line. The calculation of $g_H(\lambda, \phi; \underline{x})$ in Equation 13.75 only involves the measurements on line $L(\phi)$, which is the intersection between the detector and the plane that contains $\underline{x}$ and is subtended by the orthogonal unit vectors $\underline{\alpha}^*(\lambda, x)$ and $\underline{\theta}^{\perp}(\lambda, \phi)$.

13.4.4 Application to the Helical Vertex Path

In this section, we consider the specific problem of reconstructing f from axially truncated CB projections measured with a helical data acquisition. The vertex path is described by Equation 13.17, and a flat detector is used, with orientation given by the vectors $\underline{e}_u(\lambda)$, $\underline{e}_v(\lambda)$, and $\underline{e}_w(\lambda)$ in Equation 13.22.

Since the helix is connected, we know from Tuy's condition that TES reconstruction is possible anywhere inside the convex hull of the helix. Let $\underline{x}$ be an arbitrary point within this region. A first question to address is how to select $\Lambda(\underline{x})$. One option is to pick the smallest possible piece of the helix that will satisfy Tuy's condition for $\underline{x}$. This option, adopted here, corresponds to using the concept of π-lines and π-segments (Danielsson et al., 1997). A π-line is any segment of line that connects two source positions separated by less than 360° in λ, and a π-segment is any portion of the helix that connects the endpoints of a π-line; see Figure 13.12. It turns out that there exists one (and only one) π-line passing through $\underline{x}$ (Danielsson et al., 1997; Defrise et al., 2000), and the π-segment associated with this π-line is our choice for $\Lambda(\underline{x})$. Below, this π-segment is called C. Next comes the question of selecting $M(\lambda, \phi; \underline{x})$. For this selection, it is useful to introduce the Tam–Danielsson window (Danielsson et al., 1997; Tam et al., 1998), shown in Figure 13.13 as region T. This window is the region of the detector plane that is bounded by the CB projection of the helix turns that are directly above and below the source position. When $\underline{a}(\lambda)$ is on C, the projection of $\underline{x}$ onto the detector plane is always within this window.

Now consider Figure 13.14, which shows the detector plane for a source position on C along with (1) the boundaries of T, denoted as T^+ and T^-, (2) a dashed line, (3) the projection of $\underline{x}$, called Q, and (4) three solid lines through this projection, called L_1, L_2, and L_3. The dashed line passes through the origin $(u, v) = (0, 0)$ with a slope of $P/(2\pi R_0)$, and has the property of being a common asymptote for the upper and lower boundaries of T. Depending

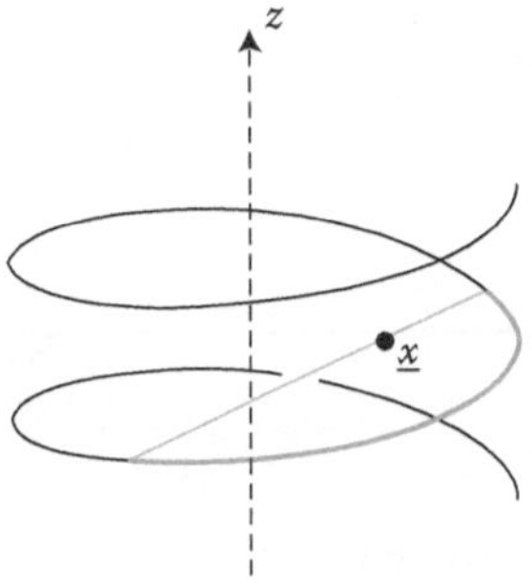

FIGURE 13.12 The π-line through a point $\underline{x}$, and the corresponding π-segment on the helix (the dark-gray piece).

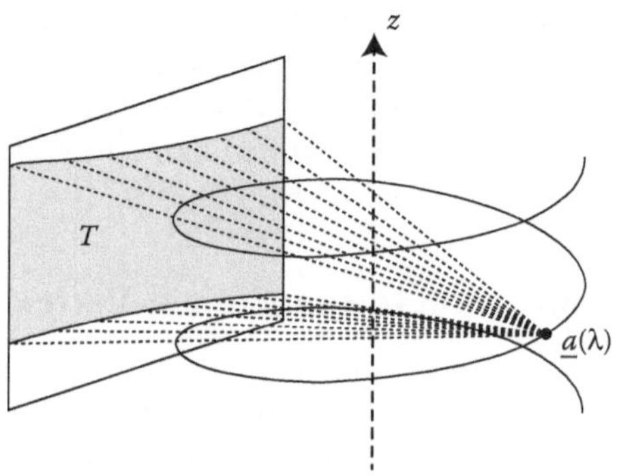

FIGURE 13.13 The Tam–Danielsson window. The intersection between the detector plane and the line that connects $\underline{a}(\lambda)$ to $\underline{a}(\lambda')$ generates a curve in the detector plane as λ' is varied over the interval $(\lambda, \lambda + 2\pi)$, and also as λ' is varied over the interval $(\lambda - 2\pi, \lambda)$. Together, the two curves delimitate the region T within the detector plane, which is the Tam–Danielsson window.

on the value of λ, Q will either be below or above this asymptote. (The figure shows it as being above.) Line L_1 is the line through the projection of the endpoints of C. Thus, Q belongs to L_1 and the projection of C can be seen as the union of two pieces: the solid portion of T^+ on the right side of the intersection between L_1 and T^+, and the solid portion of T^- on the left side of the intersection between L_1 and T^-. Line L_2 is the line parallel to the asymptote through Q, and line L_3 is the tangent to T^+ through Q. If Q were below the asymptote, L_3 would be replaced by the tangent to T^-, whereas L_1 and L_2 would be defined the same way.

Having identified the projection of C, we can draw, one after the other, all potential filtering lines, $L(\phi)$, passing through Q. Starting with $L(\phi)$ overlapping with L_1 and rotating clockwise, L_2 is first met, then L_3, and then L_1 again. Between L_1 and L_3, $L(\phi)$ intersects twice with the projection of C, whereas between L_3 and L_1 there is only one intersection. Thus, the number of intersections between C and a plane that contains $\underline{x}$ and the current source position can only be 1 or 3, and this property is valid for any source position on C. When $L(\phi)$ is between L_1 and L_2, the current source position is between the other two points of intersections. On the other hand, when $L(\phi)$ is between L_2 and L_3 these two other intersections are above the current source position.

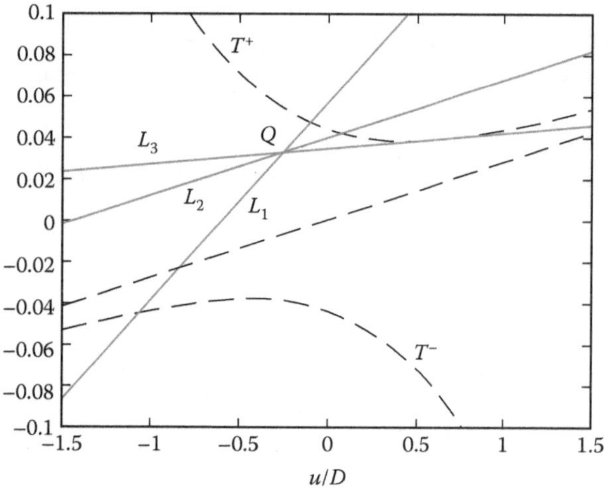

FIGURE 13.14 The detector plane for a source position on C, the π-segment through x. The three solid lines are filtering lines in the case where $M(\lambda, \phi; x)$ is selected to achieve equal weighting. See the text for further details.

A first option for $M(\lambda, \phi; \underline{x})$ is the equal weighting solution of Equation 13.57. Following the discussion above, we observe that this solution is equal to 1/3 when $L(\phi)$ is between L_1 and L_3, and equal to 1 when $L(\phi)$ is between L_3 and L_1. Hence, $q(\lambda, \phi; \underline{x})$ will change values when $L(\phi)$ overlaps with L_1 and L_3, and also when $L(\phi)$ overlaps with L_2, since L_2 is the position where sign(cos ϕ) is discontinuous. Consequently, this first option for $M(\lambda, \phi; \underline{x})$ yields an FBP formula with three filtering lines, and axial truncation is allowed because none of these lines is vertical. The required detector coverage in v is controlled by the slope of the steepest line, L_1, and is reasonably small as long as Q is close to the z-axis. (Figure 13.14 was drawn for a point far away from the z-axis to accentuate the slope of the filtering lines and allow a better visualization.)

As mentioned earlier, Equation 13.57 is not necessarily the best expression for $M(\lambda, \phi; \underline{x})$. For example, in the present case, Katsevich (2002) showed that $M(\lambda, \phi; \underline{x})$ can be selected so that there are only two filtering lines, L_2 and L_3, with the advantage that less detector coverage in v is required. The expression for $M(\lambda, \phi; \underline{x})$ that achieves this goal is equal to zero when $L(\phi)$ is between L_2 and L_3 and is equal to 1 otherwise. That is, the planes that have three intersections with C are not equally weighted: only one of the three available values of r'' on each such plane is used, namely the value that comes from the second intersection when the intersections are ranked by λ value; the other available values are eliminated from the reconstruction process by giving them a weight of zero. The mathematical expression corresponding to Katsevich's solution is

$$M(\lambda,\phi;\underline{x}) = \frac{1}{2}\left(1 + \text{sign}(\cos\phi)\,\text{sign}\left(\underline{\theta}(\lambda,\phi)\cdot\underline{\theta}^{\perp}(\lambda,\phi^{*})\right)\right), \quad (13.78)$$

where ϕ^* is such that $L(\phi^*)$ is tangent to the projection of C with $\sin\phi^* < 0$ and vectors $\underline{\theta}$ and $\underline{\theta}^{\perp}$ are given by Equations 13.54 and 13.69, respectively.

Note that the solution above is still far from being optimal because L_2 has a slope that increases with P, the helix pitch. By craftily selecting $M(\lambda, \phi; \underline{x})$, Katsevich (2004) showed that it is possible to have a single filtering line that requires no more coverage in v than that required to encompass the Tam window and is thus more practical.

13.4.5 Computational Effort

Thus far, our attention has been focussed on the issue of handling axial truncation without worrying about computational effort. However, computer resources are limited in practice, and a TES reconstruction formula is useful only when it requires moderate effort. We discuss here the cost of implementing the FBP formula presented in this section.

Let $N_x \times N_y \times N_z$ be the number of locations, where the reconstruction of f is desired. Also, let N_λ be the total number of projections and let $N_u \times N_v$ be the number of pixels on the detector. Assuming that the computation of $g_H(\lambda, \phi, x)$ in Equation 13.71 requires about N_u operations, and that each point x requires at most N_f filtering lines, the total number of operations needed to implement (Equation 13.60) is

$$C_1 = O\left(N_\lambda N_x N_y N_z N_u N_f\right), \tag{13.79}$$

where the symbol O stands for "in the order of." This number is far too large, making the FBP formula nonpractical in the form it has been presented. Fortunately, as explained hereafter, there is a way to reduce the computational effort at little cost in accuracy.

First, note that two points $\underline{x}$ and $\underline{x}'$ that lie on a same line, as illustrated in Figure 13.11, share a same value for $g_H(\lambda, \phi, \underline{x})$. Thus, there exists some shift invariance within the filtering process, which will now be highlighted. Let ρ and γ be the two parameters that define a line $L(\rho, \gamma)$ in the detector plane, and let $\underline{b}_1$ and $\underline{b}_2$ be the two unit orthogonal vectors within the plane that contains $\underline{a}(\lambda)$ and $L(\rho, \gamma)$, as drawn in Figure 13.15. Next, let

$$\hat{g}_H(\lambda,\rho,\gamma,\eta) = \int_{-\pi}^{\pi} \frac{1}{\sin(\eta-\theta)} g'(\lambda, \cos\theta\, \underline{b}_1 + \sin\theta\, \underline{b}_2)d\theta, \quad |\eta| < \pi, \tag{13.80}$$

which is an angular Hilbert transformation of the measurements on $L(\rho, \gamma)$. Using a change of variable from θ to $\theta' = \eta - \theta$, we find that

$$\hat{g}_H(\lambda,\rho,\gamma,\eta) = g_H(\lambda,\phi,\underline{x}), \tag{13.81}$$

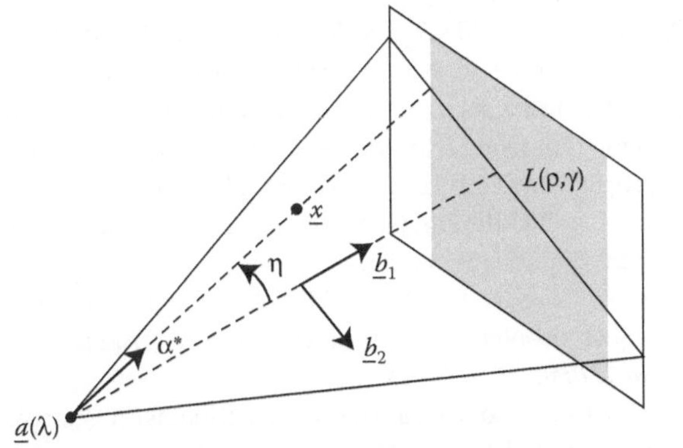

FIGURE 13.15 General filtering line: $L(\rho, \gamma)$ is a line in the detector, and $\underline{b}_1$ and $\underline{b}_2$ are two unit orthogonal vectors within the plane defined by $L(\rho, \gamma)$ and $\underline{a}(\lambda)$. The line that connects $\underline{a}(\lambda)$ to any point $\underline{x}$ inside this plane is identified by the angle η.

when the following conditions are satisfied

- $L(\phi)$ and $L(\rho, \gamma)$ are the same lines,
- $\underline{\alpha}^*(\lambda, x) = \cos\eta\, \underline{b}_1 + \sin\eta\, \underline{b}_2$,
- $\underline{\theta}^\perp(\lambda, \phi) = \sin\eta\, \underline{b}_1 - \cos\eta\, \underline{b}_2$.

Hence, the filtering step can be decoupled from the loop over $\underline{x}$ using Equation 13.81 with interpolation. This decoupling can provide a significant gain in computational effort when the filtering lines needed for all $\underline{x}$ can be described with few parameters at any given λ.

For example, in the helical case of Section 13.4.4, selecting $M(\lambda, \phi; \underline{x})$ as in Equation 13.78 implies that the filtering lines must either be parallel to the asymptote, or be tangent to T^+ or T^-. Thus, the filtering lines required for all $\underline{x}$ can be grouped into three families of lines, each described by a single parameter: the lines parallel to the asymptote, the lines tangent to T^+, and the lines tangent to T^-. For each line in each family, the Fourier transform can be used to compute Equation 13.80 in η. Then, all values of $g_H(\lambda, \phi, \underline{x})$ can be estimated by simple interpolation based on Equation 13.81. This procedure leads to a total computation cost of

$$C_2 = O\left(N_\lambda N_x N_y N_z\right) + O\left(N_\lambda N_f N_\xi N_u \log N_u\right), \tag{13.82}$$

where N_ξ is the nominal number of filtering lines within one family, which should be of the order of N_v, and $N_f = 2$. This cost is comparable to that of approximate reconstruction algorithms used in modern CT scanners, and thus makes the corresponding FBP formula competitive.

On the other hand, using the equal weighting scheme for the helix is not so advantageous because the description of lines L_1 needed for all x requires two parameters, namely one coordinate on T^+ and another one on T^-; see Figure 13.14. Thereby, the computational cost is increased to

$$C_3 = O\left(N_x N_y N_z N_\lambda\right) + O\left(N_\lambda N_\xi^2 N_u \log N_u\right), \tag{13.83}$$

which is barely practical.

In summary, the FBP formula presented in this section can be practical in terms of both detector requirement and computational effort, provided the function $M(\lambda, \phi; \underline{x})$ and the interval $\Lambda(\underline{x})$ can be appropriately chosen.

13.5 Conclusions

The intention of this chapter was to introduce the reader to fundamental tools and techniques for analytical reconstruction from CB projection data. The presentation was thus focused on Tuy's condition, Grangeat's formula, and the general FBP scheme of Katsevich. However, it should be understood that the presented material is far from representing an exhaustive review of current knowledge in CB tomography. Many other exciting results have been published, some of which are briefly reviewed hereafter.

First, there is more than one way to link the CB projection data to the 3D Radon transform. Among all existing links, the Grangeat formula has the property of being the most local, but this does not mean that the other links should be disregarded; they can still play an essential role. See Clack and Defrise (1994) and references therein for a comparative discussion on various links.

Second, there also exists more than one way to develop an FBP reconstruction method. The presented scheme has the advantage of being totally local in its formulation, and may for this reason be regarded as being more attractive than other, more global, schemes, such as those published in Defrise and Clack (1994) and Kudo and Saito (1994a). However, this flexibility requires significant input from the user to obtain an efficient reconstruction formula. The global schemes in Defrise and Clack (1994) and Kudo and Saito (1994a) are more straightforward to use and they too have been shown to be applicable to axially truncated data under various circumstances (Kudo and Saito, 1994b; Kudo and Saito, 1998; Noo et al., 1998). In these global schemes, the filtering step is shift variant; thus, it cannot be applied using a simple convolution algorithm, but Fourier techniques based on linogram coordinates have been successfully used (Axelsson and Danielsson, 1994; Axelsson-Jacobson et al., 1995; Axelsson-Jacobson, 1996).

Yet another FBP alternative to the presented FBP scheme is to use the formulas presented in (Pack et al., 2004; Pack and Noo, 2005; Ye and Wang, 2005). These formulas are less flexible, but their use is much more straightforward as their definition dictates the choice for $\Lambda(\underline{x})$ and they do not require searching for a suitable function $M(\lambda, \phi, \underline{x})$. Also, they put the selection of filtering lines at the forefront and are in that sense attractive to use in the presence of data truncation, but recall that they do not necessarily lead to the most optimal reconstruction formula, in terms of both noise and truncation handling.

Typically, efficient FBP reconstruction formulas do not allow transverse truncation. The situations under which transverse truncation is allowed are more limited, but great findings have been recently made in that direction, using a differentiated backprojection along with a finite Hilbert transform inversion in the image domain. Moreover, this methodology has also been shown to allow more axial truncation than the FBP scheme in helical CB tomography (Zou and Pan, 2004). Details on this valuable reconstruction methodology can be found in Pack et al. (2005), Ye et al. (2005), and Zou and Pan (2004) and references therein.

Both the FBP methods and the reconstruction using a differentiated backprojection requires a differentiation of the CB data with respect to the vertex path parameter, λ. Proper implementation of this differentiation is crucial to mitigate the discretization errors and resolution losses; the topic has been discussed in Noo et al. (2007). An alternative way of dealing with this differentiation step is to integrate the analytical formula by parts to eliminate it; examples of such an approach can be found in Katsevich et al. (2006), Yang et al. (2006), and Zou and Pan (2004).

An important aspect of TES reconstruction formulas that allow for truncation in the data is that they rely on a generalization of the concept of π-lines. That is, they typically assume that $\underline{x} \in \Omega$ lies on a line that contains two source positions and that there is a (connected) piece of the vertex path that links these two positions. The introduction of the reverse helix for CB imaging in radiation therapy recently illustrated that this condition can be highly inconvenient, but some new techniques allowing to circumvent the problem seem to be underway (Cho et al., 2009; Noo et al., 2009).

References

Axelsson-Jacobson, C. 1996. *Fourier Methods in 3-D Reconstruction from Cone-Beam Data.* Linköping Studies in Science and Technology, Linköping University, Sweden.

Axelsson-Jacobson, C., Defrise, M., Danielsson, P. E., Clark, R., and Noo, F. 1995. 3D-reconstruction using cone-beam backprojection, the Radon transform and linogram techniques. *IEEE Med. Imaging Conf.*, San Francisco, CA.

Axelsson, C. and Danielsson, P. E. 1994. Three-dimensional reconstruction from cone-beam data in O(N3 log N) time. *Phys. Med. Biol.*, 39, 477–91.

Burk, F. 1998. Lebesgue measure and integration: An introduction. *Pure and Applied Mathematics: A Wiley Series of Texts, Monographs and Tracts*, Wiley and Sons, USA.

Buzug, T. 2008. *Computed Tomography: From Photon Statistics to Modern Cone-beam CT*. Berlin: Springer.

Chen, G. H. 2003. An alternative derivation of Katsevich's cone-beam reconstruction formula. *Med. Phys.*, 30, 3217–26.

Cho, S., Pearson, E., Pelizzari, C. A., and Pan, X. 2009. Region-of-interest image reconstruction with intensity weighting in circular cone-beam CT for image-guided radiation therapy. *Med. Phys.*, 36, 1184–92.

Clack, R. and Defrise, M. 1994. Cone-beam reconstruction by the use of the Radon intermediate functions. *J. Opt. Soc. Am. A*, 11, 580–5.

Danielsson, P. E., Edholm, P., Eriksson, J., and Magnusson, S. 1997. Towards exact reconstruction for helical cone-beam scanning of long objects. A new detector arrangement and a new completeness condition. *1997 Meeting on Fully Three-dimensional Image Reconstruction in Radiology and Nuclear Medicine.* Pittsburgh, PA.

Deans, S. R. 2007. *The Radon Transform and Some of its Applications.* Mineola, NY: Dover Publications, Inc.

Defrise, M. and Clack, R. 1994. A cone-beam reconstruction algorithm using shift-variant filtering and cone-beam backprojection. *IEEE Trans. Med. Imaging*, 13, 186–95.

Defrise, M. and Gullberg, G. T. 2006. Image reconstruction. *Phys. Med. Biol.*, 51, R139–54.

Defrise, M., Noo, F., and Kudo, H. 2000. A solution to the long-object problem in helical cone-beam tomography. *Phys. Med. Biol.*, 45, 623–43.

Finch, D. 1985. Cone-beam reconstruction with sources on a curve. *SIAM J. Appl. Math.*, 45, 665–73.

Grangeat, P. 1991. Mathematical framework of cone-beam 3D reconstruction via the first derivative of the Radon transform. In Herman, G. T., Louis, A. K., and Natterer, F. (Eds.), *Mathematical Methods in Tomography (Lecture Notes in*

Mathematics, 1497), Berlin: Springer-Verlag.

Herman, G. 2009. *Fundamentals of Computerized Tomography: Image Reconstruction from Projections*. London: Springer-Verlag.

Hoppe, S., Noo, F., Dennerlein, F., Lauritsch, G., and Hornegger, J. 2007. Geometric calibration of the circle-plus-arc trajectory. *Phys. Med. Biol.*, 52, 6943–60.

Hsieh, J. 2009. *Computed Tomography: Principles, Design, Artifacts, and Recent Advances*. Bellingham, WA: SPIE Press Monograph.

Hubbell, J. H. 2006. Review and history of photon cross section calculations. *Phys. Med. Biol.*, 51, R245–62.

Jaffray, D. A., Siewerdsen, J. H., Wong, J. W., and Martinez, A. A. 2002. Flat-panel cone-beam computed tomography for image-guided radiation therapy. *Int. J. Radiat. Oncol. Biol. Phys.*, 53, 1337–49.

Kalender, W. A. 2006. X-ray computed tomography. *Phys. Med. Biol.*, 51, R29–43.

Katsevich, A. 2002. Theoretically-exact filtered backprojection-type inversion algorithm for spiral CT. *SIAM J. Appl. Math.*, 62, 2012–26.

Katsevich, A. 2003. A general scheme for constructing inversion algorithms for cone-beam CT. *Int. J. Math. Math. Sci.*, 21, 1305–21.

Katsevich, A. 2004. An improved exact filtered backprojection algorithm for spiral computed tomography. *Adv. Appl. Math.*, 32, 681–97.

Katsevich, A. 2005. Image reconstruction for the circle-and-arc trajectory. *Phys. Med. Biol.*, 50, 2249–65.

Katsevich, A., Taguchi, K., and Zamyatin, A. A. 2006. Formulation of four Katsevich algorithms in native geometry. *IEEE Trans. Med. Imaging*, 25, 855–68.

Kudo, H. and Saito, T. 1994a. Derivation and implementation of a cone-beam reconstruction algorithm for nonplanar orbits. *IEEE Trans. Med. Imaging*, 13, 196–211.

Kudo, H. and Saito, T. 1994b. An extended completeness condition for exact cone-beam reconstruction and its application. *Nucl. Sci. Symp. Med. Imaging Conf.* Norfolk, VA, USA.

Kudo, H. and Saito, T. 1998. Fast and stable cone-beam filtered backprojection method for non-planar orbits. *Phys. Med. Biol.*, 43, 747–60.

Lauritsch, G., Boese, J., Wigstrom, L., Kemeth, H., and Fahrig, R. 2006. Towards cardiac C-arm computed tomography. *IEEE Trans. Med. Imaging*, 25, 922–34.

Lu, Y., Zhao, J., and Wang, G. 2009. Exact image reconstruction with triple-source saddle-curve cone-beam scanning. *Phys. Med. Biol.*, 54, 2971–91.

Natterer, F. and Wubbeling, F. 2007. *Mathematical Mehods in Image Reconstruction.* SIAM Monographs on Mathematical Modeling and Computation.

Noo, F., Clack, R., and Defrise, M. 1997. Cone-beam reconstruction from general discrete vertex sets using Radon rebinning algorithms. *IEEE Trans. Nucl. Sci.*, 44, 1309–16.

Noo, F., Defrise, M., and Clack, R. 1998. Direct reconstruction of cone-beam data acquired with a vertex path containing a circle. *IEEE Trans. Image Process*, 7, 854–67.

Noo, F., Defrise, M., Clack, R., Roney, T. J., White, T. A., and Galbraith, S. G. 1996. Stable and efficient shift-variant algorithm for circle-plus-lines orbits in cone-beam CT. *The 1996 IEEE Int. Conf. Image Process., ICIP '96.* Lausanne, Switzerland.

Noo, F., Hoppe, S., Dennerlein, F., Lauritsch, G., and Hornegger, J. 2007. A new scheme for view-dependent data differentiation in fan-beam and cone-beam computed tomography. *Phys. Med. Biol.*, 52, 5393–414.

Noo, F., Wunderlich, A., Lauritsch, G., and Kudo, H. 2009. On the problem of axial data truncation in the reverse helix gemotry. *10th Int. Meet. on Fully Three-dimensional Image Reconstruction in Radiology and Nuclear Medicine*. Beijing, China.

Orth, R. C., Wallace, M. J., and Kuo, M. D. 2008. C-arm cone-beam CT: General principles and technical considerations for use in interventional radiology. *J. Vasc. Interv. Radiol.*, 19, 814–20.

Pack, J. D. and Noo, F. 2005. Cone-beam reconstruction outside R-lines using the backprojection of 1-D filtered data. In Noo, F., Zeng, G. L., and Kudo, H. (Eds.), *Eighth Int. Meeting on Fully Three-dimensional Image Reconstruction in Radiology and Nuclear Medicine.* Salt Lake City, Utah.

Pack, J. D., Noo, F., and Clackdoyle, R. 2005. Cone-beam reconstruction using the backprojection of locally filtered projections. *IEEE Trans. Med. Imaging*, 24, 70–85.

Pack, J. D., Noo, F., and Kudo, H. 2004. Investigation of saddle trajectories for cardiac CT imaging in cone-beam geometry. *Phys. Med. Biol.*, 49, 2317–36.

Pan, X., Sidky, E. Y., and Vannier, M. 2009. Why do commercial CT scanners still employ traditional, filtered back-projection for image reconstruction? *Inverse Probl.*, 25, #123009.

Tam, K. C., Samarasekera, S., and Sauer, F. 1998. Exact cone beam CT with a spiral scan. *Phys. Med. Biol.*, 43, 1015–24.

Tuy, H. 1983. An inversion formula for cone-beam reconstruction. *SIAM J. Appl. Math.*, 43, 546–52.

Wang, X. and Ning, R. 1999. A cone-beam reconstruction algorithm for circle-plus-arc data-acquisition geometry. *IEEE Trans. Med. Imaging*, 18, 815–24.

Yang, D., Ning, R., and Cai, W. 2009. Circle plus partial helical scan scheme for a flat panel detector-based cone beam breast X-ray CT. *Int. J. Biomed. Imaging*, 637867.

Yang, H., Li, M., Koizumi, K., and Kudo, H. 2006. View-independent reconstruction algorithms for cone beam CT with general saddle trajectory. *Phys. Med. Biol.*, 51, 3865–84.

Ye, Y. and Wang, G. 2005. Filtered backprojection formula for exact image reconstruction from cone-beam data along a general scanning curve. *Med. Phys.*, 32, 42–8.

Ye, Y., Zhao, S., Yu, H., and Wang, G. 2005. A general exact reconstruction for cone-beam CT via backprojection-filtration. *IEEE Trans. Med. Imaging*, 24, 1190–8.

Yu, H., Zhao, S., Ye, Y., and Wang, G. 2005. Exact BPF and FBP algorithms for nonstandard saddle curves. *Med. Phys.*, 32, 3305–12.

Zellerhoff, M., Scholz, B., Ruehrnschopf, E.-P., and Brunner, T. 2005. Year. Low contrast 3D reconstruction from C-arm data. In *Proc. SPIE*, 646.

Zeng, G. L. and Gullberg, G. T. 1992. A cone-beam tomography algorithm for orthogonal circle-and-line orbit. *Phys. Med. Biol.*, 37, 563–77.

Zhao, S., Yu, H., and Wang, G. 2005. A unified framework for exact cone-beam reconstruction formulas. *Med. Phys.*, 32, 1712–21.

Zou, Y. and Pan, X. 2004. Exact image reconstruction on PI-lines from minimum data in helical cone-beam CT. *Phys. Med. Biol.*, 49, 941–59.

14

Multimodality Imaging

Katia Passera
Mario Negri Institute

Anna Caroli
Mario Negri Institute

Luca Antiga
Mario Negri Institute

14.1 Introduction

There is an old Indian tale about six blind men who were asked to describe what an elephant looks like by touching it. The first blind man touched the elephant's leg and reported that it "looked" like a tree trunk. The second one touched the elephant's stomach and said that the elephant was a wall. The third one touched the elephant's ear and said: it was a fan. The fourth blind man touched the elephant's tail and described the elephant as a piece of rope. The fifth one felt the elephant's tusks and described it as a spear. And the sixth blind man rubbed the elephant's snout and got very scared because he thought it was a snake. All of them got into a big argument about the "appearance" of an elephant. Each blind man had touched the elephant but each of them gave a different description of the animal. So, which answer is right? All of them!

This tale, which introduces an editorial about multimodal imaging (MMI) (Azhari et al., 2007), represents the concept of MMI in a simple but effective manner.

Looking into the body with a particular imaging system is like the blind man that only touches a part of the elephant. In fact, each imaging method reports on a limited domain and has unique strengths and limitations. It is based on different physical principles, has different spatial and temporal resolutions, and is able to capture peculiar features of the part of the body under investigation. Only by integrating information from the different blind men (the different imaging systems) is it possible to have a clear view on the entire picture.

This chapter aims at providing the reader with a general overview of MMI, by first introducing MMI in clinical practice and research, and then discussing the concept of image integration through a distinction among visual, hardware, and software aspects of such integration. For the latter, the basics of multimodal registration will be described. Finally, in order to put concepts into perspective, two example applications of MMI in clinical research are described in some detail: the first example is mainly focused on clinical aspects, whereas the second delves into technical issues.

14.2 MMI in Clinical Practice and Research

Nowadays, more than one medical imaging modality is often used for diagnosis or individual therapy planning. Many clinical applications benefit from integrated visualization and combined analysis of images produced by such multiple modalities, which has justified the use of the term "multimodal imaging." MMI indeed consists of acquiring and fusing complementary information to create new images, which are more informative than each of the individual original images (Azhari et al., 2007).

MMI is not only becoming standard practice in clinics, but it is also representing a new research field aimed at providing the most complete picture of physiological and pathological states as they are achieved using the current state-of-the-art imaging techniques. To achieve this goal, different areas of research are involved. Research is indeed very active in the development of new contrast agents able to mark specific physical and biochemical features of living tissues and also being "visible" on more than one imaging modality, the so-called multimodal contrast agents. Moreover, research has addressed the issue of image integration leading to software tools for multimodal registration and visualization and to the development of new hybrid scanners permitting the simultaneous acquisition of multimodal images.

Beyond the clinics, researchers are engaged in designing and building dedicated devices, such as high-resolution scanners (microPET, microSPECT, microMRI) allowing to couple gene expression with small animal imaging. The coupling of nuclear and optical reporter genes today represents only the beginning of a whole research field with far-reaching applications (Kang and Chung, 2008; Moseley and Donnan, 2004).

14.2.1 Anatomical and Functional: From Macro to Micro

Standard imaging modalities may be roughly divided into two main classes: anatomical and functional. Structural magnetic resonance imaging (sMRI), ultrasound (US), and x-ray imaging, including computed tomography (CT), are usually used for the depiction of anatomy, whereas single-photon emission CT (SPECT), positron emission tomography (PET), and magnetic resonance spectroscopy (MRS) provide functional information, but typically present limitations in delineating the anatomy.

This classification is forcibly not precise, and continuous technology developments in this field may lead to move one imaging modality from anatomical to functional or pose it at the boundary between the two.

Relatively recent imaging techniques include molecular imaging, diffusion MRI, functional MRI (fMRI), and perfusion MRI or CT. Molecular imaging techniques include PET, SPECT, and optical imaging and allow direct imaging of biological processes *in vivo* as well as visualization of the molecular target (Tempany and McNeil, 2001). In diffusion MRI, the imaged biological tissues are weighted, with the local microstructural characteristics of water diffusion; this permits, for example, the distinction between healthy brain tissue and brain areas affected by acute stroke (Tempany and McNeil, 2001). Thanks to fMRI, it is possible to have information about neural activity in the brain by measuring the blood oxygenation-level-dependent (BOLD) change in image intensity response. In contrast to many of the mentioned imaging modalities, fMRI provides good spatial resolution (0.5–2 mm, limited only by the signal-to-noise ratio, or, equivalently, by acquisition time) of functional information in terms of increased blood oxygenation activity with a relatively good temporal resolution (0.1–10 s) (Moseley and Donnan, 2004). Finally, in perfusion MRI and CT, the anatomical area under investigation is sequentially imaged during contrast agent distribution, thus enabling study of contrast agent pharmacokinetics in particular regions of interest (ROI) and information derivation about tissue properties (e.g., microvasculature disorders due to tumor angiogenesis).

The convergence of established fields of *in vivo* imaging technologies with molecular and cell biology has led to increased interest in MMI (Kang and Chung, 2008; Moseley and Donnan, 2004).

14.2.2 Multimodal Imaging: State of Art

Literature provides a wide range of articles and reviews about MMI, which are mainly dedicated to single organs like the heart (Blankstein and Di Carli, 2010; Gaemperli and Kaufmann, 2010; Namdar et al., 2005; Rispler et al., 2007; Santana et al., 2009; Schuijf et al., 2006; Woo et al., 2008) or the brain (Blinowska et al., 2009; Liebeskind, 2009; Muzik et al., 2007; Pietrzyk et al., 1996; Saxena et al., 2007; Vince and Tülay, 2009) or to a particular imaging field (Papathanassious and Liehn, 2008; Zaidi and Prasad, 2009). This is due to the fact that each field of application (either clinical or research) adopts different imaging modalities and several combinations are possible.

In the following, we provide an overview on the use of MMI in several of the most important fields of application.

14.2.2.1 Brain Imaging

Researchers have always been fascinated by the possibility of understanding the puzzling mechanisms underlying human behavior. Since the mid-1990s, there has been an increasing interest in synchronous MMI, whereby two or more modalities are used simultaneously, which has arisen in large part from investigations on spontaneous brain activity and in particular on epilepsy (Blinowska et al., 2009; Muzik et al., 2007). Multimodal clinical nuclear medicine, PET, and MRI techniques have been involved in the growing fields of molecular and functional imaging for neuroassessment of gliomas (Floeth et al., 2005; Hsu et al., 2007), integrated stroke imaging (Chao et al., 2006; Liebeskind, 2009), and functional neuroimaging (Blinowska et al., 2009; Vince and Tülay, 2009). In addition, newer tools, such as near-infrared spectroscopy (NIRS) (Villringer, 1993) or optical imaging, that allow real-time assessment of wavelength-specific absorption of photons by oxygenated and deoxygenated tissues have been introduced (Saxena et al., 2007; Weissleder and Pittet, 2009).

With ever-improving imaging technologies and high-performance computational power, the complexity and scale of brain imaging data have continued to grow at an explosive pace, and, depending on the specific application interest, different MMI approaches have been used. A current approach to study human brain activity in physiological or pathological conditions is the integration of two-dimensional (2D) electroencephalography (EEG) and magnetoencephalography signals with imaging data (Blinowska et al., 2009; Moseley and Donnan, 2004). These signals have a direct link with neuronal synchrony as they reflect electric potential and magnetic fields resulting from synaptic transmembrane currents in neurons. Functional neuroimaging is a noninvasive way to assess brain function using MRI signal changes associated with functional neuronal activity. The most widely used method is based on BOLD, which is a particularly promising technique, as neural activity influences the metabolic demand through neurovascular coupling, and metabolic changes impact the hemodynamic response, which is dependent on physiological factors such as local cerebral blood flow, deoxyhemoglobin/oxyhemoglobin ratio, blood volume, and vascular geometry (Blinowska et al., 2009). However, it has to be stressed that inference about neuronal activity from fMRI examinations is limited by our real understanding of the coupling between BOLD signal and the real neuronal activity. For example, in epileptic spikes, simultaneous fMRI and EEG revealed a negative BOLD response beyond the stimulated regions of visual cortex, associated with local decreases in neural activity as expressed in terms of local field power below the level of spontaneous (background) activity. The relationship between neuronal inhibition and BOLD is currently under debate since inhibition could be associated with increased metabolic demand, which may be reflected as BOLD increase, while, on the other hand, most connections in the brain are excitatory and decrease of excitatory activity caused by inhibition may lead to a decrease of blood flow. Experimental results point out that both arguments may be valid. Nonetheless, integration of fMRI and EEG, with its insight into BOLD dynamics and relationships to evoked neuronal magnetic fields, has led to numerous reports and studies (Babiloni et al., 2003; Babiloni et al., 2005; Blinowska et al., 2009; Formaggio et al., 2008; Liu and He, 2008). Newer approaches using information from the NIRS signals acquired during fMRI are currently under examination (Roche-Labarbe et al., 2010; Saxena et al., 2007; Tak et al., 2010).

MMI data are used even in presurgical evaluation of patients with medically refractory epilepsy in order to define epileptogenic brain regions to be resected. The gold standard is subdural EEG recordings. However, as the accuracy of foci localization using subdural electrodes depends greatly on the location of the electrodes on the brain surface, a combination of noninvasive anatomical and functional imaging, such as MR and PET, is frequently used in order to guide their placement (Muzik et al., 2007).

Another field in which MMI plays a key role is stroke. Imaging goals in stroke typically address whether a stroke is the likely diagnosis, if the primary process is ischemic or hemorrhagic, where the lesion is situated, what is the likely mechanism, what treatments may be indicated, and what can be expected from prognosis (Liebeskind, 2009). CT and MRI, incorporating parenchymal depictions, mapping of the vasculature, and perfusion data, can provide a more complete picture of ischemic pathophysiology. For example, the initial event of ischemia, that is, obstruction of the cerebral circulation and creation of collateral circulation, can be detected by angiographic techniques (CT and MR angiography), the consequent blood flow changes can be detected by perfusion imaging through estimation of cerebral blood volume, while diffusion MRI may provide rapid delineation of cytotoxic edema.

14.2.2.2 Cardiac Imaging

Advances in cardiovascular imaging have resulted in the development of multiple noninvasive techniques to evaluate myocardial perfusion and coronary anatomy, each of which has unique strengths and limitations (Blankstein and Di Carli, 2010).

CT angiography is capable of directly visualizing the presence of atherosclerotic lesions, but cannot provide functional data and is unable to detect obstructive disease in calcified or very small vessels. SPECT or PET can perform myocardial perfusion imaging, and also cardiac MRI and CT have been used to visualize myocardial perfusion under stress and at rest. These techniques enable a physiological assessment, but have a limited capability of detecting the presence of subclinical atherosclerosis and a tendency to underestimate the extent of atherosclerosis.

In this case, combining the complementary information of coronary anatomy and myocardial perfusion imaging seems advantageous, as the weaknesses of each modality can be offset by the strengths of the other (Blankstein and Di Carli, 2010).

For this reason, the use of hybrid scanners such as SPECT-CT and PET-CT (Kaufmann, 2009; Kaufmann and Di Carli, 2009; Namdar et al., 2005) has increased, while PET-MRI scanners are currently under investigation (Judenhofer et al., 2008; Nekolla et al., 2009). In addition, several software products have been created in order to enable the fusion of imaging data acquired on different scanners; for example, stand-alone SPECT or PET images can simply be overlaid on cardiac CT scans. However, some issues, such as artifacts generated by patient movements and cardiac or respiratory motion, can still pose challenges to the process (Blankstein and Di Carli, 2010; Woo et al., 2009).

The literature presents evidence that dual-modality imaging offers superior diagnostic information compared to individual modalities (Namdar et al., 2005; Rispler et al., 2007) and that the combination of anatomical stenosis and myocardial perfusion seems to result in an incremental improvement in the prediction of adverse cardiovascular events (Gaemperli et al., 2007). In addition, the recent advent of molecular imaging seems to increase the possibility of MMI combinations. For example, it may be possible to use CT to delineate plaques, and, by using a targeted imaging probe, to reveal the location of inflammation in relation to those plaques. Molecular imaging techniques could ultimately provide a reliable means of identifying vulnerable plaques, and thus aid the selection of patients who might benefit from future, highly selective, potent therapies. In this context, the proposed integration of PET and MRI might further expand the possibilities for combining anatomical and functional data while localizing molecular targets (Blankstein and Di Carli, 2010).

There are finally other forms of MMI that aim to address more specific cardiac applications, such as the one proposed by Woo et al. (2008) for ablative heart surgery in atrial fibrillation, who combined anatomical surface models acquired by MRI/CT with localized electrical information measured by an electroanatomic mapping system in order to provide the operator with both anatomical and electrical information.

14.2.2.3 Oncology

Oncology as a field benefited to a great extent from the introduction of MMI. In clinical practice, to understand pathologic processes and their extension in individual patients, both nuclear medicine techniques and CT or MRI are used. In fact, nuclear medicine techniques, such as SPECT and PET, give insight into the metabolic and functional changes related to the pathologic process, while CT and MRI give information about anatomical changes due to the disease. Thus, the integration of information about the nature of the lesion brought by metabolic evaluation adds value to the anatomical evaluation. On the other hand, owing to the weak spatial resolution of nuclear medicine images, knowing the existence of the abnormality but not its precise location and extent may decrease their impact on the management of the disease. For this reason, many publications have focused on the role for PET/CT and SPECT/CT (Delbeke et al., 2009; Schöder et al., 2004; Townsend and Beyer, 2002).

PET/CT was shown to be very promising in many fields of oncology imaging, for staging, monitoring therapy, recurrence detection, in particular with increased specificity compared to PET alone, and sensitivity compared to CT alone (Papathanassiou and Liehn, 2008).

SPECT/CT has proved to be of great help in early detection or exclusion of bone metastases, in classifying lesions rated "equivocal" or "indeterminate" on radionuclide bone scans, in the localization of sentinel nodes in patients with melanoma and breast cancer and in the localization of lymph nodes and bone metastases in patients with differentiated thyroid cancer (Gnanasegaran et al., 2010) (Figure 14.1).

Another possibility for MMI in oncology is the combination of PET with MRI (Antoch and Bockisch, 2009). Although more technologically demanding than the combination of PET and CT, it has been proposed for both brain and body applications to introduce whole-body PET-MRI (Schmidt et al., 2007). The main advantages of PET-MRI with respect to PET-CT are the increased performance in the characterization of soft tissues, the possibility for contextual use of fMRI techniques, and the absence of ionizing radiation, allowing for more intensive follow-ups without issues related to patient safety. In particular, it may be possible to distinguish between loss of viable cells and downregulation of glucose transporter or hexokinase activity by using PET-MRI, in which diffusion-weighted MRI is used to assess the viable cell density (Brindle, 2008).

The potential of PET/CT, SPECT/CT, and PET/MRI could be further increased by using radiological contrast media, and specific CT/MR acquisition protocols for selected applications.

For example, in the last few years, perfusion imaging, in particular dynamic contrast-enhanced (DCE)-magnetic resonance imaging (MRI) and dynamic contrast-enhanced computed tomography (DCE-CT), has gained importance in oncology, thanks to the development of novel target therapies that have an antivascular effect (Brindle, 2008; Buckley and Parker, 2005; O'Connor et al., 2007; Sahani et al., 2005).

Tumor responses to treatment have conventionally been assessed from measurements of tumor size, using morphological imaging techniques such as CT and MRI. This type of measurements presents the limitation of not being able to highlight short-term effects of antivascular therapies that do not lead to a reduction of lesion size but to a decrease of tumor vascularization. For this reason, DCE-MRI measurements of tumor perfusion were used in several phase I trials of antivascular drugs, and were aimed to be employed in clinics, where they would allow the clinicians to rapidly assess the effectiveness of a new therapy

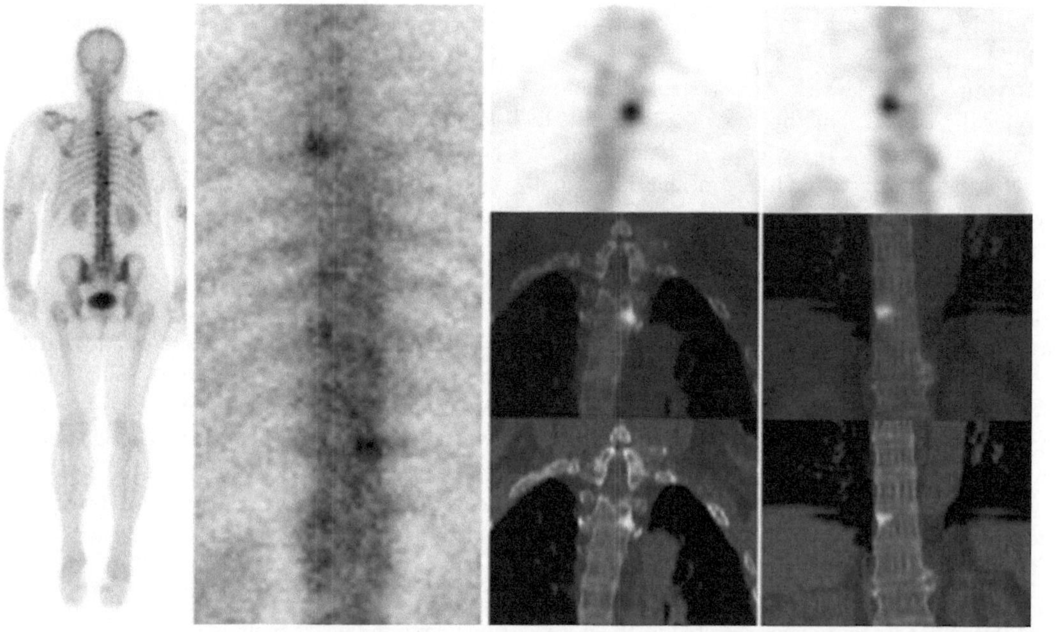

FIGURE 14.1 **(See color insert.)** Example of MMI in oncology. Bone SPECT/CT in the research for bone metastases in breast cancer; the whole body (left) and centered (second image) planar images show two abnormal foci on the spine without immediate differentiation between osteoarthrosis and metastases; the combined SPECT/CT (right columns: top: SPECT, bottom: CT; middle: fused SPECT/CT) shows without need for further imaging that the two foci correspond to malignant osteocondensation. (From Papathanassiou, D. and Liehn, J. C. 2008. *Crit. Rev. Oncol./Hematol.*, 68: 60. With permission.)

in individual patients. Ineffective treatments could be abandoned at an early stage and more effective treatments selected, with obvious benefits for the patient and cost benefits for the healthcare system (Brindley, 2008).

In brain tumors, other specific imaging modalities, such as NIRS, could be combined with structural imaging (Saxena et al., 2007).

More recently, increasing interest has been gained by molecular imaging, which provides highlights into the complexity, diversity, and *in vivo* behavior of tumors. Several researchers are now engaged in this field and a lot of recent literature deals with this topic (Cai and Chen, 2008; Ottobrini et al., 2006; Weissleder and Pittet, 2008). If at present molecular imaging systems enable doctors to see where a tumor is located in the body, the hope for the future is to visualize expression and activity, cells and biological processes that influence the behavior of tumors and/or responsiveness to therapeutic drugs, owing to MMI approaches (Weissleder and Pittet, 2008).

14.2.2.4 Molecular Imaging

The field of molecular imaging, which has flourished over the last decade, represents the next frontier in diagnostic imaging. Molecular imaging allows noninvasive mapping of cellular and subcellular molecular events, including gene expression, protein–protein interaction, dynamic cell tracking throughout the entire organism, and drug action analysis.

With respect to classical imaging technologies that can differentiate, under clinical circumstances, pathological from normal conditions but are rather nonspecific, molecular imaging techniques have led to the development of novel procedures with enhanced specificity. As a result, imaging nowadays focuses on an in-depth understanding of biological processes, early detection and characterization of diseases, and treatment evaluation based on molecular processes assessment (Nikiforidis et al., 2008; Ottobrini et al., 2006).

First developed to localize antigens in light microscopy, molecular imaging has currently been extended to several imaging modalities: molecular MRI (mMRI), MRS, optical bioluminescence imaging, optical fluorescence imaging, targeted US, SPECT, and PET.

In these modalities, sensitivity ranges from the detection of millimolar to submillimolar concentrations of contrast media with CT and MRI, respectively, to picomolar concentrations in SPECT and PET: a 10^8–10^9 difference (Zaidi and Prasad, 2009).

However, individual modalities have limitations. For example, it is difficult to accurately quantify a fluorescence signal in living subjects with fluorescence imaging alone, particularly in deep tissues; MRI has high resolution and good soft-tissue contrast but very low sensitivity; and radionuclide-based imaging techniques are very sensitive but have relatively poor spatial resolution. A combination of multiple molecular imaging modalities can offer synergistic advantages over individual modalities alone (Cai and Chen, 2008).

For this reason, different hybrid scanners (SPECT-CT, PET-MRI, fluorescence tomography combined with MR or CT) are emerging, and these multimodal platforms have improved data reconstruction and visualization (Weissleder and Pittet, 2008).

Molecular imaging always requires accumulation of contrast agent in the target site, but MMI with a small-molecule-based probe is very challenging, sometimes impossible, because of the limited number of conjugation sites and potential interference with receptor-binding affinity. The versatility of different imaging modalities has been significantly enhanced by innovative nanoparticle development (Cai and Chen, 2008; Jae-Hyun et al., 2006). These nanoprobes can be used to image specific cells and tissues within a whole organism. In fact, nanoparticles have large surface areas to which multiple functional moieties can be attached for multimodality molecular imaging. For example, in multimodal molecular imaging of tumor angiogenesis, a nanoparticle-based probe has been used for both MRI and optical imaging of integrin $\alpha_v\beta_3$ and MRI-detectable and fluorescent liposomes carrying RGD peptides were evaluated for *in vivo* tumor imaging (Cai and Chen, 2008).

14.3 Image Integration

Previous sections have highlighted the advantages and potential usefulness of an MMI approach in clinical practice and research. However, there are several technical challenges associated with MMI, in particular on the integration of anatomical and functional information obtained by the different modalities. Such integration can take place at different, nonmutually exclusive, levels: visual, hardware, and software.

14.3.1 Visual Integration

This approach is the simplest way to integrate information: physicians may visualize images coming from different stand-alone scanners side by side and mentally perform a fusion of the two. However, such a way of combining information is somewhat cumbersome and presents limitations: the position of the patient is different in the two imaged volumes, and so does the position of the inner organs; the slices thickness depends on the device used, and so does their orientation in the body. In addition, it is not an objective method to integrate information as it strongly depends on physicians' skills. Therefore, more sophisticated approaches are needed to obtain accurate spatial (and possibly temporal) alignment for image fusion.

14.3.2 Software Integration

Software registration techniques have been in use even before the advent of hybrid imaging systems in clinical research settings. In fact, the first widely attempted approach to further combine functional and anatomical images was software-based (Slomka and Baum, 2009). With this approach, image properties and tissue geometry and texture are used as clues for aligning the datasets. Alignment is thus achieved by manipulating the acquired data under certain optimization constraints or 3D models to achieve the best (most probable) match (Azhari et al., 2007).

Automated tools have recently evolved from previous research applications and can now be used in daily clinical practice. The advantages of software integration techniques are their versatility (They may be applied successfully to scans performed during different scanning sessions at different locations and timepoints.) and its flexibility in integrating PET and SPECT with other modalities. For example, PET/SPECT images acquired on stand-alone scanners can be integrated by software tools both with MRI (e.g., for brain imaging) and CT (e.g., for lung imaging) images (Slomka and Baum, 2008).

However, software integration of PET and CT or MRI images suffers from its limitations: scanners may be located in different institutions; image properties (e.g., resolution) are different and artifacts (e.g., distortion) or reference system-related differences (leading to translations and rotations) may affect the registration process; partial volume effects lead to modality-specific loss of information; and dynamic changes cannot be easily followed up in time. While for brain images software integration does not present particular issues owing to its rigid structure (apart from distortion artifacts), for other organs, especially in the abdomen or thorax, perfect spatial registration has usually been achieved with difficulty, mainly because of uncontrollable internal movements occurring between different acquisitions. In these cases, registration based on nonlinear image deformation is typically required. Even if the need for nonlinear deformation has been considered to be one of the primary limitations of software coregistration compared to hybrid imaging systems, it should be pointed out that during dual-modality hybrid imaging nonlinear deformations can occur and may need to be compensated with dedicated acquisition protocols, which include respiratory and cardiac gating (Slomka and Baum, 2009).

An aspect in favor of software integration is that it can be performed retrospectively. In cardiology, for example, MMI is typically used in subgroups of patients in which one test cannot provide a decisive answer (Slomka and Baum, 2009).

Currently, software registration is well validated for brain applications, where rigid alignment is usually sufficient (Woods et al., 1998) and specific registration tools, including nonlinear algorithms, are also applied in whole-body oncological and cardiac imaging (Rueckert et al., 1999).

It is reasonable to think that software integration will not be replaced by hardware integration but it is likely to be used in a complementary fashion with hybrid dual-modality imaging systems, particularly in cardiac and brain applications (Slomka and Baum, 2009).

Software-based registration methods have become commonly used in clinical environments. Many different approaches for achieving image registration can be found in the literature. Section 14.4 gives an overview of these approaches focusing on their application in MMI.

14.3.3 Hardware Integration

The "hardware" fusion approach utilizes a hybrid design comprising two (or more) imaging modalities that are combined in a single device. The main advantage of this approach is that the imaging modalities acquire data concomitantly within the same scanning session (Azhari et al., 2007). However, combining two or more imaging modalities into one multimodality machine is technically challenging. Imaging systems have to perform without mutual interferences while maintaining their full individual performance. Furthermore, the available space for accommodating the hardware from different scanners is often restricted. An additional challenge is that software components have to be integrated into one single analysis software package enabling an advanced workflow in routine radiology (Schlemmer et al., 2009).

A first realization of an MMI system was the combination of CT and PET; following the development of a prototype in the late 1990s, the first commercial combined PET/CT scanner was introduced in 2001 (Azhari et al., 2007; Townsend and Beyer, 2002). In addition, CT has been associated with SPECT (Delbeke et al., 2009; Kaufmann and Di Carli, 2009), which relies on the use of the widely available gamma-emitters. While at the beginning these two MMI hybrids, systems were accepted critically and many skeptical clinicians questioned about their diagnostic-added value, today they have become of common use in clinical practice and in preclinical and basic biomedical research (Cherry, 2009). In particular, the use of these techniques has shown an incremental diagnostic value compared to PET or SPECT alone or PET or SPECT retrospectively correlated with a CT, in particular in clinical tasks such as lesion detection, localization of foci of uptake (resulting in better differentiation of physiological from pathologic uptake), precise localization of the malignant foci (e.g., in the skeleton vs. soft tissue or liver vs. adjacent bowel or node), characterization of serendipitous lesions and confirmation of small, subtle, or unusual lesions (Delbeke et al., 2009; Schöder et al., 2004). In addition, PET-CT and SPECT-CT have demonstrated particular impact in monitoring response to therapy, in guiding biopsy, and in providing better maps to modulate field and dose in radiation therapy (Delbeke et al., 2009; Heron et al., 2004; Papathanassiou and Liehn, 2008). The recent incorporation of high-speed, multislice CT scanners with PET opens up the potential for applying this technology to cardiac disease as well (Azhari et al., 2007).

Another attractive modality for MMI is MRI. Although MRI imposes severe restrictions on the imaging environment, it offers a broad spectrum of scan types and image contrasts. Compared with CT, MRI offers greater soft-tissue contrast, better capability for function quantification (e.g., measurement of blood flow or tissue metabolism), potentially new types of molecule-targeted contrast agents, and no additional radiation exposure, which is of importance especially in DCE studies and repeated examinations (Azhari et al., 2007; Schlemmer et al., 2008). It has been recently demonstrated that MRI and PET can technologically be integrated into one single hybrid system, but a few doubts have been raised about this new MMI approach (Antoch and Bockisch, 2009; Judenhofer et al., 2008; Nekolla et al., 2009). In particular, it needs to be proven that a combined reading of PET and MRI improves diagnostic accuracy or efficacy of treatments via better monitoring and that the integration of both modalities is economically superior to the two individual systems. Both modalities are complex and time consuming, and the sequential application of individual modalities does have huge burdens for both the patients and the healthcare providers. Therefore, using a PET-MRI scanner may lead to reduce the overall scan time and the time needed for visualization and interpretation of separate and/or fused images (Schlemmer et al., 2008) (Figure 14.2).

Another combination that is under development is US/MRI, particularly for breast imaging. In addition, other, perhaps less obvious, combinations including CT/MRI and PET/optical are being studied (Cherry, 2009).

Considering the current trends in radiology, it can be expected that MMI devices will become increasingly available in the clinical arena.

14.4 Multimodal Registration

Image registration is a procedure that allows recovering the geometric relationship between corresponding points in multiple images of the same scene.

In the literature, a wide range of registration algorithms has been proposed, and several reviews summarize the key concepts of registration methods and provide organic classifications (Crum et al., 2004; Fitzpatrick et al., 2000; Maintz and Viergever, 1998; Slomka and Baum, 2009; Zitova and Flusser, 2003). Here, these concepts are briefly illustrated in order to point out the peculiar features of multimodal registration algorithms.

Registration algorithms can be classified according to their being point-, surface-, or intensity-based (Fitzpatrick et al., 2000).

In point-based methods, registration can be achieved by selecting a transformation that aligns a set of corresponding point pairs that are *a priori* identified for a given pair of images. Then, interpolation is used to infer correspondence throughout the rest of the image volume in a way consistent with the matched points.

Surface-based image registration methods involve determining corresponding surfaces in different images (and/or physical space) and computing the transformation that best aligns these surfaces. The surface representation can be simply a point set (i.e., a collection of points on each of the surfaces), a faceted surface (e.g., triangle set), an implicit surface, or a parametric surface (e.g., B-spline). Extraction of surfaces such as skin or bone is relatively easy and fairly automatic for head CT and MR images. Extraction of boundary surfaces between soft tissues is generally more challenging and less automatic, thus sophisticated image segmentation algorithms are usually required.

In intensity- or voxel-based methods, the transformation between images is computed using the intensities of image voxels. In this case, the registration transformation is determined by iteratively optimizing some "similarity measure" calculated from all voxel values or on a subset of voxels (a regular grid on the image or an ROI).

Intensity- methods, have recently become widely used among registration methods (Zitova and Flusser, 2003). A major attraction of intensity-based algorithms is that the amount of preprocessing or user-interaction required is lower than for point- or surface-based methods. (Sometimes they are fully automatic.) In addition, no organ segmentation is required, and therefore, registration is not affected by inevitable errors in organ segmentation (Fitzpatrick et al., 2000; Slomka and Baum, 2009).

The key elements of intensity-based registration algorithms are the similarity measure, which assesses how well two images match, and the transformation model, which specifies the way in which the source image is deformed to match the target (Crum et al., 2004).

Intensity-based approaches define matching metrics using mathematical or statistical criteria based on image intensity. The main measures of similarity include (1) squared differences

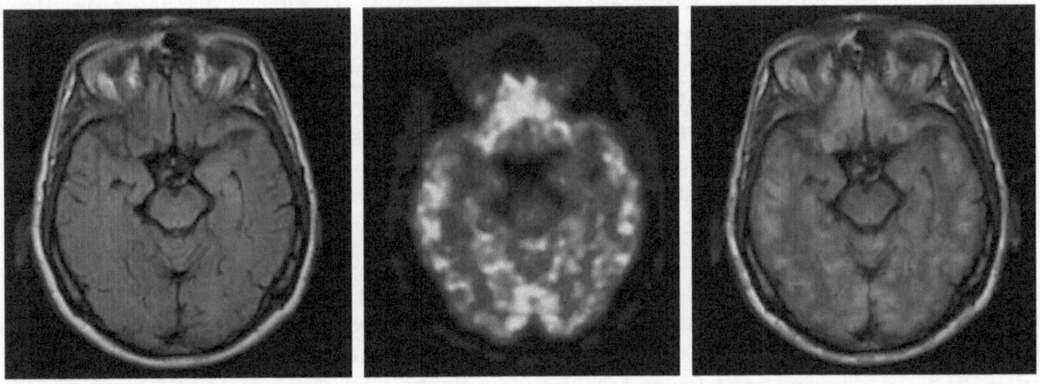

FIGURE 14.2 **(See color insert.)** MR (FLAIR sequence), (18F) FDG-PET, and fused images of the brain acquired simultaneously with an integrated MR/PET system. (From Schlemmer, H. P. et al. 2009. *Abdom. Imaging*, 34, 668–74. With permission.)

in intensities, (2) correlation coefficient, and (3) information-theoretic measures such as mutual information (MI) (Crum et al., 2004). The first two similarity measures are suitable only for monomodal registration, where the intensity characteristics are very similar among different images. For multimodal registration, similarity measures accommodate the different intensity characteristics of tissues between the different imaging modalities. In particular, MI, or relative entropy, is a basic concept from information theory, which can be considered a nonlinear generalization of cross-correlation (Pluim et al., 2003).

The maximization of MI has become the method of choice for multimodal registration in a wide range of applications. This success can be explained by its robustness, which allows for completely automated registration without need for user interaction, making the method very well suited for application in clinical practice (Maes et al., 2003; Maintz and Viergever, 1998).

As for the transformation model, three categories can be identified according to the associated degrees of freedom: rigid, affine, and nonlinear. In a rigid registration, the transformation involves only translations and rotations. This suffices to register images of rigid objects (like bones). The affine transformation preserves the parallelism of lines, but neither their lengths nor their mutual angles. It extends the degrees of freedom of the rigid transformation with scaling and shearing in each image dimension. It is an appropriate transformation model when the image has been skewed during acquisition, for example, when the CT gantry angle is incorrectly recorded.

Finally, nonlinear transformations allow the mapping of straight lines to curves. The similarity measure can be calculated globally, on the entire image, or locally, on a subimage. Smoothness of the deformation can be achieved in different ways and the deformation can be either free-form (any deformation is allowed) or guided by an underlying physical model of material properties, such as tissue elasticity or fluid flow.

In simple cases, such as rigid brain alignment, it is typically enough to use six parameters (three rotation angles and three translation parameters) to accurately align image volumes obtained at different time points by different scanners (Fitzpatrick et al., 2000). However, mainly in abdomen or thorax, nonlinear techniques are required to model deformation of organs, possibly due to respiratory effects, different bed positions, and pathological or treatment changes (Crum et al., 2004).

An important consideration for image registration concerns the fact that during iterative registration, the original image data from one scan needs to be interpolated in 3D. This step requires careful consideration of possible interpolation artifacts, especially in the case of the nonlinear registration (Slomka and Baum, 2009).

Other technical important aspects concern the optimization scheme used to adjust the transformation in order to improve the image similarity and the required computational time. In nonlinear registration applications, choosing or designing an optimizer can be difficult because the more nonlinear (or flexible) the transformation model, the more will be the parameters generally required to describe it. For the optimizer, this means that more time is required to make a parameter choice, and that the function to be optimized may present local minima which may be challenging to overcome. Anyway, which optimization scheme is suitable for a particular registration application depends on the cost function, the transformation, potential time constraints, and the required accuracy of the registration (Crum et al., 2004).

Despite their increasing use, nonlinear registration algorithms still present some difficulties that are mainly due to the selection of suitable deformations models and, to some extent, the computational cost.

14.5 Examples of Applications

In this section, two examples of MMI application, related to different clinical problems, are introduced (1) multimodality imaging in Alzheimer's disease (AD) and (2) DCE-MRI and DCE-CT in oncology. The first application is mainly focused on clinical aspects, while the second on technical aspects, in particular on spatiotemporal MMI registration.

14.5.1 Multimodality Imaging in AD

An interesting field of application of MMI is the investigation of cerebral disorders, in particular AD. This section briefly introduces the clinical issue, overviews the main neuroimaging techniques currently available to investigate cerebral alterations, points out the relevance of MMI in AD, and describes a number of studies combining different imaging techniques to improve diagnostic accuracy and prognostic performance.

14.5.1.1 Background

More than 35 million people worldwide are affected by AD, the most common form of dementia, impairing memory and cognition, and leading to death within 3–10 years after diagnosis. AD is strongly associated with age, with incidence doubling every 5 years after the age of 65. As the population is aging throughout the world, the projected disease growth is dramatic (Hebert et al., 2003; Hirtz et al., 2007), and AD thus represents a worldwide top-priority problem.

AD can be definitively diagnosed only by histopathologic examination of brain tissue (McKhann et al., 1984). Postmortem studies have identified the major hallmarks of late-stage AD, including amyloid plaques, neurofibrillary tangles, neuronal cell loss, and gliosis (Goedert and Spillantini, 2006); however, to date, the initiating mechanisms that trigger disease onset and drive its progression have not been fully understood, yet. Efforts to understand and track the early changes associated with AD will greatly increase understanding of disease-causing mechanisms, leading to the identification of novel targets for pharmaceutical intervention which could delay the course of inexorably progressive and irreversible brain damage. Furthermore, early diagnosis would enable to identify patients at the initial stages of the disease, for whom any drug therapy is more likely to be effective.

During the past 10 years, most of the AD research has been focused on finding biomarkers that could be reliably used to diagnose AD, monitor its progression, and eventually predict its onset. Among them, neuroimaging markers have been shown to play a key role (Hampel et al., 2008).

14.5.1.2 Imaging in the Brain

Advances in neuroscience and neuroimaging have led to an increasing recognition that certain neuroanatomical structures may be affected preferentially by particular disease. Neurodegenerative brain diseases mark the brain with a specific morphological "signature," detection of which may be useful to enhance diagnosis. Different imaging techniques could be used to visualize and monitor cerebral alterations occurring at different stages of AD progression.

The two main morphostructural aspects in the clinical diagnosis of AD are regional atrophy (especially assessed in the medial temporal lobe) and subcortical cerebrovascular damage. Structural imaging (mainly MRI) enables to investigate both aspects, providing markers to track the biological progression of disease (Frisoni et al., 2010). Several MR-based markers could be used to monitor the biological progression of the disease, such as Brain Boundary Shift Integral, progressive hippocampal and enthorinal cortex atrophy, cortical thickness, cortical pattern matching, and voxel-based morphometry (VBM). Most of structural imaging techniques are automated or semiautomated, and rely on different pipelines for image processing based on different mathematical procedures (Caroli et al., 2009). The most popular technique is VBM, which is based on the general linear model: MR images are registered to a global template, which allows for the removal of the global, but not the local, shape variability; registered gray matter images are smoothed, leading to normally distributed data and allowing the use of statistical parametric tools. A *t*-test is then performed on a voxel-wise basis between groups of subjects or within a group of subjects scanned at baseline and follow-up (Ashburner and Friston, 2000).

As functional alterations precede structural changes, functional imagings (mostly SPECT, FDG-PET (Jagust et al., 2007) and fMRI (Liu et al., 2008)) are playing an increasingly relevant role in detecting the dysfunction that characterizes the earliest stage of AD, providing pathophysiological information on synapse dysfunction in AD *in vivo* that cannot be detected by structural imaging. Furthermore, functional imaging provides the chance to objectively determine the extent to which clinically effective treatments attenuate or potentially compensate for disease progression. Historically, functional patterns have been defined relative to cortical areas based on scan visual inspection. Functional patterns have been further refined by the observation of functional reductions in specific ROIs other than the neocortex, relying upon within-subject coregistration of PET/MRI scans. A different approach to functional image analysis is the use of fully automated voxel-based analysis (VBA) methods performing intersubject image averaging (Minoshima, 1995). Such methods rely on spatial normalization and smoothing of the scans in order to reduce intrasubject variability and anatomically standardize all images onto a standard brain atlas while preserving metabolic/perfusion counts. Thereafter, statistical comparisons are performed on a voxel-by-voxel basis, with the resultant creation of statistical parametric maps of significant effects.

Gray matter loss assessed by structural MRI has a limited ability to capture the whole range of morphostructural changes associated with neurodegeneration in AD; it cannot discriminate neuronal from glial and axonal loss, or neuronal loss from age-associated shrinkage of healthy neurons, and it cannot appreciate white matter damage that might arise from neurofilament tau pathology. Recently, several microstructural imaging techniques probing into the finer structure of the brain have been developed, such as diffusion tensor and magnetization transfer imaging. Microstructural imaging enables investigation of the brain microstructure (Stahl et al., 2007), providing increasingly precious information to elucidate the pathophysiology of AD.

Currently, protein concentration (i.e., extracellular amyloid plaques and intracellular neurofibrillary tangles load) is assessed by lobar puncture followed by CSF laboratory analysis, an invasive procedure not suitable for all patients. Research is indeed very active in molecules labeled with radioactive isotopes that might enter the brain, bind selectively to β-amyloid, be visualized with PET scanners, and analyzed with PET imaging tools, enabling *in vivo* quantification of amyloid plaque load in AD, the compound at the most advanced stage of validation being Pittsburgh compound B (PIB) (Ikonomovic, 2008). Amyloid imaging could hopefully replace lobar puncture for the investigation of the pathological processes occurring at the cellular level, identifying *in vivo* neuroanatomic evidence of AD at a very early stage.

14.5.1.3 Relevance of Multimodality Imaging in AD

All individual neuroimaging modalities could be of help in the diagnosis of AD; however, each of them enables to investigate a single cerebral alteration. In order to investigate the overall cerebral damage, several techniques should be combined.

The combination of different imaging modalities could increase the accuracy of each modality alone. Moreover, there is considerable promise that early and specific diagnosis of AD will be rendered possible through the combination of a number of different imaging biomarkers for AD, such as medial temporal or cortical atrophy on MRI, functional defects in the temporoparietal and posterior cingulate cortex on PET, microstructural pathology on DTI, and high amyloid load on PIB-PET. A few studies have already been tried to combine biomarkers, showing an increase in prognostic power and diagnostic accuracy (Arnaiz et al., 2001; de Leon et al., 2007; Schmidt et al., 2008), but further efforts are needed.

14.5.1.4 Relationship between Structural and Functional Damage

AD is associated with widespread structural and functional brain alterations, which have been well documented, with consistent regional distribution of atrophy and hypoperfusion/hypometabolism across studies. In the very early disease stage, the atrophy and functional alteration patterns have been found to overlap only partially (Caroli et al., 2007b; Matsuda et al., 2002), suggesting that gray matter loss does not entirely explain the observed functional deficit. Through the combination of structural (MRI) and functional (PET/SPECT) imaging, it is possible to investigate the relationship between structural and functional damage, assessing the regional distribution of additional functional depression and the presence of any cerebral compensatory mechanism. A better understanding of compensatory mechanisms, taking into account the overall cerebral alteration rather than any specific structural or functional damage, could help to go deeply into the comprehension of the AD underlying pathology, hopefully opening the way to a more accurate disease marker than atrophy or perfusion and metabolism alone or suggesting novel therapeutic strategies to improve the resilience of the brain to neurodegenerative damage.

Caroli et al. investigated functional depression and compensation both in terms of cerebral perfusion (combining SPECT and MR imaging) (Caroli et al., 2010a) and glucose metabolism (combining FDG-PET and MR imaging) (Caroli et al., 2010b). SPECT and FDG-PET images were preprocessed according to a novel MR-based optimized protocol (Caroli et al., 2007a), which enables to reliably compare structural and functional loss patterns, assessed through VBM and VBA, respectively (described in Section 14.5.1.2), on a voxel-by-voxel basis. In the first study (Caroli et al., 2010a), a simple but innovative method to compare structural and functional loss was adopted: since the regions of functional compensation are denoted by significant atrophy but nonsignificant functional deficit, compensation t-map was computed masking the significant gray matter atrophy map with nonsignificant hypoperfusion; conversely, perfusion depression t-map was computed masking the significant hypoperfusion map with nonsignificant atrophy. The second study (Caroli et al., 2010b) followed a different approach, using Biological Parametric Mapping (BPM (Casanova et al., 2007)) for multimodality brain imaging analysis. BPM is based on a voxel-wise general linear model, and enables to incorporate in the analysis imaging regressors: unlike standard VBA tools based on the general linear model, in a BPM analysis each voxel has its own design matrix, with both scalar (e.g., age) and imaging voxel-specific (e.g., gray matter density) regressors. Hypometabolism detected after removing the variability due to gray matter intensity was interpreted as metabolic deficit, while gray matter atrophy detected after removing variability due to metabolism was considered as an indication of metabolic compensation. BPM is statistically more reliable than the method used in the first study which, however, is still valid for the visualization of functional compensation and depression; despite the large differences, the two methods showed pretty similar patterns of metabolic compensation and depression, further proving their validity.

Relatively preserved perfusion, indicative of compensation in the setting of neuronal loss, was found in the neocortex, likely reflecting vasodilation; conversely, perfusional depression exceeding atrophy was found in the medial temporal lobe as in this area, primarily affected by AD, the microcircle reserve was over (Caroli et al., 2010a). Metabolic compensation was mainly located in the amygdala, likely reflecting spared synaptic plasticity of the surviving neurons; metabolic depression, due both to distant effects of atrophy and to additional hypometabolism-inducing factors such as amyloid deposition, was mainly located in the posterior cingulate cortex (Caroli et al., 2010b) (Figure 14.3).

These studies point out an existing divergence between defects of perfusion and glucose metabolism, which are correlated, but not always concurrent phenomena. However, both of them found high regional variability and showed that a functional compensatory mechanism takes place against atrophy and neuropathological damage in some regions of the brain, while functional depression exceeding atrophy occurs in others.

14.5.1.5 Amyloid Toxicity

MMI provides the possibility to investigate *in vivo* the relationship between gray matter atrophy and amyloid deposition, mapping the match and mismatch of the two phenomena and thus assessing amyloid toxicity. The topographic relationship between amyloid deposition and gray matter atrophy has relevant implications on current clinical trials with antiamyloid drugs, as brain atrophy is widely used as surrogate outcome in clinical trials of drugs that might delay or arrest AD progression and, on the other hand, most current disease-modifying drugs for AD have been developed under the assumption that cognitive deterioration is due to amyloid deposition and that slowing or arresting amyloid deposition will lead to slowing or arresting cognitive deterioration.

Frisoni et al. investigated both voxel-wise and ROI-based correlation between gray matter density and (11C)-PIB uptake (Frisoni et al., 2009). Greater amyloid deposition was generally not associated with more severe gray matter atrophy except in the medial temporal lobe (Figure 14.4), in agreement with previous studies showing no relationship between synaptic integrity and amyloid deposition and pathological data showing a correlation between amyloid load in the medial temporal lobe and cognitive performance.

Medial temporal lobe was among the regions with the lowest amount of amyloid deposition, indicating that different types of amyloid may deposit in the brain, different brain regions may be differentially susceptible to its toxic effects, or amyloid may be peripheral to neurodegeneration.

14.5.1.6 Conclusions

In conclusion, this section points out the relevance and potential of multimodal analysis of neuroimaging in studying the overall cerebral damage beyond single alterations. MMI shows that structural damage, functional alterations, and protein buildup occurring at the cellular level, characterizing AD, are interrelated but neither concurrent nor colocalized phenomena. These results point out the need of using multimodal analysis techniques (combining even more than two image modalities at a time, and combining biomarkers from different fields of research) to further assess the relationship between different markers from the earliest stages, in order to better understand the mechanisms that trigger the disease onset and drive its

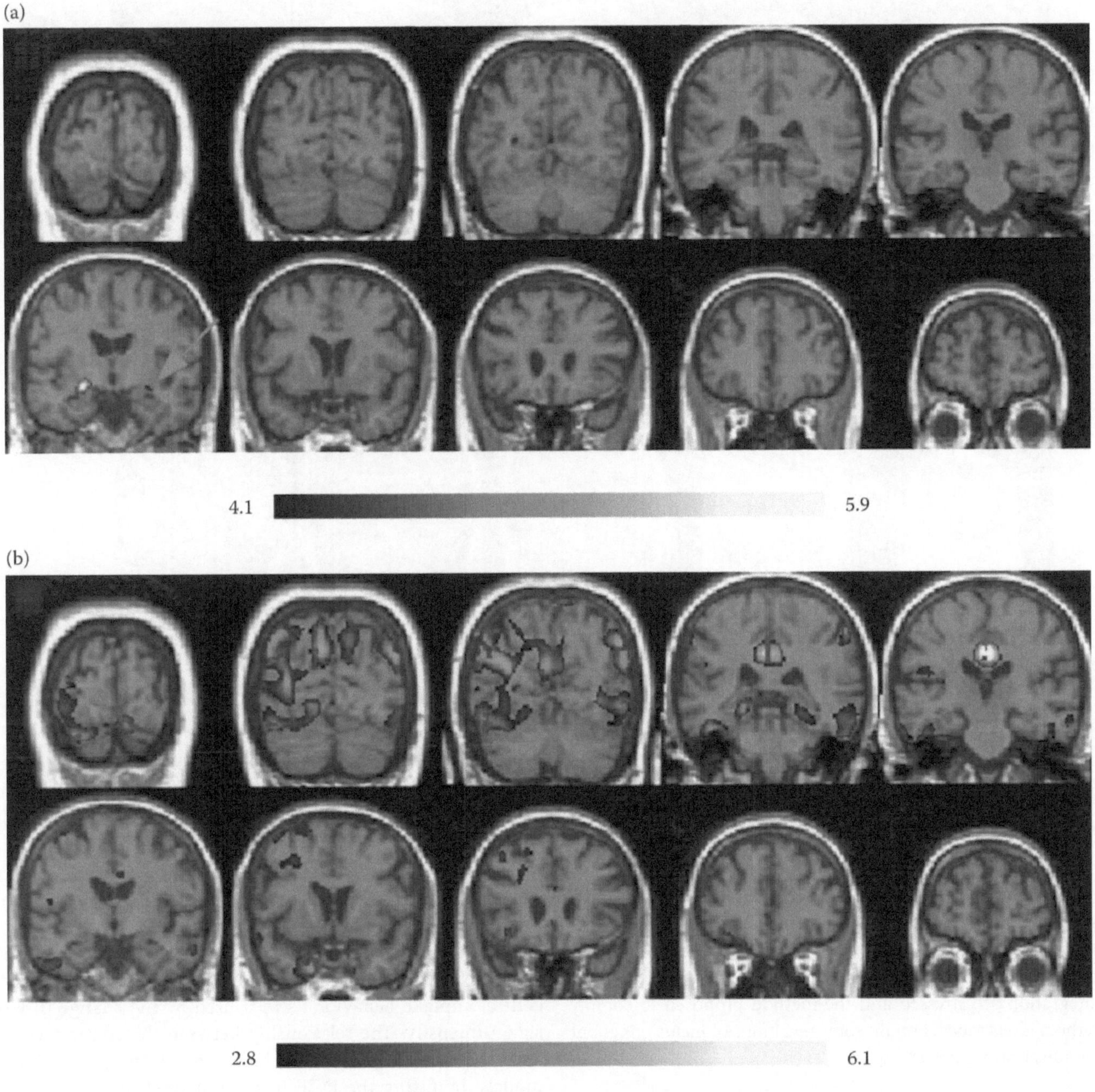

FIGURE 14.3 **(See color insert.)** (a) Metabolic compensation map, showing regions of significant relatively preserved metabolism; (b) Functional depression map, showing regions of hypometabolism exceeding atrophy in a group of 25 AD patients as compared to a group of 21 normal controls. Both maps were computed through BPM. (From Caroli, A. et al. 2010b. *Dementia Geriatr. Cogn. Disord.*, 29, 37–45. With permission.)

progression, to increase early diagnostic accuracy and to lead to the identification of novel targets for pharmaceutical intervention which could delay disease progression.

14.5.2 DCE-CT and DCE-MRI in Oncology

As seen in Section 14.4, software registration methods permit to integrate imaging modalities retrospectively, and they can be useful in the case hardware integration is not yet available. However, there are some applications where registration may be challenging. A particular case is perfusion imaging, that is, the sequential acquisition of a volume of interest (VOI) aimed to see signal changes related to contrast agent distribution in time. Both CT and MR are used for perfusion imaging in different clinical applications but mainly in oncology. In fact, from perfusion imaging it is possible to extract, after postprocessing of MR or CT signal changes in time, parameters related to tumor microvasculature. These two techniques are named dynamic contrast-enhanced computed tomography (DCE-CT) and dynamic contrast-enhanced magnetic resonance imaging (DCE-MRI). In the following, a registration

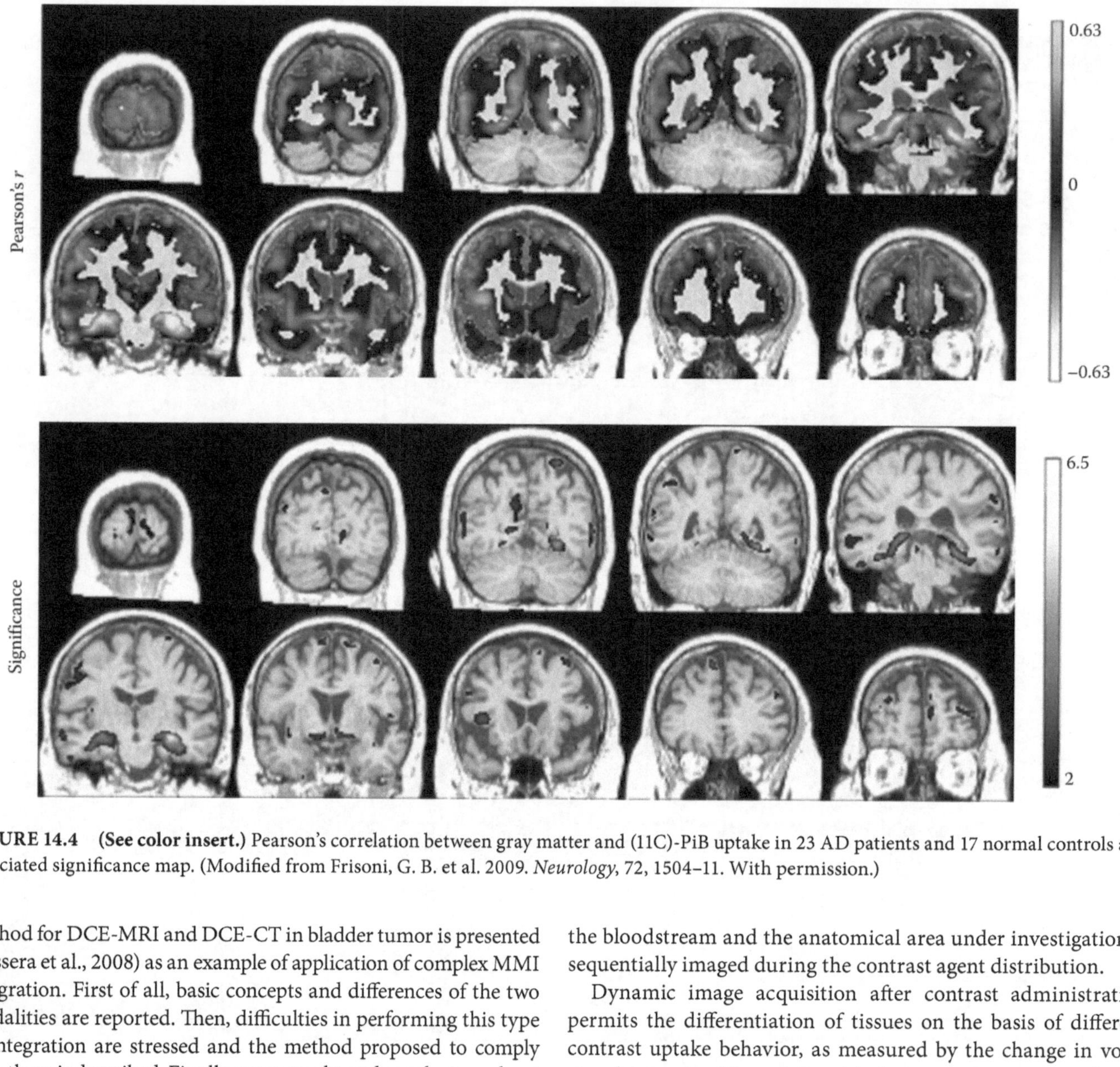

FIGURE 14.4 **(See color insert.)** Pearson's correlation between gray matter and (11C)-PiB uptake in 23 AD patients and 17 normal controls and associated significance map. (Modified from Frisoni, G. B. et al. 2009. *Neurology*, 72, 1504–11. With permission.)

method for DCE-MRI and DCE-CT in bladder tumor is presented (Passera et al., 2008) as an example of application of complex MMI integration. First of all, basic concepts and differences of the two modalities are reported. Then, difficulties in performing this type of integration are stressed and the method proposed to comply with them is described. Finally, some results and conclusions about this application are presented.

14.5.2.1 Background

It has long been known that an abnormal vasculature is an integral feature of tumors. Tumor vascularity is usually characterized by spatial heterogeneity and chaotic structure, poorly formed, fragile vessels with high permeability to macromolecules, due to the presence of large endothelial cell gaps or fenestrae, incomplete basement membrane, and relative lack of pericyte or smooth muscle association with endothelial cells, high vascular tortuosity, and vasodilatation, areas of spontaneous hemorrhage, and areas of low vascular density mixed with regions of high angiogenic activity (Padhani and Husband, 2001).

Many of the tumor angiogenesis features can be studied by perfusion imaging methods, such as DCE-MRI and DCE-CT. In both imaging methods, a contrast agent is introduced into the bloodstream and the anatomical area under investigation is sequentially imaged during the contrast agent distribution.

Dynamic image acquisition after contrast administration permits the differentiation of tissues on the basis of different contrast uptake behavior, as measured by the change in voxel signal intensity. The relationship between the contrast change and the perfusion process can be described using tracer kinetic models: by fitting the DCE imaging dataset using a compartmental model for each voxel in a VOI, it is possible to estimate different parameters related to the tumor microvascular status. (E.g., the transfer constant between plasma and extracellular extravascular space, K^{trans}, the volume of extracellular extravascular space per unit volume of tissue, v_e, and the blood plasma volume per unit volume of tissue, v_p.) The interested reader can refer to O'Connor et al. (2007) and Buckley and Parker (2005) for details about their physical meaning (Figure 14.5).

14.5.2.2 DCE-CT and DCE-MRI in the Bladder

The process of signal enhancement is different for the two modalities. In DCE-MRI, it is indirect: the passage of the bolus of paramagnetic contrast agent through a capillary bed produces

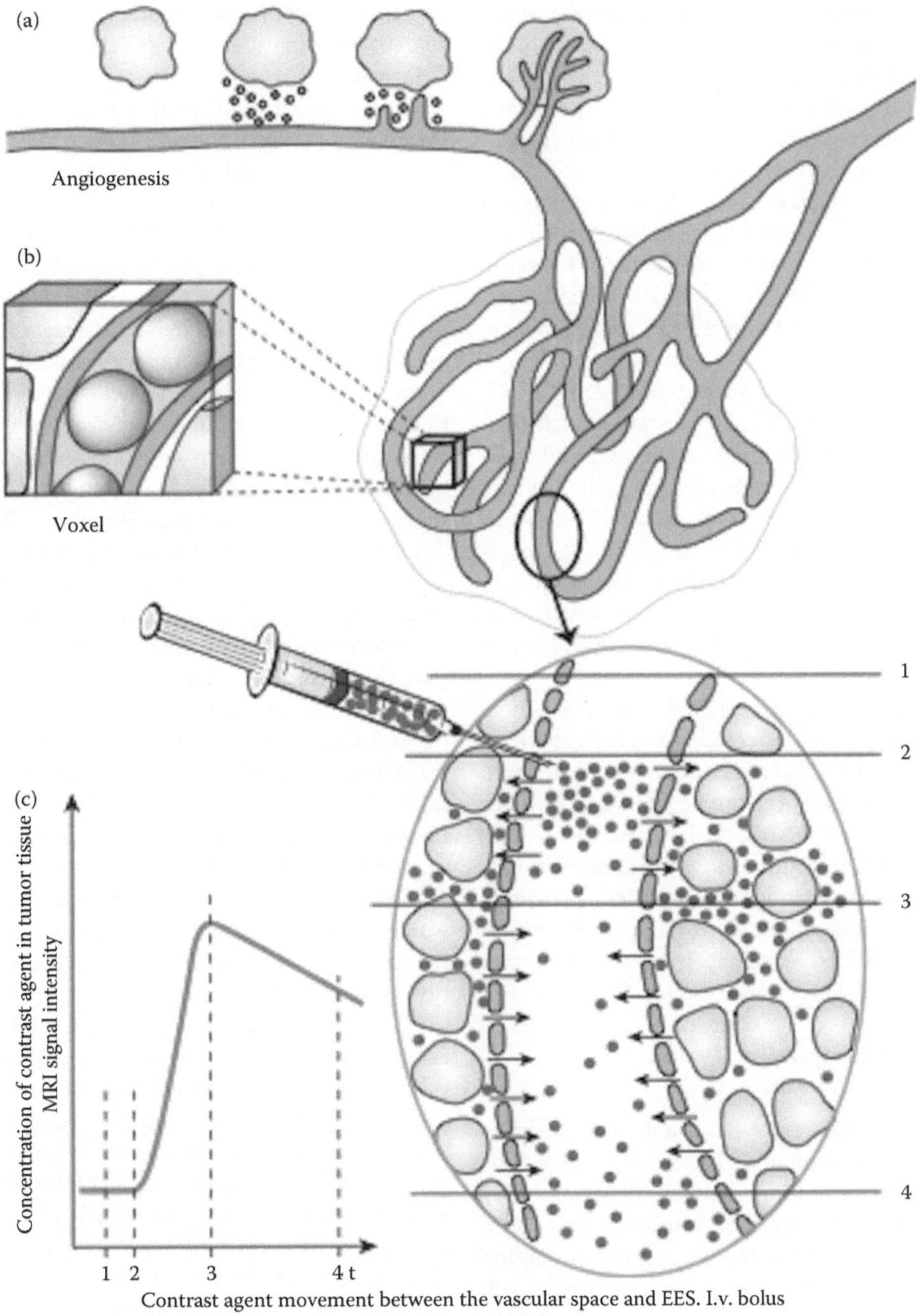

Contrast agent movement between the vascular space and EES. I.v. bolus injection at point 2.

FIGURE 14.5 To growth, tumors stimulate the formation of new vessels, that is, the angiogenesis process (a). The novel tumor microvasculature has some peculiar features (chaotic structure, high permeability, etc.). When a contrast agent arrives in this area, it leaks into the extravascular–extracellular space (EES); this phenomenon can be visualized in MR or CT images as a rapid signal increase in voxels containing tumor microvessels (b,c). Then the contrast agent accumulates in the extravascular tissue before it diffuses back into the vasculature from which it is excreted by the kidneys (c). (From O'Connor, J. P. B. et al. 2007. *Br. J. Cancer*, 96, 189–95; Buckley, D. L. and Parker, G. J. M. 2005. *Med. Radiol.*, Berlin Heidelberg: Springer-Verlag. With permission.)

local relaxation time changes that result in an increase in the signal intensity of plasma water within the capillaries, and of the tissue water in any extravascular space to where the agent may leak. In DCE-CT, the highly attenuating contrast agent itself provides the signal increase, and hence there is a direct correlation between blood signal enhancement and blood flow.

DCE-MRI is the method of choice for monitoring changes in tumor microvascular functional status due to antiangiogenic and antivascular treatments (Buckley and Parker, 2005). However, DCE-CT is an attractive alternative as it has wider clinical availability, due to which it has attracted considerable interest (Sahani et al., 2005).

DCE-MRI is known to have some advantages over DCE-CT; in particular, it is able to provide volume coverage at high temporal resolution without ionizing radiation. DCE-CT is limited in the total number of slices and number of time points during the dynamic acquisitions due to dose restrictions, which means that potentially important information from spatially heterogeneous tumors may be missed. However, DCE-MRI is thought to have some accuracy limitations due to its sensitivity to variable rates of water exchange across capillary walls (Buckley and Parker, 2005).

As both modalities have strengths and limitations, it is important to assess whether significant differences exist between them in the gross parametrization of tumor microvascular functional status, and whether these differences can be attributed to water exchange rate variability or methodological differences.

Bladder is a good candidate for this type of study as the tumor is confined and homogeneous enough. Nevertheless, bladder is a soft-tissue organ and is thus subjected to deformations, even if it is less affected by respiration movements with respect to other abdominal organs due to its lower position.

14.5.2.3 Problems for Integration

The comparison between DCE-MRI and DCE-CT is not trivial since they are acquired by different systems and, in the current best protocols, at different resolutions (both spatial and temporal). To ensure a correct voxel-by-voxel comparison, the two dynamic sets should be in the same coordinate system and a correction should be performed for any deformation inside the tumor VOI. Hence, CT and MR images should be nonlinearly registered.

Another problem related to registration is the matching of the CT/MR image registration pairs. In the dynamic datasets, image features may change during time due to the contrast distribution and this may affect the registration process, leading to different results depending on the CT-MR pairs used to calculate the transformation. Thereof a temporal registration of the dynamic set should also be obtained (Figure 14.6).

14.5.2.4 Methodologies

The registration method is composed by two steps: a temporal matching, aiming at finding a correspondence between CT/MR data-sets considering contrast agent uptake events (mainly onset of enhancement), and a spatial matching, to obtain the same spatial coordinate system and to correct any bladder tumor deformations.

First, time series of the mean signals inside the CT and MR tumor VOIs (selected by an experienced radiographer) are generated and resampled at the same temporal resolution. Then, the cross-correlation of the gradient of the mean signal intensity time course is calculated to find the temporal delay between CT and MR. The gradient was found to be a more robust measure for this task than the mean signal value (Passera et al., 2008). The original CT and MR dynamic images are then synchronized using this delay, and all temporally matched pairs are identified.

Spatial transformations may be performed by using a well-known software for image registration, which implements an intensity-based method using a hierarchical transformation model to perform a 3D registration between images using free-form deformations based on B-splines (Rueckert et al., 1999). The optimal transformation is determined by minimizing a registration cost function that represents a combination of the cost associated with the smoothness of the transformation and the cost associated with the image similarity (MI). Prior to registration, the DCE-MRI and DCE-CT data are resample at the same spatial resolution (the higher between the two) to ensure no loss of information. A first affine registration is calculated on each CT/MR pair, considering the CT as source and the MR as target image. Then, the mean MI is calculated between the CT and MR images before and after the transformation. The affine transformation that gives the maximum mean MI over all matched time points is selected as the affine transformation to apply to the whole dynamic data set. This procedure also enables to identify the image pair on which to calculate the nonlinear transformation. After affine registration to bring the CT and MR into the same coordinate system, a dilated bounding box surrounding each tumor VOI of the two images is defined. It was within these that the nonlinear transformation is calculated. Finally, the nonlinear transformation is applied to the MR volume.

After these steps, DCE-MRI and DCE-CT quantitative microvascular parameters are calculated by fitting the "extended Kety" pharmacokinetic model to both CT and MR and K^{trans}, v_e, and v_p parameters are obtained. A more extensive description of these steps may be found in Passera et al. (2008).

The complete method was applied on five DCE-MRI/DCE-CT datasets of patients with a confirmed diagnosis of primary bladder tumor.

14.5.2.5 Results

Figure 14.7 shows the effect of image registration, step-by-step. The percent overlap between the CT VOI and MR VOI was computed before and after image registration steps as the number of matching voxels between the two VOIs divided for the number of voxels of the CT VOI. Affine registration improved the mean percent overlap to 80%, and nonlinear step improved it of a further 18%. In all cases, after registration, the percent overlap was higher than 86%. These results demonstrate the importance of image registration and allow the comparison between

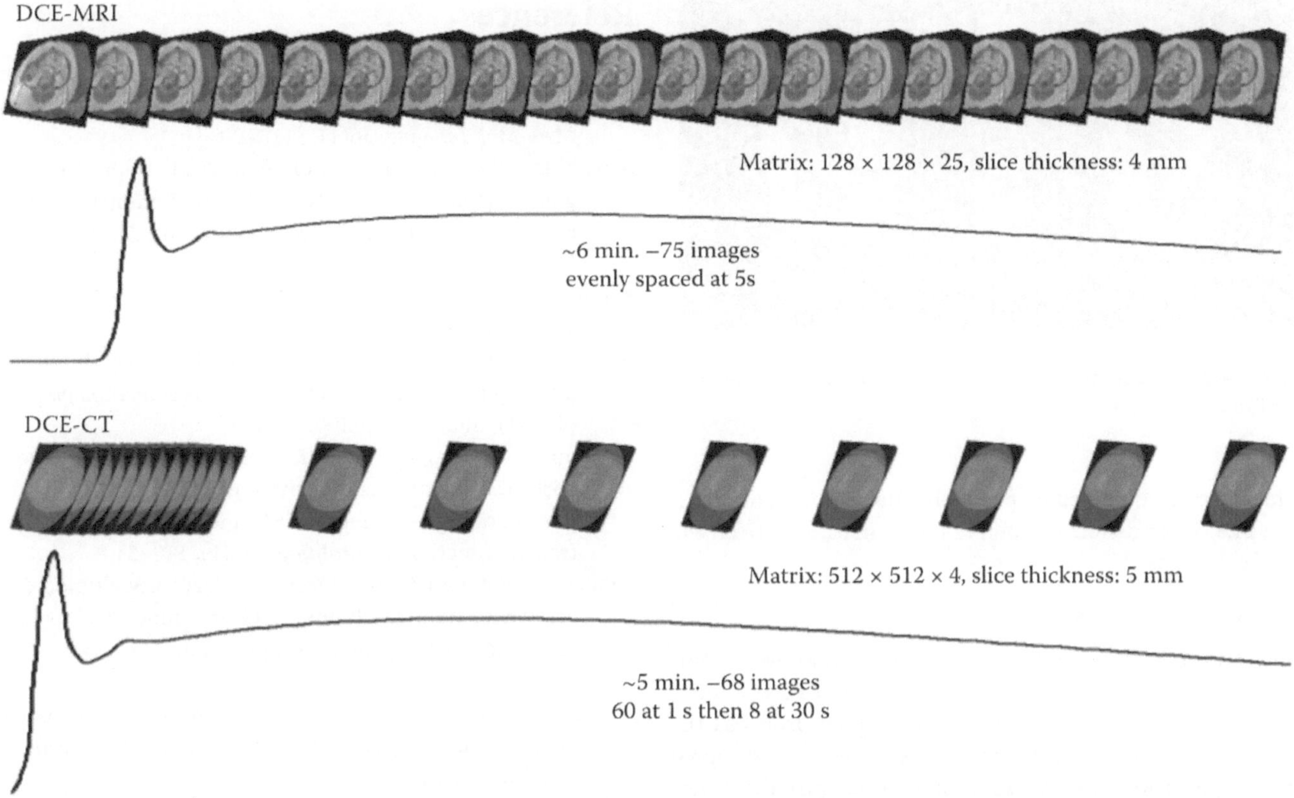

FIGURE 14.6 The two dynamic datasets. DCE-MRI protocol is different from DCE-CT protocol both in spatial and temporal resolution. In addition, the uptake process is different in the two modalities. Thereof the VOI mean signals are not aligned with respect to contrast agent arrival (pick of the curve).

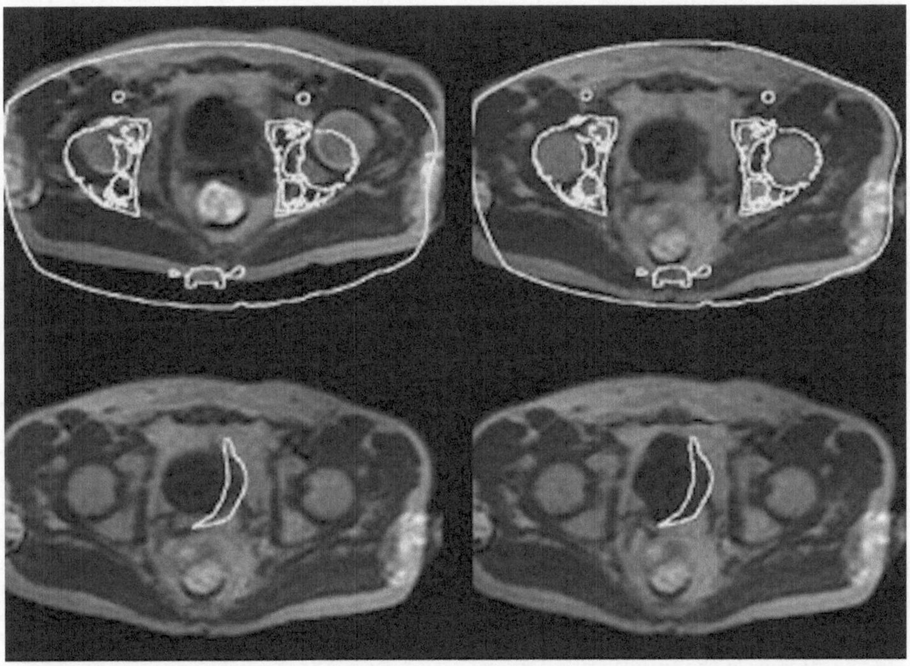

FIGURE 14.7 CT edges superimposed on MR image before (top, left) and after (top, right) affine registration. CT tumor VOI edges superimposed on MR image before (bottom, left) and after nonlinear registration (bottom, right). (Passera, K. et al. A non-linear registration method for DCE-MRI and DCE-CT comparison in bladder tumors. In *IEEE Int. Symp. on Biomedical Imaging*, pp. 1095–98. © (2008) IEEE. With permission.)

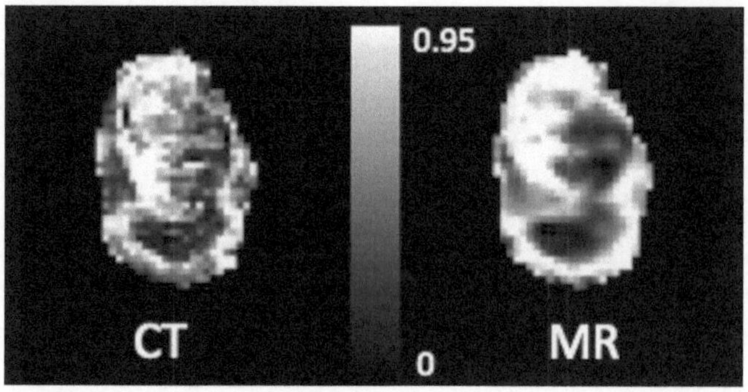

FIGURE 14.8 Parametric maps of Ktrans obtained from CT and MR for the same tumor VOI.

parametric maps obtained by processing the two modalities. In Figure 14.8, it is possible to observe the similarity between the CT and MR maps.

14.5.2.6 Conclusions

In this section, we have presented a method that allows a comparison of parametric maps generated from tracer kinetic modeling using CT and MR modalities. This comparison has the objective to investigate different processes leading to signal change in perfusion MR or CT, enabling to identify the different causes: tumor angiogenesis, water exchange rate variability, and methodological differences. The comparison may also help in choosing the most suitable imaging modality for investigating microvasculature tumor status; however, it is particularly challenging since DCE-MRI and DCE-CT differ in both spatial and temporal characteristics, and thus require both a temporal and spatial alignment. It is noteworthy that such alignment is not possible by hardware solutions as it would need a combined CT/MR system and, furthermore, a contrast agent visible in both modalities. This suggests that hardware integration cannot always replace software integration, and the latter may be of help in particular MMI applications.

14.6 Conclusions and Future Trends

In the past few years, integration of anatomical and functional images has been proved to provide added value to the pathophysiology interpretation. This chapter has reviewed the different MMI approaches, by considering the different fields of application in clinics and research. MMI is in a continuous evolution, hand in hand with the development of new imaging technologies, and new tools for integration (both software and hardware). In the field of molecular imaging, we can expect translation of the increasingly sophisticated imaging techniques, based on a thorough understanding of the underlying biology and already validated in preclinics, to the clinics. In the future, radiologists will need to take into account both biology and anatomy aiming to integrate the two types of information. The ultimate goal is to obtain a complete picture of the elephant.

References

Antoch, G. and Bockisch, A. 2009 Combined PET/MRI: A new dimension in whole-body oncology imaging? *Eur. J. Nucl. Med. Mol. Imaging*, 36(1), S113–20.

Arnaiz, E., Jelic, V., Almkvist, O. et al. 2001. Impaired cerebral glucose metabolism and cognitive functioning predict deterioration in mild cognitive impairment. *Neuroreport*, 12, 851–55.

Ashburner, J. and Friston, K. J. 2000. Voxel-based morphometry—The methods. *Neuroimage*, 14, 805–21.

Azhari, H., Edelman, R. R., and Townsend, D. 2007. Multimodal imaging and hybrid scanners. *Int. J. Biomed. Imaging*, 2007 (45353), doi: 10.1155/2007/45353.

Babiloni, F., Cincotti, F., Babiloni, C. et al. 2005. Estimation of the cortical functional connectivity with the multimodal integration of high-resolution EEG and fMRI data by directed transfer function. *NeuroImage*, 24(1), 118–31.

Babiloni, F., Babiloni, C., Carducci, F. et al. 2003. Multimodal integration of high-resolution EEG and functional magnetic resonance imaging data: A simulation study. *NeuroImage*, 19(1), 1–15.

Blankstein, R. and Di Carli, M. F. 2010. Integration of coronary anatomy and myocardial perfusion imaging. *Nat. Rev. Cardiol.*, 7(4), 226–36.

Blinowska, K., Müller-Putz, G., Kaiser V. et al. 2009. Multimodal imaging of human brain activity: Rational, biophysical aspects and modes of integration. Intell. *Neuroscience*, 2009 (813607), doi: http://dx.doi.org/10.1155/2009/813607.

Brindle, K. M. 2008. New approaches for imaging tumor responses to treatment. *Nat. Rev. Cancer*, 8, 94–107.

Buckley, D. L. and Parker, G. J. M. 2005. Dynamic contrast-enhanced magnetic resonance imaging in oncology. *Medical Radiology*. Berlin Heidelberg: Springer-Verlag.

Cai, W. and Chen, X. 2008. Multimodality molecular imaging of tumor angiogenesis. *J. Nucl. Med.*, 49 Suppl. 2, 113S–28S.

Casanova, R., Srikanth, R., Baer, A. et al. 2007. Biological parametric mapping: A statistical toolbox for multimodality brain image analysis. *Neuroimage*, 34, 137–43.

Caroli, A., Testa, C., Geroldi, C. et al. 2007a. Brain perfusion correlates of medial temporal lobe atrophy and white matter hyperintensities in mild cognitive impairment. *J. Neurol.*, 254, 1000–8.

Caroli, A., Testa, C., Geroldi, C. et al. 2007b. Cerebral perfusion correlates of conversion to Alzheimer's disease in amnestic mild cognitive impairment. *J. Neurol.*, 254, 1698–17.

Caroli, A., Geroldi, C., Nobili, F. et al. 2010a. Functional compensation in incipient Alzheimer's disease. *Neurobiol. Aging*, 31(3), 387–97. (epub June, 12 2008).

Caroli, A. and Frisoni, G. B. 2009. Quantitative evaluation of Alzheimer's disease. *Expert Rev. Med. Devices*, 6, 569–88.

Caroli, A., Lorenzi, M., Geroldi, C. et al. 2010b. Metabolic compensation and depression in Alzheimer's Disease. *Dementia Geriatr. Cogn. Disord.*, 29, 37–45.

Chao, C. P., Zaleski, C. G., and Patton, A.C. 2006. Neonatal hypoxic–ischemic encephalopathy: Multimodality imaging findings. *Radiographics*, 26(1), S159–72.

Cherry, S. R. 2009. Multimodality imaging: Beyond PET/CT and SPECT/CT. *Semin. Nucl. Medi.*, 39(5), 348–53.

Crum, W. R., Hartkens, T., and Hill, D. L. G. 2004. Non-rigid image registration: Theory and practice. *Br. J. Radiol.*, 77, S140–S53.

Delbeke, D., Schöder, H., Martin, W. H., and Wahl, R. L. 2009. Hybrid Imaging (SPECT/CT and PET/CT): Improving therapeutic decisions. *Sem. Nucl. Med.*, 39, 308–40.

de Leon, M. J., Mosconi, L., Blennow, K. et al. 2007. Imaging and CSF studies in the preclinical diagnosis of Alzheimer's disease. *Ann. NY Acad. Sci.*, 1097, 114–45.

Fitzpatrick, J. M., Hill, D. L. G., and Maurer, C. R. 2000. Image registration. In: Sonka, M., Fitzpatrick, J. M. (Eds.), Medical Image Processing. In: *Handbook of Medical Imaging*, vol. II. SPIE Press, Chapter 8, pp. 447–514. Bellingham, WA: SPIE Press, pp. 447–514.

Floeth, F. W., Pauleit, D., and Wittsack, H. J. 2005. Multimodal metabolic imaging of cerebral gliomas: Positron emission tomography with (18F)fluoroethyl-l-tyrosine and magnetic resonance spectroscopy. *J. Neurosurg.*, 102(2), 318–27.

Formaggio, E., Storti, S. F., Avesani, M. et al. 2008. EEG and fMRI coregistration to investigate the cortical oscillatory activities during finger movement. *Brain Topogr.*, 21(2), 100–11.

Frisoni, G. B., Lorenzi, M., Caroli, A., Kemppainen, N., Någren, K., and Rinne, J. O. 2009. *In vivo* mapping of amyloid toxicity in Alzheimer's disease. *Neurology*, 72, 1504–11.

Frisoni, G. B., Fox, N. C., Jack, C. R. Scheltens, P., and Thompson, P. M. 2010. The clinical use of structural MRI in Alzheimer disease. *Nat. Rev. Neurol.*, 6, 67–77.

Gaemperli, O. and Kaufmann, P. A. 2010. Multimodality cardiac imaging. *J. Nucl. Cardiol.*, 17(1), 4–7.

Gaemperli, O., Schepis, T., and Valenta, I. 2007.. Cardiac image fusion from stand-alone SPECT and CT: Clinical experience. *J. Nucl. Med.*, 48, 696–703.

Gnanasegaran, G., Adamson, K., and Barwick, T. 2010. Multislice SPECT/CT gains wider clinical acceptance. *Diagn. Imaging Eur.*, 26(1), 1–8.

Goedert, M. and Spillantini, M. G. 2006. A century of Alzheimer's disease. *Science*, 314, 777–81.

Hampel, H., Burger, K., Teipel, S. J. Bokde, A. L., Zetterberg, H., and Blennow, K. 2008. Core candidate neurochemical and imaging biomarkers of Alzheimer's disease. *Alzheimers Dement.*, 4, 38–48.

Hebert, L. E., Scherr, P. A., Bienias, J. L., Bennett, D. A., and Evans, D. A. 2003. Alzheimer disease in the US population: Prevalence estimates using the 2000 census. *Arch. Neurol.*, 60, 1119–22.

Heron, D., Andrade, R., and Flickinger, J. 2004. Hybrid PET-CT simulation for radiation treatment planning in head-and-neck cancers: A brief technical report. *Int. J. Radiat. Oncol. Biol. Phys.*, 60, 1419–24.

Hirtz, D., Thurman, D. J., Gwinn-Hardy, K., Mohamed, M., Chaudhuri, A. R., and Zalutsky, R. 2007. How common are the "common" neurologic disorders? *Neurology*, 68, 326–37.

Hsu, A. R., Cai, W., and Veeravagu, A. 2007. Multimodality molecular imaging of glioblastoma growth inhibition with vasculature-targeting fusion toxin VEGF121/rGel. *J. Nucl. Med.*, 48(3), 445–54.

Ikonomovic, M. D., Klunk, W. E., Abrahamson, E. E. et al. 2008. Post-mortem correlates of *in vivo* PiB-PET amyloid imaging in a typical case of Alzheimer's disease. *Brain*, 131, 1630–45.

Jae-Hyun, L., Yong-Min, H., Young-wook, J. et al. 2006. Artificially engineered magnetic nanoparticles for ultra-sensitive molecular imaging. *Nat. Med.*, 13, 95–9.

Jagust, W. J., Reed, B., Mungas, D. E., and Decarli, C. 2007. What does fluorodeoxyglucose PET imaging add to a clinical diagnosis of dementia? *Neurology*, 69, 871–77.

Judenhofer, M. S., Wehrl, H. F., Newport, D. F. et al. 2008. Simultaneous PET–MRI: A new approach for functional and morphological imaging. *Nat. Med.*, 14, 459–65.

Kang, J. H. and Chung, J. K. 2008. Molecular-genetic imaging based on reporter gene expression. *J. Nucl. Med.*, 49 Suppl 2, 164S–79S.

Kaufmann, P. A. 2009. Cardiac hybrid imaging: State of the art. *Ann. Nucl. Med.*, 23(4), 325–31.

Kaufmann, P. A. and Di Carli, M. F. 2009. Hybrid SPECT/CT and PET/CT imaging: The next step in noninvasive cardiac imaging. *Semin. Nucl. Med.*, 39(5), 341–47.

Liebeskind, D. S. 2009. Imaging the Future of Stroke: I. Ischemia. *Ann. Neurol.*, 66, 574–90.

Liu, Y., Wang, K., Yu, C. et al. 2008. Regional homogeneity, functional connectivity and imaging markers of Alzheimer's disease: A review of resting state fMRI studies. *Neuropsychologia*, 46, 1648–56.

Liu, Z. and He, B. 2008. fMRI-EEG integrated cortical source imaging by use of time-variant spatial constraints. *NeuroImage*, 39(3), 1198–214.

Maes, F., Vandermeulen, D., and Suetens, P. 2003. Medical image registration using mutual information. *Proc. IEEE*, 91(10), 1699–722.

Maintz, A. J. B. and Viergever, M. A. 1998. An overview of medical image registration methods. *Med. Image Anal.*, 2, 1–37.

Matsuda, H., Kitayama, N., Ohnishi, T. et al. 2002. Longitudinal evaluation of both morphologic and functional changes in the same individuals with Alzheimer's disease. *J. Nucl. Med.*, 43, 304–11.

McKhann, G., Drachman, D., Folstein, M., Katzman, R., Price, D., and Stadlan, E. M. 1984. Clinical diagnosis of Alzheimer's disease: Report of the NINCDS-ADRDA work group under the auspices of department of health and human services task force on Alzheimer's disease. *Neurology*, 34, 939–44.

Minoshima, S., Frey, K. A., Koeppe, R. A. et al. 1995. A diagnostic approach in Alzheimer's disease using three dimensional stereotactic surface projections of fluorine-18-FDG PET. *J. Nucl. Med.*, 36, 1238–48.

Moseley, M. and Donnan, G. 2004. Multimodality imaging: Introduction. *Stroke*, 35, 2632–34.

Muzik, O., Chugani, D. C., and Zou, G. 2007. Multimodality data integration in epilepsy. *Int. J. Biomed. Imaging*, 2007(13963), doi:10.1155/2007/13963.

Namdar, M., Hany, T. F., Koepfli, P. et al. 2005. Integrated PET/CT for the assessment of coronary artery disease: A feasibility study. *J. Nucl. Med.*, 46, 930–35.

Nekolla, S. G., Martinez-Moeller, A., and Saraste, A. 2009. PET and MRI in cardiac imaging: From validation studies to integrated applications. *Eur. J. Nucl. Med. Mol. Imaging*, 36(Suppl. 1), S121–30.

Nikiforidis, G. C., Sakellaropoulos, G. C., and Kagadis, G. C. 2008. Molecular imaging and the unification of multilevel mechanisms and data in medical physics. *Med. Phys.*, 35(8), 3444–52.

O'Connor, J. P. B., Jackson, A., Parker, G. J. M., and Jayson, G. C. 2007. DCE-MRI biomarkers in the clinical evaluation of antiangiogenic and vascular disrupting agents. *Br. J. Cancer*, 96, 189–95.

Ottobrini, L., Ciana, P., Biserni, A., Lucignani, G., and Maggi, A. 2006. Molecular imaging: A new way to study molecular processes *in vivo*. *Mol. Cell. Endocrinol.*, 246, 69–75.

Padhani, A. R. and Husband, J. E. 2001. Dynamic contrast-enhanced MRI studies in oncology with an emphasis on quantification, validation and human studies. *Clin. Radiol.*, 56, 607–20.

Papathanassiou, D. and Liehn, J. C. 2008. The growing development of multimodality imaging in oncology. *Crit. Rev. Oncol./Hematol.*, 68, 60–5.

Passera, K., Mainardi, L., McGrath, D. et al. 2008. A non-linear registration method for DCE-MRI and DCE-CT comparison in bladder tumors. In *IEEE Int. Symp. on Biomedical Imaging: from Nano to Macro*, Paris, May 14–17, pp. 1095–8.

Pietrzyk, U., Herholz, K., Schuster, A. von Stockhausen, H. M., Lucht, H., and Heiss, W. D. 1996. Clinical applications of registration and fusion of multi-modality brain images from PET, SPECT, CT, and MRI. *Eur. J. Radiol.*, 21, 174–82.

Pluim, J. P. W., Maintz, J. B. A., and Viergever, M. A. 2003. Mutual-information-based registration of medical images: A survey. *IEEE Trans. Med. Imaging*, 22(8), 986–1004.

Rispler, S., Keidar, Z., Ghersin, E. et al. 2007. Integrated single-photon emission computed tomography and computed tomography coronary angiography for the assessment of hemodynamically significant coronary artery lesions. *J. Am. Coll. Cardiol.*, 49, 1059–67.

Roche-Labarbe, N., Zaaimi, B., Mahmoudzadeh, M. et al. 2010. NIRS-measured oxy- and deoxyhemoglobin changes associated with EEG spike-and-wave discharges in a genetic model of absence epilepsy: The GAERS. *Epilepsia*, 51(8), 1374–84.

Rueckert, D., Sonoda, L. I., Hayes, C. et al. 1999. Nonrigid registration using free-form deformation: Application to breast MR images. *IEEE Trans. Med. Imaging*, 18, 712–21.

Sahani, D. V., Kalva, S. P., Hamberg, L. M. et al. 2005. Assessing tumor perfusion and treatment response in rectal cancer with multisection CT: Initial observations. *Radiology*, 234: 785–92.

Santana, C. A., Garcia, E. V., Faber, T. L. et al. 2009. Diagnostic performance of fusion of myocardial perfusion imaging (MPI) and computed tomography coronary angiography. *J. Nucl. Cardiol.*, 16, 201–11.

Saxena, V., Gonzalez-Gomez, I., and Laug W. E. L. 2007. A non-invasive multimodal technique to monitor brain tumor vascularization. *Phys. Med. Biol.*, 52, 5295–308.

Schlemmer, H. P., Pichler, B. J., Krieg, R., and Heiss, W. D. 2009. An integrated MR/PET system: Prospective applications. *Abdom. Imaging*, 34, 668–74.

Schmidt, G. P., Kramer, H., Reiser, M. F., and Glaser, C. 2007. Whole-body magnetic resonance imaging and positron emission tomography-computed tomography in oncology. *Top. Magn. Reson. Imaging*, 18(3), 193–202.

Schmidt, R., Ropele, S., Pendl, B. et al. 2008. Longitudinal multimodal imaging in mild to moderate Alzheimer disease: A pilot study with memantine. *J. Neurol., Neurosurg. Psychiatry*, 79, 1312–17.

Schöder, H., Larson, S. M., and Yeung, H. W. D. 2004. PET/CT in Oncology: Integration into clinical management of lymphoma, melanoma, and gastrointestinal malignancies. *J. Nucl. Med.*, 45(1), 72S–81S.

Schuijf, J. D., Wijns, W., Jukema, J. W. et al. 2006. Relationship between noninvasive coronary angiography with multislice computed tomography and myocardial perfusion imaging. *J. Am. Coll. Cardiol.*, 48, 2508–14.

Slomka, P. J. and Baum, R. P. 2008. Multimodality image registration with software: State-of-the-art. *Eur. J. Nucl. Med. Mol. Imaging*, 36(Suppl 1), S44–55.

Stahl, R., Dietrich, O., Teipel, S. J., Hampel, H., Reiser, M. F., and Schoenberg, S. O. 2007. White matter damage in Alzheimer disease and mild cognitive impairment: Assessment with diffusion-tensor MR imaging and parallel imaging techniques. *Radiology*, 243:483–92.

Tak, S., Jang, J., Lee, K., and Ye, J. C. 2010. Quantification of CMRO(2) without hypercapnia using simultaneous near-infrared spectroscopy and fMRI measurements. *Phys. Med. Biol.*, 55(11), 3249–69.

Tempany, C. M. C. and McNeil, B. J. 2001. Advances in biomedical imaging. *J. Am. Med. Assoc.*, 285, 562–67.

Townsend, D. W. and Beyer, T. 2002. A combined PET/CT scanner: The path to true image fusion. *Br. J. Radiol.*, 75, S24–30.

Villringer, A., Planck, J., Hock, C., Schleinkofer, L., and Dirnagl, U. 1993. Near infrared spectroscopy (NIRS): A new tool to study hemodynamic changes during activation of brain function in human adults. *Neurosci. Lett.*, 154, 401–04.

Vince, D. C. and Tülay A. 2009. Feature-based fusion of medical imaging data. *IEEE Trans. Inf. Technol. Biomed.*, 13(5), 711–20.

Weissleder, R. and Pittet, M. J. 2008. Imaging in the era of molecular oncology. *Nature*, 452(7187), 580–89.

Woo, J., Slomka, P. J., Dey, D. et al. 2009. Geometric feature-based multimodal image registration of contrast-enhanced cardiac CT with gated myocardial perfusion SPECT. *Med. Phys.*, 36(12), 5467–79.

Woo, J., Hong, B. W., Kumar, S., Basu Ray, I., and Kuo, C. C. 2008. Multimodal data integration for computer-aided ablation of atrial fibrillation. *J. Biomed. Biotechnol.*, 2008 (681303), doi:10.1155/008/681303.

Woods, R. P., Grafton, S. T., Holmes, C. J., Cherry, S. R., and Mazziotta, J. C. 1998. Automated image registration: I. General methods and intra-subject, intramodality validation. *J. Comput. Assist. Tomogr.*, 22(1), 139–52.

Zaidi, H. and Prasad, R. 2009. Advances in multimodality molecular imaging. *J. Med. Phys.*, 34(3), 122–8.

Zitova, B. and Flusser, J. 2003. Image registration methods: A survey. *Image Vis. Comput.*, 21, 977–1000.

15

Computer-Aided Detection and Diagnosis

Lionel T. Cheng
Singapore Armed Forces Medical Corps
Singapore General Hospital

Daniel J. Blezek
Mayo Clinic

Bradley J. Erickson
Mayo Clinic

Computers are especially suited to help the physician collect and process clinical information and remind him of diagnoses which he may have overlooked.

Robert S. Ledley and Lee B. Lusted
Science 1959, , pp. 9–21

15.1 Introduction

After the world's first general purpose, computer was built in 1940s, the idea of using computers to assist physicians in medical decisions became an increasingly enticing possibility (Ledley and Lusted, 1959). Since then, computer applications have become pervasive in daily life, and their use in the field of medicine has grown from strength to strength. Given the largely digital nature of information in images, the field of medical imaging is particularly well placed to leverage on the processing power of computers to enhance the existing work processes.

The task of medical image interpretation involves a complex series of tasks which include detection of an abnormality, description of the characteristics of a lesion, diagnosis of a specific disease entity or syndrome, and evaluation of the extent of pathology to aid prognostication. The use of computers to facilitate or perform any or all of these interpretative steps was first described in the 1960s (Lodwick et al., 1963), and continues to grow at a rapid pace, catalyzed by major advances in processing power, the widespread availability of high-resolution volumetric image data sets, improved software algorithms, and the proliferation of digital imaging technology. This exciting field was initially described as Computer-aided diagnosis (CAD), a term which has since been broadened to include the use of computers in the detection of abnormalities, characterization of lesion features, and detection of changes over time.

Early CAD work centered on developing systems could perform comparably or even better than physicians. However, it was subsequently recognized that a useful CAD system did not have to replicate or replace the physician. Instead, CAD could still add value by playing a complementary role. This paradigm shift recognized that the true value of CAD was in the synergistic combination of the clinical competence of a physician and the consistent capability of a computer, bringing along the associated benefits such as reduction in image reading time and interobserver variability. CAD is now a major research subject in the field of medical imaging (Erickson and Bartholomai, 2002; Khorasani et al., 2006; Doi, 2007).

15.2 CAD in the Medical Image Review Process

15.2.1 Overview of the Medical Image Review Process

The task of medical image interpretation begins after image data have been acquired (e.g., by a CT or MRI scanner, etc.) and appropriate quality control checks performed. The process of image interpretation by a radiologist is a complex and often iterative process during which information obtained from the images and other relevant medical data (e.g., presenting symptoms of the patient, laboratory results, past medical history) are combined to facilitate a conclusion about whether pathology exists, what the condition is, and how extensive the disease is. This in turn facilitates a clinical decision on the appropriate medical management.

The overall process of medical image interpretation, while complicated, can be simplified as shown in Figure 15.1. The first step in the image interpretation process is *detection*, which is perception of an abnormality in the image (e.g., Is there a nodule in the lung?). The next step is *description*, during which the radiologist characterizes the abnormality in order to ascertain its nature (e.g., Does the lung nodule have speculated edges? Is the nodule calcified? Where is the nodule located?). Following description, the radiologist goes on to provide a *diagnosis*, or more often, a list of differential diagnoses based on probability (e.g., spiculated lung nodule, likely malignant, most probably bronchogenic carcinoma). The final step involves providing a *prognosis* based on the likely condition, and includes the evaluation of other imaging findings in order to determine the extent, severity, and likely outcome of the condition (e.g., bronchogenic carcinoma with bilateral hilar nodes, and metastases seen in the ribs, left humerus and liver—stage 4 disease with poor prognosis). This process of image interpretation is often not as simple and linear as depicted above, and may involve several parallel or iterative steps, depending on whether multiple pathologies coexist, and whether the diagnosis is already known.

15.2.2 Types of CAD

Given the myriad potential applications of computers in image interpretation, there is understandably a correspondingly wide range of descriptive terms for CAD used in literature. For simplicity, the abbreviation "CAD" in this chapter refers generically to all the potential applications of computers to aid detection, description, diagnosis, and prognosis. Each type of CAD has different roles and capabilities, and some CAD systems are a combination of more than one functional type. Regardless of which existing CAD nomenclature one may be familiar with, it is useful to keep in mind how the various CAD systems facilitate the basic steps in the overall image interpretation process, which in turn is heavily dependent on the clinical context. Using the simplified image interpretation process outlined above, the various types of CAD systems and their potential applications may be classified as shown in Figure 15.2.

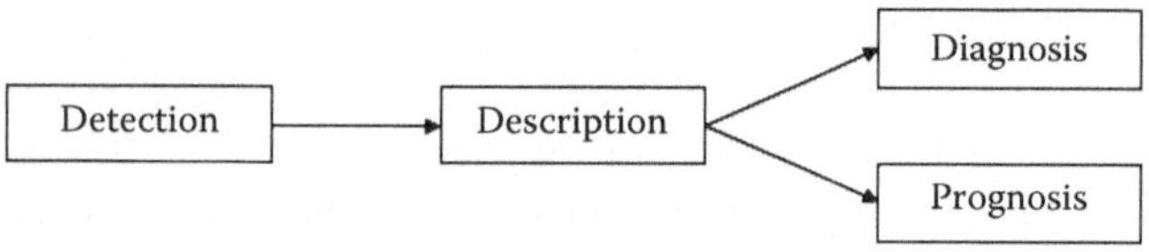

FIGURE 15.1 Simplified representation of the medical image interpretation process.

15.2.2.1 Computer-Aided Detection

Computer-aided detection (CADe) systems are designed to identify potentially abnormal findings in an image. Common examples include the use of CADe systems to identify suspicious microcalcifications in mammograms, or detect colonic polyps in CT colonography studies. Such CADe systems are typically used as prompting devices, marking locations of suspected abnormalities via an overlay on a medical image.

15.2.2.2 Computer-Aided Characterization

Computer-aided characterization (CAC) systems are designed to provide accurate and reproducible descriptions of a lesion in a medical image. These characteristics are usually obtained concurrently as part of lesion detection. For example, in the process of detecting a lung nodule on a CT scan, the size and edges of the nodule would have already been characterized, along with the volume and other features such as average Hounsfield values and relationships to adjacent structures. CAC systems are also useful in rapidly evaluating multiple complex data sets over a period of time, such as the enhancement characteristics of a breast nodule on MRI (e.g., time to peak signal intensity and the rate of washout of gadolinium). The characteristics of the lesion obtained through CAC can then be used both for diagnosis and prognostication of a disease process.

15.2.2.3 Computer-Aided Diagnosis

Computer-aided diagnosis (CADx) systems are designed to process a specific finding and describe it accurately, characterize the likelihood of a specific diagnosis (e.g., likelihood of malignancy), provide differential diagnoses in order of probability, or recommend a clinical action. In mammography, CADx is able to characterize clusters of microcalcifications based on predetermined criteria in order to provide the probability of malignancy, aiding the radiologist in deciding on the need for a tissue biopsy. Some CADx systems are able to provide a more specific and definitive diagnosis, and the utility of such CADx systems is usually restricted to clinical scenarios where a detected abnormality has a very limited list of possible differential diagnoses, for example, the presence of distinct filling defects within the pulmonary arteries on a CT pulmonary angiogram study is almost always due to pulmonary embolism.

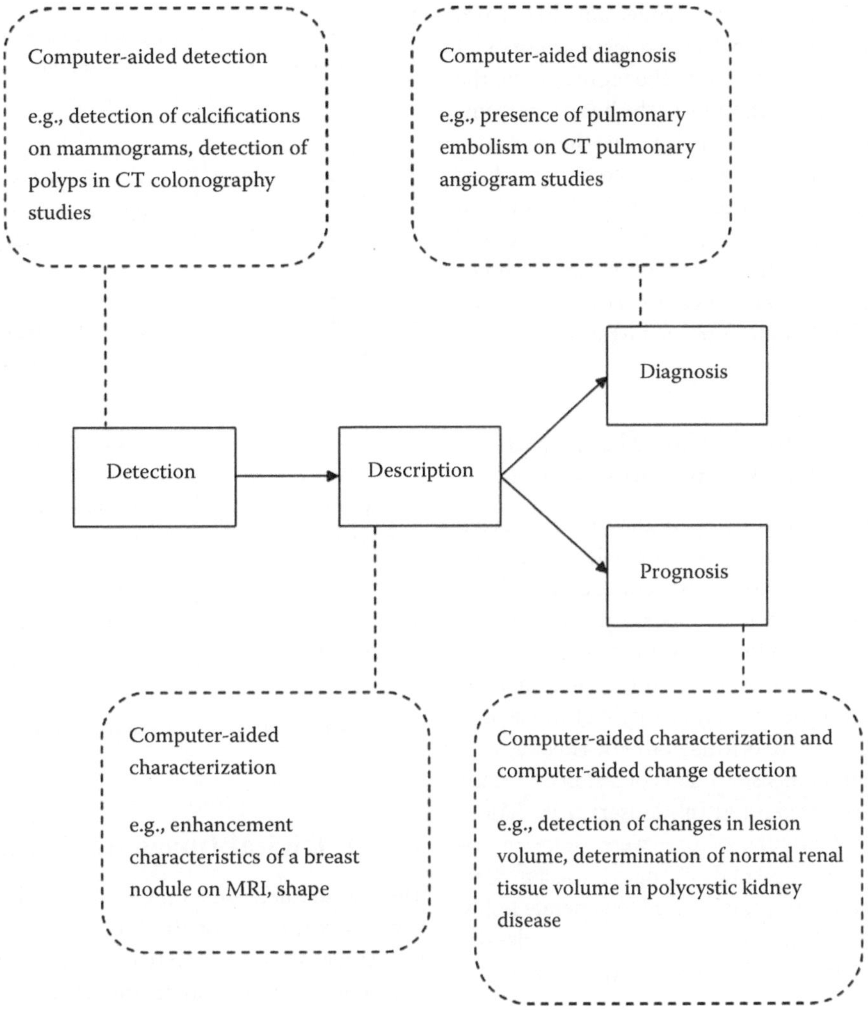

FIGURE 15.2 Potential roles of CAD in the medical image interpretation process.

15.2.2.4 Computer-Aided Change Detection

Computer-aided change detection (CACD) systems are designed to detect changes in sequential studies of the same anatomical region, quantify the amount of change, or characterize the significance of the change. An example of CACD applied to a common clinical problem is the determination of changes in brain tumor volume on MRI over time to evaluate disease status and response to treatment. More advanced CACD systems are expected to be able to evaluate variable degrees of change in multiple lesions for the same patient, and provide an overall picture of disease status. Such CACD systems have the potential to provide valuable information regarding the response to treatment in clinical trials.

15.3 Goals of CAD

Having described the overall imaging interpretation process and the potential roles of CAD in enhancing the existing workflow, the goals of CAD and the case for CAD utility will be presented. Key issues in this discussion include what value computers add to the process, and conversely, what role a radiologist can continue to play in the age of CAD. The relationship between humans and computers should ideally be based on a synergistic model, capitalizing on the strengths of each while minimizing their respective weaknesses, all for the benefit of the patient. Computers are best suited for repetitive tasks involving objective information in large data sets. On the other hand, the physician is better positioned to integrate medical information from multiple sources, recognize various patterns of clinical syndromes, and evaluate the relevance of an abnormality in a given clinical context. In other words, computers have an edge in the "science" of medicine, while physicians have the lead in the "art" of medicine.

15.3.1 Consistent Detection of Imaging Abnormalities

Humans make errors during image interpretation due to factors such as fatigue, information overload, inexperience, and environmental conditions. Computers, while not free from making mistakes, are able to rapidly process large volumes of imaging data in a more consistent fashion. The more specific the abnormality and focused the detection task, the better the computer algorithm for lesion detection is likely to be. These considerations are especially applicable to medical screening, where large numbers of imaging studies are read, the majority of which are normal. Screening examinations (e.g., mammography or CT colonography) that have standardized formats and few pathologies of interest with limited appearances are particularly suited for CADe. This area of CAD application is currently the most well-researched and developed.

15.3.2 Accurate and Reproducible Description of Imaging Features and Abnormalities

After detection, a lesion needs to be accurately characterized in terms of anatomic extent, imaging properties (e.g., size, shape, and other physical characteristics), and other features of interest (e.g., contrast enhancement, appearance of margins, etc.). Computers are better suited for performing accurate measurements and feature descriptions in a reproducible manner over an extended period of time. A clear example of this would be measurement of changes in size of lung nodules. This was previously dependent on manual measurements of lesion diameter, which was prone to interobserver and intraobserver variation. Automated calculations of nodule volume were more reproducible and better at picking up small changes in lesion size. Quantifiable lesion characteristics can be used to generate lesion descriptors that enable CAD algorithms to accurately classify lesions (diagnosis) and determine disease status (prognosis).

15.3.3 Diagnosis of Imaging Abnormalities

A final diagnosis for patient management is derived not only from a single medical imaging study, but is instead arrived at after consideration of multiple data sources, including other imaging studies, comparisons with prior imaging examinations, and other clinically relevant information (e.g., demographic data, clinical history, findings on physical examination, other test results, etc.). While computers are able to quickly search and retrieve information from vast databases, automation of this complex synthesis of medical information to arrive at a specific diagnosis remains a formidable challenge. Hence, the process of combining these multiple sources of information and deriving a list of differential diagnoses in order of probability that is customized for a particular patient within a specific clinical context is currently one better performed by a physician.

15.4 Techniques and Components of CAD

While several types of CAD systems exist, the fundamental components of a typical CAD system are similar, and are illustrated in Figure 15.3.

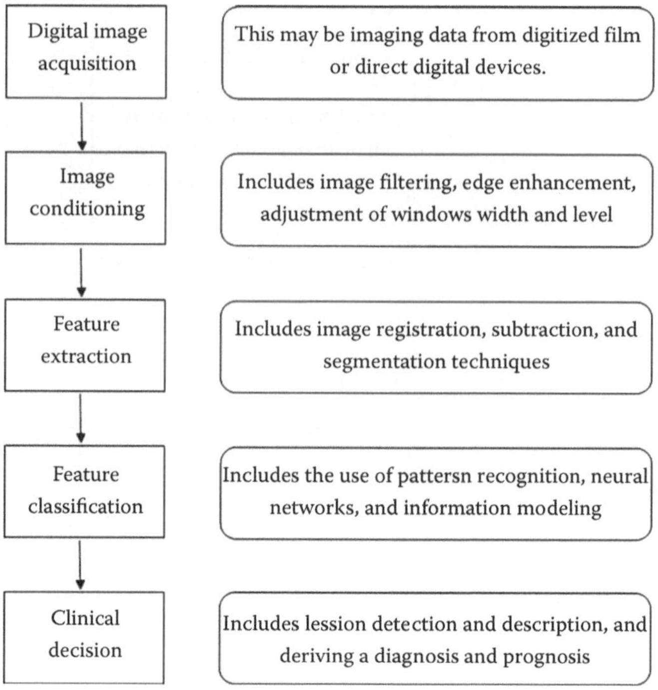

FIGURE 15.3 Functional components of a typical CAD system.

15.4.1 Digital Image Acquisition

The digital image data may be acquired from a variety of sources such as computed or digital radiography devices, ultrasound machines, gamma cameras, CT scanners, and MRI machines. Another method of obtaining usable image data includes digitizing images previously obtained with conventional radiography techniques (e.g., film-screen techniques), although some loss of information is expected through the digitization process. High-quality image data is critical to successful utility of any CAD system. Notwithstanding the different possible sources for the digital image, the basic prerequisites for the image are sufficient spatial resolution, good contrast resolution, high signal-to-noise ratio, and minimal image artifact. For volumetric data sets obtained with multidetector CTs, thin slices with isotropic voxels are preferred, although one must balance the radiation burden to the patient with the need for thin slices.

15.4.2 Image Conditioning

The aim of image conditioning is to optimize key attributes of digital images to enhance subsequent data manipulation and interpretation by the CAD system. Image conditioning techniques include the application of basic filters to image data and enhancement of features (e.g., edges) to increase lesion conspicuity. Excessive image conditioning may create artifacts in images that negatively impact CAD interpretation, resulting in increased false-positive instances.

15.4.3 Feature Extraction

Feature extraction is the process where key characteristics of an image such as an anatomical region or suspected lesion(s) are isolated from nonessential background information in the data set. Techniques available for feature extraction range from simple thresholding, segmentation, registration, and subtraction to complex shape analysis. Segmentation is a process of partitioning a digital image into various regions in order to facilitate subsequent analysis. An example would be segmentation of the colon in a CT colonography study so that only pixels related to the colon are subjected to subsequent analysis for the presence of polyps. Image subtraction highlights differences between imaging studies, and is typically used to highlight changes of a lesion over time. Image registration, the precise alignment of comparison images, is a key prerequisite for image subtraction. Modern techniques not only enable registration of images from similar studies, but also allow alignment of studies with differing contrast properties (e.g., registration of MRI and CT images).

15.4.4 Feature Classification

After relevant image features are extracted, classification ensues to derive a usable output (e.g., likelihood of malignancy, differential diagnoses) from the analysis of the features. This enables the physician to make sense of the CAD outputs to arrive at a clinical decision. The process of classification in CAD systems may incorporate rules based on *a priori* knowledge of disease features and prevalence, pattern recognition, and other probability modifiers.

15.4.4.1 Pattern Recognition and Neural Networks

Human interpretation of medical images is usually based on recognizing patterns of radiological findings and their respective associations with different diseases. Computers can mimic this pattern recognition process through the use of statistical methods and neural networks that facilitate machine learning, leveraging on vast databases of medical information (Coppini et al., 2003; Döhler et al., 2008; Suzuki, 2009).

15.4.4.2 Information Modeling and Probability

The accuracy of diagnosis is not only a function of the information in the images, but also requires consideration of the medical context of the patient who underwent the imaging study. This medical context takes into account *all* disease entities that might exist in the patient, along with their associated probabilities. Probabilities are dependent on specific information that is unique to each patient, and general information about various disease probabilities. These are factors that a physician takes into account when formulating a final impression for each imaging study. While attempts have been made to create computer algorithms to take these additional data inputs into account, the challenge remains to create CAD applications that are not only reliable in specific situations (e.g., screening), but are also applicable to diverse clinical settings.

15.4.5 Clinical Decision

After classification of the features of a detected abnormality, the CAD system provides a final usable output to facilitate the clinical decision by the evaluating radiologist. Depending on the clinical context for the particular patient who underwent the imaging study, the final output may include detection of an abnormality, description/characterization of a known abnormality, providing a diagnosis or list of differential diagnoses, or evaluation of disease progression.

15.5 Utility of CAD

15.5.1 Modes of CAD Utility

CAD devices or systems, when utilized in the image interpretation process by a physician, can be applied either as a first reader, second reader, or concurrent reader. Each mode of CAD application has its attendant advantages and disadvantages, and the optimal method of using CAD continues to be an area that requires more research.

15.5.1.1 First Reader CAD

In the first reader mode, the physician reviews only regions or findings marked by the CAD device. Imaging findings not highlighted by the CAD system are not reviewed by a physician. The main benefit of such a CAD system is an anticipated reduction in time required to review medical images, a proposition that is increasingly attractive in an era of multidetector CT, where the number of images per study can range in the thousands. However, such a CAD system would need to have a performance at least equal, if not better, than that of a physician reviewer. To date, no CAD system has been approved for first reader use.

15.5.1.2 Second Reader CAD

When CAD is used as a second reader, the physician first does a full interpretation of the imaging study without CAD, and then reinterprets the study with the CAD system. The computer provides a "second opinion," but the responsibility for the final evaluation and diagnosis is still made by the physician in all cases. This method of using CAD systems is analogous to using a spelling or grammar checking function in word processing software. This second reader mode is currently the most common clinical application of CAD systems. A key concern regarding the use of CAD in second reader mode is the extra time required to review CAD outputs, a time burden which could become excessive if the CAD system generates a large number of false-positive detections.

15.5.1.3 Concurrent Reader CAD

For the concurrent reader application of CAD, the physician performs a single-pass complete interpretation of the imaging study in the presence of CAD markings. This method of CAD utility potentially offers a way to reap the benefits of CAD while mitigating the time burden of reviewing extra CAD outputs. However, this method removes the first-cut independent review of the images by the physician, a step which is known to pick up lesions not otherwise detected by a CAD system. There is also concern that routine application of concurrent reader CAD may reduce reader vigilance by distracting the physician from performing an unbiased systematic review of the images.

15.5.1.4 Automated Computer Diagnosis

It is important to differentiate CAD from Automated Computer Diagnosis (ACD). In ACD, the final diagnosis is derived from computer algorithms alone, without human intervention. With CAD, computer performance does not necessarily have to be comparable to or exceed that of physicians, but need only be complementary. On the other hand, ACD must have a performance equal or better to that of humans in all possible clinical scenarios, as the images would not be viewed by a physician at all. To date, there is no approved software for clinical utility as an ACD system.

15.5.2 Limitations of CAD Systems

CAD performance (e.g., sensitivity, specificity, accuracy, etc.) is dependent on the underlying algorithm and type of training data set used, which are tailored for specific clinical scenarios. Inappropriate application of CAD to different clinical contexts (e.g., a patient population with different types or likelihood of disease, using a second reader CAD as a first reader, etc.) can negate the strengths of CAD and decrease physician performance. For example, a CAD system for detecting colonic polyps that was developed using CT images from a symptomatic population may not necessarily be applicable to CT images of a healthy asymptomatic population.

As with any test, there is a need to balance between sensitivity and specificity, and improving the sensitivity of CAD systems would also often increase the false-positive rate. CAD systems which routinely generate a large number of false-positive image annotations would in time be ignored by the reviewing radiologist, thus negating the expected benefits of the CAD system. Receiver Operating Characteristic (ROC) curves are a useful method to evaluate the accuracy of a CAD system and compare the results of different CAD systems. ROC curves are generated from a range of sensitivity and corresponding specificity values, and therefore can incorporate different decision thresholds that a radiologist may choose in using a CAD system. The area under the ROC curve (AUC) provides a useful summary of the accuracy of the test, and ranges in value from 0.5 (results attributable to chance, with no additional discriminatory value compared to a coin toss) to 1.0 (perfect discrimination or accuracy).

When introducing CAD systems, the learning curve and adjustment period for users (estimated to take weeks to years) need to be taken into account. Temporary changes to workflow (e.g., transiently reducing the number of reads per session) may be required to decrease the impact of this adjustment period on clinical service delivery. The expected benefits of CAD are also dependent on whether the systems are deployed for use by subspecialty expert radiologists (e.g., body imaging radiologists who routinely review large volumes of CT colonography studies and would therefore already have a high sensitivity and specificity for lesion detection) as compared to general radiologists (who may only occasionally be required to review a CT colonography examination). CAD systems may also elicit varying reactions from different users, and thus have an unexpected impact on the interpretation process and reading time.

15.6 Current Applications of CAD

15.6.1 Breast Imaging

After cancers of the skin, breast cancer is the most frequently diagnosed cancer in women. Breast cancer screening is therefore a major public health initiative globally. Mammographic screening for breast cancer involves evaluating large numbers of studies, most of which are expected to be normal, for a specific abnormality (breast cancer), which has a relatively limited range of appearances. The false-negative rate for mammographic detection of breast cancer ranges from about 10–25% (Destounis et al., 2004) and there is significant interobserver variability in the evaluation of breast imaging studies (Mussurakis et al., 1996; Skaane et al., 1997). Prior attempts to address such issues and increase the accuracy of mammographic screening include the employment of "double reads," where second radiologist independently reviews mammographic images for abnormalities, a role which may potentially be played by CAD in a second reader mode (Helvie, 2007). These characteristics of mammographic screening for breast cancer make it particularly suited for CAD deployment, and it is therefore not surprising that mammography was the first major area, where CAD usage was adopted.

The first FDA-approved CAD products were for mammography in 1998. Modern CAD systems analyze digitized or digital mammographic images to find features of breast cancer such as clustered microcalcifications, masses, and architectural distortion. The methods and challenges in detecting and characterizing such features of breast cancer have been recently discussed (Giger et al., 1994; Elter et al., 2009). Suspicious mammographic features are then highlighted to the radiologist for review. The radiologist then makes a final decision regarding the likely nature of the lesion and further management as required. It is important to note that mammographic CAD systems are currently designed for use as a second reader, which requires the radiologist to first review the images independent of the CAD system, with a subsequent "second look" with the CAD markings. The utility of current mammographic CAD systems as either a first or concurrent reader is not recommended.

To date, no randomized controlled trials have been done to document changes in patient survival from the use of CAD in mammography. Surrogate end points (e.g., cancer detection rate, stage of cancer, frequency of interval cancers, change in recall rate, etc.) have been used to evaluate CAD performance. Overall, CAD trials have generally, but not always, shown improvements in cancer detection (Helvie, 2007). Most studies show that incremental cancer detection by CAD systems are mainly restricted to ductal carcinoma *in situ* (a noninvasive type of cancer). It remains

TABLE 15.1 Summary of Results of Recent Studies of CAD Systems for Detection of Breast Cancer in Screening Mammography

Investigators (Year)	Number of Exams	Change in Cancer Detection Rate (%)	Change in Recall Rate (%)
Gromet (2008)	231,221	+1.9	+3.9
Fenton et al. (2007)	429,345	+1.2	+30.7
Dean et al. (2006)	9520	+10.8	+26.0
Ko et al. (2006)	5016	+4.7	+15.0
Morton et al. (2006)	21,349	+7.6	+10.8
Birdwell et al. (2005)	8682	+7.4	+8.2
Cupples et al. (2005)	27,274	+16.1	+8.1
Gur et al. (2004)	115,571	+1.7	+0.1
Freer et al. (2001)	12,860	+19.5	+19.0

uncertain whether such additional detections will affect overall patient outcomes in the long term. Single reads of mammographic studies with CAD have also shown promise as a viable alternative to routine double reading (Gilbert et al., 2008; Gromet, 2008). Excellent reviews of the useful role of CAD in screening mammography were recently published (Helvie, 2007; Birdwell, 2009). A summary of recent CAD trials in the detection of breast cancer in screening mammography is given in Table 15.1.

CAD has also been used to facilitate evaluation of lesions on breast MRI, including lesion morphology and the complex task of characterizing lesion enhancement kinetics (e.g., rate of enhancement and washout). CAD systems have also been developed for use with breast ultrasound in order to analyze morphologic features of breast lesions (e.g., size, shape, orientation) as well as detect lesion boundaries.

Possible future trends for CAD in breast imaging include the development of CAD systems that incorporate information from multiple mammographic images (e.g., comparisons with the contralateral breast, comparison with prior images of the same breast, correlation with different projections of the same breast region), different image sources (e.g., concurrent CAD evaluation of mammogram, ultrasound and MRI images to provide an integrated CAD output) and incorporation of newer breast imaging modalities such as digital breast tomosynthesis, contrast-enhanced digital subtraction mammography, breast PET, and breast CT.

15.6.2 Chest Imaging

The detection and evaluation of lung nodules is a very common clinical challenge encountered by radiologists. This has become a more pressing problem with the advent of multidetector CT, which has not only increased the number of images requiring review by a physician, but also resulted in a vast number of small pulmonary nodules of uncertain significance being detected on a regular basis. Furthermore, lung cancer, which accounts for the most cancer-related deaths in both men and women in the United States, often begins as a pulmonary nodule. Lung CAD has therefore been dominated by applications in pulmonary nodule detection and volumetry (Li, 2007; Marten and Engelke, 2007).

CAD systems for use with chest radiography and CT to detect lung nodules were first approved by the FDA in 2001 and 2004, respectively. Because CT has been shown to have a higher sensitivity for detection of lung nodules than conventional chest radiography, the majority of CAD development for chest imaging in recent years have focused on thoracic CT. Lung nodule CAD algorithms capitalize on the inherent high contrast between nodules and the surrounding lung tissue to facilitate detection. Overall, current lung CAD systems for nodule detection on CT have varying reported true positive rates ranging from about 70% to over 90% with a false-positive rate of about 0.1–1 nodule per section (Giger et al., 1994; Armato et al., 1999; Lee et al., 2001; Awai et al., 2004). As with breast imaging CAD, current lung nodule CAD systems should be used in a second reader capacity only. To date, no prospective outcome-based trials have been published evaluating the performance of CAD for detection of nodules in CT. A summary of recent studies on CAD utility in nodule detection is given in Table 15.2.

CAC of lung nodule features on CT such as nodule volume, lesion margins, lesion density, presence of calcifications, and enhancement characteristics allow more accurate and reproducible descriptions of lung nodules. Such characterization allows for improved follow-up of indeterminate lung nodules, and enables better evaluation of response to treatment. For example, manual two-dimensional measurements of nodules less than 5 mm in size can vary by as much as ±1 mm, which translates to a greater than 75% difference in volume. This variability is reduced by computer-aided volumetry, which would allow earlier detection of real (but small) changes in nodule size.

Current challenges in lung CAD systems for nodule detection and characterization include evaluation of lesions which do

TABLE 15.2 Summary of Results of Recent Studies Evaluating CAD Impact on Detection of Pulmonary Nodules by Radiologists for CT Thorax Studies

Investigators (Year)	Number of CT Exams	Results
White et al. (2008)	109	CAD increased AUC from 86.7% to 88.7%
Hirose et al. (2008)	21	CAD improved mean sensitivity from 39.5% to 81.0%
Rubin et al. (2005)		CAD improved mean sensitivity from 50% (individual reads) or 63% (double reads) to 76%
Marten et al. (2004)	18	CAD improved AUC from 0.71 to 0.93 (experienced reader), and from 0.49 to 0.79 (inexperienced reader)
Beigelman-Aubry et al. (2009)	54	CAD improved sensitivities of two readers by 9.6% and 23%
Brown et al. (2005)	8	CAD improved mean detection rates from 64.0% to 81.9%
Das et al. (2006)	25	CAD increased mean sensitivity from 76% to 85%
Awai et al. (2004)	50	CAD increased mean AUC from 0.64 to 0.67

Note: AUC, area under curve.

not have a uniform solid structure. This includes nodules with low density (e.g., ground glass nodules), nodules with a complex appearance (e.g., cavitatory nodules or nodules with irregular matrices), or nodules located close to other dense structures (e.g., juxta-vascular and -pleural nodules). The application of lung CAD systems in patients with co-existing lung disease (e.g., pleural effusions or interstitial lung disease) also remains a challenge as these pathologies obscure underlying nodules. The continued improvement of multidetector CT scanner technology, with the associated decreases in scan times, have reduced the impact of respiratory or cardiac motion artifacts in CT data which previously affected CAD applications.

Other areas of on-going work in lung CAD applications include the diagnosis of pulmonary embolism (Zhou et al., 2005), detection and quantification of pneumothorax (Sanada et al., 1992), quantification and characterization of interstitial lung disease (Katsuragawa et al., 1988; Arzhaeva et al., 2007), and the incorporation of multimodality information (e.g., PET and CT) to further characterize the lung lesions (Nie et al., 2006).

15.6.3 Abdominal Imaging

Colorectal cancer is the third most common cancer in both men and women. As the majority of such cancers develop over a period of time from small polyps, the detection of colonic polyps has become a key focus of reducing the morbidity and mortality from colorectal cancer. CT colonography (CTC) is an emerging technique for the detection of colonic polyps and active research and evaluation of CAD for polyp detection in CTC is underway (Yoshida and Dachman, 2005; Bielen and Kiss, 2007).

While commercial CAD systems for CTC polyp detection are already available, at the time of writing, major efforts continue to be directed at obtaining full FDA approval for CAD for CTC polyp detection. Workstations used in the evaluation of CTC studies are capable of displaying data in 2D multiplanar reconstruction views and also provide 3D virtual colonoscopic "fly through" endoluminal views. Therefore, CAD systems have been developed for use with these various viewing modes. Such CAD systems have shown good sensitivity both in isolation and as a "second reader" for detection of clinically significant polyps (Summers et al., 2005; Taylor et al., 2006; Petrick et al., 2008). However, more research is required to properly evaluate the performance and determine the optimal role of CAD in CT colonography. To date, no prospective outcome-based trials have been published evaluating the performance of CAD for detection of polyps in CT colonography. A summary of recent studies on CAD utility in polyp detection in CT colonography is given in Table 15.3.

As with other CAD systems, most of the existing CTC CAD systems are designed for use as a "second reader," although one study recently showed that "concurrent reader" application of a CTC CAD system showed better time efficiency and similar detection of polyps more than 6 mm in size when compared to a "second reader" application (Taylor et al., 2008).

TABLE 15.3 Summary of Results of Recent Studies Evaluating CAD Impact on Detection of Polyps by Radiologists for CT Colonography Studies

Investigators (Year)	Number of CT Exams	Results
Taylor et al. (2009)	50	CAD increased per-patient sensitivity from 82% to 87% for polyps 5 mm or larger
Petrick et al. (2008)	60	CAD increased average reader sensitivity by 15% for polyps 6 mm or larger
Baker et al. (2007)	30	CAD improved average sensitivity for polyp detection from 81.0% to 90.8%
Mang et al. (2007)	52	CAD increased sensitivity from 91% to 96% (expert readers) and from 75.5 % to 93% (nonexpert readers)
Halligan et al. (2006)	107	Polyp detection increased significantly with CAD; on average 12 more polyps detected per reader

Current challenges for CAD for CTC include the variable appearance of the colon (air-filled vs fluid-filled segments, residual fecal material, numerous mucosal folds), the need to integrate information from both supine and prone data sets, the detection of sessile polyps which do not conform to the typical polypoidal shape, and the use of fecal tagging and digital bowel cleansing.

15.6.4 Others

The potential applications of CAD in medical image interpretation are innumerable and continue to grow rapidly. The following are some additional examples of clinical scenarios where the use of CAD has been studied.

15.6.4.1 Cardiovascular Imaging

Within the field of cardiovascular imaging, CAD has been employed in the interpretation of myocardial perfusion SPECT studies (Garcia et al., 2001), the detection and evaluation of plaques on cardiac CT (Dey et al., 2006), and the detection of coronary artery stenoses on CT (Reimann et al., 2009).

15.6.4.2 Neuroradiology

CAD has been studied in the diagnosis of Alzheimer's disease (Brewer et al., 2009; Chaves et al., 2009), assessment of brain CTs done for head trauma (Yuh et al., 2008), detection of intracranial aneurysms on magnetic resonance angiography exams (Uchiyama et al., 2005; Arimura et al., 2006; Yang et al., 2009), detection and evaluation of ischemic brain lesions (Uchiyama et al., 2007; Yamashita et al., 2008), detection of brain tumor invasion (Jensen and Schmainda, 2009), detection of changes in brain tumor status on MRI (Patriarche and Erickson, 2007), and detection of small acute intracranial hemorrhages on CT (Chan and Huang, 2008).

15.6.4.3 Pediatric Imaging

In pediatrics, the use of CAD has been evaluated in the assessment of bone age based on skeletal radiographs (Pietka et al., 2001), identification of pulmonary nodules in pediatric oncologic patients (Helm et al., 2009), detection of childhood pneumonia on chest radiographs (Oliveira et al., 2008), and detection of therapy-induced leukoencephalopathy in pediatric leukemia patients (Glass et al., 2006). Beyond pediatric radiology, CAD has also been used in the prognostication of neuroblastoma based on digitized histological images (Sertel et al., 2009).

15.6.4.4 Musculoskeletal Imaging

CAD has also been used in the evaluation of bones and joints. Examples include the assessment of disease progression in arthritis (Duryea et al., 2000; Shamir et al., 2009), evaluation of cartilage lesions on MRI (Lee et al., 2004), detection of meniscus tears on MRI knee images (Ramakrishna et al., 2009), analysis of joints spaces in the hand (Pfeil et al., 2007), detection of vertebral body fractures on plain radiographs (Kasai et al., 2006), and change detection on successive bone scan images (Shiraishi et al., 2007).

15.7 Future of CAD

The past four decades have seen a quantum leap in the field of CAD in radiology. Although much has been achieved thus far in the field of computer-assisted evaluation of medical images, there is still some way to go before achieving widespread acceptance of routine CAD utility in medical image interpretation. The following are key areas that are expected to feature prominently in the field of CAD systems in the coming years.

15.7.1 Broadening the Scope of CAD Systems

Current CAD systems generally focus on highly specific tasks such as the detection of breast microcalcifications or colonic polyps in a screening context. Hence, such CAD systems cannot be applied to the wide range of scenarios presented to physicians in routine practice.

Future CAD systems may come as a "clinical package" to allow more general application of CAD technology to the daily practice of radiology (Doi, 2007). For example, a future "chest imaging CAD scheme" may include detection algorithms for nodules, interstitial lung disease, cardiomegaly, bone density, vertebral fractures, pneumothorax, pleural effusions, and interval changes, along with computerized classification of benign and malignant nodules and differential diagnosis of interstitial lung diseases. CAD systems, in addition to detecting lesions and providing differential diagnoses, could also automatically trawl vast image databases for similar-appearing lesions with confirmed diagnoses. Such visual comparisons would facilitate the final diagnosis by a physician.

15.7.2 Developing Multimodality Evaluation

In the process of evaluating medical images, radiologists already routinely process information from different imaging modalities in order to arrive at a diagnosis for a particular patient. Furthermore, with the rapid advances in molecular imaging, fusion imaging techniques such as PET-CT, SPECT-CT, SPECT-MR, and PET–MRI are likely to become more common in future. Enhancement of CAD systems to incorporate information from various imaging modalities (e.g., combining features from mammography, breast ultrasound, and breast MRI) may further increase the diagnostic accuracy for an individual patient.

15.7.3 Optimizing the Human–Machine Interface

More work is required in the field of human interaction with CAD systems, which will evolve with time in line with technological advances as humans become more familiar with computer-aided decision making. This key area of usability is critical to the successful incorporation of CAD into the practice of medical image interpretation. CAD systems which require a radiologist to move to a separate workstation to use the detection or diagnostic algorithms are much less likely to be accepted compared to a CAD system, which is incorporated into an existing RIS-PACS system. The manner in which CAD outputs are displayed (e.g., as key images in a separate series, or incorporating the CAD markings into the original series as annotations) is also an area where further enhancements are required. Apart from the mode of presentation, the amount of information displayed by the CAD system needs to be carefully considered, as overwhelming the display with too many annotations will likely result in the CAD outputs being ignored by the user. Future CAD systems may be customizable according to the clinical scenario and experience level of the radiologist.

15.7.4 Standardizing Evaluation and Validation of CAD

Current studies on the clinical utility of CAD systems have variable designs. Studies evaluating the clinical value of CAD can often be classified as either sequential reading studies or historical comparison studies. In the sequential reading design, radiologist performance for the same patient cohort is assessed before and after the introduction of CAD. For the historical comparison design, the performance of groups of radiologists over two periods of time is compared, and the patient cohorts and radiologists involved may not be identical for the two time periods. Beyond validation of clinical utility, further studies aimed at determining the impact of CAD on patient outcomes (e.g., improvements in survival) should also be performed to firmly establish the case for CAD in medical image interpretation.

Standardization of CAD evaluation may include clear definitions of task, patient populations, reader training with CAD, study designs, selection of ground truth, data analysis methods, identification of biases, and endpoints for assessment of success or failure. The creation of standardized image databases for different pathologies will allow different CAD systems compared against each other in an objective and reproducible manner. Such databases will also ensure that CAD systems are tested and validated across a variety of image-producing equipment and patient populations before being used in clinical practice.

15.7.5 Managing Legal Implications

As the performance of CAD systems improves, their utility in the image interpretation process may become part of the minimum standard of care. It is therefore plausible that future radiological practices could include CAD as a routine screening for all images before reports are finalized, no different from the current practice of applying "spell-check" algorithms to documents. In such a scenario, it will be necessary to decide how much of CAD information, if any, should be finally incorporated into the medical record of the patient. CAD has already been used by both plaintiff and defense lawyers to make their cases in court (e.g., mammography), a practice which could become routine in future.

15.8 Summary and Conclusion

There is increasing interest in the use of computers to facilitate any or all of the steps in medical image interpretation such as detection, description, diagnosis, and prognosis. For such CAD systems and processes to succeed, it is critical to combine physicians' clinical acumen with the technological capabilities of computers in the appropriate patient context. Key future challenges include optimizing the user interface and establishing standardized methods of evaluating these systems of disease detection and diagnosis. With extensive on-going research efforts, CAD systems will continue to evolve and improve, and are expected to become an integral part of medical image interpretation in the near future.

References

Arimura, H., Li, Q., Korogi, Y., Hirai, T., Katsuragawa, S., Yamashita, Y., Tsuchiya, K., and Doi, K. 2006. Computerized detection of intracranial aneurysms for three-dimensional MR angiography: Feature extraction of small protrusions based on a shape-based difference image technique. *Med. Phys.*, 33(2), 394–401.

Armato, S. G., Giger, M. L., Moran, C. J., Blackburn, J. T., Doi, K., and MacMahon, H. 1999. Computerized detection of pulmonary nodules on CT scans. *RadioGraphics*, 19, 1303–11.

Arzhaeva, Y., Prokop, M., Tax, D. M., De Jong, P. A., Schaefer-Prokop, C. M., and van Ginneken, B. 2007. Computer-aided detection of interstitial abnormalities in chest radiographs using a reference standard based on computed tomography. *Med. Phys.*, 34(12), 4798–809.

Awai, K., Murao, K., Ozawa, A., Komi, M., Hayakawa, H., Hori, S., and Nishimura, Y. 2004. Pulmonary nodules at chest CT: Effect of computer-aided diagnosis on radiologists' detection performance. *Radiology*, 230(2), 347–52.

Baker, M. E., Bogoni, L., Obuchowski, N. A. et al. 2007. Computer-aided detection of colorectal polyps: Can it improve sensitivity of less-experienced readers? Preliminary findings. *Radiology*, 245(1), 140–9.

Beigelman-Aubry, C., Raffy, P., Yang, W., Castellino, R. A., and Grenier, P. A. 2007. Computer-aided detection of solid lung nodules on follow-up MDCT screening: Evaluation of detection, tracking, and reading time. *AJR Am. J. Roentgenol.*, 189(4), 948–55.

Bielen, D. and Kiss, G. 2007. Computer-aided detection for CT colonography: Update 2007. *Abdom Imaging*, 35, 571–81.

Birdwell, R. L. 2009. The preponderance of evidence supports computer-aided detection for screening mammography. *Radiology*, 253(1), 9–16. Review.

Birdwell, R. L., Bandodkar, P., and Ikeda, D.M. 2005. Computer-aided detection with screening mammography in a university hospital setting. *Radiology*, 236(2), 451–7.

Brewer, J. B., Magda, S., Airriess, C., and Smith, M. E. 2009. Fully-automated quantification of regional brain volumes for improved detection of focal atrophy in Alzheimer disease. *AJNR Am. J. Neuroradiol.*, 30(3), 578–80.

Brown, M. S., Goldin, J. G., Rogers, S. et al. 2005. Computer-aided lung nodule detection in CT: Results of large-scale observer test. *Acad. Radiol.*, 12(6), 681–6.

Chan, T. and Huang, H. K. 2008. Effect of a computer-aided diagnosis system on clinicians' performance in detection of small acute intracranial hemorrhage on computed tomography. *Acad. Radiol.*, 15(3), 290–9.

Chaves, R., Ramírez, J., Górriz, J. M., López, M., Salas-Gonzalez, D., Alvarez, I., and Segovia, F. 2009. SVM-based computer-aided diagnosis of the Alzheimer's disease using t-test NMSE feature selection with feature correlation weighting. *Neurosci. Lett.*, 461(3), 293–7.

Coppini, G., Diciotti, S., Falchini, M., Villari, N., and Valli, G. 2003. Neural networks for computer-aided diagnosis: Detection of lung nodules in chest radiograms. *IEEE Trans. Inf. Technol. Biomed.*, 7(4), 344–57.

Cupples, T. E., Cunningham, J. E., and Reynolds, J. C. 2005. Impact of computer-aided detection in a regional screening mammography program. *AJR Am. J. Roentgenol.*, 185(4), 944–50.

Das, M., Mühlenbruch, G., Mahnken, A. H., Flohr, T. G., Gündel, L., Stanzel, S., Kraus, T., Günther, R. W., and Wildberger, J. E. 2006. Small pulmonary nodules: Effect of two computer-aided detection systems on radiologist performance. *Radiology*, 241(2), 564–71.

Dean, J. C. and Ilvento, C. C. 2006. Improved cancer detection using computer-aided detection with diagnostic and screening mammography: Prospective study of 104 cancers. *AJR Am. J. Roentgenol.*, 187(1), 20–8.

Destounis, S. V., DiNitto, P., Logan-Young, W., Bonaccio, E., Zuley, M. L., and Willison, K. M. 2004. Can computer-aided detection with double reading of screening mammograms help decrease the false-negative rate? *Initial Experience Radiol.*, 232, 578–84.

Dey, D., Callister, T., Slomka, P. et al. 2006. Computer-aided detection and evaluation of lipid-rich plaque on noncontrast cardiac CT. *AJR Am. J. Roentgenol.*, 186(6 Suppl 2), S407–13.

Döhler, F., Mormann, F., Weber, B., Elger, C. E., and Lehnertz, K. 2008. A cellular neural network based method for classification of magnetic resonance images: Towards an automated detection of hippocampal sclerosis. *J. Neurosci. Methods*, 170(2), 324–31.

Doi, K. 2007. Computer-Aided diagnosis in medical imaging: Historical review, current status and future potential. *Comput. Med. Imag. Graph.*, 31, 198–211.

Duryea, J., Jiang, Y., Zakharevich, M., and Genant, H. K. 2000. Neural network based algorithm to quantify joint space width in joints of the hand for arthritis assessment. *Med. Phys.*, 27(5), 1185–94.

Elter, M. and Horsch, A. 2009. CADx of mammographic masses and clustered microcalcifications: A review. *Med. Phys.*, 36(6), 2052–68.

Erickson, B. J. and Bartholomai, B. 2002. Computer-aided detection and diagnosis at the start of the third millenium. *J. Digit. Imaging*, 15(2), 59–68.

Fenton, J. J., Taplin, S. H., Carney, P. A. et al. 2007. Influence of computer-aided detection on performance of screening mammography. *N. Engl. J. Med.*, 356(14), 1399–409.

Freer, T. W. and Ulissey, M. J. 2001. Screening mammography with computer-aided detection: Prospective study of 12,860 patients in a community breast center. *Radiology*, 220, 781–6.

Garcia, E. V., Cooke, C. D., Folks, R. D., Santana, C. A., Krawczynska, E. G., De Braal, L., and Ezquerra, N. F. 2001. Diagnostic performance of an expert system for the interpretation of myocardial perfusion SPECT studies. *J. Nucl. Med.*, 42(8), 1185–91.

Giger, M. L., Bae, K. T., and MacMahon, H. 1994. Computerized detection of pulmonary nodules in computed tomography images. *Invest. Radiol.*, 29, 459–65.

Gilbert, F. J., Astley, S. M., Gillan, M. G., Agbaje, O. F., Wallis, M. G., James, J., Boggis, C. R., Duffy, S. W., and CADET II Group. 2008. Single reading with computer-aided detection for screening mammography. *N. Engl. J. Med.*, 359(16), 1675–84.

Glass, J. O., Reddick, W. E., Li, C. S., Laningham, F. H., Helton, K. J., and Pui, C. H. 2006. Computer-aided detection of therapy-induced leukoencephalopathy in pediatric acute lymphoblastic leukemia patients treated with intravenous high-dose methotrexate. *Magn. Reson. Imaging*, 24(6), 785–91.

Gromet, M. 2008. Comparison of computer-aided detection to double reading of screening mammograms: Review of 231, 221 mammograms. *AJR Am. J. Roentgenol.*, 190(4), 854–9.

Gur, D., Sumkin, J. H., Rockette, H. E., Ganott, M., Hakim, C., Hardesty, L., Poller, W. R., Shah, R., and Wallace, L. 2004. Changes in breast cancer detection and mammography recall rates after the introduction of a computer-aided detection system. *J. Natl. Cancer Inst.*, 96(3), 185–90.

Halligan, S., Altman, D. G., Mallett, S., Taylor, S. A., Burling, D., Roddie, M., Honeyfield, L., McQuillan, J., Amin, H., and Dehmeshki, J. 2006. Computed tomographic colonography: Assessment of radiologist performance with and without computer-aided detection. *Gastroenterology*, 131(6), 1690–9.

Helm, E. J., Silva, C. T., Roberts, H. C., Manson, D., Seed, M. T., Amaral, J. G., and Babyn, P. S. 2009. Computer-aided detection for the identification of pulmonary nodules in pediatric oncology patients: Initial experience. *Pediatr. Radiol.*, 39(7), 685–93.

Helvie, M. 2007. Improving mammographic interpretation: Double reading and computer-aided diagnosis. *Radiol. Clin. N. Am.*, 45, 801–11.

Hirose, T., Nitta, N., Shiraishi, J., Nagatani, Y., Takahashi, M., and Murata, K. 2008. Evaluation of computer-aided diagnosis (CAD) software for the detection of lung nodules on multidetector row computed tomography (MDCT): JAFROC study for the improvement in radiologists' diagnostic accuracy. *Acad. Radiol.*, 15(12), 1505–12.

Jensen, T. R. and Schmainda, K. M. 2009. Computer-aided detection of brain tumor invasion using multiparametric MRI. *J. Magn. Reson. Imaging*, 30(3), 481–9.

Kasai, S., Li, F., Shiraishi, J., Li, Q., and Doi, K. 2006. Computerized detection of vertebral compression fractures on lateral chest radiographs: Preliminary results with a tool for early detection of osteoporosis. *Med. Phys.*, 33(12), 4664–74.

Katsuragawa, S., Doi, K., and MacMahon, H. 1988. Image feature analysis and computer-aided diagnosis in digital radiography: Detection and characterization of interstitial lung disease in digital chest radiographs. *Med. Phys.*, 15(3), 311–9.

Khorasani, R., Erickson, B. J., and Patriarche, J. 2006. New opportunities in computer-aided diagnosis: Change detection and characterization. *J. Am. Coll. Radiol.*, 3(6), 468–69.

Ko, J. M., Nicholas, M. J., Mendel, J. B., and Slanetz, P. J. 2006. Prospective assessment of computer-aided detection in interpretation of screening mammography. *AJR Am. J. Roentgenol.*, 187(6), 1483–91.

Ledley, R. S. and Lusted, L. B. 1959. Reasoning foundations of medical diagnosis; symbolic logic, probability, and value theory aid our understanding of how physicians reason. *Science*, 130(3366), 9–21

Lee, K. Y., Dunn, T. C., Steinbach, L. S., Ozhinsky, E., Ries, M. D., and Majumdar, S. 2004. Computer-aided quantification of focal cartilage lesions of osteoarthritic knee using MRI. *Magn. Reson. Imaging*, 22(8), 1105–15.

Lee, Y., Hara, T., Fujita, H., Itoh, S., and Ishigaki, T. 2001. Automated detection of pulmonary nodules in helical CT images based on an improved template-matching technique. *IEEE Trans. Med. Imaging*, 20, 595–604.

Li, Q. 2007. Recent progress in computer-aided diagnosis of lung nodules on thin-section CT. *Comput. Med. Imag. Graphics*, 31, 248–57.

Lodwick, G. S., Keats, T. E., and Dorst, J. P. 1963. The coding of roentgen images for computer analysis as applied to lung cancer. *Radiology*, 81, 185–200.

Mang, T., Peloschek, P., Plank, C., Maier, A., Graser, A., Weber, M., Herold, C., Bogoni, L., and Schima, W. 2007. Effect of computer-aided detection as a second reader in multidetector-row CT colonography. *Eur. Radiol.*, 17(10), 2598–607.

Marten, K. and Engelke, C. 2007. Computer-aided detection and automated CT volumetry of pulmonary nodules. *Eur. Radiol.*, 17, 888–901.

Marten, K., Seyfarth, T., Auer, F., Wiener, E., Grillhösl, A., Obenauer, S., Rummeny, E. J., and Engelke, C. 2004. Computer-assisted detection of pulmonary nodules: Performance evaluation of an expert knowledge-based detection system in consensus reading with experienced and inexperienced chest radiologists. *Eur. Radiol.*, 14(10), 1930–8.

Morton, M. J., Whaley, D. H., Brandt, K. R., and Amrami, K. K. 2006. Screening mammograms: Interpretation with computer-aided detection—Prospective evaluation. *Radiology*, 239(2), 375–83.

Mussurakis, S., Buckley, D. L., Coady, A. M., Turnbull, L. W., and Horsman, A. 1996. Observer variability in the interpretation of contrast enhanced MRI of the breast. *Br. J. Radiol.*, 69(827), 1009–16.

Nie, Y., Li, Q., Li, F., Pu, Y., Appelbaum, D., and Doi, K. 2006. Integrating PET and CT information to improve diagnostic accuracy for lung nodules: A semiautomatic computer-aided method. *J. Nucl. Med.*, 47(7), 1075–80.

Oliveira, L. L., Silva, S. A., Ribeiro, L. H., de Oliveira, R. M., Coelho, C.J., and S Andrade, A. L. 2008. Computer-aided diagnosis in chest radiography for detection of childhood pneumonia. *Int. J. Med. Inform.*, 77(8), 555–64. Epub 2008 Feb 20.

Patriarche, J. and Erickson, B. 2007. Part 2. Automated change detection and characterization applied to serial MR of brain tumors may detect progression earlier than human experts. *J. Digit. Imaging*, 20(4), 321–8.

Petrick, N., Haider, M., Summers, R. M., Yeshwant, S. C., Brown, L., Iuliano, E. M., Louie, A., Choi, J. R., and Pickhardt, P. J. 2008. CT colonography with computer-aided detection as a second reader: Observer performance study. *Radiology*, 246(1), 148–56.

Pfeil, A., Böttcher, J., Seidl, B. E., Heyne, J. P., Petrovitch, A., Eidner, T., Mentzel, H. J., Wolf, G., Hein, G., and Kaiser, W. A. 2007. Computer-aided joint space analysis of the metacarpal-phalangeal and proximal-interphalangeal finger joint: Normative age-related and gender-specific data. *Skeletal Radiol.*, 36(9), 853–64.

Pietka, B. E., Pośpiech, S., Gertych, A., Cao, F., Huang, H. K., and Gilsanz, V. 2001. Computer automated approach to the extraction of epiphyseal regions in hand radiographs. *J. Digit. Imaging*, 14(4), 165–72.

Ramakrishna, B., Liu, W., Saiprasad, G. et al. 2009. An automatic computer-aided detection system for meniscal tears on magnetic resonance images. *IEEE Trans. Med. Imaging*, 28(8), 1308–16.

Reimann, A. J., Tsiflikas, I., Brodoefel, H., Scheuering, M., Rinck, D., Kopp, A. F., Claussen, C. D., and Heuschmid, M. 2009. Efficacy of computer aided analysis in detection of significant coronary artery stenosis in cardiac using dual source computed tomography. *Int. J. Cardiovasc. Imaging*, 25(2), 195–203.

Rubin, G. D., Lyo, J. K., Paik, D. S. et al. 2005. Pulmonary nodules on multi-detector row CT scans: Performance comparison of radiologists and computer-aided detection. *Radiology* 234(1), 274–83.

Sanada, S., Doi, K., and MacMahon, H. 1992. Image feature analysis and computer-aided diagnosis in digital radiography: Automated detection of pneumothorax in chest images. *Med. Phys.*, 19(5), 1153–60.

Sertel, O., Catalyurek, U. V., Shimada, H., and Gurcan, M. N. 2009. Computer-aided prognosis of neuroblastoma: Detection of mitosis and karyorrhexis cells in digitized histological images. *Conf. Proc. IEEE Eng. Med. Biol. Soc.*, 1, 1433–6.

Shamir, L., Ling, S. M., Scott, W., Hochberg, M., Ferrucci, L., and Goldberg, I. G. 2009. Early detection of radiographic knee osteoarthritis using computer-aided analysis. *Osteoarthritis Cartilage*, 17(10), 1307–12.

Shiraishi, J., Li, Q., Appelbaum, D., Pu, Y., and Doi, K. 2007. Development of a computer-aided diagnostic scheme for detection of interval changes in successive whole-body bone scans. *Med. Phys.*, 34(1), 25–36.

Skaane, P., Engedal, K., and Skjennald, A. 1997. Interobserver variation in the interpretation of breast imaging. Comparison of mammography, ultrasonography, and both combined in the interpretation of palpable noncalcified breast masses. *Acta Radiol.*, 38(4 Pt 1), 497–502.

Summers, R. M., Yao, J., Pickhardt, P. J. et al. 2005. Computed tomographic virtual colonoscopy computer-aided polyp detection in a screening population. *Gastroenterology*, 129, 1832–1844.

Suzuki, K. 2009. A supervised "lesion-enhancement" filter by use of a massive-training artificial neural network (MTANN) in computer-aided diagnosis (CAD). *Phys Med Biol.*, 54(18), S31–45.

Taylor, S. A., Halligan, S., Slater, A. et al. 2006. Polyp detection with CT colonography: Primary 3D endoluminal analysis versus primary 2D transverse analysis with computer-assisted reader software. *Radiology*, 239, 759–67.

Taylor, S. A., Charman, S. C., Lefere, P., McFarland, E. G., Paulson, E. K., Yee, J., Aslam, R. et al. 2008. CT colonography: Investigation of the optimum reader paradigm by using computer-aided detection software. *Radiology*, 246(2), 463–71. Epub 2007 Dec 19.

Taylor, S. A., Brittenden, J., Lenton, J., Lambie, H., Goldstone, A., Wylie, P. N., Tolan, D., and Burling, D. 2009. Influence of computer-aided detection false-positives on reader performance and diagnostic confidence for CT colonography. *AJR Am. J. Roentgenol.*, 192(6), 1682–9.

Uchiyama, Y., Ando, H., Yokoyama, R., Hara, T., Fujita, H., and Iwama, T. 2005. Computer-aided diagnosis scheme for detection of unruptured intracranial aneurysms in

MR angiography. *Conf. Proc. IEEE Eng. Med. Biol. Soc.*, 3, 3031–4.

Uchiyama, Y., Yokoyama, R., Ando, H., Asano, T., Kato, H., Yamakawa, H., Yamakawa, H. et al. 2007. Computer-aided diagnosis scheme for detection of lacunar infarcts on MR images. *Acad. Radiol.*, 14(12), 1554–61.

White, C. S., Pugatch, R., Koonce, T., Rust, S. W., and Dharaiya, E. 2008. Lung nodule CAD software as a second reader: A multicenter study. *Acad. Radiol.*, 15(3), 326–33.

Yamashita, Y., Arimura, H., and Tsuchiya, K. 2008. Computer-aided detection of ischemic lesions related to subcortical vascular dementia on magnetic resonance images. *Acad. Radiol.*, 15(8), 978–85.

Yang, X., Blezek, D. J., Cheng, L. T., Ryan, W. J., Kallmes, D. F., and Erickson, B. J. 2011. Computer-aided detection of intracranial aneurysms in MR angiography. *J. Digit. Imaging*, 24, 86–95.

Yoshida, H. and Dachman, A. 2005. CAD techniques, challenges, and controversies in computed tomographic colonography. *Abdom. Imaging*, 30, 26–41.

Yuh, E. L., Gean, A. D., Manley, G. T., Callen, A. L., and Wintermark, M. 2008. Computer-aided assessment of head computed tomography (CT) studies in patients with suspected traumatic brain injury. *J. Neurotrauma*, 25(10), 1163–72.

Zhou, C., Chan, H. P., Patel, S., Cascade, P. N., Sahiner, B., Hadjiiski, L. M., and Kazerooni, E. A. 2005. Preliminary investigation of computer-aided detection of pulmonary embolism in three-dimensional computed tomography pulmonary angiography images. *Acad. Radiol.*, 12(6), 782–92.

IV

Information Systems in Healthcare Informatics

16

Picture Archiving and Communication Systems

Brent K. Stewart
University of Washington

16.1 Introduction

Not to draw too many parallels between geopolitical strife and the evolution of Picture Archiving and Communication Systems (PACS), but victory for the digital operation of the radiology department has been achieved. The inherent value and necessity of digital radiology is no longer in question. Today, radiology could simply not operate without it. However, things were much different nearly 30 years ago when the concept of the digital radiology department was simply a twinkle in the eyes of some early innovators.

I was brought in on the ground floor of PACS in 1984 while a graduate student at UCLA. I had the distinct pleasure and honor of working alongside two PACS pioneers: My dear friend and mentor, Samuel J. Dwyer III, PhD and H. K. (Bernie) Huang, DSc, UCLA graduate school advisor, Image Processing Lab chief, and Biomedical Physics graduate program director. Leaving UCLA after my doctoral conferral in 1988, I returned as faculty in 1990 to work for Bernie in developing a department-wide PACS system under a newly awarded NIH grant. Sam came out to UCLA from Kansas University in 1991 to work with us for 2 years.

I have always thought of Sam as the "godfather" of PACS in his ubiquitous chalk striped suits, predictably espousing the virtues of Little's Law for bottleneck analysis (Dwyer, 2000). In the first edition of *Elements of Digital Radiology*, Bernie relates the day Sam came to visit the University of Iowa and delivered a lecture on PACS (Huang, 1987, p. ix):

> On Wednesday, November 4, 1981, Dr. Samuel J Dwyer III braved an early Midwest snowstorm, landed in Iowa City, and presented a novel seminar on PACS for Medical Imaging to the Biomedical Engineering Division and the Department of Radiology, University of Iowa. His topic was so interesting and the idea so original that Dr. Edmund Anthony Franken, Jr., Chairman of the Department of Radiology, kept discussing with me the possibility of implementing such a system in the

Radiology Department weeks after Sam returned to the University of Kansas. Since then I planned the steps to embark on such an ambitious project.

I have always personally thought of this as the beginning of the "PACS Crusade"; Sam enticing Bernie that day and Bernie later enlisting myself as a graduate student at UCLA. As with most revolutionary ideas, though it is difficult to actually determine from whence the original concept arose (Wiley, 2005).

16.1.1 A Brief History of Origins

Most in the PACS community point back to the 1st International Conference and Workshop on PACS for Medical Applications in January 1982 (SPIE, 1982) as the genesis of PACS, but there were several presenters at that meeting that had participated in previous electronic imaging meetings not focused solely on medical applications. One of these was Dietrich Meyer-Ebrecht of Philips Research Center, Hamburg. I was able to trace his writings on PACS back through 1977 (Meyer-Ebrecht et al., 1977) 5 years before the seminal 1982 SPIE meeting. The earliest I was able to trace back the PACS concept applied to medical images was a 1971 article entitled "A System for the Digitization, Storage and Display of Images" (Ausherman et al., 1971) wherein Sam Dwyer participated as a coauthor with Gwilym Lodwick, MD, who was instrumental in the initiation of many medical informatics efforts, including ACR-NEMA which led eventually to DICOM. Sam's Image Analysis Laboratory at the University of Missouri-Columbia (IAL/UMC) designed, built, and utilized this system specifically for medical radiographs. It included some very interesting equipment including an image dissecting scanner and a high-speed disc-image display system designed and implemented at IAL/UMC. This effort led eventually to Sam's more well-known work at Kansas University Medical Center, elaborated on at the early SPIE PACS conferences.

16.1.2 The Goals: Then and Now

From the very start, it was obvious that PACS, like the Platonic universe, consisted of four elements: digital imaging modalities (or digitized film), high-speed networks, an archive for image storage/retrieval, and image display systems for interpretation and review. The early goal was: could we create the PACS elements of the grand design—the totally electronic radiology department? What available technology could we integrate to create a functional system? It seemed initially like a great digital plumbing project—there certainly were a large number of schematic diagrams at the early meetings. I think the late Col. Fred Goeringer put it most succinctly a few years later: "Any image, anytime, anywhere."

As Bernie's graduate students in the UCLA Image Processing Lab, we were working on the incorporation of technologies that would enable the grand design: image compression, digital radiography (a prototype selenium detector for projection radiography), integrating an optical disk drive through the VAX 11/750 DR-11W parallel interface, integrating multiple high-resolution workstations through the General Purpose Interface Bus (GPIB), and so on.

There was still a huge gap between integrating components and building a system that demonstrated functionality in the laboratory and one that actually functioned in the clinical setting. The doctoral dissertation of my fellow UCLA graduate student, Ricky Taira, described his work in building a clinically functional Pediatric Radiology PACS.

We were helping blaze a new technology—we were PACS pioneers. We did not have many conscious higher design guides but to achieve the goal of a functional system. On presenting work at conferences in the early days it was highly descriptive: we did this and that and the result was such. The initial goal was just to get the system to operate.

We were building systems primarily, but with the 1990 NIH grant greater analysis of component and system performance commenced. When Sam came to UCLA, we began using systems modeling methodology to see if we could predict system performance under varying configurations or loading scenarios. These were applied to networks initially. Once a functional system was achieved, the vision was set on what advanced applications we could use the system to support: mining the data in PACS, natural language processing of radiology reports, devising the capability to index the content of PACS images, and so on.

16.2 PACS Elements (PACS Periodic Table)

16.2.1 PACS Elements and Work Flow

Many years ago when presenting lectures on PACS I used a graphic from a trade rag advertisement showing the various items of equipment and software that were part of what we were developing into PACS. Each item was a puzzle piece and they were shown fitting together. On reflections all these years later, it dawned on me that there were not puzzle pieces representing "people," no end users or administrators, as if all that were necessary were the hardware and software components and some lines connecting them on a schematic diagram. Honestly, the user change management issues were not on the radar back in the 1980s when we were putting many of the rudimentary elements together to get a system that actually produced the transactions we were trying to effect: digitizing an image of sufficient spatial and contrast resolution, storing the image, and displaying the image with adequate spatial and contrast resolution. I will enumerate some of these in the balance of this chapter, but change management, dealing effectively with the human component of the equation, and user education were not foremost in our minds. Demonstrating that a PACS system could function was the immediate goal.

Once the hardware and software were fairly well demonstrated, it should have been paramount for the people part of the equation to be studied, but aside from demonstrating that display monitors were as good as film using Receiver Operating Characteristic (ROC) analysis, there was no major push as I recall. People are the key ingredient in the solution of the PACS puzzle or the alchemical element in the PACS periodic table in the transmutation from hardcopy (film and paper) to entirely digital operation.

Initially, the goal of PACS development was to be a digital replication of the then ubiquitous film-based operation. I remember several analyses of the film-based operation presented at the early SPIE meetings as flow charts and initially, it was thought, in order to keep changes to the end users to a minimum, that replication of the film-based system would be the first iteration. Improvements to the departmental operation would come at a later date.

16.2.2 Informatics Standards (HL7 and DICOM)

In order for people to communicate with each other, at least verbally, it is necessary to establish common symbols, how these symbols are concatenated, how they are pronounced as syllables and what they mean as words—in other words the development of language. This is also necessary for computer systems to communicate and share information with one another, but this was not always the case.

When as a young graduate student in 1984, I was offered a position in Bernie Huang's Image Processing Laboratory at UCLA, the "rite-of-passage" was being handed a nine-track magnetic tape from a modality vendor (mine happened to be from a Diasonics MRI) and without recourse to any other source than one's "little gray cells" to decode the block structure of the tape and write a FORTRAN program so that future researchers could spool it up on our VAX 11/750 and extract the images from like tapes. At that time, magnetic tape was the only means available to share stored exam information from the various vendors imaging machines.

Health Level 7 (HL7) was founded in 1987 to develop standards for the electronic interchange of clinical, financial, and administrative information among independent healthcare-oriented computer systems; for example, hospital information systems (HIS), radiology information systems (RIS), and clinical laboratory systems. The previous version of HL7 (2.x) is a simple structured means of sending messages between healthcare-related computer systems and is still in wide use. It defines a transaction syntax for the transfer of data among the multitude of healthcare information system applications in the various environments in which healthcare is delivered. These include patient admissions/registration, discharge, or transfer (ADT) data, insurance, billing and payers, orders and results for diagnostic tests, nursing and physician observations, pharmacy orders, and resource/patient scheduling. HL7 3.x is an extension of this with transactions for exchanging information about problem lists, clinical trial enrollments, patient permissions, voice dictations, advanced directives, and physiologic waveforms.

HL7 does not try to assume a particular architecture with respect to the placement of data within applications but is designed to support a central patient care system as well as a more distributed environment, where data resides in departmental systems like Radiology and Laboratory Medicine. Message formats prescribed in the HL7 encoding rules consist of data fields that are of variable length and separated by a field separator character. Rules describe how the various data types are encoded within a field and when an individual field may be repeated. The data fields are combined into logical groupings called segments. Each segment begins with a three-character literal value that identifies it within a message (e.g., ORM—general order message). Individual data fields are found in the HL7 message by their position within their associated segments.

Newer developments in HL7 version 3 are the Reference Information Model (RIM), which defines necessary clinical and administrative contextual data as well as the unambiguous semantic and lexical relationships existing between information fields, and the clinical document architecture (CDA), specifying the encoding, structure and semantics for clinical document exchange using XML. Also see Chapter 4 in this volume for further detail.

In 1982, the American College of Radiology (ACR) and the National Electrical Manufacturers Association (NEMA) formed a joint committee to develop a standard for the transfer of medical images between imaging devices and computers. This has evolved into the Digital Imaging and Communications in Medicine (DICOM) standard (Kahn et al., 2007). The essence of the DICOM standard is that it prescribes a uniform, well-understood set of rules for the communication of digital images. This has been accomplished through defining, as unambiguously as possible, the terms it uses and in the definition of object-oriented models for medical imaging information. The DICOM standard is extremely adaptable, a planned feature that has led to the adoption of DICOM by other medical specialties that generate images in the course of patient diagnosis and treatment (e.g., cardiology, endoscopy, and ophthalmology) (Bidgood and Horii, 1996). DICOM is currently administered by the Medical Imaging & Technology Alliance (http://medical.nema.org/).

The DICOM Standard is structured as a multipart document. Each DICOM document is identified by a title and standard number taking the form "PS 3.x-year" where "x" signifies the part number and "year" the document version. There are currently 18 parts and over 150 supplements. Some of the most pertinent parts to the current discussion being Part 3: Information Object Definitions, Part 4: Service Class Specifications, and Part 6: Data Dictionary. Other specific parts are mentioned in later sections.

The fundamental functional unit of DICOM is the service-object pair (SOP). Everything implemented in DICOM is based on the use of SOP classes. The elemental units that make up the SOP are information objects and service classes. As DICOM was founded on an object-oriented design philosophy, things such as images, reports, and patients are all objects in DICOM and are termed information objects. The definition of what constitutes an information object in DICOM is called an information object definition (IOD), which is basically a structured list of tagged attributes. An example of a DICOM IOD is that of a CT image (CT Image IOD). Once the attributes (e.g., patient identification number) are "filled in," the object then becomes an information object instance.

The second elemental unit of DICOM is the service class. Information objects and the communication links between devices are not sufficient to provide functionality. It is necessary that these devices perform some operation (service) on the information objects. Some of the many DICOM services are Storage, Query/Retrieve, Print, and Modality Worklist Management. Due to the object-oriented nature of DICOM, services are referred to as service classes. In part, this is because a given service may be applied to a variety (or class) of information objects (e.g., Storage Service Class). A distinction is also made based on whether the device acts as a user or provider of a given service.

The service classes and information objects are then combined to form SOPs. The process of DICOM communication involves the exchange of SOP instances through the use of DICOM messages. An example is given in the Image Acquisition section below. Also see Chapter 4 in this volume for further details on DICOM.

16.2.3 Acquiring Necessary Exam Information

Although a radiologist looks at images, either on hardcopy film or digitally, there is a host of additional information required for the radiologist to make their interpretation as well as for archiving the exam and for billing purposes. Sam used to call this the "aside" information and it is indeed crucial. The Radiology Information System (RIS) holds this information. The RIS contains patient demographics, billing information, and exam-related information for each exam. Much of this information must be shared with PACS for PACS to function properly and with data integrity. The key item of information, is the RIS accession number, a unique number given to each exam when scheduled into RIS, either though manual entry of radiology exam requisitions, or through electronic order entry. The accession number drives all the other information systems associated with PACS, any speech recognition system and access to the image through the electronic medical record.

The RIS gets some of its information from the Hospital Information System (HIS), which may also be known as the ADT system. HIS continually sends HL7 messages to the RIS to provide authoritative data for the population of patient demographic and billing fields in RIS. In addition, orders to RIS for radiology exams are posted from computerized physician order entry (CPOE) systems. In order for the HIS HL7 messages involved to be transferred to the myriad of medical centers computer systems relying on them, a routing computer system known as an HL7 interface engine is used to examine the destination and source of HL7 messages streaming to it and route each message to its proper destination.

In essence, as the RIS speaks HL7 and the PACS speaks DICOM, it is necessary to have what was termed an HL7-DICOM broker to translate the HL7 message information from RIS into DICOM for PACS worklist functionality. The HL7-DICOM broker is itself simply another interface engine that takes in HL7 messages and transmits DICOM messages, in this case DICOM Modality Worklist (DMWL) information to the imaging modalities (e.g., computed tomography) for inclusion with the generated images.

The function of the DMWL Service Class is to push relevant patient demographic and scheduled exam information to the imaging modalities. At the imaging modalities, the radiology technologist simply selects scheduled patient exams from a worklist, rather than rely on manual entry of information (e.g., name, date of birth, accession number, etc.). Manual entry is prone to transposition, and other types of manual input error, which could lead to images being "lost" in the system. As the DMWL gets its information from an authoritative source (RIS or PACS Broker), the probability of an unspecified or missing exam is exceedingly low. The PACS Broker used to be a separate computer system from PACS, but has since been brought into the PACS (or RIS) system software architecture; yet HL7 to DICOM translation is still a necessary requirement for PACS.

The HL7-DICOM broker database also communicates with the PACS database through structured query language (SQL) transactions. This patient and exam information is required by the PACS database in order to marry this information with the images from the modality once they are received by the PACS archive through DICOM Storage Service Class messages after exam completion.

16.2.4 Image Acquisition

Once an exam is completed on a radiological modality, it is sent to PACS as a DICOM object. The DICOM object modalities include computed radiography (CR), computed tomography (CT), digital radiography (DR), magnetic resonance (MR), mammography (MG), positron emission tomography (PET), ultrasound (US), x-ray angiograms (XA), and so on. In addition, there are many nonradiological DICOM objects, including electron microscopy, endoscopy, microscopy, electrocardiograms, external camera photography, and secondary capture objects (Stewart et al., 2002). As discussed previously, each modality has its respective DICOM object structure including not only the image data, but many tagged fields or attributes associated with the image data (the DICOM "header").

To transmit this across the network requires the use of a service class, in this case the Storage Class. To send a CT exam requires the creation of an SOP, the CT Storage SOP. The Storage Service Class defines an application-level class-of-service, which facilitates the simple transfer of images between software processes on computers termed application entities (AEs). Two peer DICOM AEs implement an SOP Class of the Storage Service Class with the device that requests the image to be stored serving in the service class user (SCU) role (e.g., modality) and the device accepting the image object for storage serving in the service class provider (SCP) role (e.g., PACS archive). Storage SOP Classes are intended to be used in a variety of environments: for example, for modalities to transfer images to workstations or archives, for archives to transfer images to workstations or back to modalities, for workstations to transfer processed images to archives, and so on. Therefore, at a minimum, each digital imaging modality should act as a DICOM Storage SCU for its specific image object (e.g., CT IOD). Also see Chapter 11 in this volume for further detail.

16.2.5 Networking/Connectivity/Digital Plumbing

What do you really need to know regarding networking these days? Plug the computer into an RJ45 jack or turn on wireless networking and all you have to supply is the passphrase. No vampire tapping into a 10Base5 Ethernet coax (thicknet) yellow cable (ca. 1987) or even Windows 95 "Plug-and-Pray." The main thing to know is whether or not your application will have the requisite bandwidth to operate sufficiently. Networking using the more recent Ethernet (IEEE 802.3) variants (e.g., 100 MBps through 10 GBps) and standard media (e.g., Cat-5/6 twisted pair copper cables or single-mode/multimode fiber-optics) provides low or very low error rates. The prevalence of switched 100 MBps and 1 GBps Ethernet and possibility of fiber-optics for even higher bandwidths has relaxed early PACS bandwidth limitation issues.

Currently, 100 MBps and Gigabit switched Ethernet have become commonplace for PACS installations. Ethernet equipment purchased today uses switching and full-duplex operation to provide maximum signaling bandwidth to and from the host computer with no contention. It also makes use of the same Transport Control Protocol/Internet Protocol (TCP/IP) protocol stack, as does the larger and worldwide Internet making it possible, with DICOM to send images across the planet with ease.

The core PACS components (e.g., archive, database, and Web server) are usually colocated in an environmentally controlled and secure computer room. These core components are typically connected to an Ethernet switch with multimode fiber-optic connections that support in excess of 1 GBps transfer rates. All other PACS components, either in the computer room or outside and scanners are connected through 100 MBps Ethernet connections using Category 5 or 6 unshielded twisted pair (UTP) wiring. The PACS computer room Ethernet switch is connected to other switches and routers in the medical center through redundant gigabit fiber-optic links. The connection from these switches to the PACS workstations and to the physician desktop is usually through 100 MBps, UTP Ethernet, terminated with an RJ-45 wall jack.

Wireless local area network (WLAN) standards include the spectrum of IEEE 802.11 protocols: 802.11a: (54 MBps data rate in the 5 GHz band), 802.11b (11 MBps data rate in the 2.4 GHz band), 802.11g (54 MBps data rate in the 2.4 GHz band), and the recently ratified 802.11n, with multistreaming modulation (up to 600 MBps data rate in either 2.4 or 5 GHz bands). The 2.4 GHz spectrum is full of potential interference sources: Bluetooth devices, microwave ovens, cordless phones, and amateur radio. Most computers utilizing wireless networking are offered with multiprotocol wireless networking cards and software alleviating interaction by the user, aside from supplying the Wi-Fi Protected Access (WPA or WPA2) passphrase and possibly the Service Set identifier (SSID).

A Virtual Private Network (VPN) allows one to connect from the office, home, or on the road to an enterprise's central network using the Internet to securely transmit private data. VPNs eliminate the expense of special leased line connections and long-distance dial-up. VPNs secure private network traffic by encrypting the data stream using a variety of standard algorithms (e.g., triple Data Encryption Algorithm), authenticating incoming connections to assure the integrity of the source, and managing encryption key distribution and exchange for access control to data. Using these security measures, VPN devices construct virtual "point-to-point" tunnels through the Internet between remote users and the central network. When an authorized user logs off, the connection simply collapses.

A medical center would support one or multiple VPN servers on their subnetworks. These devices can handle hundreds of simultaneous connections. The client computer connecting to the VPN server over the Internet makes use of a program (typically freely distributed by the server vendor) to connect with the server's IP address using a password and/or digital certificates. Once the user is logged into their VPN server account, they can access systems on the subnetworks that are specified in their account authorization list. The client software automatically forwards all Ethernet packets bound for those subnetworks through the client encryption algorithm and to the VPN server, which decrypts the packets and injects them onto the medical center network. Those packets bound for other Internet entities (e.g., Google) are unmolested. Images transmitted from the medical center are encrypted by the VPN server, transmitted across the Internet to the client computer's program and decrypted and ready to use or manipulate. VPNs are also used for transmitting DICOM and other patient data across the Internet to other medical centers and sites of practice as it is heavily encrypted and authenticated. Any Internet service provider (ISP) connection to the Internet, for example, cable modem, digital subscriber line (DSL), satellite, leased line, or simple dial-up modem connections can access the VPN server.

The three things necessary to enable DICOM networking are an IP address, an Application Entity Title (AET), and an IP port. An AET is a 16-character DICOM application name for the AE. With these three simple items, one can network across the world. As a small set of port numbers are routinely used (e.g., 104, 2762, 4006), it is then incumbent on the obscurity of the AET and IP address to provide security if the called AE receiving the association request does not verify whether the calling AET is one of its authorized remote DICOM application names. In this case, the receiving unit is said to operate in promiscuous mode (Stewart, 1999a). Even so, unless a VPN is used to encrypt the message traffic over the Internet, the messages are being sent "in-the-clear." This brings us to the topic of DICOM security profiles covered in a later section. Also see Chapter 8 in this volume for further detail.

16.2.6 PACS Database

The PACS database is the core functional component of PACS. All long-term information regarding the status of exams in the system is stored in the myriad number of tables in this database. This database is typically a relational database accessed through SQL statements.

When an exam is scheduled in RIS, an HL7 message is sent to either an HL7-DICOM broker (or equivalent software process running on PACS), converted into an SQL statement and entered into various database tables so that when a new image arrives at the modality acquisition device, PACS is ready to accept the images into the archive and update the relevant database tables.

Once the exam populates a radiologist's work queue, an SQL query is generated requesting the exam location on the DICOM archive and the images are streamed over the network to the display station. How does the database know where the images are stored on the archive? A relational database has what is termed a primary key, which denotes each independent entity in the database. Remember back to DICOM and the object-oriented analysis of object relations in the radiological examination process. Each exam in the database will have a different primary key (also termed a "c-key"). Although the accession number is usually the coin of the realm with regard to identifying independent radiological exams, relational databases see the accession number as an attribute of an exam and not the sole entity identifier, which is the primary key.

Thus, associated with the exam in the radiologist's work queue is the c-key and it is this value which is sent to the database to determine the location of the appropriate images in the archive. If PACS has no prior information regarding an exam that arrives to the modality acquisition computers without an accession number, that exam is listed as unspecified. The exam is still stored in PACS, but with no accession number, but of course, the exam still has a c-key so it still appears in various queues to be displayed. For more technical information regarding how databases are used in PACS, Also see Chapter 12 in this volume for further detail.

16.2.7 Image Archive

Hierarchical storage management (HSM) systems have typically consisted of a hierarchy of digital storage media selected to fit speed/capacity/media cost criterion. Most have a front-end, short-term archiving unit with many terabytes of fast magnetic discs (Redundant Array of Inexpensive Disks–RAID level 5) that can stream requested images to display stations at a rate around 80 MBps and act as a buffer for the slower, long-term archiving units using either optical disc and/or magnetic tape libraries. HSM systems also manage the migration of images and related data that have not been accessed for a period of time to the longer-term archive media, manage the migration of long-term exam images (selected comparison exams) to the short-term RAID when either scheduled (prefetch) or requested on-demand (*ad hoc*), and proactively interrogate long-term media for reliability, rewriting this data to new media if it appears that a specific unit of media is failing.

The main limitations for the quick retrieval required for on-demand de-archiving of exam images from the long-term archive are that optical discs, although a random access media (access times on the order of 25 ms), have slow data transfer rates (max. 5 MBps) and are relatively expensive. Digital magnetic tape, on the other hand, is relatively inexpensive, has phenomenal data transfer rates, but tape being a sequential access media has mean file access times on the order of 10 s (Cecil, 2006). Smaller files (e.g., a PA and Lateral chest exam 15 MB) favor optical disc storage. Large files, like CT and MR examinations with hundreds of images or those found outside of radiology (e.g., cine fluorography and echo cardiology) involving long segments of video frames, favor storage on magnetic tape.

Several drivers have been recently transforming PACS storage. First is the consistent trend in the volume of data archived, especially with storage of submillimeter slice images from multidetector CT exams. Second, the decrease in the price of large disk array systems which has not only expanded the temporal residence of exams on fast-access magnetic disks, but also have largely made optical disc (magneto-optic disk and DVD) storage systems increasingly obsolete due to slow read transfer times and the necessity of manual intervention if the optical disk is on-shelf. Lastly, some sites have opted to either use an application service provider (ASP) to house their primary PACS archive or store a backup copy of the archive. ASP is discussed in a later section.

The use of large-scale RAID systems has flattened the storage of most PACS through the use of direct attached storage (DAS), network attached storage (NAS), storage area network (SAN), or combinations of these. However, continually adding many years of on-line disks is expensive. Although magnetic tape archiving has not been as popular in the recent past, the high transfer rates, high storage density per tape cartridge, and relatively low cost per GB still make it attractive in archive duplication for disaster recovery mechanisms.

If a slower, long-term storage is used and a currently acquired exam requires prior image(s) not on the short-term magnetic disk storage, then some type of pre-fetching algorithm needs to be invoked in order that those images necessary for the specific exam hanging protocol are resident on the short-term magnetic disk storage (Langer, 2009). In the case of disaster recovery, a strategy also needs to be worked out on how to reinstate the relevant images to the short-term storage from off-site storage media (e.g., magnetic tape cartridge) (Roy and Pressman, 2006).

Migration strategies are necessary as storage technology is evolving so rapidly that easily in a six year time period a storage technology may lose support from a PACS vendor or even the original equipment manufacturer. Thus, is it necessary to migrate old exams to new media, say from MOD to tape or MOD to magnetic disk storage. Migration is also necessary when switching magnetic disk archives due to the need for greater density per disk unit or storage technology obsolescence. Migration can take much time to complete and in consistency checking (comparing data on old media with the new), but vendors should provide means for accessing all data during the migration.

Medical image data sets are relatively large and image compression has a valuable role to play. A computed radiograph (CR) of 1760×2140 pixels and a 10-bit pixel depth requires 7.2 MB digital storage. A CT examination consisting of 2000 images (matrix size 512×512, 12-bit pixel depth) requires 1 GB. Obviously, image compression must be used judiciously as it generally trades image quality for compression ratio. With lossless compression the original images can be exactly reconstituted (bit for bit), so there is no degradation in image quality unless there is an error in the transfer of the image data—this is usually corrected by the network protocol stack (e.g., TCP/IP). Lossy compression techniques cause a varying amount of degradation in image quality, depending on image feature characteristics and the degree of compression.

Lossless compression schemes typically permit compression ratios (The numerator is the number of bytes required for storage of the original image and the denominator is the number of bytes required for storage of the compressed image.) of between 2:1 and 4:1 reduction in the number of bytes required to represent an image. Lossless compression algorithms are typically used for the PACS archive so as not to reduce the quality of images before primary diagnosis. However, some sites compress at higher compression ratios using lossy techniques after report finalization.

Techniques that claim compression ratios exceeding 4:1 are almost certainly lossy, although the term "visually" lossless has also been used. With Moore's law, the lowered cost of storage per terabyte (Nagy, 2007) and the very high-speed networks available to PACS these days, there is not the imperative to utilize lossy compression as there were only years ago. However, lossy compression still finds usefulness in providing image access through Web services. Compression ratios closer to 10:1 or 20:1 are required to have a significant practical and economic impact. Although the medical community did not readily embrace the concept of lossy data compression initially, there is sufficient evidence that such schemes can be implemented without compromising the diagnostic content of images (Aberle et al., 1993; Bolle et al., 1997; Good et al., 1994). Even though compression may cause a loss of image quality, radiographs compressed by as much as 10:1 to 20:1 and CT and MR images compressed by 5:1 to 8:1 are in most cases acceptable for postreport finalization or clinical review. Also see Chapter 9 in this volume for further detail.

16.2.8 Display Workstation Image Review/Reporting

The images from a PACS are usually displayed in either of two methods. The first is through the use of dedicated high-resolution grayscale (CT, MR, digital and computed radiography, and digital fluoroscopy) and color monitors (ultrasound, nuclear medicine, and PET). These workstations usually cost in the tens of thousands of dollars (Dwyer and Stewart, 1993; Stewart et al., 1993a). The second method is through the use of a Web server and ubiquitous and free Web browsing software, which usually require plug-in or add-on components (e.g., Active X) to provide image decompression, window-level, and other rudimentary functionality.

Dedicated PACS display workstations are usually commercial-off-the-shelf PCs using some form of the Windows or Linux operating systems. Additional memory (up to 4 GB) may be added in addition to special display cards with multiple frame buffers (e.g., $2048 \times 2560 \times 16$-bit) for use with two or four high fidelity, high-resolution LCD display monitors. If color display is used, then a single high-performance display card of sufficient resolution (e.g., $1600 \times 1200 \times 24$-bit—"true color") is used to drive one or two monitors. Basic image manipulation tools required include intensity transformation tables (automatic preset window/level, manual window/level control, and image invert), image enlargement and translation (zoom and roam), and mensuration (calibrated distances and calculated angle measurements). In addition, the ability to add and hide graphic overlays (scanner information, manually entered text annotations, arrows, or regions of interest) and rotation capabilities are important.

As the video display monitor is now the standard means of viewing radiological images through PACS, it is necessary to understand how the various myriad of display systems used in a PACS with different luminance capabilities and characteristic curves can be standardized to provide similar levels of perception and appearance for a given image. Digital driving levels (DDL) from the display card are input to the monitor and the generated luminance will depend on the characteristic curve of the monitor and the gamma curve used in the display card to modify the output luminance.

DICOM Part 14: Grayscale Standard Display Function (NEMA, 2009a) describes a mechanism, based on Barten's model of human visual system contrast sensitivity (Barten, 1999). Human contrast sensitivity is nonlinear with greater sensitivity to small changes in luminance at brighter levels of luminance than lower. Therefore, if a DDL to luminance curve could be found such that the difference in DDLs would provide the same perceptual differences in luminance (a just noticeable difference—JND) between them, then the display system would be essentially "perceptually linearized," providing similar levels of perception and appearance for a given image. This luminance curve is termed the Grayscale Standard Display Function (GSDF). In order that a monitor be "calibrated," engendering the GSDF, it is necessary to measure the precalibration luminance curve of the monitor and then apply a transformation function in order to bring the monitor luminance curve to that of the GSDF. The GSDF is defined for 1023 JNDs over the luminance range 0.05–4000 Cd/m^2.

As so much time is spent in the expert generation of images, it is absolutely critical that workstation monitors used by radiologists and technologists be calibrated to the GSDF on a periodic basis as well as other tests performed per AAPM TG18 (Samei et al., 2005). Experience tells that it is easy for these tasks to be lost through the cracks, so it is imperative to utilize whatever autocalibration/testing functionality the display system vendor provides and/or invests in a Web-based utility that keeps track of the calibration/testing routine and initiates these tests over the network when necessary.

Provision of lowered indirect lighting for viewing of images is essential as the reflected illuminance off the display monitor from room lighting sources does affect the grayscale display performance function contrast steps (Badano, 2004). With typical surface diffuse reflection coefficient materials, a room illuminance in the 75–150 lux range has been found optimal for LCDs (Chawla and Samei, 2007). Other primary concerns are the use of dark wall colors, noise reduction, temperature control, and comfortable chairs (Cohen et al., 2005). Additional concerns are height adjustable tables with adjustable monitor mounts as well as a workspace surface of adequate size to accommodate all computer peripherals, reading light, phone, and papers.

In trying to tailor the display system resolution and capabilities with end-users, usually the radiologists are given high-resolution display monitors (3–5 megapixels), the radiology technologists (a 1–1.5 megapixel monitor for QC), and most referring physicians using the Web. At our institution, there are minimum monitor requirements for viewing images on the Web (1280 × 1024 and 32-bit color). Typically, the Web application has less robust functionality than the dedicated diagnostic or review workstations in the radiology department. Some referring physicians, for example, neurosurgeons, may need more functionality than is offered with the Web-viewing application for large numbers of cross-sectional images (CT and MR). Most vendors offer a software only solution for the thick client used on the radiologist workstations that may be loaded on computers with minimum performance specifications, though the native display system may not support GSDF calibration. Then, a third-party software solution with associated photometer puck must be purchased and periodically calibrated to the GSDF.

In order that exams are read in the shortest period of time possible, it is important that the images for an exam and any associated historical images be arranged on the limited workstation display "real-estate" in the order which the radiologist wishes to see them and with the desired window/level settings. This is accomplished through a "hanging protocol," a historical name for the methodology film library personnel would use to arrange films for radiologist review on a four film-over, four film-under light alternator. This "intelligence" is encoded in the hanging protocol the radiologist creates through a series of menus on the display workstation, which are then available to them on any workstation connected with the PACS until modified (Krupinski and Kallergi, 2007). Sometimes termed default display protocols, the protocol can be setup specific to individual modalities, anatomical region, and individual or clusters of exam codes. Custom window/level settings can also be programmed into the protocol in addition to any system or user-preferred standard look up table settings.

The key end-product of the radiological exam is the radiologist interpretation. In the past, this was accomplished through either dictation into what was essentially an analog tape recorder or later into a digital voice recorder. The audio was then transcribed onto paper or into a digital text file by a radiology transcriptionist. At the beginning of this century, several companies produced speech recognition software allowing the radiologist to dictate their interpretation of an exam into a microphone, which immediately appeared as a digital text file on the display workstation. After a few rocky years, problems with limited radiological vocabulary, error rates, and lengthy training procedures were alleviated. Speech recognition software can be run in either front- or back-end modes. Due to the perceived high error rates by some radiologists, early speech recognition systems were run in a back-end mode with a transcriptionist still in the loop. Radiologists willing to work with the systems used the front-end mode, where the text appears on the display monitor as soon as it is spoken. Since then, with decreased error rates, macros (templates), and structured reporting (Hussein et al., 2004; Reiner, 2009), the front-end mode allows for the quick completion of the radiological interpretation. It is increasingly important that the interpretive report follows quickly on the heels of the image sets, which are immediately available to the referring physician through the Web (Reiner et al., 2007). Also see Chapter 10 in this volume for further detail.

16.3 PACS Operational Issues

16.3.1 Work Flow

In order to improve the way computer systems in healthcare interoperate to enable efficient workflow, a consortium of healthcare professionals and information technology experts formed the Integrating the Healthcare Enterprise (IHE) organization in 1997. IHE develops Technical Frameworks consisting of Integration Profiles, the associated actors and transactions. IHE Integration profiles model a clinical process or workflow scenario and documents how established standards (e.g., DICOM and HL7) may be used to accomplish the requisite information flow. Actors are the functional components of a distributed healthcare environment, for example, a collection of hardware and software processes that perform a particular role. Transactions are the interaction of actors in terms of a set of coordinated, standards-based transactions. Currently, Technical Frameworks have been defined for: Anatomic Pathology, Cardiology, Eye Care, IT Infrastructure, Laboratory, Patient Care Coordination, Patient Care Devices, Radiation Oncology, and Radiology.

Integration Profiles fall into three classes: (1) Content Profiles addressing the management of a particular type of content object; (2) Workflow Profiles addressing workflow process management through which content is created; and (3) Infrastructure Profiles addressing departmental issues. An example workflow integration profile is Scheduled Workflow (SWF), which covers the many transactions that (1) maintain the consistency of patient and ordering information; (2) define scheduling and imaging acquisition procedure steps; (3) determine whether images and other evidence objects associated with a particular performed procedure step have been archived and available to enable subsequent workflow steps, such as reporting; and (4) provide central coordination of the completion of processing and reporting steps as well as notification of appointments. Also see Chapter 5 in this volume for further detail.

There has been debate for several years whether or not to use PACS- or RIS-driven workflow. In PACS-driven workflow, the list of exams to be interpreted is provided to a user based on their PACS login identity. Typically, this is a limited subset of all exams taken in the department for any one radiologist. For example, a neuroradiologist would only be interested in the brain and spine CT/MR exams as well as spine exams taken using other modalities such as fluoroscopy or radiography. They do not want to find any musculoskeletal exams in their exam work queue. Getting the correct exams to that work queue depends on the exam description and/or exam code. An exam code is usually a three to eight letter abbreviation proxy for the often lengthy exam description. It is also possible to route images to specific radiologist's work queues via semiflexible rules using modality identity and time of day, for instance. With RIS-based workflow, the RIS, rather than the PACS generate the radiologist work queue.

One issue regarding work queues is the "status" of an exam. Some examples are exam scheduled, exam complete, and report finalization. In order for the RIS work queues to operate properly DICOM Modality Performed Procedure Step (MPPS), functionality must be enabled on all modalities and accepted by RIS. MPPS enables modalities to send a report regarding the status of an examination including start/end time, number of images acquired, and dose delivered (if x-rays are used). Then the RIS can discern the precise exam status from MPPS based on technologist input of the current state from the modality. This assumes that the RIS can handle MPPS reports, which is not always the case yet.

16.3.2 PACS Fault-Tolerance

With regard to PACS fault-tolerance, three major categories present themselves: redundancy and fail-over, fail-over recovery, and disaster recovery. Redundancy refers to having multiple instantiations of the same equipment/software available either as a "hot-spare" ready to assume work should one fail or in parallel in some sort of load-balancing arrangement. Failure of a unit is usually established through an arbiter (itself with redundancy), which polls the "heart-beat" of individual equipment to determine if the fail-over device requires activation. The parallel load-balancing arrangement is easier in that the process queue is served by multiple processing units and should one fail the remaining unit(s) simply keep servicing the queue. However, as the utilization of the remaining unit(s) increases as a result, the wait time increases, possibly resulting in a system bottleneck until the failed unit is brought back on line. Failure is typically due to moving parts in a server, for instance, the hard disk drive, a cooling fan or in the case of a robotic library, failure of the robotic arm. Most transaction servers are now configured in a clustered arrangement supporting internal redundancy and fail-over.

In the case, of say a database server failure, fail-over recovery can be affected through a secondary or redundant archive. The secondary database should be on a separate hardware platform than the primary database so as not to incur a single point of failure for the system. The various PACS components issuing write commands, for example, a modality acquisition server would send duplicate commands to each database. Should the primary database system fail, the secondary database realizing the state of the primary database through polling then adds the database write statements to a parallel transaction log. Thus, when the primary database server comes back on-line, this log can be reissued to the primary database server until the two are back in synchrony again wherein the parallel transaction log is cleared and the primary database server announces that it is ready to assume write/read commands from the various PACS hardware.

Ideally, in the above described scenario regarding failover recovery, the two database servers should be in different rooms if not in different parts of the country (or somewhere in between). After we built fail-over/fail-over recovery into the database of the UCLA PACS designed and implemented under the 1990 NIH contract, there was a flood in the PACS computer room (the Great Flood of '91) due to someone on the floor above leaving a backed-up sink running over the weekend. This caused the PACS computer room to flood and the primary database computer to shutdown, but the secondary database computer in the basement of the outpatient clinic half a mile away kept up with the database transactions. On Monday when staff returned to work (This was before implementation of real-time Simple Network Management Protocol agent-based monitoring.) and the problem realized, the parallel transaction log was executed when the primary database server was brought on-line and the two systems synchronized.

The above vignette points out the relevant points of disaster recovery. First, duplicate streams of data must be sent to image archives as well as database servers. Second, the disaster recovery equipment should reside off-site, preferably in a very safe location far away from the primary. For instance, it was kicked around at my current institution that the two major medical centers should mirror each other's archive and database. However, as Seattle is due for a major (9.0) earthquake sometime in the next 30 years, housing the duplicate data five miles away at the sister institution close to the fault zone is not a good idea. Preferably, the secondary copy would be housed in a different part of the country in a region not susceptible to natural disasters and in a very secure location. This is in accord with the Health Insurance Portability and Accountability Act (HIPAA) Security Rule Administrative safeguards that require that "covered entities are responsible for backing up their data and having disaster recovery procedures in place." Some healthcare organizations are more accepting of an off-site archive model (as with ASP). It is not inexpensive. In addition, HIPAA requires that "Covered entities that outsource some of their business processes to a third party must ensure that their vendors also have a framework in place to comply with HIPAA requirements," so it must be spelt out in any data outsourcing contract that the "vendor will meet the same data protection requirements that apply to the covered entity."

Providing the cost and HIPAA hurdles can be overcome, the next issue is, in the case of a disaster, how to get the PACS up and running again. As this may take several days to get damaged equipment repaired or replaced and the system operational again, an interim-gap disaster plan should be in place and tested. This may be a simple as ferrying radiographs via USB drive ("Save as DICOM") and use of DICOM software such as Osiris or Kpacs to read the images on an available computer. For cross sectional modalities such as CT and MRI, the images can be readout, if necessary, at the acquisition console.

16.3.3 PACS Image and Data Security

Encryption is the transformation of data to conceal its information content, prevent undetected modification, and/or prevent its unauthorized use. Encryption uses a key or keys. There are two main types of encryption: asymmetric encryption (also called public-key encryption) and symmetric encryption (also called private-key encryption). Encryption may be employed using hardware or software, but software is typically used in most computer applications today. Two methods are Transport Layer Security (TLS) and Secure Multipurpose Internet Mail Extensions (S/MIME).

TLS is a protocol developed for transmitting private documents via the Internet and is the successor to the Secure Socket Layer (SSL) protocol. TLS/SSL works by using a private key to encrypt data that is transferred over the TLS/SSL connection. TLS/SSL is a protocol to authenticate server to client and (potentially) client to server, to establish a "session" and to negotiate parameters for the encryption of messages exchanged during that session. These parameters include a shared "symmetric" encryption key and chosen encryption algorithm. TLS/SSL does not require any particular choice of these parameters. All Web-browsing clients support TLS/SSL and many Web sites use the protocol to obtain confidential user information. By convention, Web pages that require a TLS/SSL connection start with "https:" rather than "http:" TLS/SSL comes in multiple strengths using various length (e.g., 128, 1024, and 2048) encryption keys.

MIME is a specification for formatting non-ASCII messages so that they can be sent over the Internet. Secure MIME (S/MIME) is a protocol for the cryptographic enveloping of MIME messages. Because e-mail is asynchronous, the sender determines algorithm/key length/strength prior to sending the S/MIME message. S/MIME itself does not determine key length, merely how to securely exchange keys and algorithm information. S/MIME implementations usually support a number of algorithms but the standard only requires support for relatively weak algorithms (due to past federal export restriction and patent concerns). The sender choosing relatively strong encryption may find some recipients unable to decipher the message, while relative insecure messages will routinely be received and decrypted.

Authentication is the process of identifying an individual, usually based on a username and password, but has extended into biometric (fingerprint and retinal) identification as well as "smart cards." A smart card is a small device the size of a credit card that displays a constantly changing user ID code. A user first enters their password into the computer and then the card displays a user ID that can be used to authenticate them. Typically, the user IDs change every 1–5 min. In security systems, authentication is distinct from authorization, which is the process of giving individuals access to system functions based on their identity. Authentication merely ensures that the individual is who he or she claims to be, but says nothing about the access rights of the individual.

A certificate authority (CA) is a trusted third-party organization or company that issues digital certificates used to create digital signatures and public–private key pairs. The role of the CA in this process is to guarantee that the individual granted the unique certificate is, in fact, who he or she claims to be. Usually, this means that the CA has an arrangement with a financial institution, such as a credit card company, which provides it with information to confirm an individual's claimed identity. Certificate authorities are a critical component in authentication security on the Web because they guarantee that the two parties exchanging information are really who they claim to be. Every TLS/SSL server must have a TLS/SSL server certificate. When a Web browser connects to a Web server using the TLS/SSL protocol, the server sends the browser its public key in an ITU-T X.509 certificate.

DICOM Security Profiles (Part 15) (NEMA, 2009b) are defined very similarly to IHE Integration Profiles conceptually, employing actors and transactions between the actors. As it is assumed that the communications channel may not be trusted, the Security Profiles provide mechanisms for AEs to securely authenticate one another, to detect tampering or alteration of exchanged messages, and to protect message confidentiality while traversing a communications channel. An example is the Basic Network Address Management Profile, wherein "DHCP Client" is one of the actors and "Find and Use DHCP Server" is one of the transactions.

As healthcare information, including medical image transmission, have shifted further and further from private networks and Intranets to the Internet, enhanced security of individually identifiable patient information has become a prime concern. HIPAA mandates use of secure uniform electronic data sets for financial and administrative transactions along with a unique identifier for every participant in the health system. Along with this, the Healthcare Financing Administration (HCFA) has guidelines for the appropriate use of the Internet, though HCFA policies only officially apply to the information protected under the Privacy Act of 1974, a mandate on federal agencies. In the healthcare context, the Privacy Act of 1974 protects information about patients covered under Medicare, Medicaid, and Federal Child Insurance programs. However, the HIPAA legislation has increased the scope of this policy to apply to all patient information.

It is permissible to use the Internet for transmission of individually identifiable patient information or other sensitive healthcare data, as long as an acceptable method of encryption is utilized to provide for confidentiality and integrity of this data (HIPAA Security Rule, Administrative safeguards). Also required is the employment of authentication or identification procedures to assure that both the sender and recipient of the data are known to each other and are authorized to receive such information. Neither the HCFA policy statements nor HIPAA legislation spells out the exact mechanisms to be used for these functions as encryption and authentication technologies are moving targets.

16.3.4 PACS Problems and Bottlenecks

A PACS installation is the largest purchase a radiology department can make. These systems are extremely complex and the rapid obsolescence in the technology comprising the PACS, especially in terms of archiving, adds to the turbulence and flux (Samei et al., 2004).

All the time I knew Sam, he always had some kind of bottleneck analysis in play. Before his arrival to UCLA, we were building system components and integrating them, but were only beginning to measure the performance of these components. Using Little's law can get you to the rate at which the limiting step in a simple sequence will limit one to, but would not tell you what will happen in a more complex system with no closed state solution. So, we began playing around with simulation modeling to try to answer the question: what happens to the image delivery time to a radiologist's workstation if the imaging volume increases by say 50%? We found that even simple PACS architectures result in nonlinear response times as the loading increases and resources remain constant, especially as the utilization of any component rose above 80% (Stewart, 1993; Stewart and Dwyer, 1993b). It is indeed unfortunate that this research was never fully explored due to lack of funding, but it is probably still a great idea, if only to provide a customer an estimate of how long a disaster recovery reinstatement might take or a realistic estimate of the performance increase expected when upgrading the database server to the "latest-and-greatest."

16.4 Leveraging PACS as an Information Tool

16.4.1 PACS as an Information Tool

There is a wealth of information available in not only the PACS archive database, but also in the DICOM headers of the images in the archive. Most PACS database schemas are limited and do not include all the DICOM image tags, so in order to get to this data in the past, it has been necessary to "mine" the DICOM headers to get at this information. Most recently, the Enterprise Archive Server at our institution does keep a database (separate from the main PACS database) of the DICOM header information of the images in residence. This is very helpful as SQL queries can be posted against this database to get at more esoteric DICOM information.

Information from PACS (and RIS) can be used in decision support, quality assurance programs, and to create dashboards to support these efforts. For instance, at our institution, we have been mining the information from our PACS to monitor the exposure of computed radiography photostimulable phosphor imaging plates since 2002. These data for examinations in various radiology sections are reported out on a quarterly basis at a quality assurance (QA) meeting as a check against the proper use of manual techniques in addition to the well-known CR "dose creep" (Stewart et al., 2007). We have similar programs in places to surveil CT dose length product (DLP) and the entrance skin dose (ESD) for angiographic and interventional imaging procedures.

16.4.2 PACS as a Decision Support Tool

In terms of radiology professional and technical operation, the data in PACS (and RIS) can be used to provide periodic or integrative real-time reporting of, for example, patient wait time, radiology report turn-around time, and critical findings delivery. These results may be presented at periodic operations or QA meetings or in real-time using Web-based graphs, bubble charts, gauges, and/or treemaps (Chen et al., 2008; Nagy et al., 2009; Kruskal et al., 2006). Such data are important in decision support (Khorasani, 2006).

16.4.3 Advanced Applications Afforded by PACS

It is impossible for medical centers with appreciable volume to operate without PACS. For CT alone, multidetector CT exams, some upwards of 2500 images, are impractical to review on film, let alone the high cost involved in printing such a large number of films. The increased rate of MR image acquisition per exam and the functional imaging requirements for nuclear medicine imaging also make film obsolete. Computer-aided diagnosis (CAD) (Giger et al., 2008) and the various means of viewing large, three-dimensional image sets would be impossible without PACS (Branstetter, 2007).

16.5 PACS Acceptance and Economic Issues

So, as opposed to 10 or 15 years ago, there is no further need to justify PACS on an economic basis. It costs what it costs and one buys what one can afford (incrementally if necessary), though there are many things that can be done to keep the costs to a minimum. The driver for filmless radiology is the economic imperative of practicing radiology at a distance, providing quick service (images and associated reports) to referring physician customers for use in time-critical decision making. With the high costs of ICU care, for example, \$3500–\$8000/day (Dasta et al., 2010), shortening the stay of a patient in the ICU with time-critical information can reduce costs and improve the overall quality of patient care.

16.6 Beyond the Radiology Department

16.6.1 Enterprise PACS

It may be the case for an enterprise PACS that multiple medical centers (especially academic ones) may have different medical record number (MRN) schemes, but there is still a need for the same group of radiologists to work with images from each site. This is made possible through at least two different mechanisms. The first relies on these medical centers utilizing the same PACS vendor and operating at comparable version levels. Then it is just a matter of pointing the workstation login to the other site's PACS. If the PACS vendors are different or if the same vendors PACS version levels are not compatible, it is necessary to transfer the images from one PACS to another through a DICOM Query-Retrieve mechanism. This requires greater effort on the person requesting the exam from the other PACS and more time for the exam to migrate through the various computer systems to reside on the local archive, ready for display.

Film (laser printed or from film-screen radiography) could not be abandoned until a number of Radiology consumers, that is, referring physicians were served by another medium. The number of light boxes that used to be found all about medical centers was huge. It would have been cost-prohibitive to replace these with diagnostic or review workstations and these consumers are served through a Web server. Also, as some form of electronic medical records are being deployed in almost all medical centers, piggybacking on the computers deployed for these systems makes great economic sense.

All images on the PACS magnetic disk archive are mirrored on the Web server; however, whereas the images on the main archive are losslessly compressed, those on the Web server, as they are not meant for primary diagnosis and will most likely be displayed on lower resolution monitors, are lossy compressed. This lossy compression is upward of 10:1 so as not to overwhelm clinical networks which may not have the same high bandwidth capabilities of those in the Radiology department. Lastly, a universal resource locator (URL) is assigned to each exam.

In order to display the images from the Web server in a timely manner for most Internet connection speeds (256–1024 kbps for DSL and cable modem), compression is often used. Unless a ubiquitous compression algorithm is programmed into the native Web browser software (e.g., lossy 8-bit JPEG compression), a Java plug-in or Active X component must be utilized for image decompression (e.g., wavelet compression). The compression ratio of any specific image is a dynamic function of the modality type, and the inherent detail and contrast of each image. In addition, the plug-in allows manipulation of the full 12- and 16-bit contrast resolution of CT and MR images, respectively. The Java plug-in or Active X component also performs the following image manipulation functions: window/level, flip, invert, sort, rotate, zoom, cine-loop, and save as lossy JPEG. The required Java plug-in or Active X component is usually stored on the Web server itself for easy installation.

A Web server responds to URL requests generated by a users' Web browser through services like the Hypertext Transfer Protocol (HTTP) by returning the solicited Hypertext Markup Language (HTML) document, which is then displayed in the Web browser window after parsing the HTML document. If one wants more than simple static pages returned, then the URL can contain a Common Gateway Interface (CGI) query. The CGI is the standard by which external programs (often called gateways or gateway programs) can interface with an HTTP (Web) server. The CGI program (or script) can be designed to handle an information request to a database, take the results and assemble an HTML document from it, and return this document to the user's Web browser. Thus, CGI programs act as gateways between the HTTP server and databases, or between the server and local programs or document generators.

16.6.2 Integration into the Electronic Medical Record

An efficient mechanism for image display through an electronic medical record (eMR) is for the referring or primary care physician to find the patient list of radiological examinations they are interested in and select a hyperlink which then invokes the PACS Web server to present images from that specific examination (Stewart et al., 1999). This can be performed using a CGI query to the PACS Web server that includes the RIS accession number for that specific radiology exam report. An example of this CGI query might be: "https://pacs.uw.edu/study_list.cgi?accno=111&patid=999&namelast=TEST&namefirst=IGNORE." Note that https denotes the secure version of HTTP. As most of the DICOM images transferred from the imaging modalities have the RIS accession number imbedded in the associated DICOM header data attribute field (tag number = 0008,0050), the PACS Web server can find the relevant examination image series and display them directly to the users browser. No navigation through the PACS Web server database is required and the large volumes of images (dozens of Terabytes per year for a major teaching institution) are not strictly duplicated within the eMR database. Typically an appropriately sized Web server or cluster of Web servers can service hundreds of concurrent users.

Also see Chapter 13 in this volume for further detail.

16.6.3 Other Medical Specialties: Mini-PACS

Regarding other medical specialties, typically these have been handled through "mini-PACS" applications, for example, Cardiology cath labs or special ultrasound-based PACS (Bergh, 2006; Stewart et al., 1993c). The images of these mini-PACS systems can be archived locally on a separate DICOM archive or, providing there is sufficient capacity, the mini-PACS archiving can be rerouted to the main Radiology PACS archive. We accomplished this with my current institution's old ultrasound mini-PACS. Once the ultrasound mini-PACS archive was full, additional images were archived to the main PACS archive. The mini-PACS still retains its own database and reporting modules, but the images are stored "out there somewhere" as if in a cloud. While there is no reason why this mechanism may not be affected to accomplish a single DICOM archive for an entire medical center or enterprise, turf and budgetary silo issues can effectively stifle such efforts.

16.7 Beyond the Enterprise

16.7.1 Teleradiology and Telemedicine

Telemedicine has been variously defined, but can be broadly described as combining telecommunications technology with medical expertise for the remote delivery of medical care or education. Several telemedicine subspecialty applications have evolved: teleradiology, telepathology, teledermatology, telepsychiatry, telecardiology, medical consultations, continuing medical education, and others. Teleradiology is by far the most mature of these telemedicine subspecialties, and is therefore an obvious component of many telemedicine systems (Franken and Berbaum, 1996).

Teleradiology is the specialized use of computer technology to electronically transmit radiological images and supporting information from one location to another for the purposes of interpretation, consultation, or education. The practical use of teleradiology has become more widespread due to several converging factors. The first of these is technological: more sophisticated and ubiquitous telecommunications infrastructures and more capable computer equipment. The second of these involves the reorganization of medical practice: the emergence of managed care, outsourcing of radiological services outside of the medical center and the coverage of multiple institutions from one or more centralized locations. Teleradiology can also be thought of as Picture Archiving and Communications System (PACS) on the wide area network (Stewart, 1999b).

Several chief functional goals of teleradiology include, first, the provision of consultative and interpretative radiological services in areas of demonstrated need, for example, making radiological consultation service available to medical facilities without onsite radiology support. Teleradiology also involves the provision of timely availability of radiology images and radiological image interpretation in the support of emergent and nonemergent patient care. A further goal of teleradiology is the provision of radiological interpretation in on-call situations, such as night and weekend coverage from an attending physician residence. Other goals involve the provision of subspecialty radiology support (primary interpretation and overreading). Teleradiology is also seen as enhancing the educational opportunities of practicing radiologists, promoting efficiency and quality improvement, sending interpreted images to referring providers, supporting broader telemedicine activities, and providing direct supervision of offsite imaging studies. Also see Chapter 21 in this volume for further detail.

16.7.2 Cloud ASP

As mentioned earlier, an early PACS mantra was "any image anytime, anywhere." This was in reference to the review of the image, but with the availability of extremely large network bandwidth connectivity this can also be said not only of the storage of images ("cloud storage"), but also of the processing of large image sets ("cloud computing"). Such data services over the network are made available through an ASP. Thus, it is possible to store an off-site copy of the primary PACS archive with the ASP and in the case of disaster recovery utilize the ASP copy as the temporary primary archive until the local system is reestablished. It is also possible to utilize ASP servers to process large 3D datasets and rapidly return only the rendered projections of interest. This is made possible by rapid response times via very large network pipes. An example is the Pacific Northwest Gigapop at the University of Washington that is connected to through to the Internet2/Abilene backbone via a 10 GBps Ethernet Translight link to Chicago. A vendor ASP is used in Chicago as a backup copy of the primary archive.

16.8 Some PACS Future Directions

16.8.1 Increased Use of PACS as a Research Tool

Mining the information in PACS (as mentioned in Section 17.4) is important for the professional and technical operations of the radiology department and for hospital administration. Some future PACS directions involve the indexing of image content based on measured features, characterization of longitudinal image changes, reference atlases (e.g., brain, hand, etc.), and increased use of multispectral imaging.

16.8.2 Merging of PACS-RIS

As the PACS and RIS intercommunicate on a frequent basis with each other (e.g., HL7-DICOM Broker and MPPS), it is natural to look to merging the two systems functionality together into a single system. This simplifies the architecture of the system, reduces the number of separate computers and systems, and would theoretically cost less and function more rapidly (Kagadis et al., 2008). Bringing the PACS image database into the RIS, the DICOM archive is split off to become an image storage resource, the enhanced RIS database then keeping track of where the images are stored in the DICOM archive. With this merging, the RIS then becomes the sole information portal in the Radiology department.

16.8.3 Eventual Absorption of PACS into the eMR

As PACS, at least in a hardware and software sense is composed primarily of an image archive and database, there is no reason why it could not be eventually brought into a tight alignment with the eMR. In fact, as stated above, the various functions of PACS could be divided amongst other systems. In the era of film, film acted as the acquisition device (with the screen), a storage medium and with a light panel or alternator, a display medium. These three capabilities were in the electronic age taken up by PACS: the acquiring of images from the modalities, storage of the images in an archive and the display of images on monitors. These functions are all held together by the PACS database and controlling software programs. However, there is no reason why these functionalities cannot be split up into separate systems.

The acquisition of images and archiving of such can be formulated into a standalone DICOM archive as long as it can communicate with a database pointing to the image locations within the archive. As above, this database can reside in a RIS or mini-PACS, independent of the DICOM archive. The image display workstations then interface with their respective RIS or mini-PACS to provide work queues for the respective physicians, advanced display functionality, and reporting capabilities, be they structured reporting, speech recognition using macros or some hybrid. Thus, the DICOM archive becomes a resource for the entire medical center/enterprise and there is no reason why the images on that archive could not be directly accessed by the eMR providing the locations are shared with it by the various RIS/mini-PACS. Eventually though, the functionality of the RIS and various mini-PACS could be brought in under the eMR, but as eMR deployments have yet to generally prove themselves, such may be a ways off in the future.

References

Aberle, D. R., Gleeson, F., Sayre, J. W. et al. 1993. The effect of irreversible image compression on diagnostic accuracy in thoracic imaging. *Invest. Radiol.*, 28(5), 398–403.

Ausherman, D. A., Dwyer, S. J. III, and Lodwick, G. S. 1971. *A System for the Digitization, Storage and Display of Images*. Third Annual Houston Conference on Computer and System Sciences, Houston, TX.

Badano, A. 2004. AAPM/RSNA tutorial on equipment selection: PACS equipment overview. *Radiographics*, 24(3), 879–89.

Barten, P. G. J. 1999. *Contrast Sensitivity of the Human Eye and Its Effects on Image Quality*. Bellingham, WA: SPIE Optical Engineering Press, Technische Universiteit Eindhoven, xix, 208 pp.

Bergh, B. 2006. Enterprise imaging and multi-departmental PACS. *Eur. Radiol.*, 16(12), 2775–91.

Bidgood, W. D., Jr. and Horii, S. C. 1996. Modular extension of the ACR-NEMA DICOM standard to support new diagnostic imaging modalities and services. *J. Digit Imaging*, 9(2), 67–77.

Bolle, S. R., Sund, T., and Stormer, J. 1997. Receiver operating characteristic study of image preprocessing for teleradiology and digital workstations. *J. Digit Imaging*, 10(4), 152–7.

Branstetter, B. F. 2007. Basics of imaging informatics: Part 1. *Radiology*, 243(3), 656–67.

Cecil, R. A. 2006. PACS Archiving: A multivariate problem and solution. *JACR*, 3(1), 69–73.

Chawla, A. S. and Samei, E. 2007. Ambient illumination revisited: A new adaptation-based approach for optimizing medical imaging reading environments. *Med. Phys.*, 34(1), 81–90.

Chen, R., Mongkolwat, P., and Channin, D. 2008. RadMonitor: Radiology operations data mining in real time. *J. Digit. Imaging*, 21(3), 257–68.

Cohen, M. D., Rumreich, L. L., Garriot, K. M., and Jennings, S. G. 2005. Planning for PACS: A comprehensive guide to nontechnical considerations. *JACR* 2(4), 327–37.

Dasta, J. F., Kane-Gill, S. L., Pencina, M. et al. 2010. A cost-minimization analysis of dexmedetomidine compared with midazolam for long-term sedation in the intensive care unit. *Crit. Care Med.*, 38(2), 497–503.

Dwyer, S. J. and Stewart, B. K. 1993. Clinical uses of grayscale workstations. In Hendee, W. R. and Trueblood, J. H. (Eds.), *1993 AAPM Summer School on Digital Radiology*, pp. 241–264. Madison, WI: Medical Physics Publishing.

Dwyer, S. J., 3rd 2000. *A Personalized View of the History of PACS in the USA*. Medical Imaging 2000: PACS Design and Evaluation: Engineering and Clinical Issues. SPIE, San Diego, CA, USA.

Franken, E. A. and Berbaum, K. S. 1996. Subspecialty radiology consultation by interactive telemedicine. *J. Telemed. Telecare*, 2(1), 35–41.

Giger, M. L., Chan, H.-P., and Boone, J. 2008. Anniversary paper: History and status of CAD and quantitative image analysis: The role of medical physics and AAPM. *Med. Phys.*, 35(12), 5799–820.

Good, W. F., Maitz, G. S., and Gur, D. 1994. Joint photographic experts group (JPEG) compatible data compression of mammograms. *J. Digit Imaging*, 7(3), 123–32.

Huang, H. K. 1987. *Elements of Digital Radiology: A Professional Handbook and Guide*. Englewood Cliffs, NJ: Prentice-Hall.

Hussein, R., Engelmann, U., Schroeter, A., and Meinzer, H.-P. 2004. DICOM structured reporting. *Radiographics*, 24(3), 891–6.

Kagadis, G. C., Nagy, P., Langer, S., Flynn, M., and Starkschall, G. 2008. Anniversary paper: Roles of medical physicists and healthcare applications of informatics. *Med. Phys.*, 35(1), 119–27.

Kahn, C. E., Carrino, J. A., Flynn, M. J., Peck, D. J., and Horii, S. C. 2007. DICOM and radiology: Past, present, and future. *J. Am. Coll. Radiol.*, 4(9), 652–7.

Khorasani, R. 2006. Clinical decision support in radiology: What is it, why do we need it, and what key features make it effective? *JACR*, 3(2), 142–3.

Krupinski, E. A. and Kallergi, M. 2007. Choosing a radiology workstation: Technical and clinical considerations. *Radiology*, 242(3), 671–82.

Kruskal, J. B., Yam, C. S., Sosna, J. et al. 2006. Implementation of online radiology quality assurance reporting system for performance improvement: Initial evaluation. *Radiology*, 241(2), 518–27.

Langer, S. 2009. Issues surrounding PACS archiving to external, third-party DICOM archives. *J. Digit. Imaging*, 22(1), 48–52.

Meyer-Ebrecht, D., Lux, P., and Kowalski, G. 1977. The electronic X-ray archive: An integral approach to filing and remote retrieval of computer generated radiographs. *Medicamundi*, 22(3), 27–28.

Nagy, P. G. 2007. The future of PACS. *Med.Phys.*, 34(7), 2676–82.

Nagy, P. G., Warnock, M. J., Daly, M. et al. 2009. Informatics in radiology: Automated Web-based graphical dashboard for radiology operational business intelligence. *Radiographics*, 29(7), 1897–906.

NEMA, 2009a. Digital imaging and communications in medicine (DICOM) part 14: Grayscale display standard function. PS 3.14-2009. NEMA. Rosslyn, VA.

NEMA, 2009b. Digital imaging and communications in medicine (DICOM) Part 15: Security and system management profiles. PS 3.15-2009. NEMA. Rosslyn, VA.

Reiner, B. 2009. The challenges, opportunities, and imperative of structured reporting in medical imaging. *J. Digit Imaging*, 22(6), 562–8.

Reiner, B. I., Knight, N., and Siegel, E. L. 2007. Radiology reporting, past, present, and future: The radiologist's perspective. *J. Am. Coll. Radiol.*, 4(5), 313–9.

Roy, L. and Pressman, B. D. 2006. PACS: The long and winding road. *J. Am. Coll. Radiol.*, 3(11), 888–90.

Samei, E., Seibert, J. A., Andriole, K. et al. 2004. AAPM/RSNA tutorial on equipment selection: PACS equipment overview. *Radiographics*, 24(1), 313–34.

Samei, E., Badano, A., Chakraborty, D. et al. 2005. Assessment of display performance for medical imaging systems: Executive summary of AAPM TG18 report. *Med. Phys.*, 32(4), 1205–25.

SPIE. 1982. Picture archiving and communication systems (PACS): for medical applications; Part I; *Proceedings of the 1st International Conference and Workshop*, January 18–21, 1982, Newport Beach, CA. Bellingham, WA, SPIE—The International Society for Optical Engineering.

Stewart, B. K. 1993. Operational department wide picture archiving communication system analysis using discrete event-driven block-oriented network simulation. *J. Digit. Imaging*, 6(2), 126–39.

Stewart, B. K., Aberle, D. R., Boechat, M. I. et al. 1993a. Clinical utilization of grayscale workstations. *IEEE Eng. Med. Biol. Mag.*, 12(1), 86–100.

Stewart, B. K. and Dwyer, S. J. 3rd 1993b. Prediction of teleradiology system throughput by discrete event-driven, block-oriented network simulation. *Invest. Radiol.*, 28(2), 162–8.

Stewart, B. K., Massoth, R. J., and Thomas, S. J. 1993c. Mini-PACS. In Hendee, W. R. and Trueblood, J. H. (Eds.), *1993 AAPM Summer School on Digital Radiology*, pp. 123–156. Madison, WI: Medical Physics Publishing.

Stewart, B. K. 1999a. Networks, pipes and connectivity. In Seibert, J. A., Filipow, L. J., and Andriole, K. P. (Eds.), *Practical Digital Imaging and PACS*, pp. 259–286. Madison, WI: American Association of Physicists in Medicine by Medical Physics Pub.

Stewart, B. K. 1999b. Teleradiology. In Seibert, J. A., Filipow, L. J., and Andriole, K. P. (Eds.), *Practical Digital Imaging and PACS*, pp. 403–432. Madison, WI: American Association of Physicists in Medicine by Medical Physics Pub.

Stewart, B. K., Langer, S. G., and Martin, K. P. 1999. *Integration of Multiple DICOM Web Servers into an Enterprise-wide Web-based Electronic Medical Record*. Medical Imaging 1999: PACS Design and Evaluation: Engineering and Clinical Issues, San Diego, CA, USA, SPIE.

Stewart, B. K., Wilson, A. J., Langer, S. G., and Martin, K. P. 2002. Importing images. InNorris, T. E., Fuller, S., Goldberg, H., and Tarczy-Hornoch, P. (Eds.), *Informatics in Primary Care: Strategies in Information Management for the Healthcare Provider*, pp. 53–70. New York: Springer.

Stewart, B. K., Kanal, K. M., Perdue, J. R., and Mann, F. A. 2007. Computed radiography dose data mining and surveillance as an ongoing quality assurance improvement process. *AJR Am. J. Roentgenol.*, 189(1), 7–11.

Wiley, G. 2005. The prophet motive: How PACS was developed and sold. *Imaging Econ.*, http://www.imagingeconomics.com/issues/articles/2005-05_01.asp

17

Hospital Information Systems, Radiology Information Systems, and Electronic Medical Records

Herman Oosterwijk
OTech Inc.

17.1 Hospital Information Systems

17.1.1 Introduction

A Hospital Information System (HIS) consists of hardware and software components that support the management and operation of a healthcare institution. A HIS is essentially a hospital management system (HMS), as a matter of fact, the term HMS might have been more appropriate, and is similar in function to a Practice Management System (PMS) in the ambulatory care market. However, unlike a PMS, a HIS spans multiple departments and medical specialties across an institution.

Most facilities have a dedicated support staff for the HIS as part of their IT (Information technology) department, typically managed by a Chief Information Officer (CIO). This staff deals with implementation, training, upgrades, troubleshooting, support, help-desk functions, and interfacing the HIS to other information systems. A centralized HIS staff supports multiple clinical departments, although some health imaging-based specialties—such as radiology and cardiology—may have their own IT support staff reporting to their department supervisor.

A HIS facilitates operational, managerial, and clinical data collection and decision making. The operational aspects support direct care management, such as day-to-day operations, by collecting information during admission and providing patient tracking, inventory control, bed management, and billing functions. The managerial aspects of these systems deliver information such as the number of performed procedures in a facility, average patient stay, occupancy rate, profit and/or loss, trends, and other statistics.

The HIS is the major communication mechanism among various departments and other parties involved with patient care. In addition, it supports clinical operations by providing order entry capabilities, displays and distributes results of diagnostic tests, and interfaces with outside providers such as laboratories and suppliers. The HIS also supports healthcare professionals in their clinical decisions by providing standard protocols and templates for tests and test sets, based on patient condition and background. Last, a HIS controls patient information privacy and security by authorizing access and authentication information on the basis of a healthcare provider's role in an institution.

The term HIS is very broadly used and the functionality implied varies widely among vendors and facilities. For example, some systems do not include billing features if the institution is owned by a national healthcare organization or if it serves a certain subpopulation free of charge, such as the Veterans Administration (VA) in the United States. Most HIS systems contain departmental modules to allow a user to select certain functionalities, for example for pharmacy or surgery. In addition, other common modules are used for order entry and to provide a longitudinal patient folder or record.

Unlike a HIS, which can mix-and-match modules, imaging systems such as a Picture Archiving and Communication System (PACS [Typically used for imaging in radiology, cardiology, and other medical specialties]) are almost always standalone systems that might interface to a HIS, sometimes directly and sometimes through a department information system.

Because clinicians are becoming more dependent on paperless charts, nurses are accessing the HIS for drug administration, and multiple departments are looking for the HIS to fill orders, such as for a new lab test or radiology exam, the HIS system is a mission critical system, and redundancy, back up, and disaster recovery are very important.

The HIS is almost never the only information system in an institution, and therefore needs to communicate with other information and management systems, which in most cases is done using the HL7 (HL7 stands for Health Level 7, which defines a set of messages and protocol for healthcare informatics applications. The standard has two versions, version 2.x (i.e., 2.2, 2.3, 2.4, etc.), which is pretty much the state-of-the-art, and a completely new version, 3, which is in the process of being evaluated and sparsely implemented. More information can be found at www.hl7.org (HL7).) (Further details on HL7 can be found in Chapter 4.) messaging and communication protocol.

The HIS is the source for patient demographic information to other specialties and departments. Its role in patient information management is critical as assigns a unique and consistent set of identifiers and demographics which allows orders and results to be labeled in a coherent manner. The latter allows orders and results from different departments to be identified as belonging to a patient and accessed as such. As the HIS is considered the ultimate "source of truth" for patient demographics and identifiers, it is also where updates and changes are made to patient information. A HIS is often considered as the backbone of a healthcare institution.

17.1.2 Key HIS Features

Despite the fact that most HIS products provide similar functionality, there is a wide range of different user interfaces in various HIS software. There are also a wide range of underlying database and/or operating system (OS) architectures. Because of these differences in "cosmetics," it is critical to evaluate a HIS (and indeed any medical informatics system) on the basis of its core functionality.

One of the best ways of achieving this evaluation is to use the so-called "actor" definitions. An example would be a scheduling or order-entry actor. Actors communicate through messages, in many cases using HL7 transactions. An example of one of these transactions would be a "patient arrival," which might be exchanged between the admission actor in the HIS and the scheduling actor in a Radiology Information System (RIS). The RIS might use this message as a trigger to place the patient on a work list for a technologist, such as conducting a CT exam. Actors and their corresponding transactions have been defined by the Integrating the Healthcare Enterprise (IHE) (IHE recognizes several domains such as radiology, cardiology, eye care, and also a generic domain called the IT Infrastructure, for which specifications of use cases, actors, and transactions are defined. More information can be found at www.ihe.net (IHE).) (Further details on IHE can be found in Chapter 6.) organization in an effort to increase interoperability.

The key features that a HIS provides are given below.

17.1.2.1 Manual Data Collection

A HIS contains information that is shared by multiple organizations and departments such as patient demographic information, insurance information, clinical notes, reports, and results. Much of the patient demographic information is entered manually, for example upon admission of a patient at an institution. This information is collected by typing it in based on patient verbal information or by using a patient admission paper form as its source. Basic observations such as weight, height, and vital signs are also often manually entered by a nurse or physician assistant.

Orders are typically placed by a physician, physician assistant, or a nurse in which case there might be an approval needed from the attending physician. A centralized order-entry system would allow a practitioner to schedule procedures and/or tests at different locations based on the available schedule. The use of supplies is also typically entered manually so that the institution can bill and/or account for their usage and keep track of inventory and automatically reorder.

A common saying is "garbage in-garbage out" meaning that the information is only as good as the data that is entered. Therefore, it is critical that the people entering the data are trained properly, to limit the number of potential errors. The information entered initially is considered the "gold" standard and is the source for many tests, orders, procedures, and results. A simple misspelling of a name and/or ID can result in mismatches of clinical information to the correct patient—not only for current transactions, but also with archived results. Responsibility of data integrity is typically the purview of the medical records department of an institution. This group manages and updates patient records, especially when there are conflicts and/or mismatches.

17.1.2.2 Electronic Data Collection

Information can also be acquired electronically, not only from devices gathering clinical information, such as an EKG device, but also by scanning in documents such as authorization forms. Many institutions send their test samples to outside labs, which can be electronically connected to send back their results. In the case that there is no direct electronic connection, faxes might have to be rescanned into the HIS.

In the near future, a lot of the manual entry for patient demographics and insurance information will be replaced by automated readers. For example, instead of typing the patient name and ID, a patient might carry an identification card, as is common in several European countries, which can simply be electronically swiped to collect that information. Lastly, automated supply systems are being interfaced, which allow tracking of inventory and automatic replenishment and reordering of drugs, pharmaceuticals, and other supplies.

17.1.2.3 Information Management

Information that has been gathered needs to be managed—it requires secure and guaranteed storage and retrieval capabilities. A database management system that responds to queries and provides sophisticated search capabilities, as well as an archive to store the complete records, documents, and other patient related information, is a key component. This feature also provides the capability for management reporting and data-mining.

17.1.2.4 Clinical Messaging

An important feature of a HIS is the clinical messaging capability between an institution and other healthcare providers such as referring physicians, medical offices, and sometimes patients. This can include relevant information, such as a test result or a follow-up about a specific procedure, but also might be a reminder to get a flu vaccination. With the proliferation of intelligent personal digital assistants (PDAs) and the capability of securely accessing HIS information from the outside, consults and/or e-visits between patients and their physicians, as well as consultations with specialists can be provided. Furthermore, educational resources and tools, such as patient education information on how to deal with a diabetes problem using proper nutrition and exercise, can be accessed and made available.

17.1.2.5 Internal and External Interfacing and Communication

A HIS needs to communicate and interface with a variety of other information systems, in most cases using standardized transactions and protocols such as HL7. Core hardware components are the network with its associated routers, switches, cables, network management tools, network security, and authorization hardware and software. Most institutions also deploy an interface engine that routes messages to their appropriate destinations and is able to map application level fields and/or convert codes and transactions based on the capabilities and software version of the receiver. External communication is a little bit more involved as there are firewalls, intrusion detection systems, and in many cases a DMZ (The abbreviation DMZ literally means a De-Militarized Zone, which is typically used to indicate devices outside the institution firewall, for example to provide educational resources to patients.) involved. External access by service providers and external entities needs to be managed by gateways and routers that only allow access to certain parts of the system using VLANs. (A Virtual Local Area Network or VLAN can be configured at a network router to create subnetworks that basically lock anyone out from destinations that they should not be accessing.)

17.1.2.6 Department Management

Each department may have its own scheduling, billing, management, and regulatory requirements. A RIS has knowledge about the coding of specific procedures and results and knows the differences in resources and time required for a simple chest x-ray, CT, or a special fluoroscopic procedure that requires a physician to be present. The same applies for a lab system (LIS) that has knowledge about the battery of tests that are included in a specific order. Management tools that allow for department-specific reporting include financials, volume, performance, and other key parameters necessary for department management. Results can be entered and made available through the department system and interfaced to the main database via a HIS and/or Electronic Medical Record (EMR). In many cases there is a direct link to physicians, offices, and/or patients, such as a capability to autofax results or send secure emails.

17.1.2.7 Decision Support

Decision support systems can have a major impact on the uniform delivery of healthcare through standardized protocols based on certain patient observations, history, and tests. This can be used for prescribing drugs as well as recommended treatments and allows for the detection of potential drug interactions. With the advent of electronic PDA's, a physician can easily access these protocols and prescribe the appropriate treatment for a patient.

17.1.2.8 Portal

There are typically three different portals: a physician portal, a nursing portal, and a patient portal. Each of these portals has a different "view" into the patient information and allows for different types of actions such as updates, changes, data entry, and so on. This includes access to the longitudinal patient record, which comprises results, notes, reports, medication history, patient history, and in many cases medical images. In addition, physician and nurse portals should allow for easy data entry to place orders, enter vital signs, and observations. A patient portal allows for e-visits, e-mails, and also general educational resources.

The HIS has information available for healthcare providers and patients that should be made available in an easy-to-interpret manner, should be user friendly, and must protect patient privacy and security.

17.1.3 HIS Architecture

The features mentioned in the previous chapter are implemented as physical hardware and software components and interact with each other as shown in Figure 17.1. The main component of the HIS is the "core." The core contains the various departmental systems, with imaging as a special case, and the data repository and knowledge systems. On the left side of the figure are shown the manual input sources, electronic sources on the top, interactive communication both inside and outside the campus on the right side, and the portal on the bottom. Billing is somewhat separate as it is typically managed separately; it is not uncommon for this to be outsourced.

17.1.3.1 Core

The core of a HIS is not necessarily from the same system or vendor. It is uncommon for an institution to have a single monolithic system and a single database for all information. Some of

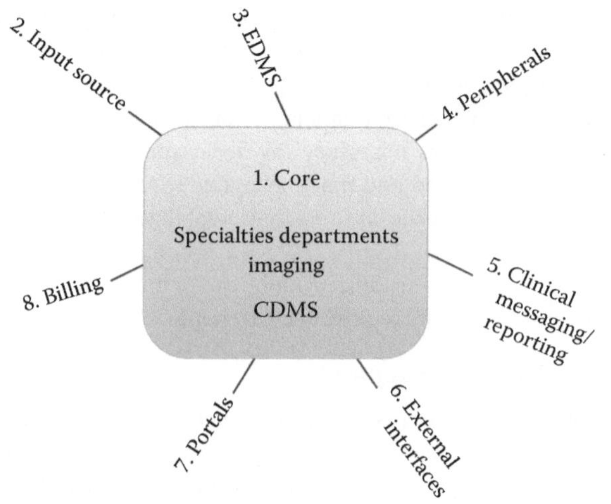

FIGURE 17.1 HIS architecture.

the clinical specialties and departments are so different that they are better served with a dedicated, specialized information system. Some specialties are also not commonly available in a typical institution, especially in a smaller hospital. An example is a radiation oncology department, which may only be present at some of the larger institutions. Other specialty surgery applications, such as transplants, often also require their own support software.

The core manages specialties and departments. Specialties typically concentrate on a particular disease and/or procedure, while departments are formed around the delivery of similar medical services. For example, the specialty "labor and delivery" might require services from the surgery, nursing, radiology, and laboratory department. Both specialties and departments need to be managed and exchange information.

Departmental information systems, such as the radiology, laboratory, pathology, surgery, deal with their own scheduling (and often billing). These systems provide a department manager the tools necessary to manage the staff and services.

PACS are very different with regard to their requirements for hardware and software due to large data volumes, performance requirements, and specialized viewing capabilities required for applications such as 3-D visualization or Computer Aided Detection (CAD). A PACS is almost always a standalone system that can be tightly interfaced with the HIS, almost always using the RIS as an intermediary, particularly for the orders, results (reports), and billing. PACS results might be directly sent to the HIS in the form of diagnostic reports, with a subset and/or sample of the images, and a link to retrieve the complete, original image data set.

A key component of the HIS Core is the Clinical Data Management System (CDMS) as shown in Figure 17.2, which consists of an archive, repositories, and data warehouses. It enables decision support capabilities, both managerial and clinical, and gets input from internal- and external-knowledge databases.

The data repository, or image manager, is the central database which indexes all information and provides query capabilities to allow users to search for the information needed. As this component is critical to the clinical operation, it is often duplicated ("mirrored") and the HIS support staff makes sure that frequent back-ups are made.

The repository also serves as a "gateway" to external repositories. For example, if a user queries for information from an external source, the repository can perform queries and, if authorized, access that information. The reverse is true as well, the repository can "register" pertinent information with an external source to make it available as needed. Note that the image manager is only a database and typically does not contain clinical results such as waveforms or images, but it does contain an index/pointer to this information stored in an archive.

The enterprise archive is used to consolidate the different information sources that exist in radiology, laboratory, pathology, and other departments. It should be able to facilitate multiple data formats such as DICOM, MPEG, JPEG clips, and documents in PDF and other formats to provide an enterprise-wide multimedia archive, which is patient-centered instead of department- or specialty-centered. Another term for this archive is a Vendor Neutral Archive (VNA), which can interface to different imaging vendors as well as multiple facilities.

An important consideration for an archive is high availability, which can be achieved by mirroring the archive or clustering the data among multiple redundant hardware components that guarantee access even if one or even more of the components fail. A good practice is to have a copy off-site, which can be managed by the institution or outsourced to a Storage Service Provider (SSP).

A data warehouse provides additional intelligence to allow for data mining and research. It is typically a separate archive so as not to interfere with the high availability and performance requirements of the main archive for clinical and operations support.

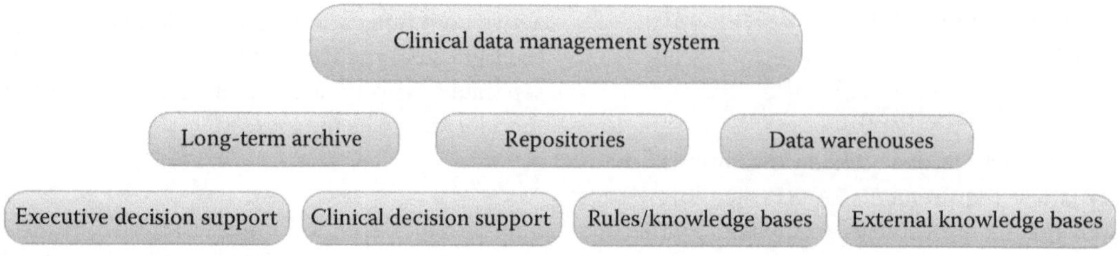

FIGURE 17.2 CDMS.

The decision support components provide the intelligence for an enterprise. Executive decision support provides management with critical data about the enterprise operation. It is typically displayed as a dashboard portal that shows trends using key performance indicators about utilization, profit centers, and rating among peers. Clinical indicators are also tracked, such as mortality rate, percentage of complications, and/or infections per procedure and department against peers and industry standard guidelines and practices. This dashboard is essentially a scorecard of how the enterprise is performing and provides the tools to make quality and performance improvements.

Clinical decision support systems focus on the clinical operation of patient care. There are two types: Passive support, which could be a warning that comes up when ordering a medication that is contraindicated for a patient; and active support, which might alert a physician if vitals from an ICU (Intensive Care Unit) patient exceed a critical threshold, or a patient has missed a necessary appointment.

The information for the decision support systems are part of a rules and knowledge base, which is based on good practices, experience, and guidelines. It contains proven protocols for certain conditions. For example, an older patient with hip pain might be a candidate for an expensive and invasive hip replacement, but a protocol might instead prescribe a certain new drug on the basis of outcomes research that has shown this to be a better approach. Treatment protocols vary widely among physicians, institutions, and even among geographic regions. For example, the number of cardiac bypasses surgeries, versus treatment in a noninvasive manner with drugs and other therapy, vary among towns and cities. There is a lot of pressure to come up with standard, proven protocols, not only to increase effectiveness and potentially reduce cost, but also to reduce infections and complications for certain procedures.

In addition to internal knowledge databases, there are also external knowledge databases, which are an important source of information. These can provide benchmarks for operational and clinical key indicators that demonstrate how an institution is performing compared with its peers. External knowledge databases are also used for drug information, including potential side effects and interactions with other medications.

Teaching files or reference data-sets are also typically provided by other parties, as they can provide a comprehensive library of images and observations collected from numerous other professionals.

The use of consistent medical terminology is also critical to improve patient care and communication among healthcare providers. For example, the term "hypertension" has multiple synonyms, and many other conditions have multiple variations as well. Medical terms are typically defined by professional organizations. The same applies for codes, such as ICD-9 or ICD-10 (International Classification of Diseases) defined by the World Health Organization (WHO) for diagnoses, and those used for procedures such as the American Medical Association's Current Procedural Terminology (CPT).

Last but not least, there is "Professor Google™" or any of the other search engines available in the public domain that aggregate a wealth of medical knowledge. These are not necessarily well organized and indexed, but for an *ad-hoc* search they are often used by healthcare practitioners. Several healthcare imaging portals also provide daily cases, which are archived for future reference.

17.1.3.2 Input Sources

The main source for data input (see Figure 17.3) is the so-called Admission, Discharge, and Transfer (ADT) system, which allows for patient demographics—including the patient history and billing information—to be entered and the patient location to be managed. Note that a change of patient status, such as "admitted" or "discharged," will have a major impact on other components, such as bed management and whether or not medications are administered. This information is typically entered by administrative support staff.

The second important source of information is entered by healthcare practitioners, that is, nurses and physicians who input notes, assessments, vital signs, and other observations. This information can be unstructured or structured, that is, as free text or drop-down menu selections, and/or barcodes or RFID (Radio Frequency Identification) sensors. The latter provides encoded information such as standard diagnosis codes, drugs, and dose information.

Information can be captured using electronic computer pads or tablets at the Point-of-Care (POC), and/or by a Computer On Wheels (COW's), which could be used in an ER (Emergency Room) setting or at a bed site. Orders can be entered using a Computerized Physician Order Entry (CPOE) system, which is used to order tests and procedures from multiple departments such as pathology and radiology. If the order is entered by a nurse, a physician can access all of the orders in their pipeline requiring their authorization.

Electronic prescribing (eRx) allows for a medication order to be sent directly to a pharmacist, either in-house or outside if it concerns an outpatient, eliminating the paper transcriptions forms which are subject to misinterpretation, especially with regard to the dosage.

Electronic Medication Administration Records (eMAR) allows for the automatic registering of medications for a specific

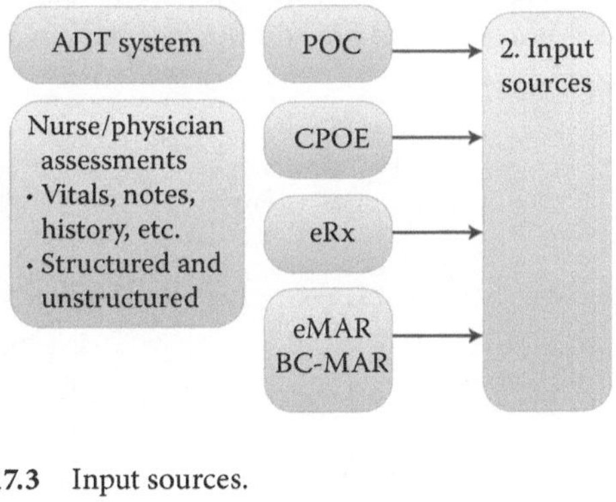

FIGURE 17.3 Input sources.

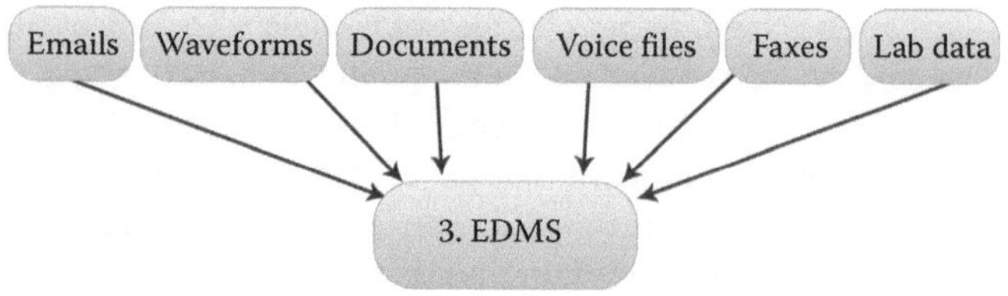

FIGURE 17.4 EDMS.

patient, using an RFID or barcode identification. This technology significantly reduces the chance for administering an incorrect drug or dosage.

17.1.3.3 Electronic Document Management System Components

The Electronic Document Management System (EDMS) (see Figure 17.4) manages documents and/or observations, forms, questionnaires, and voice clips that are entered into the system.

The EDMS is a transitional component—the goal being that all document sources will eventually be digitally integrated. In the mean time, there will be old files that might need to be scanned in, questionnaires that are filled in by hand, and results from ECG machines that create paper waveforms that need to be gathered into the data repository of the HIS. Even if all sources within an institution are interfaced, a patient could still present a paper requisition from a referring physician who is not electronically connected to the facility. In some cases, this requisition could contain critical information that needs to be made available in real time to the department or specialist. Some outside lab systems also might not be interfaced, especially if it concerns highly specialized tests that are relatively infrequent and only provide information on faxes or printed copy. These documents and digitized pieces of information need to be properly identified so they can become part of the longitudinal patient record and can be accessed by authorized healthcare providers.

17.1.3.4 Smart Peripherals

Smart peripherals (see Figure 17.5) are directly interfaced to a HIS. These devices can be at a patient home, such as monitoring systems used by diabetes patients to check their blood pressure and glucose levels. In a hospital, ICU monitors can deliver vitals in real time to physicians and make it available on their PDA's in case certain levels are exceeded.

Supply management is also highly automated with smart peripherals. Many supplies are administered using dispensing machines that keep track of inventory and can link to outside providers to automatically refill them. This can be used for pharmaceuticals, such as contrast agents, as well as supplies like the stents used for invasive cardiology. Linking the appropriate supplies to a particular patient is done by the HIS. In addition, it allows for the management of these supplies through reporting and billing. Some laboratory systems also have robots that allow results to be exchanged directly with a HIS.

External labs are sometimes interfaced, but they often require a special mapping to the HIS as many of them have their own codes for tests. These interfaces are typically routed through an interface engine, which maps the appropriate codes between the hospital and its external labs.

17.1.3.5 Clinical Messaging

Clinical information needs to be exchanged between the HIS and other healthcare providers, particularly physician offices and patients (see Figure 17.6). Results of diagnostic tests and procedures need to be reported. In addition, a HIS might have e-visit capabilities. This allows healthcare practitioners to consult patients at home or in other facilities, such as nursing homes, about health-related issues. These e-visits can use teleconferencing including video and audio consults.

In rural areas, e-visits are particularly cost effective. It also allows for chronically ill patients to be tracked in an economic and nondisruptive manner. Research has shown that when a patient is tracked with regard to their medication usage and monitored with regard to vitals, the number of ER visits and complications dramatically reduce, ultimately providing better care to the patient.

Another special application in a HIS is for the communication of critical patient results among physicians. For example, a radiologist might diagnose a condition that requires immediate attention, such as a life-threatening finding. In that case, the referring or attending physician needs to be notified immediately via every possible means, which could be paging, phone, and/or email. In case the primary physician is not available, a back-up or alternate physician needs to be alerted. Of equal importance is a function that records the receipt and acknowledgment of the critical result, so the alerting specialist knows that action is taken. It is common to use different codes for the required time that the referring physician has to be alerted, such as code red for notifying within 2 h, orange within 12 h, and yellow and green for longer time spans.

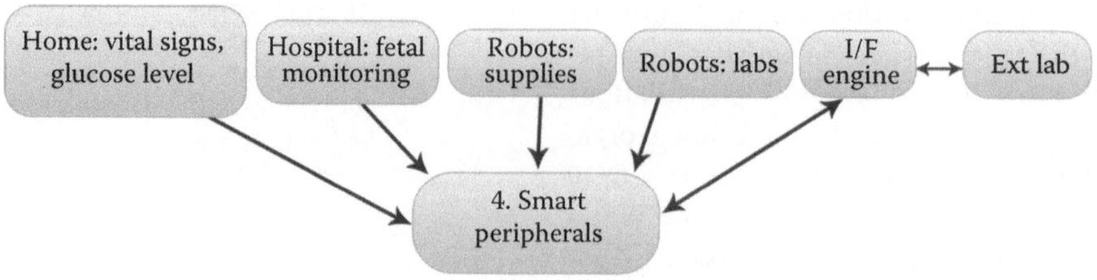

FIGURE 17.5 Smart peripherals.

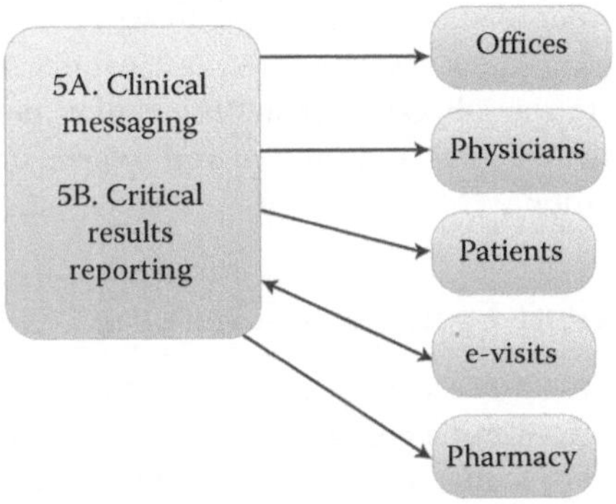

FIGURE 17.6 Clinical messaging.

Another application in a HIS is discrepancy reporting, which is a tool used when a subsequent physician disagrees with the finding or nonfinding of a primary physician. This is commonly used in the ER, where an emergency physician might discharge a patient based on a preliminary diagnosis. However, a radiologist reviewing images taken in the ER of that patient might find something different or disagree with the initial diagnosis. The primary physician needs to be notified and action should be taken depending on the urgency.

17.1.3.6 PHR/EHR and HIE Interface

An Electronic Health Record (EHR), managed by external entities, or a Personal Health Record (PHR), managed by the patient, may need to interface with the HIS (Figure 17.7). In the ideal scenario, when a patient is admitted in an institution they would authorize access to their PHR and their medical history (including vaccinations, allergies, hereditary diseases, and genomic) data can be automatically uploaded. A PHR could include dietary information, exercise logs, and vitamin supplement intake.

A PHR eliminates the need to recapture this information and fill in many pages of questionnaires. When an EHR is available from a Health Information Exchange (HIE), it can also be directly uploaded in the HIS. In case the records reside at another institution, the admitting hospital can query a central repository registry for previous images, reports, and other result information. Of critical importance is the reconciliation of patient identifiers using a Master Patient Index. Cross Document Sharing protocols using XDS (XDS is the profile defined by IHE for sharing documents and images across multiple enterprises.) allow for clinical documents and images to be exchanged.

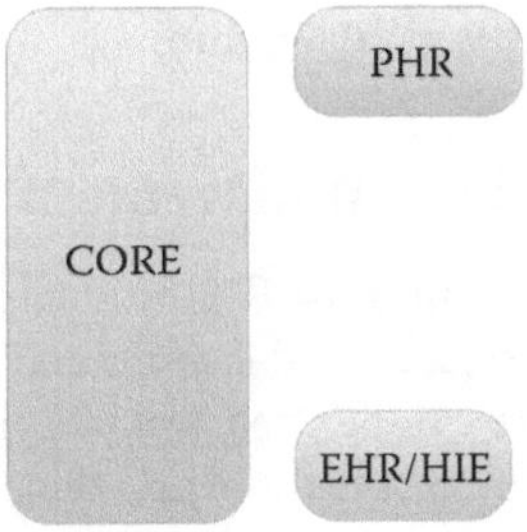

FIGURE 17.7 PHR/EHR.

17.1.3.7 Portals

The physician and/or healthcare staff will have access to the patient longitudinal health record, and also the patient electronic chart. The latter might have patient status information such as orders pending that need to be approved, the patient location, recent observations, and notes that do not necessarily end up in the permanent record, especially when it concerns transitional information.

A physician portal might also provide a window into department systems, that is, in order to provide the capability to examine diagnostic images on PACS viewer, or to sign off on the delivery of reports from a pathologist. It is not uncommon for a physician to access as many as 10 or more different applications via the portal; therefore, it is critical to have a feature that allows for a single sign-on to all applications. In addition, there must be a context exchange capability among these applications so that the information displayed by the applications always concerns the same patient.

A patient portal allows for communication between patients and healthcare providers, to access educational materials, and allows a healthcare provider to proactively send reminders and notices to a patient. It can also be linked to the patient PHR. Several commercial portals (Microsoft, Google, and others) provide a platform for a PHR.

17.1.3.8 Patient Financial Systems

The Patient Financial System takes care of the accounts receivables and submits payment claims to the appropriate organization. In many cases, eligibility is checked before services are provided, including potential copayments and/or deductibles. Claims and claim attachments are processed and exchanged with health insurance providers using standard terminology and codes. In the United States, this exchange is conducted using the X.12 protocol, as is mandated by federal regulation.

17.2 Departmental Systems: RIS

Departmental systems such as those used for laboratory, radiology, cardiology, pharmacy, and others are often part of the HIS as a "module." This allows for the sharing of its central database and some of its generic order and result screens. In some cases, these systems are more loosely connected to a HIS using an HL7 interface or a context manager, especially when the HIS vendor is different from the department system vendor.

We will discuss a typical department system using the RIS as an example/reference. Note that these systems do not differ that much, especially from a functional perspective: Each department processes and supports orders and creates results (such as a lab result, radiology report, an order to dispense medication, and so on).

A typical RIS has the following functionality:

17.2.1 Order Entry, Scheduling, and Registration

Most RIS products have the capability to register a patient, especially when used in a standalone imaging center that has no separate HIS or practice management system. It basically duplicates the ADT function of a HIS.

The same applies for orders, they can be placed using a centralized order-entry system, in many cases part of the HIS, but a RIS should also be able to place an order by itself. Scheduling is different from other departments' scheduling requirements. Each radiology procedure has its own time and resource requirement based on the exam type. A CT head study might take 10 min and need to be scheduled accordingly as it is extremely rare to have a "walk-in" clinic for CT scans. An exception to this is the ER, which may have its own CT to deal with any unexpected and unscheduled trauma cases. Other exams, such as a fluoroscopy, might take 1/2–1 h to conduct and requires that a physician and room resources are scheduled in order to perform the procedure.

Most radiology procedures have a well-defined protocol at each institution, which could occasionally change depending on regulatory guidelines with regard to patient needs, reimbursement for the procedure, and the capabilities of the device modality performing the exam. An MRI might be upgraded to include a different pulse sequence and/or coil to visualize certain body parts better, resulting in a different protocol. The same MRI might be equipped with a different table and coil to facilitate breast MRI with additional postprocessing to perform CAD (computer-aided diagnosis) in order to better visualize potential cancers. If so, the time requirement, corresponding scheduling, and reimbursement for the procedure might change.

As with other treatment protocols, diagnostic imaging protocols vary widely among physicians and institutions. This can impact the radiation dose that is received by a patient, which calls for standardization and practice guidelines.

A procedure is identified by standard procedure codes, the best known being CPT-4. For example, the code for a chest 2-view (PA and Lateral) is 71020. There is still a lack of standardization on procedure descriptions and protocol descriptions, which causes problems when managing and displaying the resulting images and reports.

For example, if a CT exam is identified as a *CT Head w and w/o contrast* and its previous study, imported from a CD that the patient brought in, is identified as *CT HD W CONTR*, it is very hard for the software to recognize that the prior study needs to be prefetched for comparison. Coding the appropriate procedures and protocols are an essential part of a department system.

17.2.2 Document Management

Unfortunately, there is still a lot of paper being used for questionnaires, release forms, and the like that might need to be scanned in order to allow a radiologist to see the reason for the procedure/order. Many systems have the capability to convert a document into the DICOM format. This allows documents to be added to the imaging study and to be displayed on the physician's workstation.

17.2.3 Patient Profile and Tracking

Sometimes patient demographic information is incorrect. This could be due to a misspelling of the patient name, a maiden name change to a married name, or other data needs to be updated. The department system will be able to issue a patient update and/or merge feature which will result in a corresponding HL7 transaction. Patient status, such as arrival status, is important to track as it could trigger placing their scheduled exam on the worklist for the modality on which the study is to be conducted.

17.2.4 Exam Status Tracking and Management

Exams are tracked with their status: scheduled, started, completed, or cancelled/aborted. Completing an exam will allow the technical component to be billed. Management information, such as the time it took for an exam to be completed, can be reviewed by an administrator. The exam status is either manually entered by a technologist or clerk, or can be automatically communicated by a modality using DICOM MPPS. (The DICOM service Modality Performed Procedure Step (MPPS), communicates status information about an exam from a digital modality to a PACS and/or RIS.)

17.2.5 Transcription/Reporting

Transcription can be part of the department information system or can be a separate application, which could reside on its own server. Traditionally, a physician would dictate a report and a medical transcriptionist would transcribe it into text, in many cases directly into the RIS.

If a physician uses speech recognition software, the voice will be converted directly into text. In most cases, the physician will correct the text themselves, but some clinicians prefer to have the editing done by a transcriptionist. When the physician does their own editing, the report can be signed off immediately, eliminating another step. Implementing speech recognition in a department has been shown to reduce report turnaround times significantly, to as little as 10–30 min.

17.2.6 Film Library Management

This feature is present in most RIS products; however, as films are becoming obsolete, it is used less often. If there are still films available, this feature can be used to track film, such as if they might be signed out of the file room and their current location. It could also be used to indicate that films are stored off-site in order to allow for a courier to pick them up if needed for comparison.

17.2.7 Billing and Inventory Control

Some departments do their own billing, as the RIS knows the specific codes to be used for reimbursement submission to payers. If billing is not done by the RIS, there is at minimum an interface to a central or external billing module that specifies the type of service, with the encoded procedure and details about the service provided, with the patient demographics. Inventory for the department needs to be managed and tracked and can be performed in many RIS products.

17.2.8 Analysis and Management Tools

A department administrator needs tools to know how their department is performing. Typical key performance indicators that are needed are the number of exams or tests, their increase or decrease over time, the time it takes to perform these, the time it takes for technologists performing exams, and so on. Reports and/or electronic dashboards are used to visualize these indicators.

17.2.9 Messaging (Critical Results, Discrepancy Reporting, Reject Analysis, Dose Reporting)

Critical results and discrepancy reporting are discussed earlier in this chapter. One item that is specific to radiology is reject analysis. For exams that use CR/DR (Computerized and Direct Digital Radiography [CR/DR]), one needs to track the number of rejects per device, and especially per technologist, along with an indication of why images are rejected.

For example, a technologist that consistently needs to retake images because of under exposure or mispositioning a patient would very likely require additional training to prevent this from recurring.

Dose reporting is also becoming very important as radiation dose used to image patients can vary widely depending on the technologist, physician, and/or institution. In addition, there have been reports of radiation overexposure due to operator error. It is very likely that dose reporting might become not only a standard, but also a legal requirement. Dose reporting may require an upgrade to the imaging modalities so it can be exchanged with the RIS, preferably using DICOM SR (Computerized and Direct Digital Radiography [CR/DR]).

17.2.10 Other: Peer Reviews, QA

Quality measures require a regular review of findings by peers. For example, every 4 weeks a random set of images will be pulled for review by a peer and discrepancies noted, reviewed, and kept.

17.2.10.1 Modality and Workstation Worklist

A digital modality needs to access its worklist to be able to select the appropriate exam to be performed while using the patient demographic information, ordering, and scheduling information from the RIS. These modalities use DICOM Modality Worklist that allows this information to be retrieved, in many cases using a bar code or patient identifier to match only a single exam record, or using a broad query to provide a worklist, for example for the MR or CT. In case that the RIS does not support this DICOM functionality, there are two options: the RIS sends the order information to a broker that then provides the DICOM worklist, or it sends it to a PACS that might be able to provide it.

A second worklist that might be provided by the RIS is a PACS workstation worklist. This is commonly referred to as a "RIS-driven worklist." A radiologist will have access to a list of exams that need to be diagnosed. The RIS keeps the status of who is reading what and updates the exam status (read, in progress, preliminary, approved). The advantage of the RIS-driven workflow is that the radiologist also has access to the requisition information such as the reason for the exam, history, and so on.

17.3 The Electronic Health Record

17.3.1 EMR and EHR

The Electronic Medical Record (EMR) is the equivalent of the traditional patient jacket or folder. It contains a longitudinal record of the patient's health events, such as lab results, imaging results, surgery notes, discharge summaries from healthcare institutions, visits with specialists, and so on.

An EMR is typically used enterprise wide, which might include multiple institutions, clinics, and other providers. In many cases, it is part of the HIS, but it can also be a standalone application that interfaces with departments using HL7 and DICOM transactions to contain the result, and with references or pointers to images and other diagnostic information. In some cases, the images are viewed through a "plug-in," which launches a PACS application to display the images.

Although an EMR spans typically patient information from a single institution and/or enterprise, an EHR is used for longitudinal record spanning across multiple institutions. Therefore, the terms EMR and EHR are used interchangeably, except that the EHR provides yet another dimension. Its functionality is basically the same.

The definition of the HER (HIMSS) is as follows:

> A longitudinal electronic record of patient health information generated by one or more encounters in any care delivery setting. Included in this information are patient demographics, progress notes, problems, medications, vital signs, past medical history, immunizations, laboratory data and radiology reports. The EHR automates and streamlines the clinician's workflow. The EHR has the ability to generate a complete record of a clinical patient encounter—as well as supporting other care-related activities directly or indirectly via interface—including evidence-based decision support, quality management, and outcomes reporting.

An EHR requires the presence of a data registry that can be used to perform a patient information query and can deliver the location of that data to a user. When crossing institutional boundaries, the Patient ID needs to be reconciled. A Master Patient Identifier (MPI) has to be assigned and corresponding ID's for specific domains have to be registered. This assumes that the registration information was submitted by the organization that created the health information, which is a requirement for the proper functioning of the EHR.

An EHR implementation varies depending on the environment in which it is used. The EHR functional definition (HL7) distinguishes three different sections: direct care, supportive care, and the information infrastructure.

17.3.2 EHR Direct Care

The direct care definition deals with the support of healthcare providers who are directly interacting with a patient. For example, a patient might be seen by a physician for a sore throat. The EHR will provide decision support functions such as signaling any contra-indications for a prescribed drug. There could also be alerts for vaccinations due, or other exams or tests based on the patient's condition.

17.3.2.1 EHR Care Management

Care Management captures demographic, clinical, administrative, and financial information, including from other EHRs and PHRs. It is basically the core of the EHR, as it should provide a summary record and different views for details. It also can manage consents and directives.

17.3.2.2 EHR Decision Support

The clinical decision support section uses protocols, plans, and guidelines to assist the healthcare professional with appropriate decisions such as standard assessment support based on a patient's medications, condition, history, age, and drug interaction, dose warnings, and recommendations.

17.3.2.3 EHR Operations Management

Operations management and communication schedules, routes, tracks, and communicates appropriate tasks involving patient care. This includes the support of clinical workflow and the communication of information with all relevant parties, including inter-providers.

17.3.3 EHR Supportive Care

EHR Supportive Care assists with administrative and financial requirements. For example, it could communicate with an immunization registry for the status of an immunization and could also communicate the relevant status when an immunization is administered. It comprises clinical support; measurements, analysis, research and reports; and administrative and financial support.

17.3.3.1 Clinical Support

This provides the provider, scheduling, and patient information at the point of the healthcare delivery. It also allows for the transfer of information to registries, the sharing information about donors and/or recipients, and the availability of healthcare resources such as beds and medical supplies.

17.3.3.2 Measurement, Analysis, Research, and Reports

This section supports the measuring and monitoring of care. For example, the outcome, performance, and accountability measurements by population, facility, and provider. In addition, it allows for both standard and *ad-hoc* queries to generate reports for decision making by providers and their patients.

17.3.3.3 Administrative and Financial

This section supports the administrative and financial aspects, such as encounter management. This will include documentation, specialized views, support data collection, billing, and could support telehealth and remote monitoring.

17.3.4 EHR Information Infrastructure

This part assists with functions such as patient safety, privacy, and interoperability. For example: Queries are performed using secure, encrypted communication using sources which are regularly backed up. The sections are security; health record information and management; registries and directories; standard terminologies; interoperability; business rules; and workflow management.

17.3.4.1 Security

This addresses the secure access to the EHR system and records and the prevention of unauthorized use of data. It requires that authentication, authorization, and audit trails are implemented in a standard manner (for example, user names); that passwords are used for access (as well as possibly RFID tags, biometrics and secure tokens and digital signatures); and that digital receipts are used to prove data integrity and confirmations.

17.3.4.2 Health Record Information and Management

Access, management and verification of accuracy, completeness and integrity of EHR information, as well as the provision of audits regarding the use and access. This entails that EHR information is retained according to regulatory rules and destroyed reliably when appropriate and that audit capabilities such as security, data, exchange, and reports are provided.

17.3.4.3 Registry and Directory Services

It is critical that information from multiple sources within and external to an EHR system is linked and made consistent. This includes information pertaining to patients, providers, payors, health plans, sponsors and employers, public healthcare agencies, and educational resources.

17.3.4.4 Terminology and Related Services

Interoperability can be achieved using standardized terminology for concepts, representations, synonyms, relationships, and definitions, such as CPT and ICD codes as well as units of measurement (LOINC) and body parts (SNOMED). Versioning (such as for ICD-9 and ICD-10) has to be facilitated.

17.3.4.5 Standards-Based Interoperability

Standards allow EHR applications to communicate with one another as well as interaction between different EHRs and regional, national, and international information exchanges. For example, communication standards such as DICOM, HL7, and X12-N should be used to perform notifications, queries, service requests, and information exchanges.

17.3.4.6 Business Rules Management

Business rules are reflected in the EHR and can be created, updated, deleted, and audited on the basis of clinical needs.

17.3.4.7 Workflow Management

This includes the management and configuration of work queues, direct work assignments, and identifies when alerts are provided such as escalations and redirection in case a patient is waiting more than "x" minutes in the ER.

17.3.5 Sample EMR Implementation: VistA

The VistA EMR is implemented by the U.S. VA (Veterans Affairs) in 175 hospitals and more than 1000 clinics and nursing homes. Because it is government funded, it is an open-source product that is available for download at no cost. Implementations of the VistA EMR are outside the VA organization, in the United States, the Middle East, northern Europe, and Central and South America.

It is considered one of the most comprehensive and complete EMRs available today. It has also been the subject of a great deal of research about the impact of an EMR on patient care. Data have shown that there is as significant reduction in medical errors that takes place (80%) and a significant reduction in postoperative deaths (30%) as a result of VistA EMR utilization.

The VistA EMR is used as an example in order to show how typical EMR functionality might look (see Figure 17.8).

A typical EMR has an overview screen, which also is called a cover sheet, and provides the most important information. It serves as an "entry" or "portal" into the EMR and is equivalent to the patient chart. Most overview screens are configurable with regard to lay-out and contents depending on the care setting (ambulatory or inpatient) and provider (physician, nurse, technologist). There are typically three sections: the demographics and provider section, selection tabs, and highlights.

17.3.5.1 Demographics

This section contains the patient demographic information, such as name, ID and birth date, and, if available, a photo of

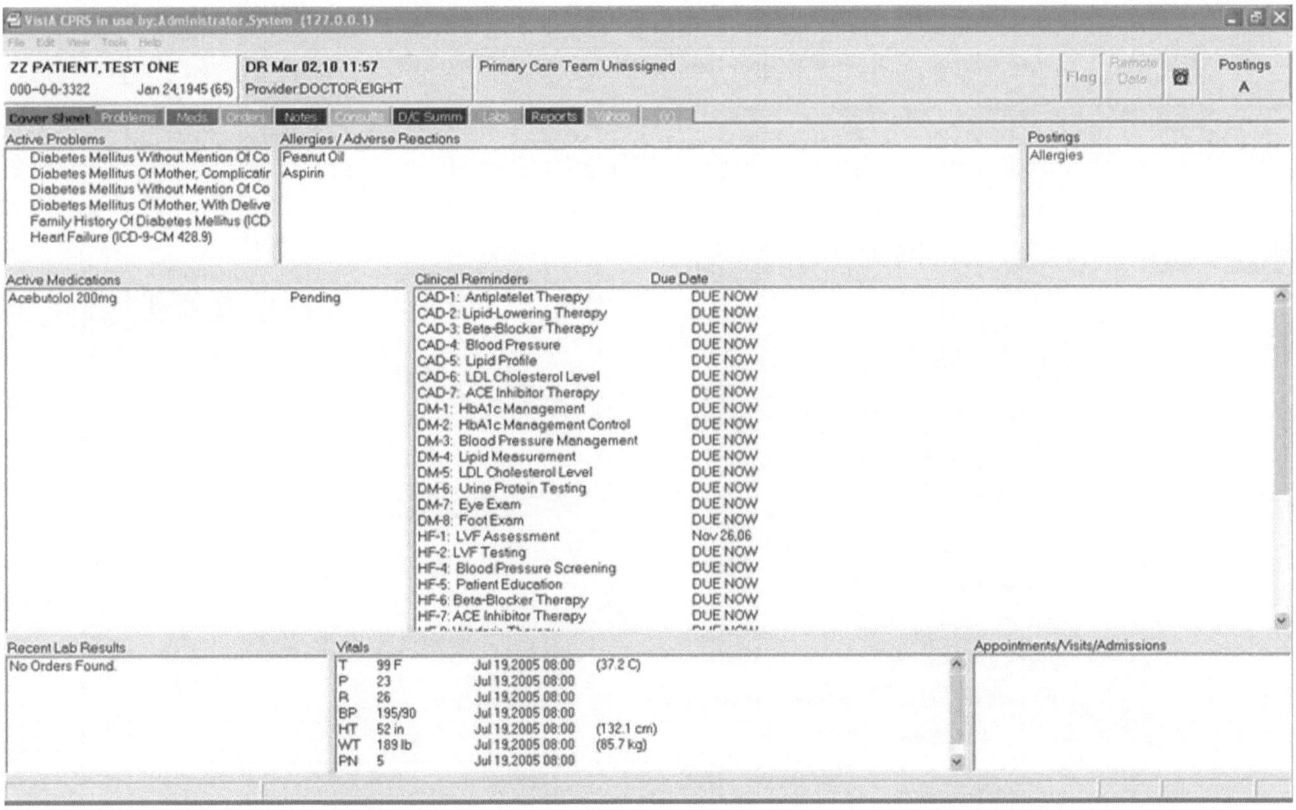

FIGURE 17.8 VistA EMR.

the patient. It also provides information about the physician and other healthcare team members. In addition, there are also alerts, which are termed "postings."

17.3.5.2 Selection Tabs

This section allows a user to "drill down" into clinical details. In this example, the options are problems, medications, orders, notes, consults, discharge summaries, labs, reports, and Internet access. Additional tabs can display and access other types of information, such as the capability to retrieve diagnostic images.

17.3.5.3 Highlights

The highlight section contains all critical information that is presented for a user, such as active problems, allergies and adverse reactions, postings, active medications, clinical reminders, recent lab results, vitals, and appointments.

At any time, a user can select either a window and/or one of the tabs to get more information or to enter additional information into the record. For example, when entering a new allergy, it would look as shown in Figure 17.9.

17.4 Conclusion

An HIS consists of hardware and software components that support the management and operation of a healthcare institution, spanning multiple departments, and medical specialties across an institution. A HIS facilitates operational, managerial, and clinical data collection and decision making. The HIS is the major communication mechanism among various departments and other parties involved with patient care; it supports clinical operations and healthcare professionals in their clinical decisions.

The HIS architecture consists of a main component, that is, the "core," which contains the various departmental systems, with imaging as a special case, and the data repository and knowledge systems. It provides manual electronic input capabilities, interactive communication to both inside and outside the campus and electronic portals as well as billing.

An EHR, managed by external entities, or a PHR, managed by the patient, may need to interface with the HIS.

A RIS is a departmental system similar to what is used for laboratory, cardiology, pharmacy, and others and is often part of the HIS as a "module." This allows for the sharing of its central database and some of its generic order and result screens.

The EMR is the equivalent of the traditional patient jacket or folder. It contains a longitudinal record of the patient's health events, such as lab results, imaging results, surgery notes, discharge summaries from healthcare institutions, visits with specialists, and so on.

An EMR is typically used enterprise wide, which might include multiple institutions, clinics, and other providers. In

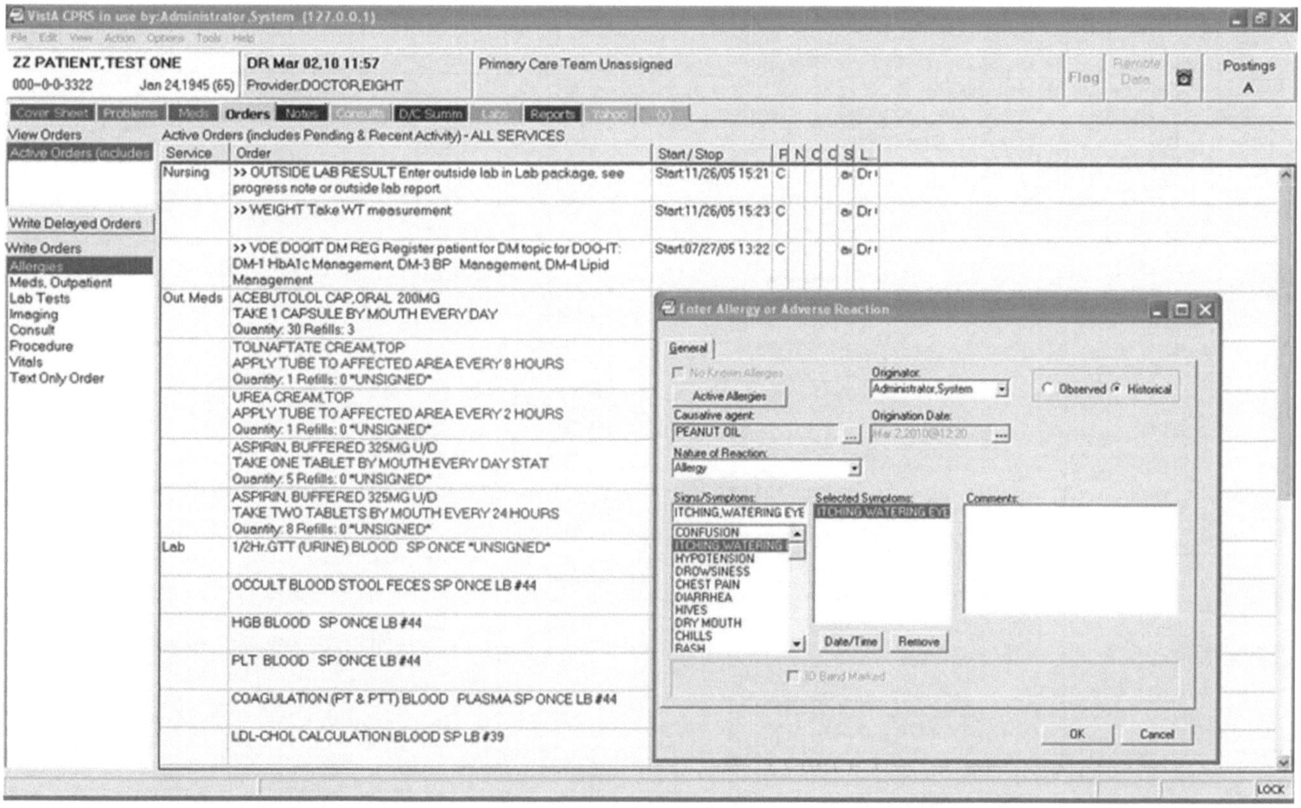

FIGURE 17.9 EMR allergy entry screen.

many cases, it is part of the HIS, but it can also be a standalone application that interfaces with departments using HL7 and DICOM transactions to contain the result, and with references or pointers to images and other diagnostic information. Although an EMR spans typically patient information from a single institution and/or enterprise, an EHR is used for longitudinal record spanning across multiple institutions. The EHR functional definition distinguishes three different sections: direct care, supportive care, and the information infrastructure.

The VistA EMR is implemented by the U.S. VA (Veterans Affairs) in 175 hospitals and more than 1000 clinics and nursing homes and is considered one of the most comprehensive and complete EMRs available today.

References

HIMSS. *HIMSS* [Online]. Available at: http://www.himss.org. Accessed September 1, 2010.

HL7. *HL7* [Online]. Available at: http://www.hl7.org. Accessed September 1, 2010.

IHE. *IHE* [Online]. Available at: http://www.ihe.net. Accessed September 1, 2010.

V

Operational Issues

18 Procurement

Boris Zavalkovskiy
Lahey Clinic

18.1 Vision for Image Distribution in the Enterprise

18.1.1 Evaluating Where You Are Today

Evaluation of the current environment and understanding where industry is headed constitute an important aspect of any enterprise. With the introduction of technology into healthcare organizations, we are presented with opportunities that had not been available before. Consider this hypothetical situation. We may start with a certain highly structured radiology department that possesses commendable human-powered processes and optimized workflows in every facet of the radiology business and provides a delivery service to an institution. This radiology department may have already had a very low percentage of film loss. The department may have already had an outstanding turnaround time in the despatch of reports to clinical peers, by having placed a transcriptionist next to a radiologist during dictation and ensuring swift film distribution to clinical practice. Perhaps each radiologist had a specific area to serve in for a specific period of time, where he/she worked with a technologist to sustain volume from specific modalities in clinical practice. However, perhaps as the department moved toward reading with greater specialization, it was recognized that those specialists that were most qualified to read certain exams may not have been in the correct location. New modalities started to produce significantly larger data sets and impeded our ability to sustain our attempt at emulation of the printed film model of professional interpretation and clinical distribution of radiological studies. Our ability to consult with colleagues was constrained by geographical proximity. Moving toward the electronic era had become inevitable. A good start for an evaluation of the current environment is a review of the state of modalities such as CT, MRI, CR, and so on and the assessment of their compliance with industry standards with regard to the modality work-list, data, and image communication. The next step is to proceed with evaluation of the environmental readiness. This is a perfect opportunity to review current reading and control rooms and examine as to how they can be altered to accommodate the new digital environment and workflows that will be in place post-PACS deployment. Validation of sufficient cooling, of appropriate illumination, of the sound-proofing of reading areas, and so on is essential. Furthermore, one should proceed with evaluation of the network infrastructure not merely in terms of the endpoint workstation's connection but with regard to the overall network requirements of the PACS system of choice, ranging from the scanners to the data centers, workstations, the final archives, and clinical distribution of the images. Some systems have distributed architecture and some have central architecture. Each system topology has its advantages; more important is the fact that each system requires a different network infrastructure and reliability.

18.1.2 Clear Definition of Desired End Points

As the department progresses with its evaluation of the current state, it is a good time to start considering as to how one may modify current practices in the new environment. The creation of an ergonomic workspace for a radiologist, such that it simultaneously permits clinical workflow, clinician collaboration, and residents' instruction, can be quite a challenge. One must understand that an electronic environment presents the opportunity of radiological studies to the specialist, no matter where this specialist physically resides (albeit with complications in legal billing). Clinical consultations with referring physicians now can be accomplished remotely with all parties examining the same images and controlling a single workspace while discussing management options in the case of a specific episode of care. The introduction of speech recognition software (assuming a high percent of recognition and quick response time) may provide some physical space to a department/institution by reducing transcription needs and costs. Distribution of radiological images to clinical departments prior to a report's delivery is also possible. This is a very interesting paradigm shift, since radiology was the gatekeeper of file films for years and could insist on providing images and reports, as a unified service, to a clinician. In contrast, with digital systems one can configure delivery of the images to critical clinical care units immediately after image acquisition. In the past, film costs would either escalate when creating copies, or film distribution to clinical areas would be limited. In a "softcopy" digital environment, one may notice that the user population that reviews diagnostics is growing.

18.1.3 Requirements Evaluation for Clinical Image Distribution

After conducting a review of the current practices, workflow, the environmental condition, and the state of infrastructure, it is useful to map out clinical image distribution requirements. As mentioned above, with a softcopy image distribution opportunity to release diagnostic studies to critical areas prior to report delivery exists. Alternatively radiology practices can assist in expediting the perusal of critical areas of reports by those in practice and thus provide higher service levels by distributing radiological studies, complete with imaging data, and professional interpretation. Understanding the image quality needs of each clinical area so as to conduct patient care is also important; the image quality will drive hardware requirements (i.e., displays and compression settings) and selection criteria in terms of PACS functionality. Some PACS vendors can deliver different qualities of images to different clinical specialties on the basis of preconfigured specifications. Furthermore, a certain type of PACS software not only understands the clinical area it is being used in, but also can evaluate the hardware of the workstation (i.e., display capabilities) and provide the appropriate image file quality.

18.1.4 Understanding Different Service Levels to Different Clinical Practices

In addition to considerations of image routing, quality, and timeliness, there are potentially more aspects in terms of migration toward a softcopy environment. The valuation of clinical practice image-processing is also very important. Some clinical environments (e.g., orthopedics) may require one to work with images postradiological interpretation, perhaps for preoperative planning. Some of these clinical practices will require the integration of third-party software products (which they are likely to already possess) into the radiology PACS. The integration of these third-party applications into the PACS vendor of choice could be crucial in clinical adoption by these departments. In many instances, a clinical practice looks forward to change only in the event of gaining efficiencies and other clinical benefits to their practice. This is one of many examples that clearly outline a need to perform a careful evaluation of each clinical area with regard to its specific requirements of image delivery, manipulation, and potential archiving of newly generated images, as a part of clinical practice, when moving toward soft copy distribution.

18.2 Building a Strategic Case for Institution and/or Market Sector

The selection of a vendor which provide images storage and distribution, and will be aligned with the strategic direction of one's healthcare organization, and at the same time will support its integration with other local, regional, and federal healthcare organizations is not a simple undertaking.

18.2.1 Current Government Directives for Information Technology in Healthcare Institutions

On April 27, 2004, President Bush issued Executive Order (EO) 13335 "to provide leadership for the development and nationwide implementation of an interoperable health information technology (IT) infrastructure to improve the quality and efficiency of healthcare." EO 13335 established the position of a National Coordinator for Health IT within the Office of the Secretary of Health and Human Services.

EO 13335 also charged the National Coordinator with development, maintenance, and direction:

> ... the implementation of a strategic plan to guide the nationwide implementation of interoperable health IT in both the public and private healthcare sectors, which will reduce medical errors, improve quality, and produce greater value for healthcare expenditures.

Accordingly the Office of the National Coordinator for Health Information Technology (ONC, 2008, p. iii) has worked across the federal government to develop this ONC-coordinated Federal Health IT Strategic Plan (the Plan), which identifies federal activities necessary to achieve the nationwide implementation of this technology infrastructure throughout the public and private sectors. The timeframe of the Plan is 2008–2012.

Plan Structure

The Plan is structured around two goals, each with four specific objectives:

Goal 1: Patient-Focused Healthcare. Enable a transformation to higher quality, more cost-efficient, patient-focused healthcare through electronic health information access and use by care providers and patients and their designees.

Goal 2: Population Health. Enable the appropriate, authorized, and timely access and use of electronic health information to benefit public health, biomedical research, quality improvement, and emergency preparedness.

18.2.2 Rationalization and More Effective Use of Technology

Meaningful use. The American Recovery and Reinvestment Act of 2009 (Recovery Act) authorizes the Centers for Medicare & Medicaid Services (CMS) to provide reimbursement incentives to eligible professionals (EPs) and hospitals that achieve success in becoming "meaningful users" of certified electronic health record (EHR) technology. The Medicare EHR incentive program will provide incentive payments to EPs, eligible hospitals, and those critical access hospitals that are meaningful users of certified EHR technology. The Medicaid EHR incentive program will provide incentive payments to EPs and hospitals for efforts to adopt, implement, or upgrade certified EHR technology or for the technology's meaningful use in the first year of their participation in the program and for the demonstration of meaningful use during each of five subsequent years. On December 30, 2009, CMS announced a notice of proposed rulemaking (NPRM) to implement those provisions of the Recovery Act that provide incentive payments for the meaningful use of certified EHR technology. The proposed rule outlines provisions governing the EHR incentive programs, including those defining the central concept of the "meaningful use" of EHR technology. CMS' aims to render the definition of "meaningful use" consistent with applicable provisions of Medicare and Medicaid law while continually advancing the contributions that certified EHR technology can make toward improving healthcare quality, efficiency, and patient safety. To accomplish this, CMS' proposed rule would phase in more robust criteria for the demonstration of "meaningful use" in three stages.

Stage I—Electronic capture of health information in a coded format; tracking key clinical conditions and communicating outcomes for care coordination; implementing clinical decision support tools to facilitate disease and medication management; and reporting outcomes for public health purposes.

Stage II—Expands on Stage I. Encourages the use of health IT to enhance computerized provider order entry; transitions in care; electronic transmission of diagnostic test results; and research.

Stage III—Expands on Stage II. Promotes improvements in quality and safety; focuses on clinical decision support at a national level by encouraging patient access and involvement; and improved population health data.

18.2.3 Movement of Care with the Patient

Health Insurance Portability and Accountability Act (HIPAA) of 1996.

The Administrative Simplification Provisions of the HIPAA of 1996 (HIPAA, Title II) require the Department of Health and Human Services to establish national standards of electronic healthcare transactions and national identifiers in the case of providers, health plans, and employers. It also addresses the security and privacy of health data. The adoption of these standards will improve the efficiency and effectiveness of the nation's healthcare system by encouraging the widespread use of electronic data interchange in healthcare. How you deploy these requirements at your institution or in your geographical area in part depends on your vendor's ability to comply with HIPAA requirements.

18.3 Understanding Technology and How It Will Change Clinical Practices

A fundamental understanding of new technology capabilities is important as it has direct implications with regard to clinical practice. This section will provide some guidelines on how to undertake a technology evaluation. This is not a plan, but includes suggested guidelines formulated on the basis of experience. PACS vendor evaluations need to be customized to an individual institution: its needs, goals, and present and intended overall electronic environment.

18.3.1 Selecting Clinical Champions

A complete understanding of clinical implications in each area may not be a realistic goal for achievement. The best that may be possible is the selection of a core group of stakeholders (technologists, IT, clinicians, and radiologists) for evaluation of the new technology and provision of input that pertains to how this new technology will impact their respective clinical areas. In any case, one needs to ensure that clinical areas are represented and that the selection committee is advised in the case of a process of vendor selection. These same individuals will be champions in their respective areas when the time for implementation arrives. These clinical liaison officers will be well-positioned to communicate with clinical practitioners and project leadership with regard to challenges and opportunities in the technology's implementation. In terms of the selection of these clinical champions, it is important that these individuals will be adept in their respective areas and enjoy a support of colleagues in that area to make technology decisions on their behalf. The question of the number of clinical areas required to achieve good representation of one's clinical environment depends on one's institution. It is recommended that one starts with areas of high image utilization, which he/she can easily identify, looking at his/her current image distribution in the environment.

18.3.2 Securing Support and Funding

As the site engages in this endeavor, it is important to secure institutional support and funding for the effort early on. One of the many ways to procure support from an institution is to show as to what this product will provide. There are many publications, articles, and books available for the extraction of this information and its application to your institutional goals. Here are some publications that will provide additional information on the justification of the PACS procurement: IEEE Xplore (ARR, 2009; Cohen et al., 2005; DICOM; Hindel, 1989; HL7; IHE; Lim, 2002; Muschlitz; Diagnosticimaging; Siegel; Trapp). There exists a range of benefits that technology offers, apart from providing good return on investment (ROI). Depending on an institution, the benefits might include fewer lost films, rapid access to diagnostic images pertaining to a specific episode of care, and availability of the prior examinations for comparison analysis in treatment. Preparation of this documentation and the communication of this project to stakeholders constitute an important part of acquiring institutional support and funding. The alignment of PACS implementation with electronic medical record (EMR) or Hospital Information Systems (HIS) is a potential benefit and could provide additional resources for planning and implementation, which will naturally improve integration opportunities with your EMR.

18.3.3 Evaluation of Different Technologies' Impact on Clinical Practice

Some technologies in the market can enable one to function without film. It is relatively inexpensive to replace your hard copy film environment with stand-alone PACS that will provide image distribution to Radiology professionals and clinical areas. However, other technologies will provide the ability to enhance clinical practice and take advantage of the newest emerging functionalities of imaging diagnostics. Some of the clinical areas that can take advantages of the radiology image acquisition to enhance their clinical treatment and diagnostics are radiation oncology, which has software that will provide treatment calculations based on radiology imaging, orthopedic surgery utilizing a templating software that provides preoperative planning, neurosurgery, endoscopy, organ transplant, and many others. Each of these clinical areas excels in adopting a different imaging software solution to enable it to provide better clinical care for its patients. Ensuring that this clinical software will be working in concert with clinical software is important and will drive clinical acceptance of the selected PACS vendor. Success of the PACS implementation in the multispecialty enterprise will be judged not merely by ability of the radiology department to streamline radiology workflow, but also by how this new imaging technology enhances other clinical practices.

18.3.4 Defining Clinical Requirements for Technology

It is only when we understand our current electronic environment that supports clinical practice, having a group of professionals who represent different specialties, that we can start drafting our requirements for new technology.

A level of integration with third-party applications needs to be clearly outlined and documented. Features like patient content need to be preserved from one application to another to ensure that the clinical test you are reviewing, for information on a specific patient, will be in fact that of the patient you are currently evaluating and not that of a prior patient or ironies patient on the system. In other words, integration of the multiple applications at the desktop level needs to be patient centric and ensure synchronization of the clinical content as physicians move from one patient to a next.

Qualities of images delivered to diverse clinical areas are necessarily not required to be the same but will require different hardware and some software functionality.

Availability of the results of a prior examination of a patient in some clinical areas is more prudent than it is in others. The speed of image loading is an important factor in the delivery of image data to the end user. We would all like images to be instantly available for review, but the creation of criteria that are realistic and achievable is important with regard to vendor selection.

Recognition of any user that logs into a system is important for security purposes, but defining different user groups with potentially different privileges on the system could be a plus an added advantage. Integration of the login process into an application is also important. No one would like to encounter multiple logins and multiple levels of authentication after initially logging on to the workstation.

18.4 Selecting Right Technology for Your Institution

At this point in time, we have a very good understanding of our current state as well as a very good idea as to how we would like to transform our current environment. How one selects vendors that will be extended an invitation to participate in on-site technology evaluations that will ultimately lead to the acquisition and implementation of this technology into one's institution is the important question. There are multiple ways to achieve this. A useful approach is to be less restrictive in the beginning of the process and allow each vendor to present his/her products and services. Engaging multiple vendors will allow the selection committee to observe, question, and learn about design differences and potentially adjust the selection criteria. A site may already have some vendors in one's institution. If these vendors have a PACS product, it might be a good practice to invite them technology an evaluation of technologies. If these vendors are modality makers, they may have better integration with their own PACS and will provide proprietary information on what is possible in modality integration to PACS. Hence, it is useful to include all those vendors currently present in one's clinical environment that have PACS products. It is also prudent to review some industry leaders in PACS technology. One can find out market leaders by evaluating the percentages of the market that are owned, identifying the owners of those percentages, and selecting the top 3–5 vendors with whom one can participate. Depending on the resources one possesses for allocation to this endeavor, it is recommended that one selects 5–10 vendors for this evaluation.

18.4.1 Vendors Evaluation

There are a number of parameters you can apply, with regard to vendor selection for your institution, depending on your goals and priorities. Here are some of the criteria that you might consider during your evaluation of vendors; adherence to industry standards, DICOM, work-list, HL7, and so on. Please review references for more information regarding HL7, DICOM, and IHE. Financial viability of the company on a market place: will this company exist 5–10 years henceforth to support its product in your environment? Depending on how your institution supports IT, that is, does it have an extensive array of IT support personnel, you may wish to evaluate the support that a vendor provides to existing customers. Furthermore, it is very important to understand the technology that you shall purchase, as technology drives all clinical requirements and integrations that you have envisioned in your enterprise.

18.4.1.1 Define Process of Technology Evaluation

Now that we know what criteria will be included in our vendor evaluation parameters, we can define a quantitative template that would outline all of the requirements from our institution and define the importance of individual criteria. Before embarking on a detailed evaluation of functions and features that are important for our integration, it is just as important to outline a process for proceedings involving different vendors' evaluations. Here is the process that we have followed in our vendor selection.

After gathering all that we needed, evaluation of the market and selection of the industry leaders, we outlined an *N* number of vendors to participate in a vendor selection process. We have created a Request for Information (RFI) and distributed the same to all *N* vendors with an idea that, after review, we would select five vendors who would be invited to present technologies available for present-day implementation and discuss what their future roadmap holds. In these demonstrations, it is important to include all stakeholders in a project including the clinical areas and support staff, if one has his/her IT shop. Soliciting quantitative feedback from participants of the demonstration and marking the one from the vendor community that will advance to the next round of an evaluation process, we have defined this process ahead of time and kept all vendors informed from the very beginning; by doing so we eliminated unnecessary calls, e-mails, and confusion regarding the successive steps in our process.

18.4.1.2 RFI and RFI Review

The creation of RFI for distribution to industry leaders is important and forms an educational part of the process of acquisition. RFIs need to include all clinical requirements for PACS functionality and performance. Here is an outline of features and the workflow that clinical practice is most concerned about:

Graphic User Interface, Work List, navigation, prior exam availability, hanging protocol customization, 3D tools, templating software, and so on.

In some healthcare institutions, IT is well developed and has strategies for data archiving, business continuance, and disaster recovery. As for institutions that have a well-developed IT department, it is important to understand how the vendor's proposed PACS hardware fits into overarching strategies of the enterprise. As for institutions that do not have well-developed IT resources in terms of hardware, support, integration, and compatibility with current IT architecture is important part of the RFI.

RFI needs to be distributed for review to an oversight group and to all disciplines in the enterprise that will be impacted by changes and can provide constructive input regarding the RFI process.

While waiting for RFI responses, it is important to create an evaluation matrix. This matrix can comprise quantitative values to individual functions, features, and hardware and software compatibility. Study an example of the matrix of RFI evaluation (Table 18.1).

As you can see in this example, we have created multiple criteria and assigned different values to individual functions, depending on their importance to our practice. At the end, you will have quantitative results that will provide you with guidance to a next round of technology evaluation. Our RFI evaluation score card consists of over 100 parameters that we used to examine all initial vendors. From these score cards, you can select 4–5 vendors to proceed to the next round.

18.4.1.3 Demonstration of Vendor Technologies

Inviting vendors for a demonstration of their technology provides an opportunity to ask questions and interact with vendor representatives. An important aspect of hosting demonstrations is planning, if there are four or five vendors, it is also important to conduct demonstrations over a sufficiently short enough period of time so that memory loss from the first presentation to the last presentation is not overly excessive. Furthermore, one should script large portions of the demonstration as "use cases" to ensure that vendors follow protocols and present comparable information to your colleagues. It is useful, in terms of these scripts, to assign different times to different specialties and to different interest groups. For example, radiologists may have a session, where vendors present tools and functions that are relevant to practicing radiologists. Another session for clinical specialties could have vendors introduce tools relevant to clinicians and exhibit integration into add on applications (e.g., the EMR

TABLE 18.1 Example of the Matrix for RFI Evaluation

Questions	Metric	1	2	3
What brand and model of hardware do you provide? What operating systems do you provide?	0–3	3	3	3
What database systems are used?	0–3, 3 = Oracle or MS-SQL	3	3	3
Does your proposal include a redundant server with redundant storage systems?	0–3	3	3	3
What security measures does your system use?	0–3	3	3	3
Does your system use Active Directory Integration for system access?	0–10, 10 = full LDAP support	10	10	10
What type of access do you require to support your system?	0–3	3	3	3
Is there antivirus on the machines? Can you use the enterprise antivirus software currently used by Lahey Clinic?	0–3, 3 = Lahey AV with auto updates	3	3	3
99.99% scheduled uptime, including prior studies. A scheduled outage for maintenance will not be counted. Preventive maintenance shall not result in clinical inoperability.	0–10, 10 = 99.99 SUT	10	10	10
Real-time remote monitoring tools	0–3, 3 = full RT monitoring	3	3	3

or Ortho templating systems). A session of IT, where a vendor presents system architecture, network requirements, hardware specifications, and opportunities for integration (EMR, HIS, etc.), is also useful. Forms created for each presentation, where attending personnel can provide feedback on a demonstration, is useful in the tabulation and reduction of the vendor pool to two or three options that would be further moved to a next round of evaluation.

18.4.1.4 Site Visits

Once the team is left with two or three options, it is useful to request a site visit, where a vendor can demonstrate his/her product implementation and clinical use in a real clinical environment. It is important to select sites that exhibit similar clinical specialties in a practice, similar sizes of clinical practice as well as similarity to our main systems that we shall use to interface our PACS system. It is important to contact the demonstration site prior to a visit and have at least one phone conference where you can agree on how a site visit will proceed. It is also important to allow time for private discussions, for example, a radiologist to another radiologist, at the visiting site. Allow the site's users to "drive" the PACS workstation, observe how a radiologist at the visiting site utilizes tools of the PACS system and determine how it helps them to achieve high productivity with a good quality. It is also important to let support personnel to meet and discuss support technical processes and workflow challenges. How did this site integrate support by vendor into their support structure, hardware and software support, issues resolution, and escalation process? What is an actual uptime of the system on this site? There should also be a postsite-visit meeting to debrief the team, collect all score cards, and add them to the evaluation process.

18.4.1.5 Selection Process

At this point, you should have a very good idea as to who your vendor of choice is. Typically, a site will have two vendors that technologically can provide a solution to an enterprise with similar results in terms of overall acceptance. It is very unusual to find that this type of selection process has revealed a vendor that stands above the rest of the industry. If you, in fact, find this to be true, then the choice is clear and obvious. In the more likely event that you have come to the end of the process and have two vendors who are similar in functionality, support model, and their ability to achieve, it would be good to step back and evaluate some factors that potentially have not been in your decision-making process during PACS selection. For example, if one of the PACS vendors is exclusively a PACS vendor, meaning it does not produce any modalities and is not in competition with some of the existing modality vendors in your environment, then it might be a good time to ask the question, "What we are trying to achieve?" One ideology is that a radiology department should always be kept away from the vendor mix perspective to avoid single source vulnerability, and, as a result, it would be beneficial to have a PACS vendor who does not produce modalities and is not in competition with the existing modality mix. Alternatively you may be already dependent on the single vendor for modalities; obtaining a PACS system from the same vendor is potentially an advantage from the perspective of the ability to integrate at least on a modality level. One needs to be careful in this type of integration as, at times, single source-vended systems have a tendency to deviate from the industry standards, making later third-party integration difficult.

18.4.1.6 Request for Proposal

We now have a very clear picture of what we would like to accomplish with this project. We have two vendors that have made it to a final selection process, and it is my recommendation to issue a request for proposal (RFP) to them; this is where you can request an actual proposal for a real system, inclusive of all required interfaces to other enterprise systems, modalities, and clinical bank trust department software integrations. It is at this time that you would like to size-up short-term storage and long-term storage so you can provide vendors with clear guidelines on the scope of the hardware requirements. RFP should be very close to your contract with regard to the deliverables pertaining to hardware, software, and services. This way you can compare these two proposed systems back-to-back from the perspective of functionality, clinical viability, capacity, and integration. I These shall become evident: the cost of initial implementation, the cost of ownership for 3–5 years and the cost of the migration, if you have any prior data that require migration to a new system. As you are returned your RFP, you will have a clear picture as your vendor of choice, and you are ready to take the next step and start contract negotiations.

18.4.1.7 Contract Negotiation

By this time, you have announced your vendor of choice to your two finalists: you have congratulated one and wished the other the best of luck. It is possible to enter into a contract negotiation with two vendors, but not advisable as it could be a lengthy and laborious process. We selected one vendor with whom we engaged in negotiations. Your RFP response from this vendor should be a template for your contract and should potentially, depending on how you constructed your RFP, include RFP as an article or exhibit to a contract. It is important not to lose the key points of that which you are attempting to achieve in this acquisition. As for us, the goal was to provide a superior reading environment to our radiologists in each subspecialty area, integration with clinical applications in clinical areas outside of the radiology, and further integration of the new system into institutional information and archiving systems. Support and "up time" of the PACS is an important part of clinical acceptance. We have included uptime of the system as a measure of compliance with support and maintenance agreement during our contract negotiations.

18.5 Conclusions

Selecting the right PACS vendor for your clinical environment is not a small undertaking. This endeavor will require resources from clinical specialties that are imaging-centric in their caregiving, IT specialists who are familiar with technologies and can provide input regarding the decision-making process on the basis of clinical requirements.

The possession of dedicated resources that will drive time line and a Project Manager are crucial to participation in a timely and successful selection process.

A clear understanding of where you are and where you need to be will provide you with apparent functions and should feature in what your negotiations with a vendor, so as to realize your vision for your institution's PACS implementation.

Most important is the support of the senior management for this project, as this is a strategic project for an institution and not a stand-along departmental endeavor.

I hope this chapter will provide you with some useful guidelines in your procurement process. Good luck!

References

ARR. 2009. *American Recovery and Reinvestment Act* [Online]. Available at: http://www.recovery.gov. Accessed September 1, 2010.

Cohen, M. D., Rumreich, L. L., Garriot, K. M., and Jennings, S. G. 2005. Planning for PACS: A comprehensive guide to nontechnical considerations. *J. Am. Coll. Radiol.*, 2, 327–37.

Diagnosticimaging. *PACS market records solid gains but Y2K may hamper growth rate* [Online]. Available at: http://www.diagnosticimaging.com/practice-management/content/article/113619/1227712. Accessed September 1, 2010.

DICOM. *DICOM Standards* [Online]. Available at: ftp://medical.nema.org/medical/dicom/2009. Accessed September 1, 2010.

Hindel, R. 1989. Methodology of PACS cost justification. *The First International Conference on Image Management and Communication in Patient Care: Implementation and Impact.*

HL7. *HL7 Standards* [Online]. Available at: http://www.hl7.org/implement/standards/index.cfm. Accessed September 1, 2010.

IHE. *IHE Standards* [Online]. Available at: http://www.ihe.net/profiles/index.cfm. Accessed September 1, 2010.

Lim, J. H. 2002. Cost justification of filmless PACS and national policy. *Medical Imaging 2002: PACS and Integrated Medical Information Systems.*

Muschlitz, L. *PAC Pays (Really)* [Online]. Available at: http://www.imagingeconomics.com/issues/articles/MI_2000-02_05.asp. Accessed September 1, 2010.

ONC. *The ONC-Coordinated Federal Health IT Strategic Plan: 2008–2012*, Executive Summary p. iii, June 3, 2008.

Siegel, E. L. *Economics of PACS: Justifying PACS* [Online]. Available at: http://www.rsna.org/Publications/rsnanews/upload/UKRC_2006_Designing_the_Radiology_Reporting_room.ppt. Accessed September 1, 2010.

Trapp, D. *House stimulus bill boosts health IT, Medicaid* [Online]. Available at: http://www.ama-assn.org/amednews/2009/02/09/gvsa0209.htm. Accessed September 1, 2010.

19

Operational Issues

Shawn Kinzel
Mayo Clinic

Steve G. Langer
Mayo Clinic

Scott Stekel
Mayo Clinic

Alisa Walz-Flannigan
Mayo Clinic

19.1 Background

19.1.1 Terms and Definitions

CCFL—Cold cathode fluorescent light, the backlight used to drive many LCD computer displays.

CR—Computed radiography, a widely used acquisition detector technology for performing portable radiography using sheets of flexible phosphors.

CRT—Cathode ray tube, an older display technology that used an electron beam to excite phosphors on a viewing screen (Bushberg et al., 2002).

DICOM—Digital imaging and communications in medicine, an international standard for medical imaging (see Chapter 4).

DR—Digital radiography, a widely used acquisition detector technology for performing fixed radiography using rigid detectors.

DMZ—demilitarized zone, when used in reference to computer systems, it is the place on a network between two firewalls, most commonly between an external Internet-facing firewall and an internal network-protecting one, creating a protected layer for external-facing applications and allowing for another layer of protection for the internal network.

GUI—Graphical user interface, as in the GUI of a computer software program.

GSDF—In DICOM, the Gray Scale Standard Display function used to perceptually linearize the gray values in an LCD display (Barten, 1992, 1993, 1999).

LCD—Liquid crystal display, the elements that make up the display mask of many modern computer displays.

LDAP—Lightweight Directory Access Protocol.

OS—Operating System, the software that runs on hardware to provide functionality.

RAID—Redundant Array of Inexpensive Disk.

Systems Administration—the practice of management computer systems.

SA—System Administrator, the person tasked with managing computer systems.

Virtualization—Technology that allows for stacking of additional Operating Systems into the same hardware footprint.

19.1.2 Introduction

The authors of this chapter bring deep experience to the several subspecialties alluded to in the major section headings: Displays, Modality Informatics Testing, and Server/Workstation maintenance. While each topic could be expanded to its own book, the material within (combined with the references) should be adequate to assist the reader in planning their own operational and support needs.

19.2 Displays

The last link in the imaging chain from patient to the eyes of the radiologist is the Picture Archiving and Communication System (PACS) electronic display. All of the visual information carefully acquired by the modality equipment, manipulated by the technologists and PACS software, and stored in the system must then make the leap from the system to the brain of the radiologist via the eyes.

The goal of a well-operating display is to convey a maximum amount of visual information as efficiently and correctly as possible. To ensure that a display is optimized for this purpose, it is necessary to understand characteristics of the display, identify, and choose between the performance attribute trade-offs, and decide on a maintenance plan.

19.2.1 Display Characteristics

The modern liquid crystal display (LCD) has largely supplanted the used of cathode-ray-based displays (Langer et al., 2004). They consist of a matrix of liquid crystal elements that act as shutters in front of a light source known as the backlight. The matrix of elements allows light through or blocks the backlight in variable degrees. On a grayscale display, there are no color filters; on a color display, there are red, green, and blue filters in front of elements and the contributions of each generate the color of the main picture element (pixel).

19.2.1.1 Pixel Resolution

In this discussion, the pixel resolution refers the matrix size of the display. There are some common matrix sizes ranging from 1024 × 768 pixels for a one megapixel (MP) display up to 2560 × 2048 pixels or greater for 5 MP mammographic displays. The pixel resolution is important when considering the image size of the modality that produced it because it determines whether a modality's image can be viewed in full-fidelity or must be down sampled. For example, a typical CT image may consist of 512 × 512 pixels and could be shown in full-fidelity on a 1 MP display. A typical CR/DR chest image, however, may be 3000 × 2500 pixels, which cannot be viewed in full-fidelity in its entirety even on a 5 MP display. While a one-image-pixel-to-one-display-pixel viewing of an image is possible by zooming in on an image via the viewing software, it can be advantageous for viewing efficiency to minimize the amount of manual zooming and panning necessary to survey the image's information. The cost of a display is often directly related to its pixel resolution.

19.2.1.2 Maximum Luminance, Luminance Ratio, Stability, and Uniformity

The LCD elements control the amount of light allowed through the panel from the backlight. When entirely open, the pixel elements display a maximum white value. When twisted all the way closed, they block their maximal amount of light and this results in a black pixel. The overall maximum luminance of a display is a function of the amount of luminance produced by the backlight. If a higher luminance is required for fully white pixels, it is necessary to increase the output of the backlight. The trade-off, however, is that because of the fixed efficiency of the panel's ability to block the backlight the black pixel luminance will also increase. This ratio of maximum white luminance to minimum black luminance is a measure of the panel's ability to generate a full range of luminance values and is known as the luminance ratio.* A greater luminance range allows for a greater difference between the individual shades of gray or color. This can enhance the contrast resolution, as discussed in the next section.

The light source for an LCD is typically a cold-cathode fluorescent light (CCFL). These sources produce a variable amount of light both in the short-term as they warm up and over the long-term as they decay. A medical-grade display incorporates a feedback circuit that serves to stabilize the luminance output of the backlight. This stabilization eliminates the luminance output variability inherent in the light source which helps to ensure that multiple displays have matching presentation characteristics.

The backlight is a nonuniform light source and the liquid crystal panel's elements have variability in response. As a result, there may be differences in the luminance output across the face of the display. Luminance Uniformity is a measure of this variability. The goal, of course, is a perfectly uniform display but in practice there will need to be some allowed variability. Some medical-grade displays have implemented individual pixel adjustments to improve the uniformity

19.2.1.3 Contrast Resolution

While the luminance range refers to the difference between the brightest white and the darkest black presented on the display, contrast resolution refers to the differences between individual shades of gray. A display will subdivide the luminance range into 256 shades for an 8-bit display, or 1024 shades for a 10-bit display. The ability for the display to produce unique shades for each of those steps is the contrast resolution. As the overall luminance range increases, the differences between the fixed individual shades of gray can also increase. This may aid in increasing the ability of the human eye to discern a difference between two adjacent pixel values and thereby increase the conspicuity of subtle findings in diagnostic images. However, perception of luminance differences by the human eye is not a linear response. Studies show that the eye is more sensitive to relative changes in luminance at the bright end of a range than in the dark end (Barten, 1992, 1993, 1999). To account for this, luminance response models have been developed that have larger relative changes between steps in the dark shades to optimize the perception of images. One such model is the DICOM Grayscale Standard Display Function (Samei et al., 2005). This fine-grain adjustment of contrast resolution is performed by software with photometric devices and test patterns and is often referred to as calibration of a display's luminance response. The importance of this displays function has been well documented (Langer et al., 2006).

19.2.1.4 Chromaticity

Another consideration for displays is the color of the shade of white. There are many variations of "white," from blue-white

* The term "contrast ratio" is commonly used to describe this characteristic. It has, however, been thoroughly subverted by marketing by the television industry, where it is common to see claims of contrast ratios greater than 1,000,000:1. We choose here to use luminance ratio as a real-world measure of the 100% white value to 0% black value.

to reddish-white, and so on. The tint exhibited by a display is a function of the backlight's color and the panel's light transmission behavior. When two displays are adjacent to each other and have different shades of white, images may appear distractingly different. A colorimeter can be used to measure the color coordinates of a display's white point and adjustments, either manual or via software tools, can be made to the display's color behavior to make displays more closely match.

19.2.2 Choosing Performance Attributes

There are a wide variety of displays and display technologies to choose from in the marketplace. Displays can be intended for general office/home use, for clinical image review, or for making critical diagnostic decisions. General-purpose displays may have lower luminance output, no stabilization of luminance output, poor uniformity, and lack the ability to be calibrated to a luminance response function but will have a lower cost. Clinical review displays typically include the ability to be calibrated to a luminance response function and color point, may include stabilization circuitry, and may exhibit better off-angle viewing due to the higher grade of LC panel technology. Diagnostic grade displays typically exhibit the highest luminance ranges, in-depth calibration offerings, uniformity corrections, and stabilization.

Even within the same model of display, there are many different attributes that can be adjusted such as maximum luminance, minimum luminance, luminance response function, white point color coordinates, "sleep mode," and others. The decisions an organization makes with respect to the display model and its performance attributes should be guided by questions such as the imaging modality to be displayed, the intended usage of the display, and budget.

19.2.2.1 Modality

The imaging modalities to be viewed on the display have an impact on the pixel resolution, luminance range, and contrast resolution. The image matrix size of a Nuclear Medicine study can be as low as 128 × 128 pixel and 256 shades (8-bits). A lower pixel resolution display could easily be used to show these images (unless an organization wanted to show multiple images simultaneously). A CR/DR chest image, however, can be on the order of 3000 × 2500 pixels and mammographic images can be even larger. A higher pixel resolution display should be used to more closely match the display's matrix size with the image matrix size so as to minimize the amount of down sampling necessary to view the entire image and the amount of zooming and panning necessary to view image details in full fidelity. Mammography often carries recommendations for very high luminance ratios and maximum luminance values (ACR, 2007a). Other modalities could certainly benefit from a very high luminance range, but the cost–benefit ratio may be much higher. Modalities which may exhibit very subtle differences in contrast may benefit from a higher luminance range because of the ability to have higher contrast resolution. Those modalities would also benefit from the ability to calibrate a display to optimize the contrast perception.

19.2.2.2 Intended Use

Displays intended for diagnostic interpretation carry the highest requirements, such as adequate luminance range, the ability to be calibrated, luminance stabilization, and uniformity correction. In nonprimary interpretation settings, displays with lesser attributes may be usable at a cost-savings over diagnostic-class displays.

The environment in which a display is used should influence the type of display used. A clinical location with high ambient light conditions warrants a display with a higher luminance output in order to provide an operating luminance range that lies above the ambient light level (Samei et al., 2005). However, it also matters greatly what use the display will be put to. Some modality displays are used simply to verify an image has been acquired and now resides in the PACS system; however, others may be used to make adjustments to the appearance of an image. These latter displays should share some performance characteristics with the diagnostic displays in terms of their luminance ratio and the contrast function chosen. By doing so, the image should be perceived in the same way by both the technologist and the interpreting physician.

19.2.2.3 Cost Implications (Warrantee and Lifetime)

Consideration should be given to the expected lifespan of a display, its luminance warranty, and the total-cost-of-ownership. Some consumer-grade displays offer a very low initial cost. However, they typically operate at a much lower luminance range and the maximum luminance will decay, perhaps rendering them unsuitable for use in some environments in a relatively short period of time. A second expenditure is then incurred because of the need for replacement.

Luminance warranties vary by the class of the display. Consumer-grade displays will typically warrant a display to maintain at least half of their initial luminance in a given time period. Clinical- and medical-grade displays often will warrant a given minimum luminance level for a period of calendar years or backlight hours of operation.

Other costs should be considered such as ongoing maintenance. The ACR specifies a threshold for luminance value for displays used in a diagnostic capacity (ACR, 2007b). If a less costly consumer-grade display is used without luminance stabilization, it will likely be necessary for more frequent checks of that display to ensure that the backlight has not decayed to the point, where the display no longer meets that requirement. A stabilized display should be able to have lower-frequency checks.

19.2.3 Guidelines for Maintenance

The goal with display maintenance is to establish a basic performance level and ensure it is maintained and that deficiencies are identified and corrected. There are numerous attributes of a display that could be measured. Fortunately, the nature of an LCD means that some tests that were standard for Cathode Ray Tube (CRT) displays are no longer necessary. For example, because of its fixed matrix, the pixel resolution of an LCD will not change and there should not be issues such as blur, tearing, or changes to the geometric accuracy or aspect ratio. (Note: Software and display firmware can change the display's behavior for aspect ratio, sometimes "stretching" an image. The end-user organization should investigate these settings and choose the one that maintains correct geometric behavior for medical imaging.) Display attributes that can change over time include the luminance output, the contrast response, display artifacts, and the cleanliness of the display face.

As discussed earlier, the backlight of a display will decay in its output. Stabilization circuitry will work to maintain the output, but toward the end of the display's useful life even that circuitry will fail to maintain the luminance output because the light source is unable to produce the required output. Establish a target luminance value for the 100% white output of the display and use a test pattern to measure it. Compare that value to an established control limit range and take action when the measured value falls outside that range. Corrective actions may include manually adjusting the display's brightness, recalibrating the display, or replacing a display whose backlight is failing.

The contrast response, or how the individual steps of gray are spread out, remains a relatively fixed attribute of the display. It is, however, an attribute that is controlled by the display's firmware and hardware. Situations such as displays being moved and reconnected to a second, uncalibrated input port, or end-users making adjustments to the display modes will cause a display to change its contrast response function. Establish control limits for mid-gray luminance levels and measure the display's values to determine if it is still operating in the expected manner. If the display has the ability to lock-out end-user manipulation of settings, consider using that ability to ensure stable, known behavior.

Display artifacts common to LCDs include dead or stuck subpixels, dust artifacts, and image retention. A stuck or dead subpixel occurs when one of the liquid crystal elements stops responding to the control signals. This may result in a bright point on a black background or a dark point on a bright pattern. Typically, there is no remediation for this occurrence. Evaluate the impact of the artifact on the imaging tasks for that display. If the artifact occurs in the imaging region of the display, it may be more unacceptable than one that occurs in a peripheral, non-imaging area of the display such as that occupied by software graphical user interface (GUI) elements.

Similarly, dust artifacts can present themselves as shadows or specks on the display's face. These can occur when dust or other foreign contaminants get into the optical stack that makes up the LCD. As with dead/stuck subpixels, the impact of the artifact should be evaluated and a decision made on remediation based on the level of impact.

Image retention, sometimes incorrectly referred to as burn-in, is also a concern for the LCD. It is most readily observed on a display that has been displaying a fixed pattern such as a GUI for a commonly used imaging application. If a blank screen is shown, it may be possible to see a faint image of the GUI or other constant elements. Suggestions for remediation vary widely from displaying a constantly varying noise pattern to turning off the display and allowing it to sit for a time roughly equivalent to the time it took to develop the retained image. If the retained image does not introduce visual elements to the imaging area of the display, it may be irrelevant. The impact should be reevaluated, however, if the imaging application changes where images are displayed or if the image may be viewed using the full area of the display.

Displays should be tested for acceptability before being used for imaging. The conformance to established luminance behavior should be evaluated as well as luminance uniformity, color coordinates, scanning for display artifacts, the physical integrity of the display, and basic inventory management data such as serial numbers and hours of operation.

After deployment, a display should be part of a routine quality assurance plan. The frequency of the plan will vary due to factors such as the display's luminance stabilization or the lack thereof. A routine inspection should include a physical cleaning of the face of the display for fingerprints and other contaminants and checks should be made for the display's white luminance, black luminance, and one or more mid-gray levels. These measurements should be compared against established control limits to verify correct performance of the display's luminance behavior. In addition, an artifact scan should be performed and a qualitative assessment of the display's color compared to any paired displays.

19.3 Modality Informatics Acceptance Testing

19.3.1 Motivation

The systems, data, and equipment in the radiology electronic environment are complex and always changing. While Information Technology (IT) and PACS specialists are invaluable to making things work smoothly, it is important that the people assigned ultimate responsibility for image quality and patient safety (be they radiologist, physicist, or technologist) understand what can be expected from the equipment. Are they knowledgeable about the capabilities of their electronic environment through the use of safeguards such as test patient (i.e., John Doe) in testing images throughout the imaging chain? It is up to the users who know what is entered into a scanner to make sure they see the expected output at various downstream viewing platforms and archives.

When new equipment is introduced into a radiology department, it is typically subject to extensive acceptance testing. Physicists assure the equipment is safe for patient use, and along with the radiologist they ensure that it provides diagnostically useful images. Radiologists are largely concerned with the user interface (Langer and Wang, 1996, 1997; Wang and Langer, 1997). But regulatory or vendor-recommended acceptance tests typically do not go farther than the acquisition workstation even though the lifecycle of an image does not end there. This is a concern because while an exam might look of diagnostic quality on the scanner, it does not follow that the image presentation state or associated information (e.g., in the DICOM header) will be preserved beyond that. For example, a correctly placed orientation label on the scanner display may end up flipped in a clinical viewer, harboring the potential for a wrong-sided surgery. The difficulties of a new installation can also create problems outside of its own workflow. For example, a new imaging system trying to send images to the PACS can bring the system to its knees by utilizing all its network resources and crippling the workflow of the department. It is not only imaging modalities that can cause problems, but also postprocessing workstations that can create or modify images in the electronic environment.

The prior examples (and for more, see Table 19.1) points out the need to perform due diligence in the procurement of imaging modalities and post-processing workstations; this includes a recommended pre-purchase evaluation and on installation perform acceptance testing of images throughout their known lifecycle. The image creators or modifiers could be scanners, postprocessing workstations, their related software upgrades, or perhaps the use of previously unused features of a device that has been around for a while. To assess the risks, it is necessary to accurately understand the electronic environment: it may consist of a radiology information system (RIS), modality worklist broker, PACS (with its associated database, short-term storage, and display), a long-term archive, and a clinical viewer. Minimally, as a patient safeguard, the image lifecycle must be understood and examined to protect the fidelity of the image and exam information.

As is the case at our institution, different modalities, vendors, and departments demand different levels of scrutiny with new installs and upgrades. Software patches and upgrades happen regularly, new research protocols are installed, new vendors' equipment comes for a trial, or a new modality type becomes integrated into the institutional PACS. For each of these situations, we seek to balance security, time, and cost in our modality informatics acceptance testing (Walz-Flannigan et al., 2007, 2008, 2009).

19.3.2 Why Do Bad Things Happen When There Is DICOM?

One might hope if all vendors complied with the DICOM standard, it would not matter if a modality was from brand X and the PACS from brand Y; images would pass smoothly from one to the other. Certainly this was the goal of the National Electrical Manufacturers Association (NEMA), one of DICOM's founders (http://medical.nema.org). However, some aspects of the

TABLE 19.1 A Snapshot of Some of the Issues that Have Been Found Since We Have Implemented Modality Informatics Acceptance Testing in Our Practice

Issue	Cause	Resolution	Related Test
PACS for the hospital CR practice was unusable	CT scanner sent in each slice of a large exam as a separate DICOM association	Scanner unplugged, sends to another destination	Evaluate unknown units in a test-system prior to evaluation in the production environment
A CAD program overlaid L/R markers on the wrong side in the clinical viewer but not in PACS or CAD workstation	Bug in vendor code	A patch was put in that burned in L/R image	Check overlays
			Check images throughout viewing chain
Different slices of an MR image were alternately set at different window/levels such when scrolling through the exam it looked like it was "flashing" from readable to dark	PACS reduces images to 12-bits when images are modified and saved. PACS displayed different bit depths with different windows/levels	Only some images in a series were re-saved and hence reduced from 16- to 12-bits.	Pretesting question: if bit-depth is greater than 12 and multislice use alternative workflow
QA issue with names being truncated of CR workstation	If demographic data was entered manually, it would be truncated. If a worklist was used, the names passed through ok	Vendor bug-fix	
Missing exams from postprocessing workstation	A new research package was installed that wiped patient demographics from the images before sending it to archive. This resulted in images not being archived	The program was pulled	Wider education about potential pitfalls so that all new workstation upgrades can be tested for effect throughout imaging chain
PACS was unusable for 1/4 of the clinical practice	Started sending tomosynthesis output from a DR unit. The size of the exams overwhelmed PACS and prevented other exams from being stored	Tomosynthesis exams were sent to an alternative storage destination than PACS	Evaluate new outputs in test-system
Images appeared as totally white in a clinical viewer	Vendor bug in LUT expression	Window/level in DICOM header were rewritten at pre-PACS processor	Check image quality in all viewer types

Note: Before, these issue would have only been uncovered in the production PACS by a radiologist, or worse in a clinical viewer by a clinician.

standard are optional, leading to various interpretations and utilizations of the standard. There is also an ongoing evolution in the DICOM standard (as technology changes) and all products may not be up to date with the most current version. Hence, a new scanner may have options that the older PACS cannot understand. Furthermore, purchasing all elements (from modality to PACS/RIS) from a single, large company does not guarantee smooth integration as modality development and PACS development divisions in large companies are often different and uncoordinated business units. Alas! while the DICOM standard may define a structure, content, and protocol for a particular task, it cannot compel the vendor's compliance to the extent needed to accomplish a typical workflow involving many tasks.

19.3.3 One Step Up: IHE Integration Profiles

Looking to bridge barriers to interoperability, a 1998 collaboration between the Radiological Society of North America (RSNA) and Healthcare Information and Management Systems Society (HIMSS) started an initiative called Integrating the Healthcare Enterprise (IHE) to specify how to use standards like DICOM and HL7 to create processes that can span many parts of a radiology image and information systems (http://www.ihe.net). IHE specifies the overall required behavior of disparate components in the department (like RIS brand X, PACS brand Y, scanner brand Z, and Archive brand Q). These components (called "actors" in IHE terminology) have defined roles in the given workflows (an IHE "integration profile") needed in the department. For example, an integration profile like "Scheduled Workflow" addresses registration, scheduling, acquisition workflow, and image content by bridging standards of HL7 and DICOM. If one desires functionality from a PACS or modality that matches up with an IHE integration profile, it is wise to make sure that it has been validated for that profile. Validation occurs during "Connectathons," where elements of disparate systems are performance tested while connected to perform the desired profile's workflow (http://www.ihe.net/Connectathon/index.cfm). If one plugs a new component into the network, the chance for success improves if that element has the relevant IHE profile validation.

Even with IHE profile validation, however, a site may still have troubles. The output of your scanner might exceed the capacity that your PACS can handle, the series structure output from your modality might not be compatible with the already designed hanging protocol, or the accumulation of compression effects from scanner to PACS to clinical viewer might lead to unwanted artifacts. It may also be the case that a device supports some IHE profiles, but not all: annotations might exist in private DICOM tags not interpretable by downstream viewers, a modality may express gray values using the modern Grayscale Softcopy Presentation State but the PACS may only understand volume of interest look-up-tables (VOI LUTs), resulting in an image rendering that is totally white with the inability to properly window/level. The possibilities for problems are endless.

19.3.4 Prepurchase Evaluation: The Connectivity Review

19.3.4.1 Intended Use

The best way to reduce compatibility problems is to not ask for them. The hottest new technology on the RSNA showroom floor might not turn out to be the latest and greatest if it cannot be integrated into your current electronic environment or department workflow. It may cost hundreds of thousands of dollars to design and provide for the workarounds that make it possible to actually use the equipment.

Pre-purchase preparation should start with extensive details from the proponent as to how the system is expected to be used. This information should be scoured to make sure that the institution's electronic infrastructure is capable of supporting the desired functionality. As an example, perhaps one of the draws for a new fluoroscopy unit is that it outputs a detailed dose report in the form of a DICOM Structured Report. If the PACS or archive cannot handle structured reporting the user would not be able to view or store this information and thus it is of little benefit. Through manuals, DICOM conformance statements or talking with vendors, it sometimes becomes clear that a proponent's hopes for equipment are not in line with what it can provide in a given environment.

19.3.4.2 Vendor Questions

Pre-purchase risk assessment is easier if you can get consistent and thorough information from vendors about their product. To encourage this at our institution, for a prospective purchase the vendor must supply answers to a lengthy connectivity questionnaire. The questions are developed from knowledge of the limitations of our electronic environment and our institutional security policy. The questionnaire is frequently revised for additional aspects of the DICOM standard and to reflect of problems we have experienced. The questionnaire covers the following areas, with sample questions give for each:

i. Security
(e.g., For Microsoft Windows based operating systems, can the device be joined to the institutional Active Directory domain?)
ii. High-Level DICOM and IHE Compliance
(e.g., Does the device support all relevant Integration Profiles currently defined for that modality?)
iii. Modality and Workstations
(e.g., Do image transfers from the modality or the workstation occur in the foreground or in the background?)
iv. Patient and Examination Data Entry
(e.g., When DICOM data elements are changed, does a new Study/Series/Instance UID get created?)
v. Image Data Transmission
(e.g., How are temporarily-unavailable (off-line) DICOM Store destinations managed?)
vi. Other DICOM Services
(e.g., DICOM Performed Procedure—Acquisition service supported by the modality?)

HIMSS and NEMA have also created a checklist to "assist professionals responsible for security-risk assessment in the management of medical device security" [NEMA].

19.3.4.3 Assessment Tools

Perhaps sample DICOM images will have been requested from the vendor by radiologist proponents in order to assess the image quality. These images can also be used to check consistency with the DICOM standard. To facilitate this analysis, a toolkit is available from the DICOM Validation Toolkit (DVTK) project for testing, validating and diagnosing communication protocols, and scenarios in medical environments (http://www.dvtk.com). The DVTK tools include a network analyzer, viewer and validator, modality emulator, RIS emulator, and others.

Alas! there is only so much one can find out from static vendor provided images and documentation. The images were not made at your institution with your patient identifiers, naming conventions, ordering and information systems, or modalities. They were not processed by your PACS. It may be likely that your exact combination of vendor systems and software revisions is a completely unique, and as of yet untested. Despite standards for interoperability, there still exists the potential for problems. This is why when a new system or element arrives there is still work to do to prepare the system for proper clinical use.

19.3.5 Informatics Acceptance Testing

One never knows how something works until its tested, and it is a good idea to try a device before depending on it in clinical practice. The motto of our lab is, "Everything not tested is assumed broken." Every unique system can bring with its own unique problems and successes, but it is wise to develop a standard set of checks to for consistent quality and to facilitate work sharing. Having a game plan is important for a task that can involve engineering, IT specialists, physicists, vendors, radiologists, technologists, and directors. What should be checked and how will depend on each installation, its features and levels of risk. At our institution, we have developed a standard set of checks for different modality types and risk level. Figuring out what to scrutinize is an ongoing process in part attained from experience with our own particular electronic environment and problems we have run into. Listed below are a number of problems that have resulted in different strategies and items for testing

19.3.5.1 System-Wide Risks

In some cases, it is unwise to add a new image creator or modifier directly into a production environment. This may be the case for new or unknown vendor equipment or for a new type of output, where there is concern for how it will interact with the electronic environment. The types of incidents which have inspired this caution include a CT scanner that sent a large series with each image sent as a separate DICOM association. The exam could not finish loading before timing out and re-sending; thus, this occupied network gateways to PACS, blocking other transmission until it was identified. As with other complex software environments, good change management practice means not testing directly in a production system. To protect production systems, risky elements are first introduced into a test electronic environment.

A test electronic environment consists of identical software to the production environment, albeit on a smaller scale and possibly with a different hardware implementation. At our institution, the test system consists of duplicates of our RIS, modality worklist broker, PACS, archive, and clinical viewer. This allows us to test the interaction of a new scanner or software and the electronic environment without jeopardizing the production systems.

Preliminary tests conducted in our test system are focused on the potential for system-wide problems, leaving more detailed tests for the production environment. As an example, here are some things to look for during test-electronic environment testing for a new image acquisition unit, accomplished with a test patient and relevant exam types:

i. Check modality worklisting.
ii. Perform all status changes on PACS/RIS (exam completion, finalization).
iii. Watch notifications between modality/PACS/RIS/Archive to make sure the status changes appropriately.
iv. Look at exam quality in PACS (Does it load?).
v. Look at exam in clinical viewer (Does it load?).
vi. Did the exam archive?

It has been a matter of experience to learn which units should first be evaluated in the test-system and which can start out with evaluation in production systems using test patients.

19.3.5.2 Image or Information Quality Risks

Once a new element has either been deemed safe enough or has passed through the test system without major issues, more thorough testing is warranted in the production environment. Before turning a unit over for clinical use, we want to verify exam details, image quality, and details of correct operation. Listed below are some specific examples of checks for informatics acceptance testing of a new scanner:

i. Modality worklisting: does it work as expected?
   a. Fetch via: last name, clinic number, barcode, manual entry
   b. Confirm demographics
ii. Send images of all different classes from modality to PACS
   a. Send duplicate
   b. Flip/rotate and send
   c. Annotate and send
   d. Reorder and send
iii. Look at the Exam in PACS
   a. Is the image count correct?
   b. Does the exam hang correctly?

   c. Is the exam information correct [time, modality, exam number, description, location, etc.] in worklist and overlay?
   d. Are Orientation Labels correct?
   e. Did annotations pass?
   f. Do window width and level adjustments behave appropriately?
iv. Look at the DICOM header
   a. Are there fields with no entries?
   b. Are values consistent with expectations?
   c. Is the image compressed?
v. Perform QA, reporting, and status changes.
vi. Look at the Exam in Clinical Viewer
   a. Is the image count correct?
   b. Is the exam information correct [time, modality, exam number, description, location, etc.] in worklist and overlay?
   c. Are Orientation Labels correct?
   d. Did annotations pass?
vii. Retrieve from Archive
   a. Look at the residual from subtraction with an image directly from the modality.
   b. What is the file size?

19.3.6 Implementation Notes

A key to introducing a new process like informatics acceptance testing is to educate people about the need for it. At our site, we persuaded people to ask for our help in evaluating new systems prior to going into production so we do not run into problems later. We request users to let us know that a new upgrade will be installed so that we can check on it. This can be politically tricky if a process is seen to slow down an installation without just cause. What helped in our case was several outages whose root causes analysis were traced back to poorly behaved modalities. It may also be important to get the backing of administrators and policies that require connectivity reviews for new purchases or informatics' sign-off before equipment is released for clinical use. It is a difficult balance to find between effort and benefit, but at the end of the day it is only the responsible thing to try to reduce problems especially those which affect patient care and safety.

19.4 Server and Workstation Uptime

19.4.1 General Systems Administration

19.4.1.1 What Is Systems Administration? Who Is the System Administrator?

Modern computer systems, while not necessarily large, are complex often-interconnected entities that require observation and on-going care. Systems, or servers, have a lifecycle that needs to be managed. In a basic sense, the lifecycle of a system can be distilled down to: design, build, configuration, deployment, maintenance, upgrade, and decommissioning.

System Administrators are tasked with managing the lifecycle of systems and components that make up the systems, which can be either a single server or a complex network of interconnecting, multitiered server farms. They also interface with users addressing issues, performing requested tasks, and ensuring the needs of the users are met.

19.4.1.2 Workstation versus Server versus Virtualization

System Administration can fall into both Workstation/Desktop support and Server support. Workstation administration covers the installation and maintenance of computer systems that are directly interfaced by users and fulfill the role of a client in client–server computing. Server administration covers the installation and maintenance of computer systems that fulfill the role of a service provider or server.

Workstations are usually located on user's desks in offices or labs. Server systems are usually in data centers or data closets.

Either or both of these classes of computers can be virtualized. Virtualization is the concept of running one or more guest Operating Systems (OS) images on top of an existing OS image and/or on top of shared or pooled hardware resources (Langer et al., 2010). Examples of virtualization include:

a. Oracle Solaris Zones and Logical Domains (Oracle Corporation, Redwood Shores, CA).
b. IBM Logical Partitions (IBM Corporation, Armonk, NY).
c. Linux KVM (Kernel Virtual Machines) guests.
d. VMware guests (VMware Corporation, Palo Alto, CA).

19.4.1.3 System Lifecycle

Considerations for designing and building a system often arise from the requirements of the intended applications that are installed and served out to clients. Considerations include: memory, Central Processing Unit (CPU) type and speed, internal storage, external storage (i.e., Storage Area Networks), network connectivity and speed, and which OS is installed. The choice of OS is often determined by the application or functionality required.

Operating Systems are tools in a toolbox. The proper one should be selected for function and reliability and specific technical requirements of the application. Budget can also impact the choice of OS. Some OSs are free such as Linux, others require licensing such as Microsoft Windows (Microsoft Corporation, Redmond, WA), and some have costs due to specialized hardware requirements such as IBM AIX (IBM *op cit*). In addition to technical and cost considerations, OS support should be considered when choosing the correct OS. Vendor Operating Systems often have supplied support either included in the cost of the OS license or for a fee. Free OS' may have no single go-to entity for support and support may need to be supplied completely in-house by the Systems Administrators, supplemented with available resources found on the Internet.

The OS to be installed dictates how a system is built. Each OS has its own install method, set of hardware requirements, and patching process. The methods for accomplishing a system build are varied and vendor dependent and beyond the scope of this chapter.

Once a system has been built and released into use, it needs to be maintained and monitored for proper functioning. Maintenance concepts are covered later in this chapter. When a system has reached the end of its effective operating life, either through replacement of functionality or depreciation of hardware, it will be either upgraded to a newer set of hardware or OS, or it will be decommissioned from active service and retired.

19.4.1.4 Environmental Considerations

Computing hardware has environmental requirements for optimal functioning. These requirements include temperature, humidity, particulates, power, and physical space. Exact requirements are dependent upon the type of hardware. Optimal air conditioning is between 68°F and –72°F and between 45% and 55% humidity (Limoncelli and Hogan, 2001). These conditions ensure proper internal cooling of memory, CPU, disk, and power supply components. The most adverse factor to optimal functioning is overheating. If the system becomes too hot, the risk for component failure increases. Efforts should be made to reduce dust and construction particulates from entering systems, where they can coat internal components and create overheating or short-circuiting situations. A filtered air supply is recommended. A controlled and monitored environment such as what is found in purpose-built data centers where power, cooling, and floor space are regulated and organized are ideal.

19.4.1.5 Documentation and Standardization

An often overlooked or poorly implemented practice of System Administration is documentation and change control. A thorough paper trail of all work planned and performed, outages taken, and faults encountered during the lifetime of a server can prove to be invaluable in helping rapid identification and resolution of future issues, to help when interfacing with vendors for support, and to aid in the planning of future upgrades. Documentation of processes and methods allows for the creation of standards and automation, which greatly speed up the implementation and repeatability of administrative duties while minimizing the effort required.

Documentation should be created with emphasis on clarity and understanding so that it is possible for others beside the author to follow the procedures. Documentation should be created in a portable format (i.e., .txt, .html) that can be read by a variety of tools and from a variety of locations such as from an office desktop, or a laptop in the data center. Standards should be clear, ordered, and as simple as possible.

Change management should be thorough (Hiaat and Creasley, 2003). Records of changes made should last as long as the system's lifetime. These records should be available to others, most notably the users of the system so they can be notified of outage or possible impact to their applications. Proposed changes should include:

- Brief description of the work to be done.
- Systems affected.
- Length and type of outage.
- Implementation description.
- Testing procedure to allow for a metric to gauge the success or failure of the change.
- Back out plans and procedures if the change is to be rolled back or fail or if the change cannot be undone, and mitigation and work-around plan to work through failures or unexpected issues.

When possible, changes should be reviewed and approved by someone other than the System Administrator performing the work. This is to increase visibility and notification of the change and to help catch any possible issues or concerns prior to the change.

19.4.1.6 Capacity Planning/Performance Monitoring

Capacity planning and performance monitoring are vital practices that allow for the optimization of system performance to maximize system potential. Maintaining historical summaries of system performance allow for the creation of metrics and benchmarks for the purposes of application tuning and system upgrades. This data can also prove vital in the diagnosis of problems as it helps to provide an over-time view of a system. This can expose conditions such as memory bottlenecks or CPU starvation that may not be visible when a system is reviewed at a specific point in time.

Various OS native and third-party tools exist that capture and log performance data. Some tools exist that can take existing performance data and build capacity models often current system operating trends that can help in the budgeting and preparation of system upgrades. Some tools can take existing data and turn them into visual models for ease of understanding. For example, the Multi-Router Traffic Grapher is a useful graphing tool that can take a variety of input sources and provides output in graphical, exportable format (MRTG http://oss.oetiker.ch/mrtg/). The tools available are OS and vendor dependent and are beyond the scope of this chapter. Consult appropriate OS vendor or online resources for more information.

Performance tuning a system is dependent upon the needs of the application. File servers may need to be tuned to help maximize disk throughput whereas Web servers will need to be tuned to maximize the number of concurrent network connections. Each application may need its own adjustments, which can make running multiple different applications on the same server problematic. In cases where multiple application need to be run on the same server, it may mean system tuning is not possible as any changes to tune for one application could have negative impact on others. These considerations need to be made before tuning a system to optimize performance.

19.4.1.7 Security Considerations

A good model for system security is the concept of the least privilege model, where users are granted access only to functionality required for their operation and maintenance of the application (Deitel, 1990). In addition, modern Operating

Systems provide a lot of features and functionalities that are often not needed for day-to-day operation. While allowing these services to run does not usually impact system performance, it does provide an increase in system vulnerability either in the form of another vector for system access or privilege escalation. Whenever possible, unneeded and unused services should be disabled. Access to the Administrator or Super User account should be limited and monitored. System security patches should be applied on a routine basis and whenever recommended by the OS or hardware vendor. Physical access to the system should be controlled and monitored to prevent unauthorized access to the machine to prevent tampering or accidental disruptions.

When a system is retired from service, care should be made to protect or destroy the data that existed on the host. Data stored on both internal and external storage can be accessed when put into other systems unless properly destroyed. Destruction of data is often based on guidelines set by information security offices of the organization in which the system lives or based off individual documentation. If security documentation does not exist, its best practices to create a document, that is publically available, that outlines user access policies, password changing schedule requirements, and data retention and disposal procedures. This serves two functions: (a) it is often required for audits and (b) it provides guidelines to use as a reference to build a decommissioning document for the disposal of retired systems. Methods of data destruction include overwriting disks with patterns of nonsensical data such as all zeros or all ones, or physical destruction of the media (such as via a shredder).

Systems that exist outside of a firewall, or are located in a DMZ should be what is known as hardened. Hardening a system varies by OS, but includes turning off all nonessential system services and restricting access to super-user and administrator accounts, thoroughly monitoring system logs, user access, and network activity. In addition, most modern Operating Systems include some type of firewall when configured correctly, help limit unwanted access while allowing proper functioning.

Keeping current on OS security patches helps promote and maintain system security. This includes installing patches for features and services that may be installed but disabled.

Identity management covers the activities of creating, authorizing, monitoring, and removing users. Management can be simple, such as allowing a small number of users on to a system via native OS methods, to the more complex such as allowing thousands of users on to a system via an identity management system such as LDAP (Lightweight Directory Access Protocol). The addition and removal of users to a system should be done via documented and controlled processes that maintain a historical record. Users should be only given the access level required to perform their tasks. Escalation of privilege should be controlled and monitored for both security and maintenance purposes. Intentional or unintentional escalation of privileges can introduce security vulnerabilities or system stability concerns. System Administrators should follow some form of privilege escalation (such as using the tool sudo) to work within a least-privileged security model and record super-user access (SUDO, http://www.gratisoft.us/sudo/).

Connection to the system via the network should be via a secured connection method such as secure shell when sensitive data such as usernames and passwords are exchanged (Barrett et al., 2005).

Many resources exist that can be referenced for more of an in-depth approach to general computing security (Garfinkel et al., 1991).

19.4.1.8 System Maintenance

System maintenance is the routine, semiroutine, often mundane activities that keep a system running from day to day. These activities can include backup of system and application data, provisioning of users to a system, archiving and rotation of log files, and the application of OS patches.

Keeping current on maintenance helps to ensure the system remains stable as well as resolving security vulnerabilities. System maintenance activities should be recorded and documented via some form of change management and release process. Ideally implementation of activities such as OS patching and testing would be performed first in a development or test environment. Once verified, with the impact of the changes understood, the maintenance activities would then be released to the active production, or user facing, systems. All changes should have well thought-out recovery processes that are clearly documented before implementation.

The majority of the system's lifetime will be spent in the maintenance phase. This is a cyclical phase that repeats until the system is upgraded or decommissioned. It is composed of elements of the previous sections encompassing security, performance monitoring, documentation, troubleshooting, and the day to day of the system's life.

19.4.2 High Availability

19.4.2.1 High Availability Considerations

High availability (also referred to as HA) is the concept of maximizing application or service availability. It can be accomplished via a variety of methods and tools. High availability can include fault tolerance, which is often referred to at a hardware level where computers are designed to resist catastrophic failure such as CPU failure, memory failures, and so on.

Most commonly, High Availability usually involves the use of multiple servers to provide a downtime resistant environment in the form of a cluster. Servers should be placed geographically such that any event that has the ability to impact one system will minimize or avoid impact to the other members of the HA environment. This can be as simple as placing systems on opposite sides of the room or as complex as locating them in different buildings. Each system that forms the HA environment should have some form of power redundancy and protection. This can be anything from a simple UPS (uninterruptable power supply) unit to a fully redundant building power backup generator system. Network connections to the servers should have some level of redundancy. Ideally the system would connect to different network switches with more than one connection per server so that any single switch failure will not impact the accessibility of environment.

19.4.2.2 Service Level Agreements

The core of every HA environment design should be the service level agreement or SLA. These agreements are meant to establish a minimum expectation of availability for service the environment will provide. Examples of an SLA could be as simple as "the application will not exceed more than 4 h of downtime a year in total." Once the statement of expectations has been established, things like budget, clustering method, environment layout, and placement can be created. Expectations will dictate cost, as the more available a service needs to be, usually the more expensive the underlying infrastructure will be to provide the level established in the expectations of the SLA.

19.4.2.3 Clustering Methods and Models

High availability can be achieved via a variety of methods. Application clustering is where an instance of an application exists on multiple different systems (physical and or virtual), where the processes and mechanics of the application manage high availability. Hardware clustering is where some piece of equipment, such as a context sensitive network switch, monitors service availability and directs workflow around detected failures or degradations to accomplish expected service levels. Another method is via clustering software; this software runs on the various members of the cluster and manages how the applications and services react to various failure or degradation conditions detected. Which method used is highly dependent upon the application or service being provided and the expectations set by the SLAs. The cost of each method in terms of both monetary and administrative overhead varies by the specific technologies chosen, the number of members in the cluster, and the expected availability of the service or application.

The method of clustering is often dictated by the model clustering chosen. In the most basic sense, clustering models can be active/active; where all members of the cluster are active and providing application functionality and in the event of a failure cluster members can be lost with no downtime, and active/passive; where one or more but not all members of the cluster are providing functionality while other members exist in a standby state, ready to take over functionality in the event of a failure or degradation situation and a brief interruption of service can occur as resources and functionality is moved to a standby member. Each model has its own pros and cons. Active/active provides the highest amount of availability, however it is also usually the most expensive, requires specialized clustering software or hardware, and requires the applications that can exist in a state where more than one instance can be active. Active/passive may seem less ideal than active/active but it is often cheaper to implement and manage and almost every application can be placed into this model, as applications are not required to be cluster aware.

19.4.2.4 Failure Testing

The final facet of High Availability is failure testing. Once service levels are written and the cluster environment designed and implemented, failure scenarios should be created and used to test the design to ensure the clustering works as expected and to make sure the functionality is maintained through a variety of impacted situations. HA failure testing should be done periodically throughout the lifetime of the cluster and done anytime changes are made to the cluster, such as new OS patch levels, application or functionality upgrades or changes, and hardware replacement or upgrade. This is to ensure the cluster remains functional and behaves as expected. Without periodic testing, high availability can be broken and not detected until the moment of an actual failure, when critical functionality is lost.

19.4.3 Data Redundancy, Disaster Recovery, and Business Continuity

Data protection is critical, and it is often overlooked. Procedures and processes should be put in place to ensure that data is always available, recoverable if lost, and repairable if damaged or corrupted. Various methods exist as insurances of data viability. Some cover data integrity, some cover data loss.

19.4.3.1 Data Redundancy

Data protection can be accomplished in multiple ways: simply duplicating files to another folder, to multiple off-site copies in different cities on archive grade media. Data, especially production data, should have at least more than one copy. At minimum a backup of the data should be done periodically and placed somewhere where it can be retrieved in the event of catastrophic system failure, user error, or corruption. One of the most common methods by which this is achieved is by placing a copy onto an external media, such as a magnetic tape or external hard drive, which is then kept separately from the system. Frequencies of backups are usually included when service level agreements are established. The process in which the data is backed up should be documented to an extent that it is possible to reliably reproduce and is testable; the same is true for the restoration and recovery process. Both should be tested on a periodic basis to ensure proper functioning of the backup and recovery methodology.

Data can also be protected via a Redundant Array of Independent Disks (RAID Levels—http://en.wikipedia.org/wiki/Standard_RAID_levels). RAID protection comes in various flavors; RAID 0 (which is also called striping) has no redundancy and is often done to improve disk access performance. RAID 1, which is also called mirroring, is for redundancy; an exact copy of one disk is made on another disk, so if one disk were to fail, the data is still available via the mirrored disk. The downside to RAID 1 is 50% of available disk space is consume, if one has two 20 GB disks, one ends up with only 20 GB of useable space—the other 20 GB is used for protection. RAIDs 2–4 are less used methods and will not be covered. RAID 5 is the next most common RAID protection next to RAID 1. RAID 5 spreads data across a system of disks via striping at the block (discrete chunk of disk) level. Each disk consists of a data portion and a parity portion. The parity portion is used for error correction and to rebuild the raid set in the event of a disk failure. RAID 5 is often used when more than two disks are present as it offers similar protection to RAID 1, but allows for greater useable space. For example if you have four 20 GB disks, you can have 60 GB of useable space with only 20 GB lost as data protection.

RAID protection can be done by either hardware or software. Specialized disk controller hardware found in most modern computers can usually provide RAID 0 or RAID 1 protection. Many modern multidisk storage arrays can offer RAID 0, 1, or 5 levels of protection. Many Operating Systems have built-in volume managers that are specialized pieces of software that can provide the same level of protection as hardware-based redundancy. The software method is often used because some form of it is present in modern operating systems and it often does not require specialized hardware yet provide the same level of protection with only a slight impact to system performance.

19.4.3.2 Disaster Recovery and Business Continuity

Disaster Recovery (DR) is the set of methods, procedures, and planning to recover an environment, often in a different location, in the event of a serious catastrophic event such as a natural disaster at the initial site (Gregory and Pothstein, 2008). DR is largely concerned with recovering data; while Business Continuity encompasses DR planning and includes the plans and processes to move both data and applications to the alternate environment and return to normal operations. Both require planning and periodic testing. Both require meticulous documentation and the storage of data off-site or in a form that is recoverable in the event the original is inaccessible or unavailable. Planning and implementation of both encompass all the concepts presented in this section so far from build procedures to data redundancy as well as security considerations. DR plans and implementation can take make forms from single system or application recovery in another room to entire production environment recovery of multiple systems, clusters, and applications as well as supportive infrastructure at special facilities in another geographic part of the world from vaulted, off-site backups. No matter the scope, documentation, planning, and testing are vital to success execution in the event of an actual disaster.

19.4.4 Operating System-Specific Considerations

19.4.4.1 UNIX/Linux Considerations

UNIX is a term that references to a specific Operating System that was created by AT&T and Bell Labs in the 1960s. It is now used more colloquially to generally refer to various flavors of vendor and open source Operating Systems derived from the original AT&T (or University of California Berkeley offshoot) such as Oracle (formerly SUN) Solaris, HP-UX, AIX, OpenBSD, and so on. Linux is a UNIX-like free open source OS that also exists in many flavors. For the purposes of this section, UNIX and Linux will be treated the same, as functionality and considerations for each is similar at a general level and for purposes of clarity, everything covering UNIX and Linux will simply be referred to as UNIX as a Systems Administrator familiar with various flavors of UNIX or Linux can easily pick up nuances of the others with minimal additional training. UNIX was designed to be a portable (meaning hardware independent) OS with flexible multitasking and time-sharing abilities. UNIX runs on a variety of platforms and CPU architectures allowing it to be adapted for specialized uses such as embedded devices, high-speed computational processing, and a variety of business applications.

UNIX has a long history and has evolved over time and its creation coincides with the development of the earliest networks. It is a powerful OS that can scale to high amounts of memory and CPUs and scales well into distributed and grid-type computing environments. UNIX has a variety of uses and features, but making full use requires specialized knowledge with a steep learning curve. Because of this, its popularity and deployment is somewhat limited in the business world compared to Microsoft Windows. UNIX systems most often fulfill the role of servers in a variety of uses such as Web servers, large-scale relational database servers, and compute engines for statistical modeling, due to its ability to scale with hardware to large amounts of CPU and memory. UNIX and Linux are not free from security vulnerabilities and the systems by default are often configured wide open from a security standpoint. The reader is advised to review some security basics (Oram and Viega, 2009; Schneier, 1996). In addition, the SANS (Sysadmin Audit Network Security) Institute posts current threat advisories (SANS Internet Storm Center—http://isc.sans.org/).

19.4.4.2 Microsoft Windows Considerations

Microsoft Windows is the general term applied to the family of Operating Systems developed by the Microsoft Corporation. Windows is the most popular Operating Systems dominating the majority of the desktop and workstation marketplaces. This is due in large part to the relative ease of installation and use, which has a much shallower learning curve than other Operating Systems such as MVS, VMS, or UNIX. This popularity has led to a large library of application created for the OS. This popularity has also led to it being the most exploited OS. Extra care must be taken deploying Windows machines that can ensure OS patch levels are kept current; some form of malware and virus protection need to be installed and configured, and extreme care must be taken if exposing a Windows machine directly to the Internet due to it being the target of choice by malicious parties. From a server perspective, Windows is no different than UNIX in terms of system administration; all of the concepts and considerations presented here equally apply to any collection of computing systems regardless of OS.

References

ACR. 2007a. American College of Radiology Mammo. ACR practice guideline for determinants of image quality in digital mammography. Resolution 35. *Practice Guidelines & Technical Standards.* Reston, VA: American College of Radiology.

ACR. 2007b. American College of Radiology. ACR technical standard for electronic practice of medical imaging. Resolution 13. *Practice Guidelines & Technical Standards.* Reston, VA: American College of Radiology. http://www.acr.org/SecondaryMainMenuCategories/quality_safety/guidelines/med_phys/electronic_practice.aspx

Barrett, D. J., Silverman, R. E., and Byrnes, R. G. 2005. *SSH: The Secure Shell (The Definitive Guide).* Sevastopol, CA: O'Reilly Media.

Barten, P. G. J. 1992. Physical model for the contrast sensitivity of the human eye. *Proc. SPIE*, 57–72, San Diego, CA.

Barten, P. G. J. 1993. Spatio-temporal model for the contrast sensitivity of the human eye and its temporal aspects. *Proc. SPIE*, 1, San Diego, CA.

Barten, P. G. J. 1999. *Contrast Sensitivity of the Human Eye and Its Effects on Image Quality.* Bellingham, WA: SPIE Press.

Bushberg, J. T., Siebert, J. A., Leidholt, E. M., and Boone, J. M. 2002. *The Essential Physics of Medical Imaging.* Philadelphia, PA: Williams & Wilkins.

Deitel, H. M. 1990. *An Introduction to Operating Systems.* Boston, MA: Addison-Wesley.

Garfinkel, S., Spafford, G., and Schwartz, A. 1991. *Practical UNIX and Internet Security.* Boston, MA: O'Reilly Media.

Gregory, P. and Pothstein, P. J. 2008. *IT Disaster Recovery Planning for Dummies.* Hoboken, NJ: Wiley Publishing.

Hiaat, J. and Creasley, T. 2003. *Change Management: The People Side of Change.* Boston, MA: Prosi Publishing.

Langer, S. and Wang, J. 1996. User and system interface issues in the purchase of imaging and information systems. *J. Digit Imaging*, 9, 113–8.

Langer, S. and Wang, J. 1997. An evaluation of ten digital image review workstations. *J. Digit Imaging*, 10, 65–78.

Langer, S., Bartholmai, B., Fetterly, K., Harmsen, S., Ryan, W., Erickson, B., Andriole, K., and Carrino, J. 2004. SCAR R&D Symposium 2003: Comparing the efficacy of 5-MP CRT versus 3-MP LCD in the evaluation of interstitial lung disease. *J. Digit Imaging*, 17, 149–57.

Langer, S., Charboneau, N., and French, T. 2010. DCMTB: A virtual appliance DICOM toolbox. *J. Digit Imaging*, 23, 681–88.

Langer, S., Fetterly, K., Mandrekar, J., Harmsen, S., Bartholmai, B., Patton, C., Bishop, A., and Mccannel, C. 2006. ROC study of four LCD displays under typical medical center lighting conditions. *J. Digit Imaging*, 19, 30–40.

Limoncelli, T. and Hogan, C. 2001 *The Practice of System and Network Administration.* Addison-Wesley.

Oram, A. and Viega, J. 2009. *Beautiful Security.* Sebastopol, CA: O'Reilly and Associates.

Samei, E., Badano, A., Chakraborty, D., Compton, K., Cornelius, C., Corrigan, K., Flynn, M. J. et al. 2005. Assessment of display performance for medical imaging systems. Report of the American Association of Physicists in Medicine (AAPM) Task Group 18.

Schneier, B. 1996. *Applied Cryptography: Protocols, Algorithms and Source Code in C.* New York, NY: John Wiley and Sons.

Walz-Flannigan, A., Charvoneau, C., Williams, K., French, T., Daly, T., and Langer, S. 2009. Using Microsoft SharePoint ofr Informatics Acceptance Testing of New Modalities in PACS. *SIIM Annual Meeting.* Charlotte, NC.

Walz-Flannigan, A., Langer, S., and Fetterly, K. 2007. Commissioning of new modalities for PACS. *Med. Phys.*, 34, 2542.

Walz-Flannigan, A., Williams, K., French, T., Daly, T., Fetterly, K., and Langer, S. 2008. Commissioning of New Modalities in PACS. *SIIM Annual Meeting.* Seattle, WA.

Wang, J. and Langer, S. 1997. A brief review of human perception factors in digital displays for picture archiving and communications systems. *J. Digit Imaging*, 10, 158–68.

20

Teleradiology

Dimitris Karnabatidis
Patras University Hospital

Konstantinos Katsanos
Patras University Hospital

20.1 Introduction

On March 10, 1876, Alexander Graham Bell, while in the process of inventing the telephone, summoned his assistant Mr Thomas Watson for medical assistance through the telephone wire, because he had spilled battery acid on himself. This incident was quoted by J. H. Thrall as the first-ever recorded case of remote emergency call giving birth to telephone-based telemedicine (Thrall, 2007a). Over the next century, step-by-step technological advancements in communications, including the wired telephone, the radiotelephone, global satellite communications, and modern Internet, in conjunction with the revolution in informatics that employed computer-based applications for acquirement, storage, postprocessing, and transmission of patient data, have fuelled an acceleration in various fields of telemedicine. Physicians are already experiencing unprecedented changes in everyday clinical practice, including services like telepathology, teledermatology, telemonitoring of the elderly, or even remote physical examination of patients through robotic interaction (Anvari et al., 2005; Balis, 1997; Della Mea, 1999; Gu et al., 2009).

In general, teleradiology belongs to the more encompassing concept of telemedicine and refers to the delivery of imaging services over a distance (Thrall, 2007a,b). Currently, there is an exponential demand for advanced medical imaging in the developed countries (Caramella et al., 2000; Pechet et al., 2010). Annual diagnostic imaging costs exceed $100 billion in the United States increasing at an unprecedented rate of as much as 20% a year (Pechet et al., 2010). The European Union (EU) eHealth initiative and action plan is the driving force toward integration of healthcare services across the EU and teleradiology represents already the most successful eHealth service available today. Radiology is the single field of medicine that inherently enjoys most of the benefits from advancements in technology and computer science. Developments in teleradiology constitute one of the leading frontiers in telemedicine and are gradually transforming the practice of traditional empirical medicine.

The ongoing growth of the teleradiology market is fostered by several factors including a shortage of radiologists, new technological developments in the field of imaging coupled with the increasing use of advanced radiological modalities, regional consolidation of healthcare services, and heightened patients expectations for timely high-quality examinations (Thrall, 2007a). The authors present a concise overview of modern teleradiology, including basic principles and operational fundamentals, current trends in teleradiology practice, and applicable business models. In addition, the authors elaborate current frameworks regulating privacy and ethical issues, and the major contribution of teleradiology into tele-education and e-learning of the whole radiology community. Finally, professional conflicts and skepticism surrounding commoditization of radiology are discussed, followed by a glimpse into the future.

20.2 Basics of Teleradiology

Teleradiology basically refers to diagnostic image acquisition in one location, transmission of image data over a distance—that is, at a local, regional, national, cross-border international or even intercontinental level, remote reading of the examination for primary diagnostic or secondary consultative purposes, and lastly communication of the finalized report back to the source location and referring physician of interest (Lundberg et al., 2010; Pechet et al., 2010; Ross et al., 2010; Thrall, 2007a). This has largely developed to the so-called "demand-push" model, where the images are pushed to a predefined fixed location and the radiologist is notified when the images become available to his site for further reading.

One of the first and simplest teleradiology applications includes remote home reporting of emergency examinations by on-call radiologists. This eliminates the need of off-the-hour traveling back to the hospital, is primarily time and cost-effective, and can be employed to consolidate referrals from different hospitals (Thrall, 2005a, 2007a). In 1999, a survey of 950 US-based radiology practices found that teleradiology constituted 5–14% of total workflow. Computed tomography studies were the most frequently interpreted exams, at 95% of the cases, followed by ultrasonography at 84%, nuclear medicine at 69%, MRI at 47%, and conventional radiographs at 43% of the cases (Larson et al., 2005). Another national survey of 114 hospital-based imaging departments performing emergency examinations reported a 25% shortage of certified radiologists and 82% nighttime teleradiology coverage (Saketkhoo et al., 2004).

The steady increase of teleradiology for interpretation of on-call exams has led to the formation of the first niche teleradiology market, otherwise referred to as "nighthawking," that is, remote off-hours coverage of emergency imaging studies (Brant-Zawadzki, 2007; Larson et al., 2005; Thrall, 2007b). The main drawback of nighthawking is that unlike an on-site radiologist, the nighthawk rarely has access to the full medical record and previous studies for comparison. In parallel, teleradiology coverage of routine daytime imaging services has also shown significant promises to optimize and expedite workflow, otherwise referred to as "dayhawking." On one hand, a general radiologist can cover for many different locations, where there is not enough work for a solo full-time radiologist, and on the other hand, a subspecialty-trained radiologist can offer his niche knowledge for specialized consultations to different practices, resulting to high-quality reporting coupled with significant corporate cost savings.

Closed-circuit interactive conventional television, initially employed for the early primitive telemedicine applications of the last century (Bird, 1972; Thrall, 2007a), has been long abandoned due to its high installation costs, poor image and spatial resolution, and cumbersome logistics. With the advent of low-cost personal computers and the Internet, telemetry techniques of medical data have made tremendous leaps forward. Nowadays, direct digital image capture and digital archiving and storage capabilities combined with computerized data compression and transmission methods have established the store-and-forward approach as the basis for the establishment and operation of most modern teleradiology links (Thrall, 2005a,b,c). However, everyday practice reveals that teleradiology is exceedingly challenging on an operational level because of lack of integration of image production, storage, management, and transmission systems.

20.3 Current Practice Trends

At the moment, Web-based distribution of radiological images remains the archetype of almost all teleradiological applications (Thrall, 2007a). However, even in the simple setting of off-hours remote reporting from your own home, there is a limited access to basic image manipulation, multiplanar reconstruction, and study navigation tools that typically all PACS workstations offer. The main technical difficulties of teleradiology applications include the general inability of integration and communication between different image management and healthcare information systems of various organizations because of incompatibility issues. For example, differences in patient registration numbers prohibit the unobstructed integration of imaging studies between two different centers, even if PACS systems from the same vendor are employed (Thrall, 2007b). Associated patient privacy and security issues produce even more limitations to remote data access. Radiological studies generally reside in PACS systems that are set-up separately from computerized archives of the rest of the patient's medical history, results of other clinical or laboratory examinations or previous radiological reports. Therefore, additional information must be collected and transmitted separately from the original hospital upon specific request from the interpreting teleradiologist as necessary. This is cost- and time-consuming and adds significantly to the complexity of information exchange before the final diagnostic report is produced.

With a view to address these problems of inter-institutional integration and interactivity, an initiative called Integrating the Healthcare Enterprise (IHE) (for further details see Chapter 6) has been established, endorsed also by the Radiological Society of North America (RSNA). The initiative seeks to enable users and developers of healthcare information technology to achieve interoperability of systems through the precise definition of healthcare tasks, the specification of standards-based communication between systems required to support those tasks and the testing of systems to determine that they conform to the proposed specifications. The work is managed by IHE committees and sponsored by various national and international bodies (Integrating the Healthcare Enterprise International, http://www.ihe.net). Adoption of IHE standardized document/image file formats (i.e., Cross-enterprise Document Sharing for Imaging XDS/XDS-I) between collaborating organizations or even states and small countries is already growing steadily in order to improve data distribution and workflow sharing (Pohjonen, 2010). Significant cost savings may be accomplished by employing "XDS-enabled" solutions that enable file compatibility between RIS/PACS systems from different vendors toward a global worklist. In this way, radiologists serving multiple healthcare facilities can remotely offer virtual reading and reporting across large geographical areas by simply logging in to their local PACS system (Pohjonen, 2010).

Contrast to cumbersome "demand-push" models that offer limited access to additional patient information, a "demand-pull" model has been put-forward. For example, a demand-pull operational teleradiology set-up in Massachusetts General Hospital, Boston, MA, USA permits the interpreting radiologist to remotely collect all the necessary studies onto the PACS monitor and also interactively pull any necessary additional info from the respective RIS system (Thrall, 2007b). Alternatively, in a typical demand-pull model, PACS may be installed off-site at the radiologist's home operating like a wide area network PACS. Historically, PACS/RIS were developed to store and handle radiological patient information on a hospital level, while teleradiology mainly includes remote image access and reporting. Today, merging of PACS/RIS with teleradiology platforms is necessary to combine local digital image archiving with the benefits of teleradiology (Benjamin et al., 2010).

While designing information systems employed in teleradiology, efficiency is one of the key parameters. Efficiency in medicine may be defined as the combination of speed and accuracy in achieving any clinical result, that is, producing and delivering a final diagnostic report to the referring source (Benjamin et al., 2010). Ideally, every organization would like to continuously increase its efficiency, that is, maximize the number of interpretations per unit time per member radiologist without sacrificing diagnostic accuracy (Benjamin et al., 2010). Effective coordination and management of teleradiology links between multiple regional or even international sites calls for an up-scaled PACS system, or alternatively for a Super-PACS that will inherently resolve most of online interactivity and file compatibility issues. Such Super-PACS systems may allow remote linking through a secure Web network of a personal computer with increased functionality of image manipulation, while integrating customized access to patient files through PACS/RIS interconnections (Benjamin et al., 2010). A "SuperPACS" system would simultaneously serve multiple sites merging them into one virtual site, and provide one and the same virtual desktop interface to all cooperating parties to efficiently complete all reports. However, several tedious tasks of functionality and interactivity have to be solved before such applications become widely available.

First of all, the SuperPACS should merge all studies from the originating sites into one virtual global worklist before further distribution for reporting (Benjamin et al., 2010). This global worklist will be the sum of all local worklists. Second, global access to studies and relevant supporting patient data must be granted upon reporting, so that the whole study file can be "locked" during interpretation and "unlocked" soon after delivery of the finalized report back to the originating institution to avoid conflicts between different competing groups (Benjamin et al., 2010). Third, the SuperPACS should provide a desktop client for remote login of "roaming users." This client will supply all image manipulation and reconstruction tools to allow accurate and efficient reporting, for example volumetric postprocessing of thin-section cross-sectional studies, "demand-pull" access to supporting patient info and of course a full array of reporting tools like dictation software, speech recognition, structured formats, and more (Benjamin et al., 2010). Log-in privileges and level of security clearance will depend on the specific profile and professional contract of each "roaming" user. To ensure protection of patient data from unauthorized access according to the Health Insurance Portability and Accountability Act of 1996 (HIPPA), the SuperPACS system must include properly structured access control gates. For example, one individual with postgraduate fellowship training in neuroradiology may be granted access only to "nighthawk" trauma emergency exams and in parallel he may be also assigned to all spine and head-and-neck studies within the SuperPACS system. Further details about the elegance of workflow management and increased reporting productivity offered by SuperPACS systems may be found elsewhere (Benjamin et al., 2010).

Cross-border teleradiology aims to improve workflow, cover on-call services, and reduce costs and patient waiting lists with the exchange of services between countries in different time zones. However, a magnitude of organizational, technical, legal, security, and most importantly linguistic issues may hinder the delivery of cross-border teleradiology services (Ross et al., 2010). Currently, there are only a few commercial agents offering cross-border teleradiology services in the EU with an annual turnaround of approximately 100,000–200,000 examinations compared to more than 500 million exams in the whole of EU. Various medicolegal and financial matters remain to be solved before widespread implementation of teleradiology on a European level (Pattynama, 2010; Ross et al., 2010). In addition, contracts and agreements about licensing, professional accreditation, liability, and reimbursement of services have to be negotiated. Cross-border networks may involve two organizations or a wider area set-up serving multiple institutions (Radiology eMarketplace).

Lessons from two partially EU-funded telemedicine projects—Baltic e-Health and R-Bay (Ross et al., 2010)—including 649 cross-border cases of first- or second-opinion reporting underscore that language is the biggest problem for workflow sharing across borders of different nations or continents. For that reason, special tools for multilingual structured reporting have been developed (Pattynama, 2010; Ross et al., 2010). The idea of structured multilingual reports is to offer automatic translation to multiple foreign languages. For example, a pilot multinational partnership between the Czech Republic, Denmark, Estonia, Finland, Lithuania, and the Netherlands tried to develop a special structured reporting application based on anatomical semantics for plain x-ray examinations (Ross et al., 2010). Today this tool supports four different languages (English, Danish, Estonian, and Lithuanian) covering two anatomical regions of the skeleton, the hip, and the knee. A reporting template and multiple choice pull-down menus are employed by radiologists for assembly of final diagnosis, which is then translated to the language of the referring institution. Translation is based on semantics of findings and anatomy. Quality assurance was achieved with on-site visits and person-to-person familiarization for exchange of ideas and build of trust.

Experience with these two projects showed that irrespective of increased costs because of language barriers, lack of PACS/RIS integration and labor-intensive manual transfer, translation and anonymization of studies and referral letters, significant added value for both the customer side and the provider side was accomplished. The set-up of an international cross-border radiology eMarketplace can achieve mutual benefits by taking care of all security, privacy, payments, and contracting issues. It may open up a whole customer market allowing open competition between reports, pricing, niche subspecialties, 24/7 availability, and response times (Ross et al., 2010). To the end of efficient multilingual interoperability, design and development of robust and accurate semantic translation software and user-friendly structured reporting tools based on anatomy and/or pathology are the fundamentals for development of a successful cross-border teleradiology platform. Of note, structured reporting systems may help reduce missed findings during routine image interpretation.

20.4 Modern Business Models

Patient–physician personal interaction remains the cornerstone of empirical medicine. Given the integral role of imaging in diagnosis and therapy, development of teleradiology has been groundbreaking in terms of medicine practice and new business models. There are now three to four distinct teleradiology business models that could not have been imagined even a few years ago (Benjamin et al., 2010; Thrall, 2009).

The first basic model involves offer of cost-effective teleradiology services to public district hospitals or other healthcare providers. In the United States, up to 50% of current radiology practice is estimated to be outsourced to commercial teleradiology providers (Kaye et al., 2008). Both "nighthawking" and "dayhawking" services may be included in this category (Benjamin et al., 2010; Boland, 2009). The main use of "nighthawking" is for emergency radiological exams, so that resident radiologists can avoid off-hours on-call duties and improve their quality of life (Larson et al., 2005). The primary use of "day-hawking" is to expedite routine workflow and achieve faster turnaround of subspecialty reporting in regional hospitals with shortage of appropriately trained radiologists. The second approach involves delivery of specific subspecialty radiological reports by recruiting radiologists trained in relevant accredited fellowship programs. Fortunately, the majority of new radiologists coming into practice are already subspecialized (Thrall, 2009). In a typical scenario, a private radiology group contracts to offer complete and timely on- and off-hours reading services, including subspecialty reporting, to several sites covering a large geographical region (Brant-Zawadzki, 2007).

The third model, which is probably most widely used in tertiary academic hospitals that incorporate experience and expertise, involves blending of remote with on-site consultative and reporting service (Benjamin et al., 2010). Second-opinion interpretation service, which was initially promoted in the 1980s, is another option taking advantage of the experience and unique subspecialization of academic practice (Jarvis and Stanberry, 2005). The fourth model involves reciprocal teleradiology coverage between two or more radiology facilities on a rotating basis. Such a model represents a coalition of different groups and would effectively diminish the need of off-hour reporting while maintaining a high level of trust between cooperating parties. All sides will eventually enjoy improved quality of work life and economies of scale on running costs and subspecialty training.

20.5 Ethics, Privacy, and Security

Teleradiology continues to be an underdeveloped legal frontier with myriads of medicolegal issues surrounding each and every conceivable business model (for further details see Chapter 22). Professional licensure, both at a national and local state level, is a major filter for quality control and point of accountability for delivered healthcare services (Thrall, 2007a). Professional licensure establishes a threshold for determining the appropriate qualifications to practice medicine and represents states' vested interest in regulating the practice of medicine in the best interest and safeguard of their citizens (Moore et al., http://www.acr.org).

In the United States, the American College of Radiology (ACR) published the *ACR Standard for Teleradiology* as early as 1994 (American College of Radiology, 1994). The ACR standard explicitly stated that all physicians providing official services of diagnostic teleradiology should hold and maintain appropriate licensure at both the initiating and receiving sites and should have staff credentials relevant to the hospital of origin of the reported examinations. In keeping with the above standards, radiologists currently offering teleradiology services within the United States must obtain a medical license for every state from which they receive diagnostic examinations for remote reporting (Thrall, 2007a).

The ACR Task Force on International Teleradiology published a statement on interpretation of radiology images outside the United States (Moore et al., http://www.acr.org). In general, though, radiologists offering imaging services in the United States but living overseas have to comply with the same standards, that is, obtaining licensure and credentials from every state where a cooperating hospital resides. New legislation to regulate teleradiology is already under development to cope with legal challenges on a European level (Pattynama, 2010). The proposed EU Directive on cross-border Healthcare was first accepted by the European parliament in April 2009. With respect to the provision of teleradiology services, the Directive states that "when healthcare is provided in a Member State other than the state where the patient is insured, such healthcare must comply with the legislation, standards and guidelines of the Member State of treatment." In addition, e-Health and other telemedicine services must fully "adhere to the same professional medical quality and safety standards as those in use for nonelectronic healthcare provision" (Pattynama, 2010, p. 27). In all cases, European legislation acts in addition to national health laws.

Reimbursement for various telemedicine services is also a debatable issue between insurance providers. For example, the United States Center for Medicare and Medicaid Services (CMS) considers that a physician may charge for a service only when the physician himself examines the patient or is able to visualize some aspect of the patient's condition without the interposition of a third's person judgment (American Telemedicine Association, http://www.atmeda.org/news/library.htm). Thus, reimbursement of international teleradiology imaging services is justified provided that the person rendering the examination report is the one interpreting and submitting the finalized authenticated report, is licensed in the respective state and credentialed as a member of the medical staff at the institution performing the examination and receiving the report, and the reporting member is also available for further consultation (Moore et al., http://www.acr.org).

Information privacy is a well-acknowledged human right and as such, teleradiology security issues include but are not limited to confidentiality, integrity, and accountability (White, 2004). Privacy and security of patient data handled and exchanged by teleradiology networks must fulfill some of the most stringent requirements. Security and privacy enhancing tools like strong user authentication, data encryption, nonrepudiation services, and audit logs constitute core elements of modern teleradiology applications (Ruotsalainen, 2010; White, 2004). Proposed principles of basic information privacy include free informed consent for the collection of personal data, fair and lawful processing of personal data compatible with the purpose for which they were initially acquired, appropriate technical measures to protect against any unauthorized or unlawful data processing and against any data loss, error, abuse or misuse, and proven accountability of data processing to prevent wrong-doings (Ruotsalainen, 2010). Several international guidelines and ISO standards (ISO/IEC 27799 and ISO/IEC 27001) for protection of personal health data are available (HIPAA documentation, www.hipaaadvisory.com). Of note, the EU Data Protection Directive (95/46/EC) defines the legal environment for protected data transfer between different jurisdictions in case of cross-border teleradiology practice.

In teleradiology, the whole network of data exchange is considered a security-sensitive domain and administrators must assign data controlling and data-processing privileges (Ruotsalainen, 2010). For example, the data controller may be the site where a radiological exam has been performed and the medical images have been acquired and archived, while the data processor is the radiologist consultant who interprets the films and writes the final report. From a privacy and security standpoint, all information exchange between data controllers and data processors may be subject to unauthorized manipulation or unlawful use. Undoubtedly, data de-identification and encryption is the least security measure any over-the-Web teleradiology network has to undertake. Data controllers and data processors must be protected against malignant viruses, worms, and Trojans crawling the Web at all times. Privilege management and access control services ensure every participating group member is readily certified and granted the appropriate level of clearance. Regular audit-logs have to be produced by data controllers to check when and how imported data has been used by data processors. Anonymization should be used in all cases so far it does not influence clinical decision making (Ruotsalainen, 2010).

Briefly, proposed security and privacy services for trusted teleradiology networks include (1) identification and authentication service of all participating users and entities, (2) encryption of transferred information, (3) e-signing of transferred data for integrity, (4) pseudoanonymization and anonymization, (5) audit-logs for accountability, and (6) nonrepudiations services for network safety.

20.6 Tele-Education and E-Learning

Tele-education, peer review, and quality assurance of the whole discipline of radiology practice are also key fields that may benefit from teleradiological networking (Thrall, 2007a). Electronic learning (e-learning) is an affordable user-friendly service that employs the Internet to provide learners with remote access to virtually unlimited knowledge from their home office or computer. E-learning may also combine immediate assessment of student's understanding and peer guidance as necessary (Kumar and Krupniski, 2008). Digitization of radiological images has produced a wealth of teaching files storming the Internet. These online resources include digital image files, lecture Web-casts, online conferences or just interesting everyday cases and are for the whole radiological community. The vast majority of this educational material is usually available free of charge to your laptop or mobile smart phone and is poised to play a major role in continued lifelong training of physicians ranging from student to resident and attending level. The material comes into modular categories usually arranged by anatomy or pathology and may be supported by extensive links into standard textbooks or other Web resources to which someone may refer on real-time, for example while consulting or reporting a difficult case with complex differential diagnostics. However, in spite of its tremendous impact and major cost and time savings, tele-education in radiology misses the invaluable person-to-person mentoring, whose value cannot be overstressed.

On the other hand, every institution must periodically evaluate the quality of its teleradiology services, both from a medical and an operational standpoint. Engagement of different radiology groups into reciprocal quality review may identify sensitive or inadequate areas of their services and establish a positive feed-back of continuous improvement. In addition, continuous investment to new computer technologies is mandatory to further optimize workflow and expedite productivity. Web-based collaborative clinical teleradiology can be considered as a set of quality assurance services that enable expertise sharing and promote health education opportunities (Kumar and Krupniski, 2008). For example, interventional radiology is a field that may benefit the most in terms of tele-education from innovative telemedicine technologies.

Taking advantage of modern computer technology and advances in medical physics and engineering of miniature endovascular catheters and instruments, interventional radiology is an innovative discipline involving the performance of minimally invasive image-guided procedures for diagnosis or therapy with lower morbidity and mortality. By incorporating teleradiology systems, collaboration of remotely located practitioners is feasible with a view to allow real-time scientific consultation during real-world interventional practice and improve patients' outcomes (Kumar and Krupniski, 2008). Synchronous and asynchronous Web-based modules can be used as adjuncts to traditional face-to-face training for supplementation and enrichment of the educational experience (Varga-Atkins and Cooper, 2005).

Interactive communication of voice and image over the Web during execution of interesting cases of interventional radiology is the simplest method of synchronous or real-time learning that may be used to convey the experience of high-volume interventional centers to groups of students, residents, or postgraduate fellows during training symposia and workshops. Remote learners can thus study the performed techniques in real-time with improved outcomes in cognition and understanding. Alternatively, collaborative consultation may be performed between an experienced high-volume center and a less experienced low-volume one for mentoring and effective proctoring of a challenging or novel interventional procedure. In such a scenario, use of high-bandwidth cable or satellite audio and video streaming technologies is of utmost importance to allow unobstructed zero-latency communication between collaborating physicians round the globe. Modern Web-based collaborative applications can also operate in asynchronous mode with storage, transfer, and retrieval of archived digital medical videos and images upon request (Kumar and Krupniski, 2008).

20.7 Risks and Opportunities

The business of teleradiology raises several risks and opportunities for the involved radiology community at the same time. First of all, lack of access to necessary patient information or even patient–physician physical interaction may lead to lower quality or incomplete reporting. At the same time, this reflects a direct risk also to the reputation and professional good-standing of the radiology group that outsources the examination.

The risk of commoditization of radiology services is by far the biggest risk for radiologists outsourcing or remotely offering their interpretations. Radiologists may abolish their professional role as highly regarded consultants while commercial teleradiology companies raid the field of general imaging with night- and day-hawking. As open market rules dictate, demand and supply will drive pricing of teleradiology services like any other commodity. Hence, a substantial part of the income of individuals who do not offer subspecialty consulting services will be put at stake. On the other hand, subspecialty-trained radiologists may benefit from increased referrals through links requesting either primary reporting or second-opinion expertise. Radiology associates in the strongest position to strive in the modern era of teleradiology are those who provide outstanding imaging services with periodic quality assurance, avoid outsourcing while taking on responsibility 24-h-7-day-a-week themselves and combine high-quality up-to-date subspecialty consultation (Thrall, 2007b). Understandably, creative incorporation of high-quality teleradiology services, either on the side of the outsourcing or on the side of the receiving organization, may in parallel improve the financial well-being and quality of work life for all involved individuals. For example, small practices could merge together to offer round-the-clock coverage on a regional level and combine subspecialty services in different fields. Second-opinion readings may further increase quality and thereby satisfaction of patients and referring physicians.

Commoditization of radiology consultative services has already led to the appearance of the so-called eMarketplaces in the EU and North America that serve as online open market platforms for implementation of business-to-business deals between teleradiology providers and healthcare customers. In such integrated platforms, radiology groups can offer their services worldwide and stakeholders can compare prices, fields of imaging and subspecialty services, or even perform auctions to achieve better pricing (Boland, 2009; Thrall, 2009). In an online open market, the opportunities for business deals are indeed endless. Notwithstanding the dark side of the inevitable commoditization of imaging, teleradiology is fostering a major paradigm shift in overall radiology practice with consolidation of services and introduction of corporate strategy in a highly competitive globalized environment. Outcomes may depend on the ratio of remote to on-site covered studies. The higher the fraction of outsourced examinations, the greater the risks of pure commoditization of the radiology discipline.

On the other hand, radiologists act as professionally trained physicians and should therefore be engaged in a variety of value-added activities that are not expendable and cannot be duplicated over Web networks. Such value-added activities are exceptionally important for the public recognition of radiologists and distinguish between the actual "clinical radiology practice" versus "remote computer-based image interpretation" (Patti et al., 2008, p. 1050). All practitioners have the ethical and moral obligation to incorporate proposed value-added activities into their everyday practice in order to escape expendability and enhance their own self-esteem. Several regulatory, educational, and practical issues have to be considered toward a reform of radiology practice in the next decade (Berquist, 2010). Self-conscious radiologists need to engage in lifelong learning and engage in maintenance of their subspecialty certification. They need to proactively resolve quality, quantity, appropriateness, and outcome measures for imaging procedures and develop strategic relationships with other clinical parties (Berquist, 2010). Patient safety and quality of care with appropriate use of imaging when and wherever necessary should be our motto.

20.8 Glimpse into the Future

Teleradiology is destined to become one of the key stakeholders in the development of remote cost-effective healthcare services and the advancement of telemedicine in general. Within the future pervasive healthcare environment, teleradiology will soon represent one of the most successful domains. Dynamic, mobile, and location-independent general or subspecialized teleradiology applications with 24 h/day availability to anyone, anytime, and anywhere under on-the-fly access control scrutiny by autonomous computer systems have been envisioned (Ruotsalainen, 2010). Teleradiology will be also of integral value for the delivery of highly specialized imaging services in case of emergency within the frame of dedicated high-risk manned explorative missions in rough environments far away in the sea, deep down in the oceans, in remote deserted continents like the Antarctica or above all deep in space. The United States National Aeronautics and Space Administration (NASA) have a long expressed interest in telemedicine for remote monitoring of astronauts health, as well as remote delivery of invasive therapy if deemed necessary.

Of interest, NASA has pioneered the first extraterrestrial teleradiology links (Fincke et al., 2005). First of all, the agency tested the ability of a nonexperienced minimally trained crewmember to perform musculoskeletal ultrasonography of the shoulder by following a just-in-time training algorithm and using real-time remote guidance aboard the International Space Station (ISS). The crew used special positioning techniques for subject and operator to facilitate ultrasonography in a microgravity environment and initial probe placement was aided by common anatomic reference points. An extraterrestrial teleradiology link was set up for transmission of real-time ultrasound video of the examined shoulder to experienced sonologists in Telescience Center at Johnson Space Center for remote complete rotator cuff evaluation (Fincke et al., 2005). NASA has also reported performance of focused assessment with sonography for trauma (FAST) aboard the International Space Station from crewmembers with modest training in hardware operation, sonographic techniques, and remotely guided scanning (Sargsyan et al., 2005). To facilitate the real-time telemedical ultrasound examination, identical reference cards showing topologic reference points and hardware controls were available to both the crew member and the ground-based expert who guided the examination. The investigators reported that the anatomic content and fidelity of the ultrasound video were excellent and would allow adequate clinical decision making despite a 2s satellite communication latency for both video and audio (Sargsyan et al., 2005). Similarly, a nonphysician crewmember astronaut recently performed the first ultrasound examination of the genitourinary tract and the retroperitoneum in space under remote ground-based real-time tutoring and guidance by voice commands from experienced, earth-based sonographers stationed in Mission Control, Johnson Space Center, Houston. The crewmember achieved to acquire all of the necessary target ultrasound images that were received and further analyzed at Johnson Space Center in Houston in real time (Jones et al., 2009).

The above projects have provided essential proof-of-evidence that just-in-time training, combined with remote experienced physician guidance, may allow performance of complex medical tasks by nonexperienced personnel in a variety of remote settings, especially under microgravity in the interest of current and future space programs. Taking into consideration, the recent explosive growth of image-guided minimally invasive procedures, that is, interventional radiology that has transformed almost all fields of traditional surgery and internal medicine, teleradiology will also be a key player for the remote delivery and performance of radiologically guided diagnostic and therapeutic procedures with the aid of robotic interaction. In parallel, merging of synchronous and asynchronous e-learning scenarios will gradually lead to the evolution of live, dynamic, and interactive e-learning platforms that will provide collaborators and practitioners with unlimited flow of mentoring and consultation during a procedure in real-time as needed. Widespread use of powerful handheld computers and smartphones along with high-bandwidth telecommunications will probably fuel the next phase of teleradiology with on-demand live subspecialty teleconsultation and on-the-road mobile dynamic multilingual structured reporting. Dissemination of telemedicine will transform the practice of traditional empirical medicine and teleradiology will be spearheading the progress.

References

American College of Radiology. 1994. *ACR Standard for Teleradiology.* Reston, VA: ACR.

American Telemedicine Association. (http://www.atmeda.org/news/library.htm). Medicare reimbursement for telemedicine. Accessed May 2010.

Anvari, M., Broderick, T., Stein, H., Chapman, T., Ghodoussi, M., Birch, D. W. et al. 2005. The impact of latency on surgical precision and task completion during robotic-assisted remote telepresence surgery. *Comput. Aided Surg.*, 10(2), 93–9.

Balis, U. J. 1997. Telemedicine and telepathology. *Clin. Lab. Med.*, 17(2), 245–61.

Benjamin, M., Aradi, Y., and Shreiber, R. 2010. From shared data to sharing workflow: Merging PACS and teleradiology. *Eur. J. Radiol.*, 73(1), 3–9.

Berquist, T. H. 2010. Healthcare reform: Radiology in the next decade. *AJR Am. J. Roentgenol.*, 194(1), 1–2.

Bird, K. T. 1972. Cardiopulmonary frontiers: Quality healthcare via interactive television. *Chest*, 61(3), 204–5.

Boland, G. W. 2009. Teleradiology for auction: The radiologist commoditized and how to prevent it. *J. Am. Coll. Radiol.*, 6(3), 137–8.

Brant-Zawadzki, M. N. 2007. Special focus—Outsourcing after hours radiology: One point of view—Outsourcing night call. *J. Am. Coll. Radiol.*, 4(10), 672–4.

Caramella, D., Reponen, J., Fabbrini, F., and Bartolozzi, C. 2000. Teleradiology in Europe. *Eur. J. Radiol.*, 33(1), 2–7.

Della Mea, V. 1999. Telepathology and other telemedicine fields: Lessons to learn. *Adv. Clin. Pathol.*, 3(4), 107–9.

Fincke, E. M., Padalka, G., Lee, D., van Holsbeeck, M., Sargsyan, A. E., Hamilton, D. R. et al. 2005. Evaluation of shoulder integrity in space: First report of musculoskeletal US on the International Space Station. *Radiology*, 234(2), 319–22.

Gu, J., Wolters, R., and Gustafsson, U. 2009. Temporal matching in endoscopic images for remote-controlled robotic surgery. *Int. J. Telemed. Appl.*, 2009, 627625.

HIPAA documentation. (www.hipaaadvisory.com). Retrieved Accessed June 2010.

Integrating the Healthcare Enterprise International. (http://www.ihe.net). Cross-enterprise Document Sharing for Imaging. Retrieved June 2010.

Jarvis, L. and Stanberry, B. 2005. Teleradiology: Threat or opportunity? *Clin. Radiol.*, 60(8), 840–5.

Jones, J. A., Sargsyan, A. E., Barr, Y. R., Melton, S., Hamilton, D. R., Dulchavsky, S. A. et al. 2009. Diagnostic ultrasound at MACH 20: Retroperitoneal and pelvic imaging in space. *Ultrasound Med. Biol.*, 35(7), 1059–67.

Kaye, A. H., Forman, H. P., Kapoor, R., and Sunshine, J. H. 2008. A survey of radiology practices' use of after-hours radiology services. *J. Am. Coll. Radiol.*, 5(6), 748–58.

Kumar, S. and Krupniski, E. A. (Eds.). 2008. *Teleradiology*. Berlin and Heidelberg: Springer-Verlag.

Larson, D. B., Cypel, Y. S., Forman, H. P., and Sunshine, J. H. 2005. A comprehensive portrait of teleradiology in radiology practices: Results from the American College of Radiology's 1999 Survey. *AJR Am. J. Roentgenol.*, 185(1), 24–35.

Lundberg, N., Wintell, M., and Lindskold, L. 2010. The future progress of teleradiology—An empirical study in Sweden. *Eur. J. Radiol.*, 73(1), 10–19.

Moore, A. V., Allen, B. Jr, Campbell, S. C., Carlson, R. A., Dunnick, N. R., Fletcher, T. B. et al. (http://www.acr.org). Report of the ACR task force on international teleradiology. Charlotte, NC: Charlotte Radiology PA. http:\\www.vanmoore@aol.com. Accessed May 2010.

Patti, J. A., Berlin, J. W., Blumberg, A. L., Bryan, R. N., Gaschen, F., Izzi, B. M. et al. 2008. ACR white paper: The value added that radiologists provide to the healthcare enterprise. *J. Am. Coll. Radiol.*, 5(10), 1041–53.

Pattynama, P. M. 2010. Legal aspects of cross-border teleradiology. *Eur. J. Radiol.*, 73(1), 26–30.

Pechet, T. C., Girard, G., and Walsh, B. 2010. The value teleradiology represents for Europe: A study of lessons learned in the U.S. *Eur. J. Radiol.*, 73(1), 36–9.

Pohjonen, H. 2010. Changing the European healthcare IT procurement market. *Eur. J. Radiol.*, 73(1), 1–2.

Ross, P., Sepper, R., and Pohjonen, H. 2010. Cross-border teleradiology—Experience from two international teleradiology projects. *Eur. J. Radiol.*, 73(1), 20–5.

Ruotsalainen, P. 2010. Privacy and security in teleradiology. *Eur. J. Radiol.*, 73(1), 31–5.

Saketkhoo, D. D., Bhargavan, M., Sunshine, J. H., and Forman, H. P. 2004. Emergency department image interpretation services at private community hospitals. *Radiology*, 231(1), 190–7.

Sargsyan, A. E., Hamilton, D. R., Jones, J. A., Melton, S., Whitson, P. A., Kirkpatrick, A. W., et al. 2005. FAST at MACH 20: Clinical ultrasound aboard the International Space Station. *J. Trauma*, 58(1), 35–9.

Thrall, J. H. 2005a. Reinventing radiology in the digital age. Part II. New directions and new stakeholder value. *Radiology*, 237(1), 15–8.

Thrall, J. H. 2005b. Reinventing radiology in the digital age. Part III. Facilities, work processes, and job responsibilities. *Radiology*, 237(3), 790–3.

Thrall, J. H. 2005c. Reinventing radiology in the digital age: Part I. The all-digital department. *Radiology*, 236(2), 382–5.

Thrall, J. H. 2007a. Teleradiology. Part I. History and clinical applications. *Radiology*, 243(3), 613–7.

Thrall, J. H. 2007b. Teleradiology. Part II. Limitations, risks, and opportunities. *Radiology*, 244(2), 325–8.

Thrall, J. H. 2009. Teleradiology: Two-edged sword or friend of radiology practice? *J. Am. Coll. Radiol.*, 6(2), 73–5.

Varga-Atkins, T. and Cooper, H. 2005. Developing e-learning for interprofessional education. *J. Telemed. Telecare*, 11(Suppl. 1), 102–104.

White, P. 2004. Privacy and security issues in teleradiology. *Semin Ultrasound CT MR*, 25(5), 391–5.

21

Ethics in the Radiology Department

William R. Hendee
Medical College of Wisconsin

21.1 Introduction

The term "Ethics" refers to an agreed-upon set of rules that define acceptable behavior in a society or social environment. There are several subdivisions of Ethics. In this chapter, we are concerned about Applied Ethics. This branch of ethics describes rules of acceptable behavior in an organization, profession, or setting. In particular, we focus on the ethics (i.e., rules of acceptable behavior) appropriate to the profession of radiology.

Often there is confusion between the terms "Ethics" and "Morality." One exhibits moral behavior if he or she adheres to the rules consistent with a particular set of religious beliefs or religion. That is, morality and moral behavior are dictated by religious doctrine and canons. If I am Catholic, for example, then I exhibit morality and am thought of as a moral person if my behavior is consistent with the teachings of Catholicism. One exhibits ethical behavior, on the other hand, if he or she adheres to rules of behavior that are agreed-upon by the society in which one lives and the profession in which one works. Morality is driven by religious beliefs, whereas ethics is driven by secular rules of behavior. In a certain sense, one might say that morality is ethical behavior conditioned by religious beliefs. In this chapter, we focus on ethics and ethical behavior, and do not venture into morality and moral behavior.

There are agreed-upon and understood rules of ethical behavior in all professions, including medicine, and these rules extend to everyone working in a healthcare discipline. Radiology is such a discipline, and we will examine the ethics and rules of ethical behavior appropriate for radiology and a radiology department. These rules, however, are equally applicable to other healthcare disciplines such as medical physics, radiation oncology, and many others.

21.2 Conflicts of Interest

21.2.1 Financial Conflict of Interest

A conflict of interest occurs when an individual has the authority to make a decision for an administrative unit such as a radiology department that could yield benefits to the individual.

Often conflicts of interest are financial, because the decision could yield some amount of financial remuneration to the individual making the decision. A "kickback" or financial payment from a company to the individual making a decision that favors the company is an obvious financial conflict of interest. In fact, such an overt conflict of interest is simply a bribe intended to influence the decision. Offering or accepting such a kickback is strictly unethical. Financial conflicts of interest can be far more subtle, however. For example, having a financial investment (e.g., holding stocks) in a company that benefits from your decision to favor the company with a purchase is a conflict of interest, because the purchase potentially enhances the value of the financial investment. A conflict of interest also exists for an individual if members of the individual's immediate family (parents, spouse, children) have financial interests in the company. In most situations, there is a financial limit to the investment held by an individual or family member, below which the amount is considered trivial and does not constitute a financial conflict of interest. In many situations in healthcare, the financial limit is $10,000. If a person is in a financial conflict of interest with regard to purchase decisions in a department, he or she should request recusal from the decisions so that the conflict of interest may be avoided. A similar conflict of interest arises if an investigator is conducting research on a product from a company in which the investigator has financial investments. In cases of financial conflicts of interest, the conflicts should either be avoided or should be made known to the individual's institution and, in the case of research, whenever results of the research are presented or published. A challenging conflict of interest occurs when one buys or sells stock in a company based on knowledge of an impending product release from a company, or perhaps of a pending investigation of a company's product by the U.S. Food and Drug Administration. An action of this sort may be subject to civil and criminal penalties through the use of "insider" information about the company's products. For more information on financial conflicts of interest, the reader is referred to the literature (Davis and Stark, 2001; U.S. Department of Health and Human Services-NIH, 2010).

21.2.2 Other Forms of Conflict of Interest

Many forms of conflict of interest other than financial may be encountered in the radiology setting. For example, research that yields positive results may lead to a publishable manuscript, whereas negative results are usually not publishable. In an academic setting, an individual's recognition and ascension through academic ranks depend in part on grants and publications. Therefore, one may be tempted to tweak research results so that they are publishable. This temptation reflects a conflict of interest for the individual, and requires diligent attention to ensure that research results are always interpreted objectively and without bias that could favor the individual making the interpretation.

Many of us are in positions where we supervise students, residents, and fellows. These individuals may perform research, teach other students, and provide clinical services, all of which benefit the radiology department. We monitor the progress of these individuals and have an influence on when they are declared to have completed their training. This influence places us in a conflict of interest, because the department (and perhaps us as well if the individuals are conducting research in our laboratories) benefits from the presence of the individuals, whereas the individuals benefit from the completion of their training. In this situation, we must guard against making decisions that benefit the department or ourselves at the detriment of the individuals whom we supervise as students, residents, or fellows.

Most of us work in a competitive environment where funding for our research programs or compensation for our clinical work is based in part on how well we compete with our colleagues. At times, this competition can interfere with collegiality and collaboration in research and clinical practice. When this happens, we are in a conflict of interest because our welfare in terms of research funding or personal compensation is interfering with maintenance of a cooperative atmosphere essential for the efficient working of the department. Whether we are able to put the welfare of the department above our individual well-being is a challenge for each of us in such a conflict of interest.

We usually work in groups to accomplish our clinical work, teaching responsibilities, and research objectives. Each of these groups succeeds in achieving its goals through the cooperative efforts of individuals in the group. Yet each of these individuals has personal expectations and desires that may not always be in agreement with the objectives of the group as a whole. In this situation, the individuals are in a conflict of interest, and must decide how to resolve the conflict so that the efforts of the group can flourish without total subjugation of the rights and desires of the individual.

The descriptions above are just examples of conflicts of interest that we may encounter routinely in a radiology department. They are neither avoidable nor necessarily bad. They are simply the result of the environment in which we work. What is important is to recognize that they exist and to conduct ourselves in a manner so that they are addressed in an ethical and transparent manner.

21.2.3 Disclosure of Conflicts of Interest

When a financial conflict of interest exists, it should be disclosed in certain situations. For example, many journals require disclosure of financial conflicts of interest in papers submitted for possible publication. Often sponsors of scientific, educational, and professional meetings require that speakers disclose financial conflicts of interest. Service on a board of directors or a committee in an organization frequently requires disclosure of financial conflicts of interest. Even within an institution or organization, disclosure of financial conflicts of interest may be required if the conflict amounts to more than a preset amount (e.g., $10,000). Each situation is different, and we should be aware of and abide by disclosure requirements when we encounter them (Association of American Universities, 2010).

21.3 Patient Confidentiality

A healthcare provider is in an especially sensitive position with regard to information about an individual patient. This information is highly private and should be shared only with those individuals who have a need to know because they are caring for the patient. Otherwise the information must remain confidential, and measures must be taken to prevent its inadvertent disclosure to persons who do not need it. There exists a bond of trust between a patient and a healthcare provider (e.g., physician, nurse, medical physicist, technologist), and this trust is violated if confidentiality is compromised by disclosure of patient information beyond the small group of individuals who need it because they are directly involved in the patient's care. Each patient is an autonomous individual who deserves the respect of every person in the healthcare team. This principle is parallel with the first (respect for individuals) of the three Belmont principles discussed in Section 21.8 concerning research with human participants. A violation of confidentiality outside those with a need to know is disrespectful to the patient and a violation of the patient's rights and autonomy as an individual (enotes.com, 2010).

21.3.1 Health Insurance Portability and Accountability Act

The Health Insurance Portability and Accountability Act (HIPAA) (P.L.104–191) was enacted by the U.S. Congress in 1996 (U.S. Department of Health and Human Services, 1996). It was originally sponsored by Sen. Edward Kennedy (D-Mass.) and Sen. Nancy Kassebaum (R-Kan.). The Act is intended to ensure continuity of health insurance coverage for workers and their families when they change or lose their jobs. Title II of HIPAA, known as the Administrative Simplification (AS) provisions, requires the establishment of national standards for electronic healthcare transactions and national identifiers for providers, health insurance plans, and employers. The AS provisions also address the security and privacy of patient data through the Privacy Rule that became effective in 2003 (U.S. Department of Health and Human Services, 2010).

The Health Information Technology for Economic and Clinical Health (HITECH) Act, enacted as part of the American Recovery and Reinvestment Act of 2009, promotes the adoption and meaningful use of health information technology. Subtitle D of the HITECH Act addresses the privacy and security concerns associated with the electronic transmission of health information, in part through several provisions that strengthen the civil and criminal enforcement of HIPAA rules. Section 13410(d) of the HITECH Act establishes:

- Four categories of violations that reflect increasing levels of culpability.
- Four corresponding tiers of penalty amounts that significantly increase the minimum penalty amount for each violation.
- A maximum penalty amount of $1.5 million for all violations of an identical provision.

The HIPAA Privacy Rule regulates the use and disclosure of certain information held by "covered entities" (e.g., healthcare clearinghouses, employer-sponsored health plans, health insurers, and healthcare providers that engage in certain types of transactions). The Rule establishes regulations for the use and disclosure of Protected Health Information (PHI), defined as any information held by a covered entity that concerns the health status, provision of healthcare, or payment for healthcare that can be linked to an individual. This definition is interpreted rather broadly and includes any part of an individual's medical record or payment history. Noncompliance with HIPAA regulations can result in civil and criminal penalties, depending on whether the noncompliance was willful or simply a result of neglect, and on the degree to which patient information was disclosed to those without a need to know (U.S. Department of Health and Human Services, 2010).

21.4 Behavior of Others

21.4.1 Unethical Conduct

In the work environment and in our personal lives, each of us engaged in healthcare is expected to behave in an ethical manner. There are circumstances in which a person's behavior is such an egregious violation of ethics that it must be identified and corrective action taken. This section provides a few examples of egregious behaviors (De Blois, 2007).

A healthcare provider who commits a crime violates the confidence entrusted in him or her by colleagues and patients. Criminal behavior is considered unethical conduct, and the privileges accorded to the individual as a member of the healthcare team should be revoked. These privileges may be restored once the individual has endured the penalties associated with the crime.

An individual who participates in a fraudulent or deceptive practice violates the ethics of the healthcare profession. Such practices could include fraudulent billing for healthcare services, research misconduct (defined in Section 21.6), deceptive advertising, or other activities outside the boundary of ethical behavior. An individual who is guilty of fraudulent or deceptive practice violates the basic ethical tenets of the healthcare profession and should lose the right to be part of the profession.

Mental impairment can occur as a consequence of aging (e.g., dementia), a disease such as a brain tumor, an accident with head injuries, or no known cause. Healthcare providers who are mentally impaired may have lost their ability to care for patients, and should remove themselves from direct responsibility for patients. If this is not done voluntarily, then the institution or department must act to remove the impaired individual from the healthcare setting. It is simply unfair to patients, and puts them at risk, if an impaired individual is allowed to provide care.

Radiology is a dynamic field in which the technologies of imaging are changing and improving at a very rapid pace. It is a challenge for the radiologist, medical physicist, and radiographer to keep abreast of these changes and remain skilled in the use of the technologies for patient care. Initiation in 2002 of Maintenance of Certification for radiologists and medical physicists was a strong move by the American Board of Radiology to encourage radiologists and physicists to continue to be competent practitioners in an era of rapid technological evolution. In spite of this initiative, not all radiologists or medical physicists remain competent to practice. These individuals should discontinue their practice in areas of imaging where they are no longer competent. If withdrawal from practice is not done voluntarily, then the individuals should lose their privileges to practice in areas of imaging where they are no longer competent. It is less traumatic if the department acts to remove the privileges rather than to have the problem addressed at the level of the hospital or its medical board.

Substance abuse can occur in any profession, and medicine is no exception. When an individual engages in alcohol or drug abuse, his or her ability to practice competently and dependably is compromised. When a person is suspected of substance abuse, the department chair should speak frankly with the individual to determine if the suspicions are grounded in fact. If they are, then counseling should be arranged and the person should be removed from contact with patients until the counseling is completed. Recidivism in substance abuse is common and if it happens, should be reported to the hospital medical board and to the licensing board of the state.

Improper sexual behavior is also a type of unethical conduct that must be dealt with by the department or institution. Sexual engagement with a patient is forbidden, and engagement with a subordinate employee is also considered undesirable. Flagrant violation of the sexual mores of society also is considered improper for a professional in a healthcare setting.

21.4.2 Reporting Unethical Conduct

A person who witnesses or suspects unethical conduct by an employee is obligated to report the conduct to an appropriate individual. Usually, the person who should receive the information is the departmental chair or division head. If action is taken to correct the problem, then the reporting responsibility has been fulfilled. A quandary arises when no action is taken, because then the person witnessing or suspecting unethical conduct must decide whether to report it to someone higher in the institutional hierarchy. There is no easy answer to this quandary, but ignoring the problem when patient care could be compromised is not an acceptable answer. If the problem has reached the departmental chair and no action has been taken, the next reporting step should be to the medical board of the hospital or the licensing board of the state. The reporting individual should recognize that there is a risk in proceeding in this manner, because in spite of "whistle-blower protection" laws, the atmosphere of the department toward the individual may become unwelcoming or even hostile. Nevertheless, the worst solution is to do nothing and watch patients be put at risk as a consequence (Yamey and Roach, 2001).

21.5 Peer Review

Peer review is a hallmark of every profession. Each of us benefits from having our work reviewed by professionals who are our peers and who can assess the quality and quantity of our work productivity. There are several conditions that should be met for a peer review process to be useful and credible. The first condition is that peer review must be conducted by one or more individuals who are truly peers of the person under review. These individuals should be competent professionals in the area of work performed by the person under review, and should have a level of experience and maturity that permits them to judge the quality and quantity of the work. They should be independent of the reviewed individual, not be an employee of the institution or organization for which the reviewed individual works, and have no personal or overt professional connections to the individual. A former mentor or research collaborator of the reviewed individual should not be chosen as a peer reviewer, for example, nor should a person who has ties to the reviewed person directly through family or indirectly through marriage.

Peer review should be conducted in a manner that preserves the integrity and respect of the individual under review. The person being reviewed must know that the review is occurring, and should be given ample opportunity to interact with the reviewers and to present all evidence of the quality and quantity of his or her work. When the review is completed and a draft written report is available, the reviewed individual should be permitted to correct any factual errors in the report. However, he or she should not be allowed to alter the findings, conclusions, or recommendations in the written report. The final report with factual errors corrected should be shared among only those persons with a need to know, and a copy of the report should be provided to the reviewed individual.

A person undergoing peer review is in a vulnerable position, and the sensitivity of this position should be recognized and understood by all those engaged in the peer review process. Peer review should be entirely objective, honest, and fair. It is not appropriate to surface personal grudges or hard feelings about the individual during the peer review process. Of course, there is never an appropriate time to engage in such theatrics, either in the workplace or elsewhere. Peer review is intended to help an individual become more productive and work at a higher level of quality, and to benefit the department as a consequence. It is not intended as an opportunity to denigrate an individual, and never should be used in this manner.

21.6 Research Misconduct

There are many ethical issues important to the conduct of biomedical research. Each of the issues could be greatly expanded from what is presented here, and some of them (e.g., research misconduct and research with human participants) are the subjects of books in themselves. This section provides an introduction to the topic of research misconduct; the interested reader is encouraged to pursue the topic in greater depth through study of the literature (Bell, 1992; Goodstein, 2010).

There are certain violations of research ethics that are so egregious that they are categorized separately as research misconduct. These violations are data fabrication, data falsification, and plagiarism. Data can be fabricated, falsified, or plagiarized only by willful action of the person exhibiting unethical behavior. That is, fabrication, falsification, or plagiarism of research data are not done accidentally; they are done intentionally, and penalty for such willful actions should be severe. In general, someone who is guilty of such actions loses credibility as a researcher, often is unable to secure future research funding, and frequently loses his or her position as an academic or industry researcher. These penalties are so severe that one wonders why anyone would commit such ethical violations. Still, they happen from time to time, probably because the perpetrators believe that their violations will go undetected. It is also true that biomedical research is an extremely competitive endeavor, and that acquiring research funding and publishing papers are keys to success as a researcher. A struggling researcher might be tempted to commit research misconduct as a last-ditch effort to salvage his or her research career because he has been unsuccessful in competing for funding or publishing articles in peer-reviewed journals.

Fabrication of data occurs when a researcher makes up data that supports the hypothesis of his or her investigation. This is analogous to "dry-labbing" experiments in an introductory chemistry course, except that the consequences are much more severe. A student who dry-labs a chemistry experiment puts the grade received in the course at risk. A researcher who fabricates data in a research experiment, and then publishes the fabricated data, puts an entire career at risk. Once the fabrication is discovered, the paper containing the data must be withdrawn from publication. If the paper has already been published, an erratum must be published in the same scientific journal declaring that the paper and the data it contains are invalid, and the paper must be expunged from the electronic form of the journal. If the fabricated data were part of a research project funded by an outside agency, the agency must be notified of the fraudulent character of the research conducted with the funds. In many cases such notification ensures that committed but unused research funds will be withdrawn and no future funding will be provided to the investigator. In most cases, an investigator guilty of fabricating data loses his or her position and finds it impossible to continue in a research career.

Falsification of research data is similar to data fabrication, except that data collected in an experiment are altered so that they support the investigator's research hypothesis. That is, the data are not "made up," but the data are changed ("tweaked") in a manner that makes the data fraudulent. The net result is the same—data that the investigator knows are incorrect are submitted for publication and for continuation of research funding. Data fabrication and data falsification are equally unethical, and the penalties are not different between them.

Perhaps the most grievous offense is plagiarism. Plagiarism occurs when an investigator copies the data of another researcher and presents it in a publication as his or her own data. Although not a common experience, most editors of scientific journals have examples where entire sets of data have been lifted from one publication (perhaps a relatively obscure one) and copied into a paper submitted to a different journal. There are even examples where an entire published paper has been copied and submitted to a different journal, with only the authors and the institutions changed to the plagiarizers' names and institutions. Often these abuses are caught because the editor or a reviewer of the article containing the plagiarized material recognizes that the material came from a different already-published article. When one is working in a highly specialized area of research, the community of scholars working in the same area is usually quite small. This community furnishes both the researchers and the reviewers of articles, and it is not uncommon that the same individual serves as reviewer for both the original publication and the article containing the plagiarized information. Sometimes the reviewer of the article with plagiarized material is an author of the earlier article from which the material was stolen.

Of course, plagiarizing published research data is a violation of copyright, but the offense is more grievous than that. Plagiarism simply is thievery, and the penalty should be severe. Plagiarism is usually dealt with at the level of the editors of scientific journals, and these individuals have no authority to invoke a penalty other than to barricade the journal against future articles from the plagiarizing individuals. Some journal editors notify the institutions employing the individuals who have committed plagiarism, but the threat of legal action against the journal is a disincentive for such notification.

21.7 Ownership of Research Data

In discussions of research data, the question sometimes arises of who owns the data. In academics, research is usually supported by funds from an outside source such as a federal agency (e.g., the National Institutes of Health or the National Science Foundation), a foundation (e.g., the American Heart Association or the Kaufman Foundation), or a commercial vendor (e.g., General Electric or Varian). The funds are made available to the researcher through a contract between the funder and the researcher's institution. Under the contract, the institution is responsible for the judicious use of research funds and for fulfillment of the terms of the contract, including delivery of any data, products or services called for in the contract. That is, the institution is the responsible party in the contract, not the researcher. Further, the institution is committed to ensure that the research is conducted in a scientific manner and that the data generated by the research are true and accurate.

Because of the obligation of the institution to the contractor supporting the research, many institutions claim ownership of any and all data generated by the research. Only in this manner can they assure access if a claim is levied in the future about the credibility of the research or the accuracy of the data. When the institution claims data ownership, it usually leaves the data in the hands of the researcher with the understanding that the researcher is the caretaker, not the owner, of the data.

If the researcher leaves the institution, usually he or she may take the data with the provision that it must be maintained in a secure and unaltered manner, and that the institution has the right of future access to the data upon request.

The issue of data ownership is challenging, and researchers should be informed of the policies of their institution or organization at all times, and certainly before moving to a new employer or before any data are destroyed, since the data are not their own to move or to destroy without permission.

21.8 Research with Human Participants

Research with humans as experimental subjects is a particularly sensitive activity laden with many rules and regulations. In the United States, the foundation for research with human participants is the Belmont Report and its three governing principles: (1) Respect for individuals; (2) Beneficence; and (3) Justice. The principles of the Belmont Report are derived from the Declaration of Helsinki.

The Declaration of Helsinki was developed in 1964 by the World Medical Association (WMA) as a set of ethical principles for experimentation with human subjects. It is widely regarded as the cornerstone document for countries developing guidelines for human subjects research (World Medical Association, 2000). The Declaration of Helsinki is not a legally binding instrument in international law, but instead draws its authority from the degree to which it influences national or regional legislation and regulations (Human and Fluss, 2001).

The Belmont Report (National Commission for the Protection of Human Subjects of Biomedical and Behavioral Research, 1979) was created in 1979 by the former United States Department of Health, Education, and Welfare (now renamed the United Stated Department of Health and Human Services). The title of the Belmont Report is "Ethical Principles and Guidelines for the Protection of Human Subjects of Research." The report, an especially important historical document in the field of medical ethics, derived its name from the Belmont Conference Center where the document was drafted. The conference center is in Elkridge, Maryland, 10 m south of Baltimore.

From the Belmont Report, the three fundamental ethical principles for all research in which human participants (subjects) are involved can be summarized as

1. Respect for Individuals: protecting the autonomy of all people and treating them with courtesy and respect and requiring informed consent of the persons before they participate in research.
2. Beneficence: maximizing the benefits of the research project to the participants and/or to society while minimizing the risks to the research subjects.
3. Justice: ensuring whenever possible that reasonable, nonexploitive, and well-considered procedures are administered fairly to all potential participants, and that costs and benefits are distributed fairly to *potential* research participants without regard to race, creed, or gender.

The Belmont Report is an essential foundational reference for institutional review boards (IRBs) that review research proposals at the institutional level when the research involves human participants. Every IRB must include representatives of the public as well as experts in various areas of research involving human subjects. IRB approval is required for all research involving human participants that is funded by any agency of the Department of Health and Human Services (DHHS). Most institutions follow the General Rule, which states that all research involving human participants must obtain IRB approval, and not just that funded by agencies of the DHHS (Office of Human Research Protections, 2009).

Persons who participate as human subjects in research must understand that their participation is entirely voluntary and that no penalties will occur if they choose not to participate. They must be fully informed of the reasons for the research and of the benefits and risks of their participation, and they must sign a form indicating their Informed Consent (Berg et al., 2001) to participate in the study. For certain individuals who are unable to grant Informed Consent (e.g., children and mentally impaired individuals), a guardian may sign for the individuals. In most cases, prisoners, students, and institutional employees are not considered appropriate participants in human research because they may feel coerced into participation because of their subordinate status to an investigator.

Research involving human subjects that is funded by agencies of the DHHS is overseen by the DHHS's Office of Human Research Protections (OHRP). Rules of the OHRP are based on the Belmont Principles, and violations are subject to fines, suspension of DHHS funding of the research project and the investigator, or, in extreme cases, withdrawal of all DHHS funding from the institution.

Research involving human subjects is also overseen by the U.S. Food and Drug Administration, another agency of the DHHS, when the research involves the use of experimental drugs in a clinical trial. This research usually is performed under an FDA-approved application termed an Investigational New Drug application, and is monitored by the FDA to ensure that the research (which often involves multiple institutions) is being conducted according to the approved protocol.

Many institutions conducting research with human participants have sought and been awarded accreditation by the Association for the Accreditation of Human Research Protection Programs Inc. (AAHRPP) (Association for the Accreditation of Human Research Protection Programs, Inc. 2010). Accreditation involves both an extensive institutional self-study of its research program involving human subjects, and a site visit by expert volunteers who review the institutional IRB process for approving and monitoring research involving human participants. All individuals in an institution receiving NIH funds, including investigators, nurses, and other personnel who are involved in human subjects research, are expected to have completed a course on the protection of the rights and welfare of persons participating as subjects in research.

21.9 Research with Animals

Animals have been used in research for more than two centuries, and over this period many discussions and disagreements have arisen concerning the ethics of research with animals. The controversy over the use of animals in research is at times more heated than that over research with human participants, because animals have no voice in whether or not they will be used as research subjects. Some who favor the use of animals in research argue that humans have dominion over all other creatures on earth, and cite passages in the Bible and other religious teachings as justification for their belief. Others believe that all life is sacred and humans have no right to exploit other species in research or in other ways (e.g., hunting, wearing animal skins, or eating animals). There is a wide spectrum of opinions between these two extremes, and almost everyone has an opinion on the subject that he or she feels strongly about.

The use of animals in research is justified on the basis that they are good surrogates for humans, and that many scientific and technological advances have resulted from animal research that would not have occurred if research animals had not been available. When animals are used in this fashion, they must be treated as humanely as possible consistent with the goals of the research. Proposals for the use of animals in research must have the approval of an institutional committee referred to as the animal care committee or, often, the Institutional Animal Care and Use Committee (IACUC). Approval is granted only if the committee finds that the research is meritorious, the use of animals is necessary, the number of animals to be employed is consistent with the goals of the study, and the animals will be treated as humanely as possible. Once the IACUC has approved a study, it is responsible for periodically monitoring the study to ensure that the study's objectives are being pursued as proposed, and that the animals are treated with compassion and care.

Policies for ethical experimentation with animals have been established worldwide. Examples of these policies are those developed by the International Council on Laboratory Animal Science (ICLAS, 2010); Council for International Organizations of Medical Sciences (CIOMS, 2010); Institute for Laboratory Animal Research—National Research Council (ILAR-NRC, 1991); Scientists Center for Animal Welfare (SCAW, 2010); and the Canadian Council on Animal Care (CCAC-CCPA, 2010).

The Animal Welfare Act was passed by the U.S. Congress in 1966 (USDA—APHIS, 2010). The current version of this act requires that alternatives to the use of animals be considered if painful procedures are proposed, the use of proper anesthetics and analgesics if painful procedures are performed, and requires that institutions have an IACUC and an attending veterinarian with laboratory animal experience. The act includes specific rules for the exercise of dogs and the establishment of an environmental enrichment program to promote the psychological well-being of nonhuman primates. The Act specifically excludes birds, rats, and mice bred exclusively for use in research. The Animal Welfare Act is enforced by the United States Department of Agriculture.

The Association for Assessment and Accreditation of Laboratory Animal Care International (AAALAC) is a private, nonprofit organization that promotes the humane treatment of animals in science through programs of voluntary accreditation and assessment (AAALAC, 2010). AAALAC was started in 1965 by veterinarians and researchers concerned with the humane treatment of animals in research. More than 700 organizations worldwide have received AAALAC accreditation. Along with meeting all applicable local and national regulations, AAALAC-accredited institutions must also demonstrate that they are achieving the standards outlined in the Guide for the Care and Use of Laboratory Animals written by the National Research Council of the U.S. National Academy of Sciences (National Research Council, 1996).

A number of organizations have attempted to prevent or impede the use of animals in research. These organizations are referred to collectively as animal rights organizations, and include the Animal Defense League (ADL), Animal Liberation Leagues (ALL), Animal Liberation Front (ALF), National Anti-Vivisection Society (NAVS), People for the Ethical Treatment of Animals (PETA), Physicians Committee for Responsible Medicine (PCRM), and many others. Some of these groups have been responsible for entering institutional laboratories, freeing research animals, and damaging property. In a few cases, investigators using animals in research have been harassed and threatened by members of more militant animal rights groups. (DeGrazia, 2002) (Sunstein and Nussbaum, 2004)

The use of animals in research (and in medical education) has become an issue of concern to the general public, and an institution in which animals are used must ensure that the animals are handled in the most humane way possible consistent with the objectives of the research. The IACUC must be diligent in its processes of approving and monitoring animal research protocols, and the public affairs and security offices of the institution must be prepared for demonstrations and even possible invasions of the premises by animal rights groups.

21.10 Vendor-Sponsored Research

A substantial fraction of the research conducted in radiology is sponsored by commercial organizations such as manufacturers and vendors of imaging equipment, devices, and supplies used in radiology procedures. This research support is important to the advance of medical imaging, but is accompanied by several issues that should be recognized and addressed before support is accepted. Some of the more important issues are outlined in this section.

The acceptance of research funds from a vendor of equipment or supplies should be independent of decisions by the radiology department to purchase equipment or supplies from the vendor. If this condition is not met, the research funds could be perceived as (or could actually be) a bribe to influence the purchase decisions of the department. Meeting this condition is tricky, because certain forms of research may require the use of the vendor's equipment, supplies or devices. In this situation, the researchers and the researcher's institution must be careful to ensure that the research to be conducted is legitimate and well-designed to produce useful results, and that the researcher is free to conduct the research and publish the findings without any interference from the vendor. The research contract should establish milestones of progress in the research, and progress reports should be submitted periodically to demonstrate that the milestones have been achieved or, if not, the reasons why progress has been less than anticipated.

Research sponsored by a commercial entity should never restrict the rights of the researcher to publish research findings. A researcher and his or her institution have an obligation to share the results of research with other investigators through the processes of scientific presentation and publication, and they must not be prevented from fulfilling that obligation. Sometimes a research contract will allow the vendor a period of time to review manuscripts before publication to determine if there is intellectual property that should be protected. The time for review should be limited to 30 or so days, after which the vendor must either release the article for publication or request a longer time (2–3 months at most) to file a patent application. The conditions of research support from industry are relevant to investigators in academic institutions, hospitals, and other independent agencies. They do not pertain to researchers who work for manufacturers and vendors. In these cases, the employer has every right to inhibit publication of research results if the results are considered proprietary to the employer, because the researcher is an employee of the company.

Another issue that often arises in negotiations over research contracts with industry is the matter of indemnification. If a researcher is using a drug or device experimentally and a research participant is injured because of adverse effects of the drug or failure of the device, then who is responsible—the manufacturer of the drug or device, or the investigator who is using it? Similarly, an operator error or improper use of a drug or device may harm a research participant even though the drug or device is functioning normally—who is responsible in this situation? These questions give rise to the question of indemnification—who shoulders the blame and covers the cost of an accident or injury that happens with an experimental drug or device? Usually this question is resolved in the following way: If the accident or injury results from drug or device failure, the manufacturer is responsible, whereas if the untoward event occurs because of operator error or carelessness, the researcher and his or her institution are liable. This separation of responsibilities should be written in clear and unarguable terms into the research agreement between the manufacturer or vendor and the researcher's institution. Even if this is done, however, an unfortunate occurrence in which someone is injured can result in finger-pointing between the vendor and the researcher in terms of who is to blame and responsible. Some research institutions have taken a hard line and refuse to accept indemnification clauses of any sort, irrespective of who is at fault if an accident occurs. This posture can compromise the ability of the institution to successfully negotiate research contracts with industry.

Guidelines have been prepared by agencies and institutions to assist institutions and commercial organizations in negotiating research contracts that serve the mutual interests of both parties. Examples of these guidelines are provided in the references (University of California, 1989; Pharmaceutical Research and Manufacturers of America, 2009).

Support of research from a company is viewed as a conflict of interest for the investigator when he or she presents or publishes the results of the research, if the results describe properties of a drug or device furnished by the company. There is nothing inherently wrong about a conflict of interest, and as described earlier, it is virtually impossible to escape having a conflict of interest of some form when conducting research. Still, a conflict of interest must be declared whenever results are reported or published from research supported by a commercial entity.

An issue that has received considerable adverse publicity recently is the occurrence of ghost-written scientific articles. In these situations, an investigator or clinician allows an article to be written about a company's product and then agrees to submit the article for publication as an author, often the first and sometimes the only author. The investigator or clinician may or may not have conducted research or acquired clinical experience with the product. The person who writes the article is referred to as a "ghost-writer" or "ghost," and often is not acknowledged in the published article. Usually the ghost-writer is an employee of the company, or someone the company has hired to write the article. Ghost-written articles are frowned upon in the scientific community, and most editors will categorically reject a paper if it is apparent that it has been written by someone other than the author(s).

In the past, commercial organizations have sponsored meetings, often in lavish surroundings, and hosted dinners and entertainment evenings for researchers and clinicians. These activities are rapidly fading into oblivion as the research and clinical communities have become sensitized to the financial waste and poor image they project both to themselves and to members of the public. Healthcare has entered a new era of public accountability, and some of the abuses of the past have ended, certainly none too soon.

21.11 Publication Ethics

Researchers have an obligation to publish the results of their research so that other investigators can benefit from knowledge of the results. Increasingly, there is an interest in making research findings available to the public when public funds have been used to support the research through grants from federal agencies. For example, articles supported by agencies of the DHHS must be posted on PubMed, a public-access website within 12 months of their appearance in a scientific journal (PubMed, 2010).

The privilege of serving as an author on a scientific publication carries certain obligations that may include (1) Contributing to the conception and design of the research reported in the publication; (2) Participating in the acquisition of data during the research project; (3) Analyzing and interpreting the data collected during the research study; (4) Composing and revising the manuscript prior to its submission for publication; and (5) Approving the manuscript for submission to a journal. Most journals require involvement in at least three of these five activities for each person included as an author on a scientific publication. An author must also be involved in revising and resubmitting a manuscript after it has been returned following scientific review.

Articles have various types of authors. The first author of an article is the individual who has been most involved in the research project and the preparation of the scientific article. Often the first author is a graduate student or postdoctoral fellow who has spent virtually full-time on the project and the paper preparation. The first author on an article is recognized as the individual most responsible for the article. The article may have a corresponding author who may or may not be the first author. Often the corresponding author is an individual in whose laboratory the research was conducted. He or she will be accessible to correspondence about the article even if the first author has moved to another institution, which is often the case if the first author is a student or fellow. The article may also have a senior author who is preeminent in the field of research and deserves recognition because he or she has mentored the first author and satisfies the conditions for authorship outlined above. It sometimes happens that the first author, corresponding author, and senior author are the same individual, but more frequently they are different individuals.

In the past, it was considered a courtesy to list a senior individual as an author even though he or she contributed little if anything to an article. This might be done as a courtesy to the individual, or perhaps the other authors hoped to curry favor with the senior individual. Often the person added without justification as an author on an article was a departmental chair or other individual with authority over the other authors. Today courtesy authorship (sometimes called guest authorship) is frowned upon, and some journals require that each author sign a statement that he or she has contributed in a meaningful way to the research and the article. Other journals make it clear that courtesy authorship is undesirable in their instructions to authors.

A hallmark of scientific publication is the peer review process. In this process, an article submitted to a scientific journal is reviewed by individuals who are knowledgeable and have experience in the area of research described in the article. Serving as a reviewer of an article submitted to a journal is a major responsibility. A reviewer must have demonstrated expertise in the area of research. He or she must be an experienced researcher who has maturity in the discipline and is known to be timely and thorough in fulfilling the responsibilities of a reviewer. He or she must be totally objective and fair in the review, and not be influenced by his or her own interests in the field of research. The point of the review process is to help the authors by guiding them to ways to strengthen the article and improve its clarity and comprehensiveness. Sarcasm or cavalier comments have no place in a scientific review. Finally, the reviewer must keep the article and the research results it reports confidential, and should not exploit knowledge of the results to his or her own advantage. Announcing or using the results of others obtained by serving as a reviewer is an abridgement of the obligations assumed when one agrees to serve as a reviewer.

21.12 Education Ethics

21.12.1 Obligations of the Teacher

Serving as an educator in molding the minds of students is one of the highest privileges that can be acquired, but it is a privilege accompanied by great obligations. Teachers have many responsibilities beyond the simple one of guiding students to knowledge and enlightenment. Teachers are a model for students, and all of us who teach can point to examples of effective teaching that have influenced us over the course of our education. Perhaps the greatest compliment a teacher can receive is to learn that his or her former students view the teacher as a model educator and someone to emulate in their own educational endeavors.

The first obligation of a teacher is to respect students as individuals and to help them acquire the information and skills that ultimately will lead to knowledge, and hopefully, wisdom. This simply cannot be done unless the students feel that the teacher respects them and that they are in a physically and mentally safe environment. The environment must provide an atmosphere of intellectual freedom so that students can express their thoughts and questions without fear of ridicule. The environment must be nondiscriminatory, and all students must be treated equally irrespective of ethnicity, creed, color, political persuasion, or other distinguishing characteristics. The teacher must instill trust in the students and respect the confidentiality of sensitive information disclosed by a student in confidence. A teacher must evaluate each student fairly and make the reasons for an evaluation transparent to the student. The teacher must respect the work of the students and also display respect for other teachers and for researchers in the teacher's discipline. The goal of each student should be to complete the course of study and ultimately graduate from the educational program, and the teacher should help the student achieve that goal in all possible ways. Consensual relationships between students and teachers are discouraged in academic institutions, and harassment of students has no place whatsoever in such institutions or anywhere else (Plaut 2010; University of Pennsylvania, 2010).

21.12.2 Obligations of the Student

Students in an academic institution also have a number of obligations. They are expected to abide by the policies of the institution, and violation of these policies may constitute grounds for dismissal. Students should be informed of policies that affect them, and they should conduct themselves accordingly. Furthermore, students must respect and care for the property of the institution, and should be held responsible for purposeful damage to property. Students are expected to be honest and conduct themselves with integrity, and to be present when their attendance is expected unless there is a clear reason for their absence. Students should respect their teachers and each other, and should always acknowledge the contributions of others in papers and articles that they author. They should not reveal information given to them in confidence. Students are in an institution to learn, and they should take advantage of every opportunity to gain knowledge and insight into the discipline that they are studying.

21.13 Employment Ethics

Employees are expected to function with honesty and integrity at all times, and to be responsive to the mission and goals of the institution or organization. Employees should be responsive to the priorities of the employer, and not attempt to foster discontent among other employees. When negotiating an employment agreement, the potential employee should be straightforward about his or her expectations in the job and should fully understand the expectations of the employer for the person in the job. It is better to decline an offer of employment if there is a major disparity between these two sets of expectations. It is unfair to the employer and to other potential employees for an individual to deliberate over an offer for an extended period of time without making a decision. It is equally unfair to everyone if a person offered a job uses the offer to better the situation with his or her current employer. Although this is a common tactic, it is fundamentally unethical because the individual is interviewing for the new position under false pretenses.

When interviewing for a position or a promotion, an individual should recognize that others may be affected by the action. These persons should be treated with courtesy and respect. If someone will be replaced by the person assuming the position, the interviewee should not accept the position until he is sure that the incumbent knows he or she will be replaced. When vacating a position, an employee should leave the workplace prepared for the person who will take his or her place, and should make every effort to complete tasks and not leave work undone for the replacement.

21.14 Attributes of Professionals and Organizations

Individuals working in a professional manner exhibit a number of laudatory characteristics that are identified below. When any of these characteristics is compromised, the individual falls short of what it means to be a professional. Among the characteristics of a professional are

Honesty: In all transactions and interactions, a professional is honest and acts with dignity and trustworthiness.

Forthrightness: In all transactions and interactions, a professional is forthright and does not keep features of the transactions or interactions hidden.

Diligence: A professional is diligent in his or her work and responsibilities, and can be trusted to provide an intelligent, thorough, and accurate work product.

Respect for Persons: A professional has respect for others and respects the dignity and autonomy of all individuals with whom he or she works.

Service to Others: A professional recognizes that the greatest privilege one can attain is to be of service to others, and works toward that objective in his or her employment and life.

Patient Welfare: Individuals serving as professionals in healthcare acknowledge through their words and actions that the care and safety of patients has an extremely high priority in their work.

Organizations that serve professionals also should exhibit professional attributes that help their members function in a professional manner. Among these attributes are

Standards of Conduct: An organization for professionals should establish standards of conduct and "rules of the road" to guide its members in professional attitudes and behaviors.

Compliance: Organizations with established Standards of Conduct should be able to enforce compliance with the standards, and should void the membership of individuals who are unable to meet the standards.

Encourage Conduct: An organization for professionals should encourage ethical conduct and skilled practice of its members that exceed the minimum standards of conduct of the organization, and should provide opportunities for members to achieve a higher level of performance.

Education: An organization for professionals should provide educational opportunities for its members to hone their ethical behavior, knowledge base, and practice skills.

Service: Organizations for professionals should accentuate the service aspects of the profession to society and, if in healthcare, to patients.

Public: Every professional organization should recognize its accountability to the public for the profession it serves, and should pursue ways to demonstrate such public accountability.

21.15 Conclusions

Everyone involved in healthcare, including individuals engaged in research, education, and administration as well as those actually providing care to patients, is subject to rules of behavior that are derived from ethical principles. These principles encompass a variety of human activities, ranging from personal relationships to publication of research results. The ethical principles are straightforward, and many of them simply reflect the Golden Rule: Do unto others as you would have them do unto you. Nevertheless, each person should understand the principles and applications of ethics in specialties of healthcare, including radiology, to ensure that unintentional violations of ethics do not occur and that willful violations of ethics are detected and corrected quickly.

References

Association for Assessment and Accreditation of Laboratory Animal Care International. AAALAC International web site. Accessed April 16, 2010. www.aaalac.org

Association for the Accreditation of Human Research Protection Programs, Inc. AAHRPP—Web Site. Accessed April 16, 2010. www.aahrpp.org/www.aspx.

Association of American Universities. AAU Report on Individual and Institutional Conflict of Interest. Accessed April 12, 2010. http://www.aau.edu/research/COI.01.pdf

Bell, R. 1992. *Impure Science: Fraud, Compromise and Political Influence in Scientific Research*. New York, NY: Wiley.

Berg, J. W., Appelbaum, P. S., Parker, L. S., and Lidz, C. W. 2001. *Informed Consent: Legal Theory and Clinical Practice*. Oxford: Oxford University Press.

Canadian Council on Animal Care. CCAC web site. Accessed May 21, 2010. www.ccac.ca/.

Council for International Organizations of Medical Sciences Publications. CIOMS web site. Accessed May 21, 2010. www.cioms.ch/frame_publications.htm

Davis, M. and Stark, A. 2001. *Conflict of Interest in the Professions*. Oxford: Oxford University Press.

De Blois, J. 2007. *A Primer for Healthcare Ethics: Essays for a Pluralistic Society*. Washington, DC: Georgetown University Press.

DeGrazia, D. 2002. *Animal Rights: A Very Short Introduction*. Oxford: Oxford University Press.

Enotes.com. *Doctor–Patient Confidentiality: Encyclopedia of Everyday Law*. Accessed April 12, 2010. http://www.enotes.com/everyday-law-encyclopedia/doctor-patient-confidentiality

Goodstein, D. 2010. *On Fact and Fraud: Cautionary Tales from the Front Lines of Science*. Princeton, NJ: Princeton University Press.

Human, D. and Fluss, S. S. 2001. The World Medical Association's Declaration of Helsinki: Historical and Contemporary Perspectives, 5th draft. World Medical Association.

Institute for Laboratory Animal Research—National Research Council. 1991. *Companion Guide to Infectious Diseases of Mice and Rats*. Washington, DC: National Academy Press. www.nap.edu/openbook.php?record_id=1540

International Council on Laboratory Animal Science. ICLAS web site. Accessed May 20, 2010. www.iclas.org

National Commission for the Protection of Human Subjects of Biomedical and Behavioral Research. The Belmont Report. pdf. *Ethical Principles and Guidelines for the Protection of Human Subjects of Research*. April 18, 1979. http://or.org/pdf/BelmontReport.pdf

National Research Council. 1996. *Guide for the Care and use of Laboratory Animals*. Washington, DC: National Academy Press.

Office of Human Research Protections (OHRP, HHS). Code of Federal Regulations, Title 45–Public Welfare. Part 46—Protection of Human Subjects 2009. www.hhs.gov/ohrp/humansubjects/guidance/45cfr46.html

Pharmaceutical Research and Manufacturers of America (PHRMA): *Principles on Conduct of Clinical Trials and Communication of Clinical Trial Results*. April 2009. http://www.phrma.org/sites/default/files/105/042009_clinical_trial_principles_final.pdf

Plaut, M. *Advocate Web Teacher–Student Relationships—Boundary Issues in Student–Teacher Relationships*. Accessed April 12, 2010. http://www.advocateweb.org/home.php?page_id=79

PubMed. U.S. National Library of Medicine. Accessed April 12, 2010. www.ncbi.nlm.nih.gov/pubmed

Scientists Center for Animal Welfare. SCAW web site. Accessed May 20, 2010. www.scaw.com

Sunstein, C. R. and Nussbaum, M. C. 2004. *Animal Rights: Current Debates and New Directions*. Oxford: Oxford University Press.

U.S. Department of Agriculture, Animal and Plant Health Inspection Service. USDA-APHIS-Animal Welfare. Accessed April 16, 2010. www.aphis.usda.gov/animal_welfare/publications_and_reports.shtml

U.S. Department of Health and Human Services, Office of Extramural Research, National Institutes of Health. Conflict of Interest (COI) Page. Accessed April 12, 2010. http://grants.nih.gov/grants/policy/coi/

U.S. Department of Health and Human Services. P.L. 104–191. August 21, 1996. aspe.hhs.gov/admnsimp/pl104191.htm

U.S. Department of Health and Human Services. The Privacy Rule. Accessed April 12, 2010. www.hhs.gov/ocr/privacy/hipaa/administrative/privacyrule/index.html

U.S. Department of Health and Human Services. *Summary of the HIPAA Privacy Rule*. Accessed April 12, 2010. www.hhs.gov/ocr/privacy/hipaa/understanding/summary/index.html

University of California. *Guidelines on University–Industry Relations*. June 1989. www.ucop.edu/raohome/cgmemos/89-20.html

University of Pennsylvania. *Sexual Harassment Handbook: A Guide for Faculty and Staff*. Accessed April 12, 2010. www.upenn.edu/affirm-action/shisnot.html

World Medical Association. 2000. Declaration of Helsinki: Ethical principles for medical research involving human subjects. *JAMA*, 284: 3043–5.

Yamey, G., Roach, J. 2001. Witnessing unethical conduct: The effects. *West J. Med.*, 174, 355–56.

VI

Medical Informatics beyond the Radiology Department

22

Imaging Informatics beyond Radiology

Konstantinos Katsanos
Patras University Hospital

Dimitris Karnabatidis
Patras University Hospital

George C. Kagadis
University of Patras

George C. Sakellaropoulos
University of Patras

George C. Nikiforidis
University of Patras

22.1 Introduction to Imaging Informatics

There is a rapid pace of discovery in modern imaging technologies and physicians are challenged by the explosion of bioimaging information available in everyday clinical practice. Imaging informatics has developed to a distinct subspecialty of imaging sciences, mainly radiology, and endeavors to improve the efficiency, accuracy, and reliability of imaging services within the medical enterprise (Branstetter, 2007a,b; Harrington, 2006a,b). Basically, informatics studies the flow and distribution of information between different places or individuals and analyzes the various means that are employed for data processing and manipulation (Branstetter, 2007a; Hersh, 2009; Kagadis et al., 2008). Van Bemmel has conceptualized six different structural levels of computer applications in health and medicine. This model essentially applies across all disciplines involving informatics in everyday patient care. Level 1 is *Communication and Telematics* and deals with data acquisition and transfer. Level 2 is *Storage and Retrieval* and deals with information storage in a data repository. Level 3 is *Processing* and involves a computer program to process the data as necessary. Level 4 is *Diagnosis* and combines data processing with human interaction to assist with decision making. Level 5 is *Treatment* and finally level 6 is *Research* (van Bemmel, 1984).

Most of the excitement and groundbreaking applications of informatics in medicine have nourished within the field of radiology that has been traditionally the major stakeholder of imaging. Of note, a distinct training pathway in Imaging Informatics with an accredited subspecialty fellowship program is now available in the United States, as well as the internationally recognized Society of Imaging Informatics in Medicine (SIIM) that supports a formal structured curriculum. Most commonly used tools of informatics can be classified into digital imaging, telehealth practice, Web-based tools, electronic data mining, and digital teaching resources (Gabril and Yousef, 2010).

Although at the moment radiologists are admittedly the major beneficiaries of imaging informatics (Branstetter, 2007a), it is becoming clear that almost all personnel involved in the delivery of medical services get an increasing exposure to the field and stand to gain the most from its understanding and wider adoption. The ramifications of imaging informatics applications such as Picture Archiving and Communication Systems (PACS), Radiology Information Systems, Health Information Services (HIS), Electronic Medical Records, and Computer Aided Detection and Diagnosis, which are already felt strongly throughout the radiology departments (Nikiforidis et al., 2006), are also drifting to other image-based disciplines including but not limited to pathology, dermatology, and ophthalmology. Dermatologists obtain color pictures of suspicious pigmented skin lesions, endoscopists take color pictures of the gut, pathologists capture high-resolution images of histological slides, cardiologists remotely check electrocardiograms, and ophthalmologists take digital pictures of the retina for clinical and educational purposes.

In addition, step-by-step technological advancements in communications, including the wired telephone, the radiotelephone, global satellite communications, and modern Internet, in conjunction with the revolution in informatics that employed computer-based applications for acquirement, storage, postprocessing, and transmission of patient data have fuelled an acceleration in various fields of telemedicine. Olsen et al. (2000) have identified three structural levels of computer applications necessary for the provision of telemedicine and telecare services. Level I involves the plain transfer of alphanumeric information in addition to teleconferecing. Level II involves the transfer of image-based data, including image-postprocessing capabilities, to a site distant to where the images were produced. Level III applies to the performance of real-time operations using images from remote workstations. The first two levels are already widely implemented in projects of home-based telecare and the teleradiology enterprise. The latter envisions the robotic delivery of healthcare services to remote locations with extended patient–clinician interaction under augmented reality concepts.

Delivery of telemedical services is nowadays a strong impetus for further advancement of imaging informatics and is destined to become a key player with decisive impact to future developments and other pioneering applications. The authors present a concise overview of the framework and modern applications of imaging informatics outside the radiology department, including basic principles and operational fundamentals, current trends in clinical services, and specific applications in the context of telepathology and virtual microscopy, teledermatology, telecardiology, and teleophthalmology. In addition, the authors elaborate on security regulations and ethical issues, and touch upon the major contribution of imaging informatics into tele-education and electronic learning (e-learning) in general. Finally, the role of imaging informatics in research and the development of telemedicine and e-Health in general is analyzed followed by the authors' perspectives about the future.

22.2 Imaging Informatics beyond Radiology

Physicians and other scientists aiming to get involved into imaging informatics either as end users of some software application or within their managing and administrative duties should first gain a wide knowledge of PACS and its basic components and supporting software and next familiarize themselves with more advanced applications designed to optimize workflow, improve ergonomics, perform quality assurance and control, and augment person–machine interactivity such as speech recognition software and image processing and computer-aided-diagnosis elements. Today, if the radiologist is the brain of the radiology department, then the electronic environment is the lifeblood, facilitating the transmission, storage, presentation, processing, and finally interpretation of medical images. Radiologists, technologists, physicists, and other clinicians interface with PACS through the hardware and software responsible for image processing, manipulation, and display. Image and data format standards such as the DICOM file type (Digital Imaging and Communications in medicine Standard) have greatly contributed to the success of PACS, and today most institutional vendors are highly invested in the DICOM image format.

Based on the experience and know-how of radiology informaticists, the same principles apply for all other disciplines outside radiology aiming to capture, record, transfer, analyze, and process digital images of a particular anatomy or organ function (Krupinski, 2010). A typical PACS comprises output ports from the various image production modalities, computer networks to transfer, communicate and distribute the information, multiple workstation monitors to enable image viewing, reading and processing, storage devices large enough to archive and back up the vast amounts of generated data, and finally a maintenance team of appropriately trained computer personnel (Branstetter, 2007a,b).

Workstation medical displays are an essential component with technology shifting from traditional cathode-ray tube displays to newer liquid crystal display (LCD) technologies. Albeit their higher cost, LCD displays have a flat-panel design, thereby requiring significantly less desk space and improving versatility when designing workspace ergonomics. Medical displays must have adequate brightness (luminance), much higher than average home monitors, and sufficiently high resolution, in the order of a few megapixels (the reader is referred to Chapter 20 for further information). However, specific display characteristics will vary depending on the end application the display is designed for. For example, 1–2 megapixel resolution may be enough for display of typical 512×512 cross-sectional computed tomography (CT) and magnetic resonance images, whereas up to 5–10 megapixels are required for standard digital mammographic interpretation (Branstetter, 2007a,b). However, in case of more demanding applications such as virtual computerized microscopy where automated slide scanners acquire digital datasets with image resolution in the range of gigapixels, extreme display settings will be necessary to accommodate the images and allow sufficiently high virtual magnification (Weinstein et al., 2001, 2009). Standards for digital radiological imaging have strictly confined image display into displays with gray-scale settings because of their superior brightness and less blurriness (Hirschorn et al., 2002). However, display of nonradiological medical images, such as digital pathology slidesets or dermoscopic or ophthalmoscopic microphotographs, will routinely require a color monitor, since color is an indispensable feature of such images (Weinstein et al., 2009). Recommendations for settings' adjustment and assessment of display performance by the American Association of Physicists in Medicine task force may serve as an excellent consulting document during design and implementation of imaging informatics systems for nonradiological purposes also.

Wired or wireless computer networks between servers and clients are necessary to facilitate communication and exchange of data. The applicable standards for topological design of modern computer networks for medical purposes, whether local (local area networks [LAN]) or wide (wide area networks [WAN]), would be generally the same regardless of the origin of the data and/or the modalities applied for image acquisition—that is, a CT machine or a virtual slide scanner or a high-resolution digital camera. However, depending on the volume of generated data, certain elements of the computer networks may be adapted as deemed necessary, like for example the network bandwidth or the image compression algorithms or the security protocols so as to better serve the end-user application. One network model that is of special interest not only to diagnostic radiologists but also to pathologists and other image interpretation specialties is the virtual private network (VPN). The VPN will ordinarily allow users, who are logged in computers that reside outside the hospital LAN, to have access to sensitive medical imaging information with the same level of user authentication and data security (Branstetter, 2007a). VPNs are therefore an elegant low-cost solution for the realization of remote image interpretation services.

An important feature of modern computer networking, which is of special interest to telemetrical and telemedical applications, is the incorporation of synchronous or asynchronous communication tools. Synchronous communication refers to simultaneous real-time contact between two or more parties logged in the network. Asynchronous communication refers to creation and transmission of data that is received and processed by another party at a later time. The classical store-and-forward approach used in most of modern telemedical applications is an example of asynchronous communication. On the other hand, the practice of remotely controlled robotic telepathology requires implementation of synchronous communication protocols. In general, asynchronous communication is cheaper, less technically demanding to implement, and enhances efficiency and productivity of cooperating parties by improving flexibility in use of time and resources (Branstetter, 2007a). Several other features of computer networks, including but not limited to network bandwidth, network security, online and offline means of data storage, and image compression algorithms, are of utmost importance in maximizing data transfer capacity and workflow efficacy, while maintaining patient confidentiality (Branstetter, 2007a). Archiving and safe storage of adequate backup copies of sensitive medical imaging data using protocols with redundant computer links; for example, a redundant array of inexpensive spinning disks or Redundant Array of Independent Disks (RAID); is crucial in the extreme case of a disastrous event.

Apart from PACS, which nowadays serve as the supporting skeleton of every image distribution network, several additional tools of informatics have been developed to further expedite the efficiency and accuracy of the image interpretation process and improve the working conditions of the end user. For example, development of speech recognition software engines allows radiologists to dictate in a natural cadence with decreased transcription costs and above all significantly less turnaround time between execution and reporting of the examinations (Houston and Rupp, 2000; Langer, 2002a,b). Although modern speech recognition software have their core built around English radiology-specific language, they may be easily adapted to other terminology more suitable for pathologists, dermatologists, or ophthalmologists, or even cross-border multilingual telemedical platforms.

Image processing techniques such as multiplanar reformatting, three-dimensional volume rendering, and virtual fly-through endoscopy are now commonplace in radiology. Similar techniques may be applied for the manipulation of a stack of thin pathology sections to evaluate three-dimensional features of whole-mount histological analysis (Vlad et al., 2010). Computer-aided detection to enhance image interpretation has been studied extensively in screening mammography, as well as in the detection of lung nodules, colonic polyps, and pulmonari emboli (Branstetter, 2007a; Engelke et al., 2010). In principle, computer-aided interpretation refers to implementation of sophisticated image processing and segmentation algorithms in order to detect tissue features, patterns, or geometries correlating to specific pathologies. However, the significant rate of false-positive results and the challenges in computer algorithm development and training has limited to certain radiological applications like mammography. A similar framework applies for computer-aided analysis of digital pathology images. Image analysis algorithms have been developed for data mining of the extremely rich digital histopathology datasets (Daskalakis et al., 2007a,b; Georgiadis et al., 2008; Kostopoulos et al., 2007, 2008; Nikiforidis et al., 2008). Over the last decade, manifold learning and nonlinear dimensionality reduction schemes have emerged as popular and powerful machine learning tools to cope with pattern recognition problems during interpretation of digitized histopathology images (Madabhushi et al., 2010).

Finally, more advanced informatics tools and technologies can improve tasks such as workflow analysis and optimization, quality assurance and control, and workplace ergonomics (Branstetter, 2007b). Workflow analysis tools deconstruct complex everyday tasks into a list of simpler tasks, calculate the time necessary to perform each of them, and classify them according to a hierarchy of importance in order to eliminate interruptions, remove bottlenecks, remove redundancies, and incorporate out-of-band tasks. Reading workstations may be equipped with simple applications with plain alert messages or more sophisticated visual tools, such as traffic light metaphors where different colors may correspond to different volumes of outstanding examinations or to different states of reports (e.g., unsigned, signed, verified, etc.), have been proposed to improve report turnaround times and patient care (Morgan et al., 2008). Quality assurance refers to the assessment of the quality of the process by which the image interpretation services are performed and delivered, whereas quality control evaluates the quality of the deliverable product, that is, the produced reports themselves. Imaging informatics provides all the necessary tools for rapid collection, analysis, and monitoring of all quality control data. Finally, workspace ergonomics are crucial for improving comfort and efficiency of physicians working in front of tiresome computer workstations. For example, adjustment of lighting architecture and layout modifications of the reading "cockpit" room may help prevent repetitive stress injuries and reduce eyestrain (Thrall, 2005a,b,c).

22.3 Digital Pathology, Virtual Microscopy, and Whole Slide Imaging

With radiology leading the way in the digital era, the discipline of pathology soon followed the transition to the digital environment (Krupinski, 2010). Digital pathology has several advantages over traditional optical microscopy, but also significant challenges remain to be solved. A generation of pathology trainees who prefer digital pathology imaging over hands-on light microscopy is already under the way (Weinstein et al., 2009). The reader must be introduced to relatively new terms such as virtual "whole slide imaging," "virtual microscopy," and "virtual slide telepathology" (Kayser et al., 2006). Whole slide imaging includes first the creation of digital images of the entire area of a glass histopathology or cytopathology slide using virtual slide scanners and second the viewing of such large digital image slides using virtual slide viewers (Yagi and Gilbertson, 2005, 2008). Virtual microscopy refers to technologies that emulate traditional hands-on light microscopy by manipulating digitized virtual slide images on a computer screen using appropriate microscope emulator software or a so-called virtual slide viewer (Dee and Meyerholz, 2007; Kumar et al., 2004; Weinstein et al., 2001, 2009). Operators of virtual slide viewers have the ability to pan and zoom, explore the slide, and make independent observations more easily compared to digital photomicrographs (Dee and Meyerholz, 2007). For example, the operator may select a range of digital magnifications, that is, ×0.6 to ×40, instead of interchanging objective lenses in the actual microscope. After selecting the magnification of choice, he can further pan the image to zoom in other areas of the slide just by simple drag-and-drop-like mouse functions. The end result is the substitution of cumbersome light microscopy with a highly versatile and user-friendly graphics-user-interface emulator. Virtual slide telepathology refers to the process of distal transfer and remote interpretation of the digital virtual slides by a specialist located in a distant site (Kayser, 2000; Weinstein et al., 1987, 2009). Nowadays, digital pathology and virtual whole slide imaging encompasses a series of operational steps such as tissue sample fixation and embedding, sectioning of the specimen, glass mounting and appropriate staining in the local histological laboratory, whole slide image acquisition and digitization using a virtual slide scanner, image postprocessing and storage, image compression and distribution across computer networks, and finally reconstitution and display of the digital image file on the interpreting pathologist's computer display (Kayser et al., 2006). In terms of informatics, the whole process is almost identical to image acquisition, storage, and transmission across PACS within radiology departments.

Digital pathology standards have to be developed for widespread dissemination of telepathology networking and manufacturing of interoperable pathology workstations (Kayser et al., 2006, 2008). In radiology, the workstation is often referred to as the "cockpit" with a "digital dashboard" and the reading room as the "control room." Several groups have tried to incorporate available digital tools into the pathologist's "digital dashboard" (Krupinski, 2010). The "digital dashboard" is a portal to biomedical information, including medical images and relevant patient data. Like in radiology, appropriate use of informatics is crucial for the communication, management, and monitoring of data flow. Pathologists' workstations may be designed to seemlessly integrate with other information systems such as HIS and summarize key workflow metrics to augment productivity and facilitate informed decision making by the responsible physicians (Krupinski, 2010). Today, several workstations designs or custom in-house products are available for the average practicing pathologist. However, standardized metrics and guidelines are missing and pathology informaticists have yet to overcome the challenge of designing the optimal pathology "cockpit" (Weinstein et al., 2001). A major difference between radiology and pathology workstations is the necessity of color displays in the latter. This results in increased complexity and difficulty in calibrating them properly (Fetterly et al., 2008). In addition, the staining technique employed by the local laboratory technicians is another important factor that has to be dealt with under the rubric of color management. Using spectral analysis and proper calibration, accurate reproduction of color settings and standardization of stains for digital imaging may be achieved (Kayser et al., 2006). The *Macbeth Color Checkers* are routinely applied for color adjustment and standardization of workstation displays used in digital pathology. The *Macbeth Color Checker* is used for precise color balance in digital photography and consists of an array of 24 printed color squares that include spectral simulations of various color shades (Weinstein et al., 2009). In contrast to radiology being confined to gray-scale imaging, color fidelity is of paramount importance in digital imaging of other specialties, including pathology and dermatology. Informaticists must be familiar with color adjustments of the digital camera or slide scanner when acquiring the raw digital image data, of the computer monitor used for display of the image file and of the ambient lighting settings in the room where image interpretation takes place. In addition, light illumination optics, light filters, glass stability, camera focus, and compression algorithms during digital slide scanning may permanently influence the quality of the final slide images (Kayser et al., 2006; Weinstein et al., 2009).

Development of suitable standards about color resolution, file compression, and other video monitor specifications would greatly enhance interoperability between pathology workstations of different commercial vendors, facilitate file sharing and exchange, and accelerate workflow. The DICOM file format has been developed by the American College of Radiology and the National Electrical Manufactures Association in the United States and is the universally accepted standard for image sharing in a clinical context (Kayser et al., 2006). However, the DICOM standard has been designed for gray-scale radiological images and is not optimal for color-coded clinical images. Ever since 2005, efforts have been made by pathologists working groups to develop DICOM file extensions suitable for digital pathology and telepathology imaging in general. The approach taken involves dividing the digital raw data of its whole slide image into a lot of smaller "tiles." The location of each smaller "tile" is indexed in relation to its neighboring "tiles" and the result is a long, indexed stack of image "tiles," similar to the way CT data are treated. Each image is also accompanied by information describing the acquisition process, including the source of illumination, light filers, original magnification, and lenses used, color encoding, compression, and more (Weinstein et al., 2009). However, according to Weinstein et al., radiological DICOM image standards suffer from two significant limitations that have to be overcome for successful virtual slide telepathology. First, pixel dimensions of DICOM image objects are stored as unsigned 16-bit integers, for a maximum value of 65 K, which is not enough for virtual slide imaging. Second, data sizes of DICOM image objects are stored as unsigned 32-bit integers, for a maximum value of 2 GB, which again does not suffice for high-resolution whole slide imaging. Furthermore, the time has come for implementation of pathology PACS platforms at a local institutional or even WAN-based national level to streamline workflow and update pathologists' practice in line with their colleague radiologists.

There have been several efforts, mostly by researchers led by Dr. Elizabeth and A. Krupinski who pioneered research into digital pathology imaging, to standardize the design of telepathology workstations, based on experience transferred and lessons learned from teleradiology (Krupinski, 2009). In fact, choosing a workstation for daily use in the interpretation of digital pathology images can be a very daunting task. There is actually no "one size fits all" workstation, so users must consider a variety of hardware and software aspects when choosing a workstation for viewing medical images (Krupinski, 2009). Most importantly, unlike radiology, pathology requires consideration of multiple information systems (anatomical, clinical, molecular, etc.) (Krupinski, 2010). Key features such as unsigned report status, number of unread studies, average report turnaround times, number and type of interpreted cases, and other functions are equally essential in radiology and pathology workstations.

22.4 Teleradiology and Telepathology

In order to better understand developments in image-based telemedical applications like telepathology, teledermatology, and more, the reader must be familiar with the example of teleradiology first. Given the integral role of imaging in diagnosis and therapy, development of teleradiology has been groundbreaking in the medical enterprise producing innovative business models. Teleradiology belongs to the more encompassing concept of telemedicine and refers to the delivery of imaging services over a distance (Thrall, 2007a,b). Teleradiology has actually set the overarching principles of informatics involved in modern e-Health and may be referred to as the patriarch of all telehealth models.

In almost parallel with teleradiology, telepathology has shown steady progress since 1986. Today, pathology services may be offered at a long distance employing new-generation so-called virtual slide telepathology systems that allow computerized scanning and digitization of the specimen slides (Weinstein et al., 2009). Hundreds or even thousands of virtual slide scanners have already been installed by commercial vendors around the world enabling rapid virtual microscopy, digital whole slide imaging, and remote interpretation of digitized slides, that is, telepathology. To be exact, telepathology was originally defined as the practice of pathology at a long distance (Weinstein, 1986; Weinstein et al., 1987). Long distance meant a distance of many miles, but in the era of modern digital pathology instrumentation it means any distance between the virtual slide scanner machine and the interpreting pathologist, ranging from a few meters between different rooms to thousands of miles between different institutions around the globe (Dunn et al., 2009; Evans et al., 2009).

The ongoing growth of the teleradiology market highlights the fact that developments in telemedicine are fostered by several factors including a shortage of appropriately trained clinicians at the point of care, new technological developments in all fields of imaging, regional consolidation of healthcare services, and heightened patients expectations for timely high-quality examinations (Thrall, 2007a). Nowadays, direct digital image capture and digital archiving and storage capabilities combined with computerized data compression and transmission methods have established the store-and-forward approach as the basis for the establishment and operation of most modern telediagnosis links (Thrall, 2005a,b,c). The same technologies have been adopted for other image-based telemedicine operations such as telepathology and teledermatology.

Several centers in the United States have pioneered in the institutionalization of sustainable robotic and virtual slide telepathology services in the early 1990s (Weinstein et al., 2009). In practice, telepathology may involve remote interpretation of static slide images or even remote control of a motorized robotic light microscope (Weinstein et al., 2001). Dunn and colleagues have run a robotic telepathology program (Apollo Telepathology System) serving the Milwaukee region for more than a decade using key-enabling technologies such as a remotely controlled robotic interface and videoconferencing for teleconsultation. Over the years, Dunn's group has carefully documented several interesting parameters of their everyday practice like differences in practice patterns between novice and seasoned pathologists, viewing and report turnaround times, learning curves and diagnostic accuracy (Agha et al., 1999; Dunn et al., 1997, 2000, 2001, 2009). In fact, the experience amassed by has been integral for the design of future studies examining the diagnostic accuracy of telepathology programs.

Again, several business model applied today in telecare are based on similar ones developed within the field of teleradiology. One of the first and simplest teleradiology applications includes remote home reporting of emergency examinations by on-call persons. The steady increase of teleradiology for interpretation of on-call exams has led to the formation of the first niche teleradiology market, otherwise referred to as "nighthawking," that is, remote off-hours coverage of emergency imaging studies (Brant-Zawadzki, 2007; Larson et al., 2005; Thrall, 2007b). Based on this concept, cross-border teleradiology has been proposed in order to improve workflow, cover on-call services, and reduce costs and patient waiting lists with the exchange of services between countries located in different time zones. In parallel, teleradiology coverage of routine daytime imaging services has also shown significant promises to optimize and expedite workflow, otherwise referred to as "dayhawking." On the one hand, a general radiologist can cover for many different locations, where there is not enough work for a solo full-time radiologist, and on the other hand, a subspecialty-trained radiologist can offer his niche knowledge for specialized consultations to different practices, resulting to high-quality reporting coupled with significant corporate cost savings. "Nighthawking" and "dayhawking" businessmodels are generally applicable in all image-based disciplines suffering from a shortage of consultants in order to cut costs and expedite report turnaround times.

Modern telepathology services encompass viewing and interpreting the digital slide images, remote access to relevant patient data when necessary, generating a properly written report, teleconsultation with other physicians involved in patient care like surgeons and oncologists, and of course internal quality control and assurance of the whole process to ensure accuracy and promote efficiency of the service (Kayser, 2000; Kayser et al., 2004; Weinstein et al., 2001). Interestingly, setup of a telepathology link may also provide a low-cost solution for immediate second-opinion readings of challenging cases in surgical pathology (Graham et al., 2009). For example, Dr. Evans and colleagues have used a telepathology link in Toronto covering a demanding frozen section service of neuropathology just a 15-min walk apart. Users initially employed a dynamic robotic telepathology link, but the service soon shifted to a modern system of virtual slide telepathology (Evans et al., 2009, 2010). Dr. Weinstein and colleagues have created and run an international second-opinion static image telepathology service in Arizona since 1993 (Halliday et al., 1997; Weinstein et al., 1995). They also studied the diagnostic accuracy and identified the limitations of static image telepathology. The overall concordance between written reports of static image telepathology and after rereview of the respective glass slides on light microscopy was 88% and up to 96% when looking for clinically significant diagnoses (Weinstein et al., 1997). Researchers have realized that most of the false diagnostic conclusions in telepathology reporting were due to the glass slide fields selected and photographed during original light microscopy by the local pathologists. On rereview of the same glass slides, more important fields were recognized that were initially missed (Weinstein et al., 2009). A quality control audit of 3064 cases referred for robotic telepathology second-opinion readings by Arizona rural hospitals showed a very high overall diagnostic accuracy of 98–99%. This is a very good example of a telepathology model offering university-quality pathology second readings in remotely located rural communities (Bhattacharyya et al., 1995; Graham et al., 2009; Weinstein et al., 2004, 2009). Another lesson from international telepathology services was the offering of false hope to patients in cases where the referring country could not have access to the therapy proposed by the diagnosis of the second-opinion telepathologist. Such cross-border disparities in the availability, level, and quality of healthcare services may lead to significant frustration of both the physician who cannot offer and the patient who cannot receive the recommended therapy (Weinstein et al., 2009).

Another model, which is most widely used in tertiary academic hospitals that incorporate experience and expertise, involves blending of remote with onsite consultative and reporting services (Benjamin et al., 2010). Second-opinion interpretation services are another option in order to take advantage of the experience and unique subspecialization of academic practice (Jarvis and Stanberry, 2005). Along with teleradiology and other telehealth endeavors, telepathology is already transforming the face of modern medicine producing innovative laboratory services. Remote offering of second-opinion STAT-quality assurance (QA) surgical pathology and telehealth-enabled rapid breast care services bundling telemammography, telepathology, and teleoncology services into a single-day process are excellent examples of the paradigm shift within the discipline of pathology itself (Lopez et al., 2009a; Weinstein et al., 2009). The so-called *UltraClinics Process* service integrated the core services of four different physical locations with a virtual telepathology link. The virtual slide telepathology quality assurance program of STAT second-opinion readouts on newly diagnosed breast cancer cases found only a small number of significant diagnostic discrepancies, less than 10%, all of which were resolved the same day with increased patient satisfaction and physician confidence (Lopez et al., 2009b).

22.5 Teledermatology, Teleophthalmology, and Telecardiology

Dermatology is another visual specialty, particularly suited for development of image-based telemedical applications to cut long waiting lists, deal with consultant shortages, and offer high-level subspecialty services (Eedy and Wootton, 2001; English and Eedy, 2007; Massone et al., 2008). Teledermatology has already demonstrated successful practice models and generally high levels of concordance in patient diagnosis and management between remote and face-to-face consultation (Eedy and Wootton, 2001; Hockey et al., 2004; Oakley et al., 2000, 2001; Wootton et al., 2000). It has been used for screening programs, primary diagnosis, and second-opinion readings. Informatics systems employed for image transfer in teledermatology may be the classical store-and-forward approach for acquisition and transmission of static images or more expensive and cumbersome real-time videoconferencing techniques that have the added advantage of dynamic patient interaction (Massone et al., 2008). Digital acquisition and over-the-Web, wired or wireless mobile transmission of dermoscopic images for evaluation of pigmented skin lesions, the so-called teledermoscopy, has shown around 90% concordance between actual physical examination and remote viewing and has been proposed as a suitable triage syst, which is a combination of remote practice of pathology and dermatology for interrogation of melanoma in suspicious excised skin lesions, has been reported with high accuracy in images captured with a robotic microscope (Weinstein et al., 1997). Of note, a novel optical imaging system employing a three-dimensional camera and disposable optical marker has been recently reported for remote assessment of diabetic foot ulcers. This innovative project showed that a physician viewing only the three-dimensional images produced by telemetry could accurately and conveniently measure and assess a diabetic foot wound from a remote setting (Bowling et al., 2011).

Mobile teledermatology using low-resolution cameras installed in cellular personal digital assistant (PDA) and smartphones is another elegant development in image-based telehealth applications (Ebner et al., 2008; Massone et al., 2005b, 2006). Cellular phones are now abundantly available, even among the lower socioeconomic levels of society. They are almost universally equipped with average-resolution digital cameras and may serve as the ideal input device for mobile on-the-road acquisition of biomedical image data. Photographs of skin lesions may be taken by the primary care physician or even by the patient himself either for triage of suspicious lesions or for consultative reading of emergent skin lesions (Braun et al., 2005; Massone et al., 2005b, 2006). Mobile teledermoscopy has also been applied for remote evaluation of pigmented skin moles and early diagnosis of melanoma (Massone et al., 2007). In this way, mobile teledermatology introduces two novel concepts in practice of modern medicine. First, it engages patients themselves in home-based health data acquisition, and second it overcomes the barriers of traditional wired networks. Thus, development of secure and ultrahigh-bandwidth networks for mobile telediagnosis of body images and other biometric data is a challenge for future informaticists, especially within the context of patient–physician interactivity and interoperability between different systems. At the moment, Web-based teleconsultation is a rather new trend in the field of teledermatology. These so-called DermOnline Communities are based on motivated users to create and expand their content, and promote the concept of Open Access Teleconsultation in Dermatology (Massone et al., 2008). Such online platforms may allow physicians to seek expert advice in dermatology easily and quickly from a predefined pool of subspecialized consultants and discuss challenging cases. An up-to-date list of Web resources, training opportunities, online atlases, and electronic textbooks in dermatology may be found elsewhere (Massone et al., 2008).

In a similar way, teleophthalmology has been applied for evaluation of the retinopathy of prematurity (Murakami et al., 2008; Richter et al., 2009), as well as for diabetic retinopathy screening in a primary care setting (Andonegui et al., 2011). Retinopathy of prematurity is a vasoproliferative disorder affecting low birth weight infants and remains the leading cause of treatable childhood blindness throughout the world. Timely diagnosis and treatment can significantly reduce the risk of severe late complications. However, current management strategies of retinopathy of prematurity suffer from a shortage of adequately trained ophthalmologists at the point of care, which coupled with extensive travel and logistical coordination requirements for subspecialty-trained retinal specialists or neonatal ophthalmologists, drove the development of teleophthalmology services offered by a remote expert (Richter et al., 2009).

Asynchronous store-and-forward telemedicine systems have been developed for early diagnosis and prompt treatment of retinopathy of prematurity. Images are captured by appropriately trained neonatal personnel and subsequently interpreted by a remote neonatal retinal specialist. This can significantly reduce travel time for ophthalmologists and greatly improve accessibility to expert care for neonate patients. In principle, teleophthalmology can be either used to report comprehensive retinal findings in retinopathy of prematurity cases or as a screening test to identify infants with equivocal or moderate to severe findings before referral for a more thorough ophthalmoscopy (Trese, 2008). In addition, serial retinal imaging may provide a method for more objective documentation of disease findings and progression (Richter et al., 2009). Several pilot studies have validated the accuracy, intergrader reliability, and intragrader reliability of telemedicine as a tool for timely diagnosis of retinopathy of prematurity. It has been also found that in contrast to standard ophthalmoscopy with routine scleral depression, digital retinal photography causes less physiologic stress to the infants (Mehta et al., 2005; Mukherjee et al., 2006). These have led to the implementation of pilot telemedical networks in the United States that involve screening premature babies at risk of developing blindness by using trained nonexpert personnel to capture images and transfer data to remote ophthalmologists (Richter et al., 2009).

Of note, digital ophthalmoscopic imaging has been also applied in other challenging pediatric retinal conditions such as retinoblastoma and shaken baby syndrome (Hussein et al., 2004; Lee et al., 2004). Another study examined the performance of four general practitioners (GPs) trained to assess nonmydriaticretinography images of patients with diabetes. In total, 1223 patients were screened for diabetic retinopathy and all positives ($n = 297$) and a group of normal negatives ($n = 120$) were referred for standard ophthalmoscopy (Andonegui et al., 2011). The specificity of GPs for detecting diabetic retinopathy by nonmydriaticretinography was 83%, the overall sensitivity was 90.9%, and the sensitivity for detecting treatable lesions was 99.2%. The investigators concluded that adequately trained GPs can screen for treatable lesions of diabetic retinopathy with a very high level of reliability using nonmydriaticretinography (Andonegui et al., 2011).

Finally, telemedicine has been also widely tested for telemetric cardiac monitoring including multiple physiological parameters such as multichannel electrocardiography (ECG), blood pressure, and oxygen saturation (Hilbel et al., 2008). Continuous measurement of oxygen saturation is necessary for the remote monitoring of patients with cardiac pacemakers, whereas 12-lead ECG systems with diagnostic quality are advantageous in monitoring patients with chest pain syndromes. Home-based systems for remote monitoring of implantable cardiac devices with telemetric functionalities are becoming increasingly popular because they allow prompt remote diagnosis of any device dysfunction and also convenient remote optimization of the device settings (Hilbel et al., 2008).

22.6 Teaching in the Digital Era and E-Learning

Virtual microscopy has proved to be a transformational technology for academic pathology departments (Weinstein, 2008). Some of the first killer applications of modern digital technologies belong to education and virtual slide microscopy is considered one of them (Weinstein et al., 2009). The trend at medical schools in the United States is to discard light microscopes and produce purely digital pathology courses by building virtual slide libraries (Dee, 2009; Dee and Meyerholz, 2007; Weinstein et al., 2009). Pathology courses have been developed based on virtual microscopy technologies in order to teach pathology of human disease for bioscience graduate students, cytology education, a comparative pathology research resource, and histology and histopathology for veterinary medicine. Advantages of virtual microscopy over traditional microscopy include accessibility and efficiency of learning and the ability to integrate virtual microscopy with computer-assisted interactive learning (Dee, 2009; Dee and Meyerholz, 2007).

Soon, digital pathology took also advantage of digital learning technologies like e-learning. E-learning is an affordable user-friendly service that employs the Internet to provide learners with remote access to virtually unlimited knowledge from their home office or computer. Within this context, high-quality teaching sets of virtual pathology slides are already available over the Internet as online courses. Apart from digital image file sets, other online resources include supplemental text courseware, lecture Web-casts, online conferences or just interesting everyday cases. The vast majority of this educational material is usually available free of charge to your laptop or mobile smartphone and is poised to play a major role in continued lifelong training of physicians ranging from student to resident and attending level.

E-learning may also combine immediate assessment of student's understanding and peer guidance as necessary (Kumar and Krupniski, 2008). For example, Dee and colleagues have developed a 20-question competency assessment test for residents in surgical pathology using virtual microscopy slide sets as test materials (Bruch et al., 2009). The idea was to develop an efficient and reliable performance-based virtual slide competency examination in general surgical pathology that objectively measures pathology resident's morphologic diagnostic skill. Results from a pilot implementation of the test showed that the learning curve was much steeper in early training days. The researchers also proposed that digital technologies like, for example, virtual microscopy and computer accessibility have advanced to the point that similar online competency assessment tests could have applicability across multiple residency programs (Bruch et al., 2009). In line with this, development of a whole library of Web-based competency assessment tools for academic purposes has been put forward (Gorstein and Weinstein, 2001; Weinstein et al., 2009). Using such tools, competency-based curricula for residency training could be developed that would be adapted to each individual's needs and progress (Gorstein and Weinstein, 2001).

In parallel, merging of synchronous and asynchronous e-learning scenarios will gradually lead to the evolution of live, dynamic, and interactive e-learning platforms that will provide collaborators and practitioners with unlimited flow of mentoring and consultation during a procedure in real-time as needed. Widespread use of powerful handheld computers and smartphones along with high-bandwidth telecommunications will probably fuel the next phase of telecare with on-demand live subspecialty teleconsultation and on-the-road mobile dynamic multilingual structured reporting.

22.7 Data Privacy and Network Security

Modern medical informatics tools merge the disciplines of medicine, law, and computing while exploring solutions to problems such as connectivity and secure networking, preservation of data integrity and patient confidentiality, data mining, and quality assurance policies (McCullagh et al., 2008). Network security is routinely referred to as the Achilles heel of many hospitals infrastructures, despite extensive national and international security regulations. Information privacy is a well-acknowledged human right and as such, telemedicine security issues include but are not limited to confidentiality, integrity, and accountability (White, 2004). Privacy and security of patient data handled and exchanged by teleradiology networks must fulfill some of the most stringent requirements. Security and privacy enhancing tools such as strong user authentication, data encryption, non-repudiation services, and audit logs constitute core elements of modern teleradiology applications (Ruotsalainen, 2010; White, 2004). Encryption is now widely used in protecting sensitive systems such as hospital records, Internet e-commerce, mobile phone networks, and bank ATMs. Security measures may be basic like plain access control with a username and password, or enhanced requiring use of a smart card and card reader (McCullagh et al., 2008). Identification of user privileges based on unique biometric data such as fingerprint scanning is another example of technologies that may be employed to ensure network security. Moreover, speech recognition may also include voice recognition authentication protocols to enhance security during user log-in.

Undoubtedly, data de-identification or encryption is the least security measure any over-the-Web teleradiology network has to undertake. Data controllers and data processors must be protected against malicious viruses, worms, and Trojans. Privilege management and access control services ensure that every participating group member is readily certified and granted the appropriate level of clearance. Regular audit-logs have to be produced by data controllers to check when and how imported data have been used by data processors. Screen blanking and automatic user log-out after a configurable period of inactivity must be incorporated into the basic security protocols of each individual reading workstation. Anonymization should be used in all cases as far as it does not influence the clinical decision making (Ruotsalainen, 2010). Briefly, proposed security and privacy services for trusted telemedicine networks include (1) identification and authentication service of all participating users and entities, (2) encryption of transferred information, (3) e-signing of transferred data for integrity, (4) pseudoanonymization and anonymization, (5) audit-logs for accountability, and (6) nonrepudiations services for network safety (Langer and Stewart, 1999).

Several international guidelines and ISO standards (ISO/IEC 27799 and ISO/IEC 27001) for protection of personal health data are available (HIPAA documentation, www.hipaaadvisory.com). The HIPAA legislation has been into enforcement by the United States Department of Health and Human Services since 1996. The Society of Imaging Informatics in Medicine has also published a guidebook describing some of the technical methods used to secure sensitive health-related electronic data (Dwyer et al., 2004). Of note, the EU Data Protection Directive (95/46/EC) defines the legal environment for protected data transfer between different jurisdictions in case of cross-border teleradiology practice and may serve as the basis for the development of a similar framework regulating telepathology, teledermatology, and other telemedical applications entailing remote image interpretation.

Cross-border image-based telediagnosis aims to improve workflow, cover on-call services, and reduce costs and patient waiting lists with the exchange of services between countries in different time zones. However, a magnitude of organizational, technical, legal, security, and most importantly linguistic issues may hinder the delivery of cross-border healthcare services (Ross et al., 2010). The proposed EU Directive on cross-border Healthcare was first accepted by the European parliament in April 2009. With respect to the provision of teleradiology services, the Directive states that "when healthcare is provided in a Member State other than the state where the patient is insured, such healthcare must comply with the legislation, standards and guidelines of the Member State of treatment." In addition, e-Health and other telemedicine services must fully "adhere to the same professional medical quality and safety standards as those in use for nonelectronic healthcare provision" (Pattynama, 2010, p. 27). In all cases, European legislation acts in addition to national health laws.

Telepathology and other image-based telehealth models, just like teleradiology, continue to be an underdeveloped legal frontier with myriads of medicolegal issues surrounding each and every conceivable business model (Leung and Kaplan, 2009; Tsuchihashi et al., 2000). Of interest, some medicolegal issues surrounding telepathology depend on the type of telepathology system used. In an active diagnostic system, where the telepathologist or telecytopathologist can control a remotely located robotic microscope, the interpreting pathologist assumes most of the diagnosis-related responsibility. On the other hand, in a modern passive diagnostic system where the telepathologist reviews the microscopic images that have been selected, scanned, and transmitted by someone in the remote hospital, shared responsibility has to be assumed by both the originating and the interpreting sites (Tsuchihashi et al., 2000). An interesting collection of selected medicolegal and reimbursement issues in the practice of telepathology may be found elsewhere (Leung and Kaplan, 2009).

22.8 Informatics, Research, and E-Health

Medical informatics lies at the intersection of information science, computer science, and healthcare enterprise (Kagadis et al., 2008). The healthcare business is expected to generate an exponentially increasing amount of medical information, and imaging will be responsible for the vast majority of it. Different specialties and disciplines who have vested interests in medical imaging need to develop and agree upon the standards and metrics necessary for future workstations and informatics tools in order to handle the vast amount of biomedical data. Informatics methods can assist interpreting consultants to recognize important findings in images, whether radiological, histological, dermatological, or other, then combine their findings with other clinical data in multiparametric decision models in order to decide the best plan of therapy, and finally engage patients themselves in the management and provision of their own healthcare services (Rubin, 2009).

Again, teleradiology serves as an excellent example for many pioneering informatics applications that may be adopted by other telemedical image-based disciplines. The authors and others envision dynamic, mobile, and location-independent image-based telediagnosis applications with 24 h/day availability to anyone, anytime, and anywhere under on-the-fly access control scrutiny by autonomous computer systems (Ruotsalainen, 2010). While designing information systems employed in telecare, efficiency is one of the key parameters. Efficiency in medicine may be defined as the combination of speed and accuracy in achieving any clinical result, that is, producing and delivering a final diagnostic report to the referring source (Benjamin et al., 2010). Ideally, every organization would like to continuously increase its efficiency, that is, maximize the number of image interpretations per unit time per member consultant without sacrificing diagnostic accuracy (Benjamin et al., 2010). Effective coordination and management of telemedicine links between multiple regional or even international sites calls for an up-scaled PACS system, or alternatively a so-called Super-PACS that will inherently resolve most of online interactivity and file compatibility issues while transfering and exchanging various types and formats of image files originating from different hospital departments and distributed to local and remote locations (Benjamin et al., 2010).

Since 1989, the European Union (EU) has continuously increased funding to support research and development of information and communication technologies for healthcare. This

EU so-called eHealth initiative is gradually becoming a global reality and will be instrumental in strengthening European industrial competitiveness and tackling new health and social problems originating from the free movement of people in the EU (Iakovidis et al., 2007; Iakovidis and Purcarea, 2008). Otherwise put, eHealth aims to combine health telematics, telemedicine, biomedical engineering, and bioinformatics to the edge for enhanced healthcare services (Blobel and Zvarova, 2010). The e-Health action plan is the driving force toward integration of healthcare services across the EU and teleradiology represents already the most successful eHealth service available today (Baardseng, 2004; Palsson and Valdimarsdottir, 2004; Reponen et al., 2008). Radiology is the single field of medicine that inherently enjoys most of the benefits from advancements in technology and computer science.

Telemedical applications such as mobile teledermatology include informatics tools that open up new horizons and enable delivery of patient-centered health services (Gatzoulis and Iakovidis, 2007). In fact, one of the cardinal points of the European e-Health initiative is research toward prevention and management of health problems using "Personal Health Systems," that is, patient-centered systems that combine diagnosis and treatment and are handled and managed by the patient himself (Blobel and Zvarova, 2010). This is a groundbreaking concept that aims to empower citizens to adopt a proactive role in controlling their own health status by employing easy-to-use computer tools for implementation of preventive medicine screening programs and acceleration of early diagnosis in emergent situations.

The science of informatics has an elemental role in supporting the practice of patient-centered medicine. With the appropriate tools and methods, informatics are engaged throughout the imaging workflow spanning the steps of selecting an appropriate imaging examination, interpreting the acquired biomedical images, communicating the results to the patient, and reinforcing a shared decision-making model (Rubin, 2009). Therefore, informatics lies before the great challenge of putting the patient into the center of his own well-being and care. Development of personal health systems, as envisioned by the eHealth initiative, will greatly promote the level and quality of telediagnosis, telecare, and telemedicine in general.

References

Agha, Z., Weinstein, R. S., and Dunn, B. E. 1999. Cost minimization analysis of telepathology. *Am. J. Clin. Pathol.*, 112(4), 470–8.

Andonegui, J., Serrano, L., Eguzkiza, A., Berastegui, L., Jimenez-Lasanta, L., Aliseda, D. et al. 2011. Diabetic retinopathy screening using tele-ophthalmology in a primary care setting. *J. Telemed. Telecare*, 16(8), 429–32.

Baardseng, T. 2004. Telemedicine and eHealth in Norway: Administration and delivery of services. *Int. J. Circumpolar Health*, 63(4), 328–35.

Benjamin, M., Aradi, Y., and Shreiber, R. 2010. From shared data to sharing workflow: Merging PACS and teleradiology. *Eur. J. Radiol.*, 73(1), 3–9.

Bhattacharyya, A. K., Davis, J. R., Halliday, B. E., Graham, A. R., Leavitt, S. A., Martinez, R. et al. 1995. Case triage model for the practice of telepathology. *Telemed. J.*, 1(1), 9–17.

Blobel, B. and Zvarova, J. 2010. eHealth: Combining health telematics, telemedicine, biomedical engineering and bioinformatics to the edge. *Methods Inf. Med.*, 49(2), 121–2.

Bowling, F. L., King, L., Paterson, J. A., Hu, J., Lipsky, B. A., Matthews, D. R. et al. 2011. Remote assessment of diabetic foot ulcers using a novel wound imaging system. *Wound Repair Regen.*, 19(1), 25–30.

Branstetter, B. F. t. 2007a. Basics of imaging informatics. Part 1. *Radiology*, 243(3), 656–67.

Branstetter, B. F. t. 2007b. Basics of imaging informatics: Part 2. *Radiology*, 244(1), 78–84.

Brant-Zawadzki, M. N. 2007. Special focus—Outsourcing after hours radiology: One point of view—Outsourcing night call. *J. Am. Coll. Radiol.*, 4(10), 672–4.

Braun, R. P., Vecchietti, J. L., Thomas, L., Prins, C., French, L. E., Gewirtzman, A. J. et al. 2005. Telemedical wound care using a new generation of mobile telephones: A feasibility study. *Arch. Dermatol.*, 141(2), 254–8.

Bruch, L. A., De Young, B. R., Kreiter, C. D., Haugen, T. H., Leaven, T. C., and Dee, F. R. 2009. Competency assessment of residents in surgical pathology using virtual microscopy. *Hum. Pathol.*, 40(8), 1122–8.

Daskalakis, A., Cavouras, D., Bougioukos, P., Kostopoulos, S., Georgiadis, P., Kalatzis, I. et al. 2007a. Genes expression level quantification using a spot-based algorithmic pipeline. *Conf. Proc. IEEE Eng. Med. Biol. Soc.*, Lyon, France, pp. 1148–51.

Daskalakis, A., Cavouras, D., Bougioukos, P., Kostopoulos, S., Glotsos, D., Kalatzis, I. et al. 2007b. Improving gene quantification by adjustable spot-image restoration. *Bioinformatics*, 23(17), 2265–72.

Dee, F. R. 2009. Virtual microscopy in pathology education. *Hum. Pathol.*, 40(8), 1112–21.

Dee, F. R. and Meyerholz, D. K. 2007. Teaching medical pathology in the twenty-first century: Virtual microscopy applications. *J. Vet. Med. Educ.*, 34(4), 431–6.

Dunn, B. E., Almagro, U. A., Choi, H., Sheth, N. K., Arnold, J. S., Recla, D. L. et al. 1997. Dynamic-robotic telepathology: Department of Veterans Affairs feasibility study. *Hum. Pathol.*, 28(1), 8–12.

Dunn, B. E., Choi, H., Almagro, U. A., and Recla, D. L. 2001. Combined robotic and nonrobotic telepathology as an integral service component of a geographically dispersed laboratory network. *Hum. Pathol.*, 32(12), 1300–3.

Dunn, B. E., Choi, H., Almagro, U. A., Recla, D. L., and Davis, C. W. 2000. Telepathology networking in VISN-12 of the Veterans Health Administration. *Telemed. J. E Health*, 6(3), 349–54.

Dunn, B. E., Choi, H., Recla, D. L., Kerr, S. E., and Wagenman, B. L. 2009. Robotic surgical telepathology between the iron mountain and Milwaukee Department of Veterans Affairs Medical Centers: A 12-year experience. *Hum. Pathol.*, 40(8), 1092–9.

Dwyer, S. J. I., Reiner, B. I., and Siegel, E. L. 2004. *SIIM U Primer 5-Security Issues in the Digital Medical Enterprise*, 2nd edition. http://siimweb.org/index. Retrieved December 2010.

Ebner, C., Wurm, E. M., Binder, B., Kittler, H., Lozzi, G. P., Massone, C. et al. 2008. Mobile teledermatology: A feasibility study of 58 subjects using mobile phones. *J. Telemed. Telecare*, 14(1), 2–7.

Eedy, D. J. and Wootton, R. 2001. Teledermatology: A review. *Br. J. Dermatol.*, 144(4), 696–707.

Engelke, C., Schmidt, S., Auer, F., Rummeny, E. J., and Marten, K. 2010. Does computer-assisted detection of pulmonary emboli enhance severity assessment and risk stratification in acute pulmonary embolism? *Clin. Radiol.*, 65(2), 137–44.

English, J. S. and Eedy, D. J. 2007. Has teledermatology in the U.K. finally failed? *Br. J. Dermatol.*, 156(3), 411.

Evans, A. J., Chetty, R., Clarke, B. A., Croul, S., Ghazarian, D. M., Kiehl, T. R. et al. 2009. Primary frozen section diagnosis by robotic microscopy and virtual slide telepathology: The University Health Network experience. *Hum. Pathol.*, 40(8), 1070–81.

Evans, A. J., Kiehl, T. R., and Croul, S. 2010. Frequently asked questions concerning the use of whole-slide imaging telepathology for neuropathology frozen sections. *Semin. Diagn. Pathol.*, 27(3), 160–6.

Ferrante, F. E. 2005. Evolving telemedicine/ehealth technology. *Telemed. J. E Health*, 11(3), 370–83.

Fetterly, K. A., Blume, H. R., Flynn, M. J., and Samei, E. 2008. Introduction to grayscale calibration and related aspects of medical imaging grade liquid crystal displays. *J. Digit. Imaging*, 21(2), 193–207.

Gabril, M. Y. and Yousef, G. M. 2010. Informatics for practicing anatomical pathologists: Marking a new era in pathology practice. *Mod. Pathol.*, 23(3), 349–58.

Gatzoulis, L. and Iakovidis, I. 2007. Wearable and portable eHealth systems. Technological issues and opportunities for personalized care. *IEEE Eng. Med. Biol. Mag.*, 26(5), 51–6.

Georgiadis, P., Cavouras, D., Kalatzis, I., Daskalakis, A., Kagadis, G. C., Sifaki, K. et al. 2008. Improving brain tumor characterization on MRI by probabilistic neural networks and non-linear transformation of textural features. *Comput. Methods Programs Biomed.*, 89(1), 24–32.

Gorstein, F. and Weinstein, R. S. 2001. Rethinking pathology residency training and education. *Hum. Pathol.*, 32(1), 1–3.

Graham, A. R., Bhattacharyya, A. K., Scott, K. M., Lian, F., Grasso, L. L., Richter, L. C. et al. 2009. Virtual slide telepathology for an academic teaching hospital surgical pathology quality assurance program. *Hum. Pathol.*, 40(8), 1129–36.

Halliday, B. E., Bhattacharyya, A. K., Graham, A. R., Davis, J. R., Leavitt, S. A., Nagle, R. B. et al. 1997. Diagnostic accuracy of an international static-imaging telepathology consultation service. *Hum. Pathol.*, 28(1), 17–21.

Harrington, D. P. 2006a. Imaging and informatics at the National Cancer Institute, part 1. *J. Am. Coll. Radiol.*, 3(2), 88–9.

Harrington, D. P. 2006b. Imaging and informatics at the National Cancer Institute, part 2. *J. Am. Coll. Radiol.*, 3(3), 169–70.

Hersh, W. 2009. A stimulus to define informatics and health information technology. *BMC Med. Inform Decis. Mak.*, 9, 24.

Hilbel, T., Helms, T. M., Mikus, G., Katus, H. A., and Zugck, C. 2008. [Telemetry in the clinical setting]. *Herzschrittmachertherapie Elektrophysiol.*, 19(3), 146–54.

HIPAA documentation. (www.hipaaadvisory.com). Accessed June 2010.

Hirschorn, D., Eber, C., Samuels, P., Gujrathi, S., and Baker, S. R. 2002. Filmless in New Jersey: The New Jersey Medical School PACS Project. *J. Digit. Imaging*, 15(Suppl 1), 7–12.

Hockey, A. D., Wootton, R., and Casey, T. 2004.. Trial of low-cost teledermatology in primary care. *J. Telemed Telecare*, 10(Suppl 1), 44–7.

Houston, J. D. and Rupp, F. W. 2000. Experience with implementation of a radiology speech recognition system. *J. Digit. Imaging*, 13(3), 124–8.

Hussein, M. A., Coats, D. K., and Paysse, E. A. 2004. Use of the RetCam 120 for fundus evaluation in uncooperative children. *Am. J. Ophthalmol.*, 137(2), 354–5.

Iakovidis, I., Le Dour, O., and Karp, P. 2007. Biomedical engineering and eHealth in Europe. *IEEE Eng. Med. Biol. Mag.*, 26(3), 26–8.

Iakovidis, I. and Purcarea, O. 2008. eHealth in Europe: From Vision to Reality. *Stud. Health Technol. Inform*, 134, 163–8.

Jarvis, L. and Stanberry, B. 2005. Teleradiology: Threat or opportunity? *Clin. Radiol.*, 60(8), 840–5.

Kagadis, G. C., Nagy, P., Langer, S., Flynn, M., and Starkschall, G. 2008. Anniversary paper: Roles of medical physicists and healthcare applications of informatics. *Med. Phys.*, 35(1), 119–27.

Kayser, K. 2000. Telepathology in Europe. *Anal. Cell. Pathol.*, 21(3–4), 95–6.

Kayser, K., Gortler, J., Goldmann, T., Vollmer, E., Hufnagl, P., and Kayser, G. 2008. Image standards in tissue-based diagnosis (diagnostic surgical pathology). *Diagn. Pathol.*, 3, 17.

Kayser, K., Kayser, G., Radziszowski, D., and Oehmann, A. 2004. New developments in digital pathology: From telepathology to virtual pathology laboratory. *Stud. Health Technol. Inform*, 105, 61–9.

Kayser, K., Molnar, B., and Weinstein, R. S. 2006. *Digital Pathology Virtual Slide Technology in Tissue-Based Diagnosis, Research and Education*. Berlin: VSV Interdisciplinary Medical Publishing, pp. 1–93.

Kostopoulos, S., Cavouras, D., Daskalakis, A., Bougioukos, P., Georgiadis, P., Kagadis, G. C. et al. 2007. Colour-texture based image analysis method for assessing the hormone receptors status in breast tissue sections. *Conf. Proc. IEEE Eng. Med. Biol. Soc.*, Lyon, France, pp. 4985–8.

Kostopoulos, S., Cavouras, D., Daskalakis, A., Kagadis, G. C., Kalatzis, I., Georgiadis, P. et al. 2008. Cascade pattern recognition structure for improving quantitative assessment of estrogen receptor status in breast tissue carcinomas. *Anal. Quant. Cytol. Histol.*, 30(4), 218–25.

Krupinski, E. A. 2009. Virtual slide telepathology workstation-of-the-future: Lessons learned from teleradiology. *Semin. Diagn. Pathol.*, 26(4), 194–205.

Krupinski, E. A. 2010. Optimizing the pathology workstation "cockpit:" Challenges and solutions. *J. Pathol. Inform*, 1, 19.

Kumar, R. K., Velan, G. M., Korell, S. O., Kandara, M., Dee, F. R., and Wakefield, D. 2004..Virtual microscopy for learning and assessment in pathology. *J. Pathol.*, 204(5), 613–8.

Kumar, S. and Krupniski, E. A. (Eds.). 2008. *Teleradiology*. Berlin: Springer-Verlag.

Langer, S. 2002a. Radiology speech recognition: Workflow, integration, and productivity issues. *Curr. Probl. Diagn. Radiol.*, 31(3), 95–104.

Langer, S. and Stewart, B. 1999. Aspects of computer security: A primer. *J. Digit. Imaging*, 12(3), 114–31.

Langer, S. G. 2002b. Impact of tightly coupled PACS/speech recognition on report turnaround time in the radiology department. *J. Digit. Imaging*, 15(Suppl 1), 234–6.

Larson, D. B., Cypel, Y. S., Forman, H. P., and Sunshine, J. H. 2005. A comprehensive portrait of teleradiology in radiology practices: Results from the American College of Radiology's 1999 Survey. *AJR Am. J. Roentgenol.*, 185(1), 24–35.

Lee, T. C., Lee, S. W., Dinkin, M. J., Ober, M. D., Beaverson, K. L., and Abramson, D. H. 2004. Chorioretinal scar growth after 810-nanometer laser treatment for retinoblastoma. *Ophthalmology*, 111(5), 992–6.

Leung, S. T. and Kaplan, K. J. 2009. Medicolegal aspects of telepathology. *Hum. Pathol.*, 40(8), 1137–42.

Lopez, A. M., Graham, A. R., Barker, G. P., Richter, L. C., Krupinski, E. A., Lian, F. et al. 2009a. Virtual slide telepathology enables an innovative telehealth rapid breast care clinic. *Semin. Diagn. Pathol.*, 26(4), 177–86.

Lopez, A. M., Graham, A. R., Barker, G. P., Richter, L. C., Krupinski, E. A., Lian, F. et al. 2009b. Virtual slide telepathology enables an innovative telehealth rapid breast care clinic. *Hum. Pathol.*, 40(8), 1082–91.

Madabhushi, A., Doyle, S., Lee, G., Basavanhally, A., Monaco, J., Masters, S. et al. 2010. Integrated diagnostics: A conceptual framework with examples. *Clin. Chem. Lab Med.*, 48(7), 989–98.

Massone, C., Di Stefani, A., and Soyer, H. P. 2005a. Dermoscopy for skin cancer detection. *Curr. Opin. Oncol.*, 17(2), 147–53.

Massone, C., Hofmann-Wellenhof, R., Ahlgrimm-Siess, V., Gabler, G., Ebner, C., and Soyer, H. P. 2007. Melanoma screening with cellular phones. *PLoS One*, 2(5), e483.

Massone, C., Lozzi, G. P., Wurm, E., Hofmann-Wellenhof, R., Schoellnast, R., Zalaudek, I. et al. 2005b. Cellular phones in clinical teledermatology. *Arch. Dermatol.*, 141(10), 1319–20.

Massone, C., Lozzi, G. P., Wurm, E., Hofmann-Wellenhof, R., Schoellnast, R., Zalaudek, I. et al. 2006. Personal digital assistants in teledermatology. *Br. J. Dermatol.*, 154(4), 801–2.

Massone, C., Wurm, E. M., Hofmann-Wellenhof, R., and Soyer, H. P. 2008. Teledermatology: An update. *Semin. Cutan. Med. Surg.*, 27(1), 101–5.

McCullagh, P. J., Zheng, H., Black, N. D., Davies, R., Mawson, S., and McGlade, K. 2008. Section 1: Medical informatics and eHealth. *Technol. Healthcare*, 16(5), 381–97.

Mehta, M., Adams, G. G., Bunce, C., Xing, W., and Hill, M. 2005. Pilot study of the systemic effects of three different screening methods used for retinopathy of prematurity. *Early Hum. Dev.*, 81(4), 355–60.

Morgan, M. B., Branstetter, B. F. t., Lionetti, D. M., Richardson, J. S., and Chang, P. J. 2008. The radiology digital dashboard: Effects on report turnaround time. *J. Digit. Imaging*, 21(1), 50–8.

Mukherjee, A. N., Watts, P., Al-Madfai, H., Manoj, B., and Roberts, D. 2006. Impact of retinopathy of prematurity screening examination on cardiorespiratory indices: A comparison of indirect ophthalmoscopy and retcam imaging. *Ophthalmology*, 113(9), 1547–52.

Murakami, Y., Jain, A., Silva, R. A., Lad, E. M., Gandhi, J., and Moshfeghi, D. M. 2008. Stanford University Network for Diagnosis of Retinopathy of Prematurity (SUNDROP): 12-month experience with telemedicine screening. *Br. J. Ophthalmol.*, 92(11), 1456–60.

Nikiforidis, G. C., Kagadis, G. C., and Orton, C. G. 2006. Point/counterpoint. It is important that medical physicists be involved in the development and implementation of integrated hospital information systems. *Med. Phys.*, 33(12), 4455–8.

Nikiforidis, G. C., Sakellaropoulos, G. C., and Kagadis, G. C. 2008. Molecular imaging and the unification of multilevel mechanisms and data in medical physics. *Med. Phys.*, 35(8), 3444–52.

Oakley, A., Rademaker, M., and Duffill, M. 2001. Teledermatology in the Waikato region of New Zealand. *J. Telemed. Telecare*, 7(Suppl 2), 59–61.

Oakley, A. M., Kerr, P., Duffill, M., Rademaker, M., Fleischl, P., Bradford, N. et al. 2000. Patient cost-benefits of realtime teledermatology—A comparison of data from Northern Ireland and New Zealand. *J. Telemed. Telecare*, 6(2), 97–101.

Olsen, D. R., Bruland, S., and Davis, B. J. 2000. Telemedicine in radiotherapy treatment planning: Requirements and applications. *Radiother. Oncol.*, 54(3), 255–9.

Palsson, T. and Valdimarsdottir, M. 2004. Review on the state of telemedicine and eHealth in Iceland. *Int. J. Circumpolar Health*, 63(4), 349–55.

Pattynama, P. M. 2010. Legal aspects of cross-border teleradiology. *Eur. J. Radiol.*, 73(1), 26–30.

Piccolo, D., Smolle, J., Wolf, I. H., Peris, K., Hofmann-Wellenhof, R., Dell'Eva, G. et al. 1999. Face-to-face diagnosis vs telediagnosis of pigmented skin tumors: A teledermoscopic study. *Arch. Dermatol.*, 135(12), 1467–71.

Reponen, J., Winblad, I., and Hamalainen, P. 2008. Current status of national eHealth and telemedicine development in Finland. *Stud. Health Technol. Inform*, 134, 199–208.

Richter, G. M., Williams, S. L., Starren, J., Flynn, J. T., and Chiang, M. F. 2009. Telemedicine for retinopathy of prematurity diagnosis: Evaluation and challenges. *Surv. Ophthalmol.*, 54(6), 671–85.

Ross, P., Sepper, R., and Pohjonen, H. 2010. Cross-border teleradiology-experience from two international teleradiology projects. *Eur. J. Radiol.*, 73(1), 20–5.

Rubin, D. L. 2009. Informatics methods to enable patient-centered radiology. *Acad. Radiol.*, 16(5), 524–34.

Ruotsalainen, P. 2010. Privacy and security in teleradiology. *Eur. J. Radiol.*, 73(1), 31–5.

Thrall, J. H. 2005a. Reinventing radiology in the digital age. Part II. New directions and new stakeholder value. *Radiology*, 237(1), 15–8.

Thrall, J. H. 2005b. Reinventing radiology in the digital age. Part III. Facilities, work processes, and job responsibilities. *Radiology*, 237(3), 790–3.

Thrall, J. H. 2005c. Reinventing radiology in the digital age: Part I. The all-digital department. *Radiology*, 236(2), 382–5.

Thrall, J. H. 2007a. Teleradiology. Part I. History and clinical applications. *Radiology*, 243(3), 613–7.

Thrall, J. H. 2007b. Teleradiology. Part II. Limitations, risks, and opportunities. *Radiology*, 244(2), 325–8.

Trese, M. T. 2008. What is the real gold standard for ROP screening? *Retina*, 28(Suppl 3), S1–2.

Tsuchihashi, Y., Okada, Y., Ogushi, Y., Mazaki, T., Tsutsumi, Y., and Sawai, T. 2000. The current status of medicolegal issues surrounding telepathology and telecytology in Japan. *J. Telemed. Telecare*, 6(Suppl 1), S143–5.

van Bemmel, J. H. 1984. The structure of medical informatics. *Med. Inform (Lond.)*, 9(3–4), 175–80.

Vlad, R. M., Kolios, M. C., Moseley, J. L., Czarnota, G. J., and Brock, K. K. 2010. Evaluating the extent of cell death in 3D high frequency ultrasound by registration with whole-mount tumor histopathology. *Med. Phys.*, 37(8), 4288–97.

Weinstein, L. J., Epstein, J. I., Edlow, D., and Westra, W. H. 1997. Static image analysis of skin specimens: The application of telepathology to frozen section evaluation. *Hum. Pathol.*, 28(1), 30–5.

Weinstein, R. S. 1986. Prospects for telepathology. *Hum. Pathol.*, 17(5), 433–4.

Weinstein, R. S. 2008. Time for a reality check. *Arch. Pathol. Lab Med.*, 132(5), 777–80.

Weinstein, R. S., Bloom, K. J., and Rozek, L. S. 1987. Telepathology and the networking of pathology diagnostic services. *Arch. Pathol. Lab Med.*, 111(7), 646–52.

Weinstein, R. S., Bhattacharyya, A., Yu, Y. P., Davis, J. R., Byers, J. M., Graham, A. R. et al. 1995. Pathology consultation services via the Arizona-International Telemedicine Network. *Arch. Anat. Cytol. Pathol.*, 43(4), 219–26.

Weinstein, R. S., Descour, M. R., Liang, C., Barker, G., Scott, K. M., Richter, L. et al. 2004. An array microscope for ultrarapid virtual slide processing and telepathology. Design, fabrication, and validation study. *Hum. Pathol.*, 35(11), 1303–14.

Weinstein, R. S., Descour, M. R., Liang, C., Bhattacharyya, A. K., Graham, A. R., Davis, J. R. et al. 2001. Telepathology overview: From concept to implementation. *Hum. Pathol.*, 32(12), 1283–99.

Weinstein, R. S., Graham, A. R., Richter, L. C., Barker, G. P., Krupinski, E. A., Lopez, A. M. et al. 2009. Overview of telepathology, virtual microscopy, and whole slide imaging: Prospects for the future. *Hum. Pathol.*, 40(8), 1057–69.

White, P. 2004. Privacy and security issues in teleradiology. *Semin. Ultrasound CT MR*, 25(5), 391–5.

Wootton, R., Bloomer, S. E., Corbett, R., Eedy, D. J., Hicks, N., Lotery, H. E. et al. 2000. Multicentre randomised control trial comparing real time teledermatology with conventional outpatient dermatological care: Societal cost-benefit analysis. *BMJ*, 3207244., 1252–6.

Yagi, Y. and Gilbertson, J. R. 2005. Digital imaging in pathology: The case for standardization. *J. Telemed. Telecare*, 11(3), 109–16.

Yagi, Y. and Gilbertson, J. R. 2008. The importance of optical optimization in whole slide imaging (WSI) and digital pathology imaging. *Diagn. Pathol.*, 3(Suppl 1), S1.

23

Informatics in Radiation Oncology

George Starkschall
University of Texas

Peter Balter
University of Texas

In this chapter, we will describe the role of informatics in radiation oncology (RO). We will begin with a brief history of the development of RO informatics. This history will be followed by a case study illustrating the information flow in a large, contemporary RO clinic. We will then present some of the standards used in RO informatics that are, in many cases, extensions of the corresponding standards in diagnostic imaging. Finally, we will conclude the chapter with some speculation about future developments in RO informatics.

23.1 History of Informatics in RO

Informatics in RO developed from the 1970s to meet the three needs: (1) to obtain and process the information needed to generate treatment plans for patients receiving radiation therapy (RT), (2) to set the parameters on the radiation delivery machine to match those determined by the treatment plan, and (3) to create a permanent record of the treatment machine parameters and any images acquired during treatment delivery. We have presented these actions in the order of dataflow for the patient, but they were developed and implemented into RO informatics systems in the reverse order. Table 23.1 summarizes these developments.

The first issue addressed leading to the study of RO informatics was the development of the record and verify (R & V) system, developed during the 1970s to reduce potential errors in the establishment and recording of radiation treatment parameters (Cederlung et al., 1976; Chung-Bin et al., 1976; Dickof et al., 1984; Frederickson et al., 1979; Kipping and Potenza, 1976; Mohan et al., 1984; Rosenbloom et al., 1977). These R & V systems confirmed that the machine parameters, which were set manually, were correct, and verified that the delivered treatment was consistent with the machine settings. If the treatment was interrupted unexpectedly, the R & V system automatically stored the data related to the partial treatment delivered, and then resumed the treatment from the point of interruption.

In the 1980s, R & V systems became a key component of computer-controlled RT systems (Seelentag et al., 1987; Takahashi et al., 1987), which were needed to handle the more complicated treatment delivery techniques that were being developed. For instance, the setting of multileaf collimators was too laborious a task to be performed manually and, because of its complexity, was prone to human error. With the coupling of the radiation treatment-planning computer to the computer-controlled linear accelerator, R & V systems became instrumental for verifying computer-generated reference multileaf collimator settings. The issues in generating and verifying complicated machine settings were further exacerbated with the development of dynamic therapies such as intensity-modulated RT.

R & V systems also evolved potentially reducing the likelihood of medical errors in the planning and delivery process (Bates et al., 2001). An Institute of Medicine report published in 1999 identified several errors that occur in the healthcare setting and concluded that a reduction of such errors is required for a safer health system (Kohn et al., 1999). A goal of the R & V system was to reduce the frequency of radiotherapeutic errors (Patton et al., 2003). Recently, several investigators have demonstrated that the use of an R & V system to transfer the data from the treatment planning system (TPS) to the linear accelerator does indeed reduce the frequency of errors in RO (Klein et al., 2005; Yeung et al., 2005). The true improvements in patient safety due to R & V systems may be underestimated, as prior to the introduction of these systems, many errors were probably made and never identified.

The second issue of RO informatics was the need to computerize the RO medical record. Initiatives to develop an electronic RO medical record began in the 1980s with the efforts of investigators

TABLE 23.1 Development of RO Informatics Systems

Approx. Dates	Development	Purpose
1970s	R & V systems	Reduce potential errors in setting; record radiation treatment parameters.
1980s	Computer-controlled RT systems	Deliver complex radiation treatments, for example, multileaf collimators, dynamic delivery.
1980s	Computerized medical record	Improve access to medical record and radiation treatment parameters.
1990s	RT-PACS	Store, access, and display RO images.

at The University of North Carolina (Sailer et al., 1997; Salenius et al., 1992). The early electronic RO medical records were text-based systems that stored patient-specific information in a commercial relational database (Gfirtner et al., 1994). The database could also store treatment-specific information extracted from an R & V system. By the mid- to late 1990s, several electronic RO recording systems were commercially available.

The third issue was the need to store, access, and display the images used in RO. In the early 1990s, several research groups developed the Radiation Therapy Picture Archival and Communications System (RT-PACS) to fill this need, using the successful model of the development of the radiology PACS (McGee et al., 1995; Starkschall, 1997; Takenaka and Hosaka, 1987). During the early development of the RT-PACS, several functional differences between the radiology PACS and the RT-PACS were identified (Law and Huang, 2003). For example, the high resolution typically required of a radiographic image was not necessary for an RO image, as it was not necessary to identify the pathologies in the latter because they had already been identified in the former. Consequently, no high-resolution display was required for the RT-PACS viewing station, nor were high bandwidth and high-volume image transfer and storage capabilities. However, the RT-PACS still had to be capable of removing the geometric distortion that sometimes appeared in radiographic images because RO images were used for radiation treatment planning, which requires accurate geometries. Perhaps the key difference between the radiology PACS and the RT-PACS was the flow of data. In the radiology PACS, data flowed in only one direction, from the imaging device to the PACS. Once on the PACS, the image could be viewed, but nothing more could be added. However, in the RT-PACS, data flowed in two directions. First, data would be transferred from an imaging device to the RT-PACS. Once on the RT-PACS, the image could be transferred to a radiation TPS, where a treatment plan could be developed and added to the image, after which the entire dataset could be transferred back to the RT-PACS. Because of the bidirectional flow of information, new objects needed to be developed to handle the evolution of the patient information with time.

The primary function of the early RT-PACS was to enable the comparison of simulation images and digitally reconstructed radiographs (DRR) to portal images (Starkschall et al., 1994), but as the RT-PACS evolved it was combined with the radiation TPS and interfaced with the R & V system (Becker et al., 1994; Hyodynmaa et al., 1994).

23.2 Information Flow in the RO Process

Three types of information are used in the RO process: patient-, treatment-, and machine-specific. To identify the information used in the process and to trace its flow, we will describe the information flow associated with a single patient treated in a large, contemporary RO clinic. In several instances, the description of the information flow will be specific to our facility, but elsewhere the description will be generic.

The patient enters the RO clinic with both digital and non-digital data stored in a hospital-wide electronic medical record (EMR). Here, we will not discuss nondigital data, such as clinical notes and pathology reports, that are stored in the EMR, although this information is relevant to the patient's treatment. Digital data from previous imaging studies, treatments, and medical interventions administered in the institution are all stored in the EMR, but they are not necessarily transferable into the RO information system.

By the time the patient enters the RO clinic, he or she has been evaluated by the radiation oncologist, who has already determined the tumor extent based on images stored in the EMR. In a fully integrated information system, patient information, such as tumor site and disease stage, previous and concurrent treatments, and demographics, has been entered into a hospital-wide database prior to the patient's arrival at the RO clinic.

After arriving at the RO clinic, a computed tomography (CT) data set is created and the patient is simulated. A coordinate system must be established on the CT data set to accurately transform beam information from the radiation TPS to the linear accelerator for treatment delivery. The coordinate system can be either embedded in the image or transferred as a separate object. If the coordinate system is embedded in the image (e.g., by using external imageable markers such as BBs and tattoos), verifying the patient position on the linear accelerator is relatively straightforward; however, if the coordinate system is transferred as a separate object, verifying patient position is more complicated since it is necessary to first verify that patient markings actually correspond to their location on the digital image. In addition to creating the CT data set, at this time, one normally adds information, such as photographs and setup instructions, to assist the radiation therapist in ensuring the accuracy and reproducibility of the patient setup.

Next, treatment-specific data, such as the treatment plan, are generated using the TPS. In currently available technologies, patient demographic information must be entered manually into the TPS, but eventually, this information will be directly transferable from the EMR to the TPS. In this step, additional digital data are merged with the simulation data. For this merger to be successful, the orientation of the simulation CT scan and that of the additional images must be consistent. Serious errors in the delivery of radiation can occur if the patient is scanned in one configuration, for example, in the supine position with the feet toward the scanner, but the TPS reads the data assuming that the patient was imaged in another configuration, for example, in the supine position with the head toward the scanner. Quality assurance (QA) procedures are necessary to verify that the tag position that identifies patient orientation during data acquisition is identical to that entered into the TPS. This information is often entered manually. Additional imaging data from various sources may also be used to develop the treatment plan. The transfer of the metadata associated with these additional images into the TPS must also be verified.

Machine-specific data are then incorporated into the treatment plan. Such data may include the beam model (the set of parameters that characterize the dose distribution), the geometric capabilities of the treatment delivery machines (e.g., maximum and minimum collimator settings, multileaf collimator leaf widths, etc.), the coordinate system conventions of the treatment machines, and the CT number conversion tables (CT to density for photon beam calculations and CT to stopping power for particle beam calculations). Because this information was most likely entered into the TPS at the time of commissioning of the TPS, it has already been validated. However, additional validation may be necessary to ensure that the information has not been altered or corrupted.

The next step in the planning process is the segmentation of the target volume. Although in some cases the tumor is clearly visible on the simulation CT scan, additional imaging information, for example, positron emission tomography, single-photon emission CT, or magnetic resonance images, may also be used to aid tumor segmentation. If additional images are used, they may have been viewed on the hospital's PACS, so the registration of these image data sets may have been previously done. However, if these additional data sets are input directly into the TPS, image registration, either rigid-body or deformable, becomes an issue that needs to be addressed.

The treatment plan is then developed. Data input into the treatment plan include the prescribed dose as well as various treatment planning constraints, which are typically entered manually by the treatment planner or read by the TPS from a file that lists standard-of-practice guidelines for the treatment of specific tumor sites. In either case, the accuracy of the data input into the treatment plan must be validated by the treatment planner.

After the treatment plan is generated, it is reviewed and approved by the attending physician. The treatment plan can be reviewed either locally, that is, on the treatment planning workstation on which the plan was developed, or remotely. If the plan is reviewed remotely, it is necessary to ensure that the attending physician is seeing the same display as the treatment planner. While this may be straightforward in a homogeneous (single-vendor) environment, it is likely to be a more complicated in a heterogeneous (multivendor) environment.

Once the treatment plan has been approved, the information travels in three directions. First, the treatment plan is archived in the TPS. All the information that was used to generate the treatment plan needs to be stored in the event that the treatment plan needs to be retrieved at a later date. The beam geometry, patient images, and tumor and normal structure contours must be stored in the TPS, as well as the beam model, the CT voxel-to-electron density conversion table, and the version of the TPS software used, so that, in the event of retrieval, the retrieved information is identical to that used during the treatment planning.

Second, all the information required to drive the linear accelerator is transferred from the TPS to the R & V system. Although the R & V system may be manufactured by a different vendor from either the TPS or the linear accelerator, a seamless and accurate transfer of information from the TPS and the R & V system must take place and be verified (Siochi et al., 2009a). Serious harm can be inflicted on the patient when the transfer of information from the TPS to the R & V system is flawed (Bogdanich, 2010).

Third, the information is sent to the EMR. Not all of the treatment planning information needs to be sent to the EMR, but an appropriate abstract of the information should be stored in the EMR and that information must be sufficient to enable one to correlate the radiation treatment response with the radiation dose and fractionation delivered. Ideally, the EMR could directly access the treatment planning archive so that all the treatment plan information would be accessible through the EMR, but this may not be feasible. QA procedures are therefore necessary to ensure that the data stored in all three locations—the TPS archive, the R & V system, and the EMR—are consistent.

Next, the patient's treatment schedule must be added, most likely into the R & V system. At this point in the RO process, all the information needed to deliver the radiation treatment to the patient is now in the R & V system, and the patient comes in for treatment. At the time of treatment, it is necessary to verify the beam geometry, multileaf collimator settings (whether static or dynamic), radiation modality and energy, and treatment duration, as well as the appropriate digital devices, for example, the immobilization device, beam modifiers, and treatment machine. A treatment session with the treatment parameters is then delivered. After the patient is treated, the parameters that were used in the actual treatment, including any reference images, are recorded. Finally, event flags that prompt actions such as changes in the treatment fields and the termination of the treatment need to be added to the R & V system.

During the course of treatment, independent QA procedures are required to ensure that the treatment is being delivered as planned. These procedures include the checking of charts and images, and these procedures need to be recorded. QA procedures are also required to ensure that the treatment records are archived correctly and, even more important, that they can be retrieved correctly as well.

Clearly, not only must patient, machine, and planning data be transferred accurately from the source to the user to the archive, but metadata, such as the coordinate systems, beam model, and TPS version must as well. Although this information flow may be relatively straightforward in a homogeneous environment, it is likely to be more complicated in a heterogeneous environment, which is more likely to develop in an RO clinic that is seeking to maximize the flexibility and utilization of its equipment. Consequently, the fast and accurate transmission of data among the various components of the treatment planning, delivery, and verification systems may be difficult to achieve.

Additional machine-related information is also acquired to maintain the QA program in the RO clinic. For example, regular (e.g., daily, monthly, annual) measurements of beam characteristics can be stored and retrieved for later analysis. With such information, trends in machine behavior as well as individual events can be identified.

Finally, all patient-, treatment-, and machine-related information should be retrievable for case review, patient follow-up, and clinical studies.

23.3 Information Standards

Many vendors develop and manufacture RO equipment. Vendors often claim that open standards are not adequate for the transfer of information between systems that a homogeneous environment with proprietary data formats will result in a more accurate transfer of information. However, this argument is weakened by the fact that even individual vendors manufacture different products that do not communicate with each other. Moreover, as an individual vendor's products evolve, proprietary standards tend to change. Thus, open information standards are necessary to enable interoperability among data sources and users regardless of the environment. Not only do new open standards need to be developed, but also the transfer of information among equipment adhering to these standards must be demonstrated. In this section, we describe the most accepted standard for data transfer in RO, Digital Imaging and Communications in Medicine (DICOM) Supplement 11, often referred to as DICOM-RT, as well as the steps that are being taken to demonstrate connectivity among equipment manufactured by different vendors adhering to this standard.

23.3.1 DICOM and DICOM-RT

The DICOM-RT standard was developed as an extension of the DICOM Version 3.0 standard to handle RO information. (For further details, please see Chapter 5.) Its development began in 1994, when a "DICOM-RT *ad hoc* Working Group" was established by various vendors of RO equipment (Neumann, 2003). The International Electrotechnical Commission (IEC) was simultaneously developing an analogous RO standard, and the two groups entered into collaboration in 1995. The major vendors of RO equipment are represented in the DICOM-RT Working Group.

The RT extension to the DICOM standard defines five additional objects associated with patient-specific studies: RT Image, RT Plan, RT Dose, RT Structure Set, and RT Treatment Record. These additional objects are described below:

The RT Image object includes all planar images used in RO including projection images, such as simulation and portal images, and virtual images, such as DRRs. Image specifications include pixel spacing on the imaging plane, the location of the treatment isocenter with respect to the imaging plane, exposure sequences for multiple-exposure images and cine images, and descriptions of beam-limiting devices, electron applicators, and blocks.

The RT Plan object is used for several purposes. It is primarily used to transfer geometric data and machine parameters from the radiation TPS to the R & V system and from the R & V system to the radiation delivery device. It may also be used to communicate and archive derived data in the treatment plan such as dose–volume histograms (DVH), dose prescriptions, and dose levels by fraction. Some of these data are located in the RT Dose object described below. The RT Plan object includes planned geometric and dosimetric data for a course of RT, either external-beam RT or brachytherapy, including tolerance tables, fractionation schemes, and patient setup information. The RT Plan object may also include a reference to the dose distribution, as specified in the RT Dose object; (a reference to the geometric frame of reference; treatment plan relationships, such as versions, prior treatment plans, and alternative treatment plans; and control points for dynamic therapy). Dose prescription information, such as minimum dose, prescription dose, maximum dose, and an under dose volume fraction, is also included in the RT Plan object. Patient setup information in the RT Plan object includes fixation devices, shielding devices, and setup technique. Beam information includes beam identification, treatment unit description, identification of and information about wedges, compensators, bolus, blocks, and applicators, as well as control point information for dynamic treatment deliveries.

The RT Dose object includes radiation dose data generated during treatment planning, such as dose matrices, point doses, isodose curves, and DVH. This object also may contain either a cumulative dose from a set of radiation beams or dose matrices identified for individual beams. These matrices are always referenced to a three-dimensional data object such as a CT image data set. DVH information includes differential and cumulative DVHs, as well as minimum, maximum, and mean doses to regions of interest (ROIs). Dose distributions specify dose units and dose values, as well as normalization points and normalization values.

The RT Structure Set object includes patient-related ROIs and points of interest, such as dose points, all referenced to a three-dimensional data object. The RT Structure Set object also includes objects that are not patient structures, such as bolus or brachytherapy applicators. The algorithm used to generate the ROI is also specified, whether the ROI has been generated automatically, semiautomatically, or manually. It should be noted that ROIs, though three-dimensional, are expressed as contours in a series of two-dimensional parallel, transverse planes. Other methods of representing three-dimensional structures exist, such as triangulated surface tiles or bitmaps, but these are not supported in the present DICOM-RT standard.

The final object, the RT Treatment Record, includes all treatment session data for external-beam and brachytherapy treatments, summaries of recording information, dose calculations, and dose measurements. For each beam, the date, time, and fraction number, as well as the number of monitor units, both specified as well as delivered, are all identified in the RT Treatment Record object.

As in DICOM 3.0, many implementations support only a subset of the objects in DICOM-RT. For example, an external-beam radiation TPS may import RT Image, RT Structure Set, and RT Plan objects, and export RT Plan and RT Image objects.

23.3.1.1 Coordinate Systems

When communicating radiation treatment data from one entity to another, a consistent geometry is essential. Serious errors in radiation delivery have been caused by even small errors in coordinate system transformations (IAEA, 2008). One way that geometric ambiguity can be minimized is by using a common coordinate system. Both DICOM and the IEC (IEC, 1997) have defined coordinate system conventions, but there is an important difference between the patient coordinate system of DICOM and that of the IEC. Both coordinate systems are right-handed with an arbitrary origin, but the DICOM coordinate system is based on a set of transverse images, whereas the IEC coordinate system is based on a three-dimensional representation of the patient. In particular, in the DICOM coordinate system, the +X direction is to the right of a transverse image, with the +Y direction to the bottom of the image; in the IEC coordinate system, the +X direction is to the right of the patient, with the +Y direction to the patient's head.

23.3.2 Integrating the Healthcare Enterprise in RO

Although open information standards such as DICOM-RT have been developed, RO equipment manufacturers still need to adopt these standards and to demonstrate that these standards allow the transfer of information across different equipment platforms. Integrating the Healthcare Enterprise (IHE) is designed by healthcare professionals and industry to promote coordinated use of established standards (e.g., DICOM-RT) to facilitate the implementation, communication, and more effective use of information across different equipment platforms. IHE in RO (IHE-RO) is the initiative that specifically addresses RO. IHE-RO is sponsored by the American Society for Radiation Oncology, with collaboration from the American Association of Physicists in Medicine, the Radiological Society of North America, and the Healthcare Information and Management Systems Society.

IHE-RO performs its tasks by developing and testing use cases, called IHE Integration Profiles (IHE, 2008a). These integration profiles describe solutions to specific integration problems, document the roles of the components of the system being integrated, and document standards and design details for implementers to use in developing systems that cooperate to address the specific integration problem. In designing the integration profiles, IHE-RO group defines actors and transactions. Actors are information systems or system components that produce, manage, or act on information, whereas transactions are the interactions between actors that communicate the required information through standards-based messages. Vendors then support the integration profile by implementing appropriate actors and transactions.

The use of an integration profile can be illustrated by means of an example, the Normal Treatment Planning-Simple. This example illustrates the flow from the acquisition of CT images to the review of dose distributions. Note, in particular, the similarities between this profile and the case study described earlier in this chapter. This profile consists of six actors, one of which is an archive. Each actor performs a specific set of tasks that interact with a specific set of tasks performed by other actors via a specific set of transactions.

The first actor in the Normal Treatment Planning-Simple profile is the Image Acquirer. The Image Acquirer is typically a CT scanner, which acquires the CT data set that becomes the basis for the treatment plan. Once the Image Acquirer has obtained the data set, it stores the data set in the archive. The specification for storage of the CT data set has already been developed in IHE-Radiology, another IHE initiative (IHE, 2008b).

The second actor is the Contourer, which retrieves the image set from the archive. It may resample the CT data set and/or combine the CT data set with additional image data sets, such as previous CT data sets, positron emission tomography data sets, or magnetic resonance imaging data sets, which may be used to assist in contour delineation. The Contourer enables the user to delineate anatomical structures, thus creating the RT Structure Set object. Finally, the Contourer stores both the resampled image data set and the RT Structure Set object in the archive.

The third actor is the Geometric Planner, which retrieves the CT image set and the RT Structure Set object from the archive. The Geometric Planner enables the user to define the geometry of the treatment plan, specifying, for example, the isocenter, beam angles, and field sizes to create the Geometric Plan. Finally, the Geometric Planner stores the Geometric Plan in the archive.

The fourth actor is the Dosimetric Planner, which retrieves the CT image set, the RT Structure Set object, and the Geometric Plan from the archive. It allows the user to define the dosimetric properties of the treatment plan including the dose prescription, dose matrix, and beam calculation algorithm, thus creating the Dosimetric Plan. The Dosimetric Planner then calculates the dose based on the Geometric Plan and the Dosimetric Plan to create the RT Dose object, and it stores both the Dosimetric Plan and the RT Dose in the archive.

The final actor is the Dose Displayer, which retrieves the CT image set, the RT Structure Set object, the Dosimetric Plan, and the RT Dose object from the archive and displays the dose in a clinically useful manner, for example, as isodose distributions or DVH.

Normal Treatment Planning-Simple is one IHE-RO profile; other IHE-RO profiles include Multimodality Registration for RO, which shows how RO TPSs integrate positron emission tomography and magnetic resonance imaging data into the contouring and dose review process, and Treatment Workflow, which integrates daily imaging with radiation treatments using the workflow.

Use cases are developed to test these profiles. Participants in IHE-RO typically meet every year to test these use cases in what are called "Connectathons." In addition, public demonstrations of connectivity are regularly held, typically at meetings of the American Society for Radiation Oncology.

23.4 Future Development of RO Informatics

In this final section, we speculate on several issues in the future directions that the development of RO informatics might take.

The first direction is the integration of the RT-PACS with the radiology PACS and the radiation TPS. Two methods are commonly used to view radiology objects in current PACS, dedicated review stations and Web-based viewers; neither of these options is readily implemented in the RO environment. Whereas TPS allow the downloading of radiotherapy objects and browsing of dose distributions in three dimensions, few currently allow the importation of DICOM dose objects and therefore require recomputation of the dose, provided an appropriate beam model exists. Nor are any commercially available Web viewers available for RT objects. If we were to follow the imaging model, the RT dose distribution and/or plan could trigger the automatic downloading of its associated volumetric dataset and the two could be automatically overlaid for viewing either on a dedicated workstation or through a Web-based viewer.

The second direction is the development of a searchable EMR. The need for a medical record that can be queried for key pieces of information is not disputed; however, the format of the searchable medical record is open for debate. One method for creating a searchable medical record would be to place the information in a relational database, since it would improve the accessibility of the record. Closely related to a formalized database would be a structured report such as that being developed for radiologic examinations through the RadLex project (Langlotz, 2006). In the RadLex context, clinical information from radiologic examinations is stored using standardized terms in a structured format rather than as text narratives. A specific finding is thus described using a unique set of identifiers, with the goal of reducing ambiguity and facilitating searches. The downside of using either a database or a structured report is that they are not backward compatible, so all reports generated prior to the initiation of the searchable medical record would not be accessible unless converted into the searchable format. An alternative method for creating a searchable medical record would be to use a Google-type search engine in a free-form, text narrative report. Such a search engine would be capable of searching for specific words or concepts in a set of text narratives, but it is unclear how complete such a search would be or how well the search engine could filter through nonrelevant material.

Another issue regarding the development of an electronic RO record is the heterogeneity of the record. It is not uncommon for a patient to receive radiation treatments at more than one institution. How would treatment information from various institutions be combined? In addition to the computational challenges of combining multiple treatment plans, there is the potential informatics challenge of combining treatment planning information generated from various RO records. The solution would be to have DICOM dose treatment records, along with the associated image sets, for all of the patients care available to all institutions providing care, much like is being developed for radiology objects.

A final challenge is mining data from the RO records of a large cohort of patients. Data mining is necessary for many patient studies. However, finding the most effective way to represent these data is a formidable challenge.

23.5 Conclusions

In conclusion, RO informatics is a relatively new field. Although some issues faced in RO informatics are related to those encountered in radiology informatics, the nature of the data encountered in the RO process poses challenges that are unique to the discipline. Fortunately, there exists an open standard in DICOM to facilitate this challenge. The knowledgeable RO physicist is well equipped to handle informatics, bridging the gap between clinical RO and information technology (Siochi, 2009b).

References

Bates, D. W., Cohen, M., Leape, L. L., Overhage, J. M., Shabot, M. M., and Sheridan, T. 2001. Reducing the frequency of errors in medicine using information technology. *J. Am. Med. Inform. Assoc.*, 8, 299–308.

Becker, G., Mack, A., Jany, R., Major, J., and Bamberg, M. 1994. PACS and networking systems in radiotherapy. In Hounsell, A. R., Wilkinson, J. M., and Williams, P. C. (Eds.), *Proceedings of the Eleventh International Conference on the Use of Computers in Radiation Therapy*, pp. 46–7. Amsterdam: North-Holland.

Bogdanich, W. 2010. Radiation offers new cures and ways to do harm. *New York Times*, January 23, 2010. Available at: http://www.nytimes.com/2010/01/24/health/24radiation.html

Cederlung, J., Lofroth, R.-O., and Zetterlund, S. 1976. An attempt to check radiation treatment parameters with a mini-computer. In Sternick, E. S. (Ed.), *Computer Applications in Radiation Oncology, Proceedings of the Fifth International Conference on the Use of Computers in Radiation Therapy*, pp. 60–2. Hanover, New Hampshire: University Press of New England.

Chung-Bin, A., Kartha, P., Wachtor, T., and Hendrickson, F. 1976. Development and experience in computer monitoring and verification of daily patient treatment parameters. In Sternick, E. S. (Ed.), *Computer Applications in Radiation Oncology, Proceedings of the Fifth International Conference on the Use of Computers in Radiation Therapy*, pp. 57–9. Hanover, New Hampshire: University Press of New England.

Dickof, P., Morris, P., and Getz, D. 1984. Vrx: A verify-record system for radiotherapy. *Med. Phys.*, 11, 525–7.

Frederickson, D. H., Karzmark, C. J., Rust, D. C., and Tuschman, M. 1979. Experience with computer monitoring, verification and record keeping in radiotherapy procedures using a Clinac-4. *Int. J. Radiat. Oncol. Biol. Phys.*, 5, 415–8.

Gfirtner, H., Kropf, F., and Schenk, G. 1994. A check and recording system based on the relational data base Sybase realized on NeXT workstations. In Hounsell, A. R., Wilkinson, J. M., and Williams, P. C. (Eds.), *Proceedings of the Eleventh International Conference on the Use of Computers in Radiation Therapy*, pp. 84–5. Amsterdam: North-Holland.

Hyodynmaa, S., Aalto, J., and Pitkanen, M. 1994. A computer network for transferring radiotherapy images and treatment set-up data. In Hounsell, A. R., Wilkinson, J. M., and Williams, P. C. (Eds.), *Proceedings of the Eleventh International Conference on the Use of Computers in Radiation Therapy*, pp. 54–5. Amsterdam: North-Holland.

Integrating the Healthcare Enterprise (IHE). 2008a. *IHE-Radiation Oncology Technical Framework Volume 1—Integration Profiles*. Available at: http://wiki.ihe.net/images/8/8a/IHE_RO_TF_3.0_Volume1.pdf

Integrating the Healthcare Enterprise (IHE). 2008b. *IHE-Radiology Technical Framework Volume 1—Integration Profiles*. Availanle at: http://www.ihe.net/Technical_Framework/upload/ihe_tf_rev9-0ft_vol1_2008-06-27.pdf

International Atomic Energy Agency (IAEA). 2008. *Module 2.10: Accident Update—Some Newer Events*. Available at: http://rpop.iaea.org/RPOP/RPoP/Content/Documents/Training AccidentPrevention/Lectures/AccPr_2.10_Accident_update1_WEB.ppt.

International Electrotechnical Commission (IEC). 1997. *IEC 61217: Radiotherapy Equipment—Coordinates, Movements, and Scales*. Geneva: IEC.

Kipping, D. and Potenza, R. 1976. The CART system: Automated verification, recording, and controlled accelerator setup. In Sternick, E. S. (Ed.), *Computer Applications in Radiation Oncology, Proceedings of the Fifth International Conference on the Use of Computers in Radiation Therapy*, pp. 63–75. Hanover, New Hampshire: University Press of New England.

Klein, E. E., Drzymala, R. E., Purdy, J. A., and Michalski, J. 2005. Errors in radiation oncology: A study in pathways and dosimetric impact. *J. Appl. Clin. Med. Phys.*, 6, 81–94.

Kohn, L. T., Corrigan, J. M., and Donaldson, M. S. 1999. *To Err Is Human: Building a Safer Health System*. Washington, DC: National Academy Press.

Langlotz, C. P. 2006. RadLex: A new method for indexing online educational materials. *Radiographics*, 26, 1595–7.

Law, M. Y. and Huang, H. K. 2003. Concept of a PACS and imaging informatics-based server for radiation therapy. *Comput. Med. Imaging Graph*, 27, 1–9.

McGee, K. P., Das, I. J., Fein, D. A., Martin, E. E., Schultheiss, T. E., and Hanks, G. E. 1995. Picture archiving and communications systems in radiation oncology (PACSRO): Tools for a physician-based digital image review system. *Radiother. Oncol.*, 34, 54–62.

Mohan, R., Podmaniczky, K. C., Caley, R., Lapidus, A., and Laughlin, J. S. 1984. A computerized record and verify system for radiation treatments. *Int. J. Radiat. Oncol. Biol. Phys.*, 10, 1975–85.

Neumann, M. 2003. *Educational Course in DICOM-RT*. Switzerland: Neuchatel. Available at: http://www.sgsmp.ch/dicom/neumann1.pdf

Patton, G. A., Gaffney, D. K., and Moeller, J. H. 2003. Facilitation of radiotherapeutic error by computerized record and verify systems. *Int. J. Radiat. Oncol. Biol. Phys.*, 56, 50–7.

Rosenbloom, M. E., Killick, L. J., and Bentley, R. E. 1977. Verification and recording of radiotherapy treatments using a small computer. *Br. J. Radiol.*, 50, 637–44.

Sailer, S. L., Tepper, J. E., Margolese-Malin, L., Rosenman, J. G., and Chaney, E. L. 1997. RAPID: An electronic medical records system for radiation oncology. *Semin. Radiat. Oncol.*, 7, 4–10.

Salenius, S. A., Margolese-Malin, L., Tepper, J. E., Rosenman, J., Varia, M., and Hodge, L. 1992. An electronic medical record system with direct data-entry and research capabilities. *Int. J. Radiat. Oncol. Biol. Phys.*, 24, 369–76.

Seelentag, W. W., Lutolf, U. M., and Heinze-Assmann, R. 1987. Dynaver for treatment verification and recording. In Bruinvis, I. A. D., van der Giessen, P. H., van Kleffens, H. J., and Wittkamper, F. W. (Eds.), *The Use of Computers in Radiation Therapy, Proceedings of the Ninth International Conference on the Use of Computers in Radiation Therapy*, pp. 379–82. Amsterdam: North-Holland.

Siochi, R. A., Balter, P., Bloch, C. D. et al. 2009b. Information technology resource management in radiation oncology. *J. Appl. Clin. Med. Phys.*, 10(4), 16–35.

Siochi, R. A., Pennington, E. C., Waldron, T. J., and Bayouth, J. E. 2009a. Radiation therapy plan checks in a paperless clinic. *J. Appl. Clin. Med. Phys.*, 10(1), 43–62.

Starkschall, G. 1997. Design specifications for a radiation oncology picture archival and communication system. *Semin. Radiat. Oncol.*, 7, 21–30.

Starkschall, G., Bujnowski, S. W., Wong, N. W. et al. 1994. Implementation of image comparison methods in a radiotherapy PACS. In Hounsell, A. R., Wilkinson, J. M., and Williams, P. C. (Eds.), *Proceedings of the Eleventh International Conference on the Use of Computers in Radiation Therapy*, pp. 186–7. Amsterdam: North-Holland.

Takahashi, T., Sakamoto, K., and Kikuchi, A. 1987. Computer controlled verification of irradiation condition and its recording in multiple irradiation apparatuses. In Bruinvis, I. A. D., van der Giessen, P. H., van Kleffens, H. J., and Wittkamper, F. W. (Eds.), *The Use of Computers in Radiation Therapy, Proceedings of the Ninth International Conference on the Use of Computers in Radiation Therapy*, pp. 383–5. Amsterdam: North-Holland.

Takenaka, E. and Hosaka, R. 1987. Radiation therapy PACS. In Bruinvis, I. A. D., van der Giessen, P. H., van Kleffens, H. J., and Wittkamper, F. W. (Eds.), *The Use of Computers in Radiation Therapy, Proceedings of the Ninth International Conference on the Use of Computers in Radiation Therapy*, pp. 213–7. Amsterdam: North-Holland.

Yeung, T. K., Bortolotto, K., Cosby, S., Hoar, M., and Lederer, E. 2005. Quality assurance in radiotherapy: Evaluation of errors and incidents recorded over a 10 year period. *Radiother. Oncol.*, 74, 283–91.

Index

E

I

N

O

P

Q

R